ISNM
International Series of
Numerical Mathematics
Vol. 118

Edited by
K.-H. Hoffmann, München
H. D. Mittelmann, Tempe

Control and Estimation of Distributed Parameter Systems: Nonlinear Phenomena

International Conference in Vorau (Austria), July 18–24, 1993

Edited by

W. Desch
F. Kappel
K. Kunisch

Springer Basel AG

Editors

W Desch and F Kappel
Institut fur Mathematik
Universitat Graz
Heinrichstrasse 36
A-8010 Graz
Austria

K Kunisch
Fachbereich Mathematik
Technische Universitat Berlin
Strasse des 17 Juni 136
D-10623 Berlin
Germany

A CIP catalogue record for this book is available from the Library of Congress, Washington D C , USA

Deutsche Bibliothek Cataloging-in-Publication Data
Control and estimation of distributed parameter systems:
nonlinear phenomena , international conference in Vorau
(Austria), July 18–24, 1993 / ed by W Desch – Basel ,
Boston , Berlin Birkhauser, 1994
(International series of numerical mathematics , Vol 118)
ISBN 978-3-0348-9666-5 ISBN 978-3-0348-8530-0 (eBook)
DOI 10.1007/978-3-0348-8530-0
NE Desch, Wolfgang [Hrsg], GT

Originally published by Birkhäuser Verlag in 1994
Camera-ready copy prepared by the editors
Printed on acid free paper produced from chlorine-free pulp
Cover design Heinz Hiltbrunner, Basel

ISBN 978-3-0348-9666-5

9 8 7 6 5 4 3 2 1

Preface

The International Conference on Control and Estimation of Distributed Parameter Systems (Nonlinear Phenomena) was held at the Chorherrenstift Vorau (Austria), July 18 – 24, 1993. The meeting formed a platform for the exchange of new ideas and recent results in control, identification and optimization of nonlinear infinite dimensional systems. It was designed to provide a melting pot for a broad variety of viewpoints, including theoretical aspects like the Maximum Principle, relaxation and stabilizability as well as numerical algorithms for optimization and applications to elastic structures, flow control and population dynamics.

About 45 participants from 9 countries have attended. We thank all of them for their contribution to the success of this meeting. Once again we have enjoyed the pleasant and stimulating atmosphere of the Bildungshaus Chorherrenstift Vorau. Our thanks go to the very helpful staff and in particular to Director P. Riegler who have created an optimal environment for our conference.

We have received funding for this conference from Amt der Steiermärkischen Landesregierung, Graz, Bundesministerium für Wissenschaft und Forschung, Wien, Christian Doppler Laboratorium für Parameter Indentifikation und Inverse Probleme, Graz, United States Air Force European Office of Aerospace Research and Development and United States Army Research Development and Standardization Group, London. We greatly appreciate the support from these institutions.

We thank in particular Mrs. G. Krois, whose efforts have been an indispensable help in hundred details of administration work, and who has prepared the TEX manuscript for these proceedings.

Graz, June 1994

W. Desch, F. Kappel, K. Kunisch

List of Participants

H.T. Banks	North Carolina State University
V. Barbu	Universitatea "Al.I.Cuza"
J. Blum	Université Joseph Fourier Grenoble I
J.A. Burns	Virginia Polytechnic Institute & State University
C.I. Bynres	Washington University
E. Casas	Universidad de Cantabria
W. Desch	Universität Graz
H.O. Fattorini	University of California
B.G. Fitzpatrick	North Carolina State University
D.S. Gilliam	Texas Tech University
K.B. Hannsgen	Virginia Polytechnic Institute & State University
S.W. Hansen	Iowa State University
M. Heinkenschloss	Virginia Polytechnic Institute & State University
M.A. Horn	University of Minnesota
L.S. Hou	Simon Fraser University
W. Huyer	Universität Graz
K. Ito	North Carolina State University
F. Kappel	Universität Graz
V. Komornik	Université Louis Pasteur
W. Krabs	Technische Hochschule Darmstadt

M. Kroller	Universität Graz
K. Kunisch	Technische Universität Berlin
J.E. Lagnese	Georgetown University
G. Leugering	Universität Bayreuth
R. Ober	University of Texas at Dallas
R. Peer	Universität Graz
G. Peichl	Universität Graz
W. Prager	Universität Graz
G. Propst	Universität Graz
W. Ring	Universität Graz
I.G. Rosen	University of Southern California
T. Roubíček	Czech Academy of Sciences
E.W. Sachs	Universität Trier
F. Santosa	University of Delaware
C. Scherer	Universität Würzburg
O. Scherzer	Universität Linz
G. Schmidt	McGill University
M.E. Tessitore	Università di Roma "La Sapienza"
D. Tiba	Romanian Academy
F. Tröltzsch	Technische Universität Chemnitz-Zwickau
C. Wang	University of Southern California
R.L. Wheeler	Virginia Polytechnic Institute & State University

M. Yamamoto University of Tokyo

B.-Y. Zhang University of Cincinnati

E. Zuazua Universidad Complutense de Madrid

Contents

H.T. Banks, F. Kappel and C. Wang:
A semigroup formulation of a nonlinear size-structured distributed rate population model 1

H.T. Banks and Y. Wang:
Damage detection and characterization in smart material structures 21

M. Bergounioux, T. Männikkö and D. Tiba:
Optimality conditions for non-qualified parabolic control problems 45

C.I. Byrnes, D.S. Gilliam and V.I. Shubov:
Convergence of trajectories for a controlled viscous Burgers' equation 61

P. Cannarsa and M.E. Tessitore:
Optimality conditions for boundary control problems of parabolic type 79

E. Casas:
Pontryagin's principle for optimal control problems governed by semilinear elliptic equations 97

H.O. Fattorini:
Invariance of the Hamiltonian in control problems for semilinear parabolic distributed parameter systems 115

B.G. Fitzpatrick:
Rate distribution modeling for structured heterogeneous populations 131

S.W. Hansen:
A model for a two-layered plate with interfacial slip 143

M. Heinkenschloss and E.W. Sachs:
Numerical solution of a constrained control problem for a phase field model 171

M.A. Horn and I. Lasiecka:
Uniform stabilizability of nonlinearly coupled Kirchhoff plate equations 189

K. Ito:
Boundary temperature control for thermally coupled Navier-Stokes equations 211

J. Kazimir and I.G. Rosen:
Adaptive estimation of nonlinear distributed parameter systems 231

V. Komornik:
Decay estimates for the wave equation with internal damping 253

W. Krabs:
On the controllability of the rotation of a flexible arm 267

J.E. Lagnese:
Modeling and controllability of interconnected elastic membranes 281

G. Leugering:
On feedback controls for dynamic networks of strings and beams and their numerical simulation 301

T. Roubíček:
Various relaxations in optimal control of distributed parameter systems 327

F. Tröltzsch:
Convergence of an SQP–method for a class of nonlinear parabolic boundary control problems 343

M. Yamamoto:
Conditional stability in determination of densities of heat sources in a bounded domain 359

B.-Y. Zhang:
Boundary stabilization of the Korteweg-de Vries equation 371

E. Zuazua:
Controllability of the linear system of thermoelasticity: Dirichlet-Neumann boundary conditions 391

A SEMIGROUP FORMULATION OF A NONLINEAR SIZE-STRUCTURED DISTRIBUTED RATE POPULATION MODEL

H T BANKS F KAPPEL AND C WANG

Center for Research in Scientific Computation
North Carolina State University

Institut fur Mathematik
Universitat Graz

Department of Mathematics
University of Southern California

ABSTRACT A variation of the Sinko-Streifer model in population dynamics where besides the observable characteristics size in addition a non observable characteristics responsible for variations in growth rates for individuals of the same size is investigated It is shown that the model can be formulated as an abstract Cauchy problem in an appropriate Banach space We proof wellposedness of the abstract linear problem and also for a nonlinear perturbation of the model

1991 *Mathematics Subject Classification* 92D25 47D03, 47H20

Key words and phrases Structured populations, Sinko-Streifer model C_0-semigroups Lumer Phillips theorem, semilinear abstract Cauchy problems

1. Introduction

In most of the vast literature on population models (see [10] for an overview), each individual (or each individual with the same age or size) is assumed to follow the same pattern of development (e g , growth, mortality, fecundity) However, it is well known that in addition to the observable characteristics such as age or size of individuals, non-observable genetic characteristics may often play a critical role in the development of the individuals In the classical models proposed by F R Sharp, A Lotka, A G McKendrick, J W Sinko, and W Streifer and the

Research supported in part(H T B and C W) by the U S National Science Foundation under the grant NSF INT-9015007 and in part (F K) by the Fonds zur Forderung der Wissenschaftlichen Forschung (Austria) under Grant P8146-PHY and in part by the Commission of the European Communities under the SCIENCE Program "Evolutionary Systems, Deterministic and Stochastic Evolution Equations, Control Theory and Mathematical Biology"

numerous nonlinear variations of these models, the growth rate of each individual is assumed to be a non-increasing function of the individual's age or size. As a consequence, there can be no dispersion of the density of the population in age or size (see [1] for more detailed discussions). Therefore the Sharp-Lotka-McKendrick class of models is in conflict with most of the field data collected by experimental biologists. Alternative models with diffusion mechanisms derived from random walk models (see [7] for examples) are not based on convincing biological explanations (an individual can not choose to stay young simply because there are too many older individuals). Thus, in such models some of the parameters do not lend themselves to clear biological interpretations. In [1], we first proposed that the total population should be most properly considered to be a mixture of subpopulations of individuals with specific biological characteristics assigned to a given subpopulation. For each subpopulation of individuals with identical biological characteristic, the growth pattern of the subpopulation could thus be modeled by a Sharp-Lotka-McKendrick-Sinko-Streifer type model with specific growth parameters. It was demonstrated in [1] through simulations that the solutions of the resulting heterogeneous population model exhibit dispersion similar to that observed in field data. In subsequent investigations [2], this modeling approach was given a sound theoretical foundation in the context of aggrete population probability distributions for rate parameters and parameter identification techniques for the rate distribution on the heterogeneous population model were developed. These ideas were extended in [5] where it was shown that a semigroup of linear operators on a space of vector-valued measures can be used to formulate a heterogeneous population model. A somewhat different extension of these modeling ideas was proposed in [6], where the model equation is given by

$$\begin{aligned} \partial_t u(t,x,q) &= -\partial_x(g(x;q)u(t,x;q)) - \mu(x;q)u(t,x;q), \\ & t>0,\ x\in(0,x_{\max}(q)),\ q\in Q, \end{aligned} \tag{1.1}$$

$$g(0;q)u(t,0;q) = \int_Q \int_0^{x_{\max}(\hat{q})} K(q,\hat{q},x)u(t,x;\hat{q})\,dx\,dm(\hat{q}),\ t>0,\ q\in Q, \tag{1.2}$$

where $q \in Q$ represents the biological characteristic of an individual, $x_{\max}(q)$ is the maximum size attainable by individuals with characteristic q, and m is a probability measure on the space of characteristics. The model (1.1) - (1.2) is further generalized in [6] by including a nonlinear mortality rate μ which depends on the total population

$$P(t) = \int_Q \int_0^{x_{\max}(q)} u(t,x;q)dx\ dm(q). \tag{1.3}$$

The global existence of solutions of (1.1) – (1.3) is established in [6] using the method of characteristics and a fixed point theorem. No results related to semigroups are given. In this paper, we develop a semigroup formulation for a heterogeneous population model similar to that of (1.1) – (1.3) and its nonlinear perturbation wherein the mortality depends on the total population. For the linear model, our semigroup approach is similar to that for the Sinko-Streifer type size-structured model studied in [3]. Development of such a functional analytical framework for these problems is desireable since this allows one to use powerful techniques for qualitative investigations as well as approximation studies leading to numerical solutions.

In Section 2 below, we investigate the proposed solution space. In Section 3, a semigroup formulation of the linear heterogeneous population model is given while in Section 4 we discuss a nonlinear perturbation of the linear model.

2. The Solution Space

As we indicated in Section 1, the set Q represents the set of possible biological characteristics vectors. It is assumed in this paper that Q is a compact metric space with a probability measure $m(\cdot)$ defined on the Borel algebra of subsets of Q. Let $\mathcal{H} = C(Q; L^1(0,1))$ with a norm defined by

$$\|u\|_{\mathcal{H}} := \sup_{q\in Q} \|u(\cdot;q)\|_{L^1(0,1)} = \max_{q\in Q} \|u(\cdot;q)\|_{L^1(0,1)}. \tag{2.1}$$

It is not difficult to see that $\mathcal{H}$ is a Banach space. We define a subspace H of $\mathcal{H}$ as follows:

$$H = \{u \in \mathcal{H} \mid u(x;q) = 0 \text{ a.e. on } (x_{\max}(q), 1) \text{ for all } q \in Q\}$$

where $x_{\max}(\cdot)$ is a given continuous function from Q into $(0,1]$. For a given $q \in Q$, $x_{\max}(q)$ represents the maximum potential size of an individual with biological characteristic q. It is not difficult to verify that H is a closed subspace of $\mathcal{H}$, therefore H is also a Banach space with $\|\cdot\|_{\mathcal{H}}$ in (2.1) as the norm. Since Q is compact, for each $u \in H$, the collection of L_1 functions given by $\{u(\cdot;q) \mid q \in Q\}$ is a compact subset of $L_1(0,1)$. The following lemma states a few important properties of compact subsets of $L_1(0,1)$.

Lemma 2.1. a) *For any $u \in H$ and any $\epsilon > 0$, there exists a constant $\delta_0 > 0$ such that*

$$\int_{-\infty}^{\infty} |u(x+\delta;q) - u(x;q)|\, dx \le \epsilon$$

for all $\delta \in [0,\delta_0)$ and all $q \in Q$ where $u(\cdot;q)$ is extended by zero outside of the interval $[0,1]$.

b) *For any $u \in H$ and $\epsilon > 0$, there exists a $\delta_0 > 0$ such that*

$$\int_A |u(x;q)|dx < \epsilon \quad \textit{for all } q \in Q$$

provided A is a measurable subset of $[0,1]$ with meas $A < \delta_0$.

Proof. Statement (a) is a part of Kolmogoroff's compactness criterion for subsets of $L^1(0,1)$ (see, for instance [4], Theorem IV.8.20). To establish (b), note that from the continuity of u in q and the compactness of the space Q, we can find a family of values $\{q_i\}_{i=1}^n \subset Q$ such that for all $q \in Q$, there exists q_i with

$$\int_0^1 |u(x;q) - u(x;q_i)|\, dx \le \frac{\epsilon}{2}.$$

For each of the vectors q_i, there exists a constant δ_i such that for any set A with meas $A \le \delta_i$,

$$\int_A |u(x;q_i)| dx \le \frac{\epsilon}{2}.$$

Let $\delta_0 = \min_{1\le i\le n}\{\delta_i\}$. Then we have that for all $q \in Q$,

$$\int_A |u(x;q)|\, dx \le \int_0^1 |u(x;q) - u(x;q_i)| dx + \int_A |u(x;q_i)|\, dx \le \frac{\epsilon}{2} + \frac{\epsilon}{2} = \epsilon.$$

□

In order to establish dissipativity of operators on H, it is useful to derive a more explicit representation of the left-sided directional derivative τ_- of the norm $\|\cdot\|_{\mathcal{H}}$. Recall that for any $u, v \in \mathcal{H}$,

$$\tau_-(u,v) = \lim_{h\uparrow 0} \frac{1}{h}\big(\|u + hv\|_{\mathcal{H}} - \|u\|_{\mathcal{H}}\big).$$

Lemma 2.2. *Let $\tau_-^{L_1}$ denote the left-hand directional derivative of the norm in $L^1(0,1)$ and for any $u \in \mathcal{H}$, let $\Omega(u) \subset Q$ be defined as*

$$\Omega(u) = \{q \in Q \mid \|u(\cdot;q)\|_{L^1(0,1)} = \|u\|_{\mathcal{H}}\}.$$

Then we have

$$\tau_-(u,v) = \inf_{q\in\Omega(u)} \tau_-^{L^1}\big(u(\cdot;q), v(\cdot;q)\big), \quad u, v \in \mathcal{H}.$$

Proof. We first show

$$\tau_-(u,v) \le \inf_{q\in\Omega(u)} \tau_-^{L^1}\big(u(\cdot;q), v(\cdot;q)\big). \tag{2.2}$$

Consider a sequence ϵ_n of negative real numbers such that $\lim_{n\to\infty} \epsilon_n = 0$. Let q_0 and $\{q_n\}$ be arbitrarily chosen such that $q_0 \in \Omega(u)$, $q_n \in \Omega(u + \epsilon_n v)$. As a result,

$$\begin{aligned}\frac{1}{\epsilon_n}\big(\|u + \epsilon_n v\|_{\mathcal{H}} - \|u\|_{\mathcal{H}}\big) &= \frac{1}{\epsilon_n}\big(\|u(\cdot;q_n) + \epsilon_n v(\cdot;q_n)\|_{L^1(0,1)} - \|u(\cdot;q_0)\|_{L^1(0,1)}\big)\\ &\le \frac{1}{\epsilon_n}\big(\|u(\cdot;q_0) + \epsilon_n v(\cdot;q_0)\|_{L^1(0,1)} - \|u(\cdot;q_0)\|_{L^1(0,1)}\big).\end{aligned}$$

Taking the limit as $n \to \infty$, we obtain (2.2). On the other hand, for any sequence of negative real numbers ϵ_n with $\lim_{n\to\infty} \epsilon_n = 0$, let $q_n \in \Omega(u+\epsilon_n v)$. By compactness of Q, there exists a subsequence, again denoted by q_n such that $\lim_{n\to\infty} q_n = q_0$ for some $q_0 \in Q$. For any $q \in Q$, we have

$$\|u(\cdot;q)+\epsilon_n v(\cdot;q)\|_{L^1(0,1)} \leq \|u(\cdot;q_n)+\epsilon_n v(\cdot;q_n)\|_{L^1(0,1)} \leq \|u(\cdot;q_n)\|_{L^1(0,1)}+|\epsilon_n|\,\|v\|_{\mathcal{H}}.$$

By taking the limit as $n \to \infty$, we obtain

$$\|u(\cdot;q)\|_{L^1(0,1)} \leq \|u(\cdot;q_0)\|_{L^1(0,1)}$$

and hence $q_0 \in \Omega(u)$. But

$$\begin{aligned} \frac{1}{\epsilon_n}\big(\|u+\epsilon_n v\|_{\mathcal{H}} - \|u\|_{\mathcal{H}}\big) &= \frac{1}{\epsilon_n}\Big(\|u(\cdot;q_n)+\epsilon_n v(\cdot;q_n)\|_{L^1(0,1)} - \|u(\cdot;q_0)\|_{L^1(0,1)}\Big) \\ &\geq \frac{1}{\epsilon_n}\Big(\|u(\cdot;q_n)+\epsilon_n v(\cdot;q_n)\|_{L^1(0,1)} - \|u(\cdot;q_n)\|_{L^1(0,1)}\Big). \end{aligned}$$

The right side of the above inequality is equal to

$$\frac{1}{\epsilon_n}\int_0^1 \big(|u(x;q_n)+\epsilon_n v(x;q_n)| - |u(x;q_n)|\big)\,dx.$$

Since $\{u(\cdot;q_n)\}$ and $\{v(\cdot;q_n)\}$ converge to $u(\cdot;q_0)$, $v(\cdot;q_0)$ in $L^1(0,1)$, respectively, as $n \to \infty$, we can find a subsequence, again denoted by q_n such that $\{u(\cdot;q_n)\}$, $\{v(\cdot;q_n)\}$ converge to $u(\cdot;q_0)$, $v(\cdot;q_0)$ almost everywhere. Let $\Sigma_1(u(\cdot;q_0))$ and $\Sigma_0(u(\cdot;q_0))$ be two subsets of the interval $[0,1]$, defined by

$$\begin{aligned} \Sigma_1\big(u(\cdot;q_0)\big) &= \{x \in [0,1] \mid u(x;q_0) \neq 0\}, \\ \Sigma_0\big(u(\cdot;q_0)\big) &= \{x \in [0,1] \mid u(x;q_0) = 0\}. \end{aligned}$$

Then we can argue that

$$\lim_{n\to\infty} \frac{1}{\epsilon_n}\big(|u(x;q_n)+\epsilon_n v(x;q_n)| - |u(x;q_n)|\big) = \operatorname{sign}\big(u(x;q_0)\big)v(x;q_0)$$

almost everywhere on $\Sigma_1(u(\cdot;q_0))$. Hence by dominated convergence, we have

$$\begin{aligned} \lim_{n\to\infty} \int_{\Sigma_1(u(\cdot;q_0))} &\frac{1}{\epsilon_n}\big(|u(x;q_n)+\epsilon_n v(x;q_n)| - |u(x;q_n)|\big)\,dx \\ &= \int_{\Sigma_1(u(\cdot;q_0))} \operatorname{sign}\big(u(x;q_0)\big)v(x;q_0)\,dx. \end{aligned}$$

On the subset $\Sigma_0(u(\cdot;q_0))$, we have

$$\frac{1}{\epsilon_n}\big(|u(x;q_n)+\epsilon_n v(x;q_n)| - |u(x;q_n)|\big) \geq -|v(x;q_n)|.$$

As a result, we find

$$\liminf_{n\to\infty} \frac{1}{\epsilon_n} \int_{\Sigma_0(u(\cdot;q_0))} \big(|u(x;q_n)+\epsilon_n v(x;q_n)| - |u(x;q_n)|\big)\,dx \geq -\int_{\Sigma_0(u(x;q_0))} |v(x;q_0)|\,dx.$$

Combining the above results, we obtain

$$\begin{aligned}\tau_-(u,v) &= \lim_{n\to\infty} \frac{1}{\epsilon_n}\big(\|u+\epsilon_n v\|_{\mathcal{H}} - \|u\|_{\mathcal{H}}\big)\\ &\geq \int_{\Sigma_1(u(\cdot;q_0))} \operatorname{sign}\big(u(x;q_0)\big)v(x;q_0)\,dx - \int_{\Sigma_0(u(\cdot;q_0))} |v(x;q_0)|\,dx\\ &= \tau_-^{L^1}\big(u(\cdot;q_0), v(\cdot;q_0)\big).\end{aligned}$$

□

3. Well-Posedness of the Linear Model

In this section, we rewrite equations (1.1) – (1.2) as an evolution equation in H of the form

$$\frac{d}{dt}u(t) = Au(t), \quad t \geq 0.$$

To this end, we make some standing assumptions on the growth, mortality, and fecundity functions in (1.1) – (1.2). We assume:

(A1) The growth rate function satisfies $g \in C(Q, W^{1,\infty}(0,1))$ and for each $q \in Q$ we have $g(x;q) > 0$ on $[0, x_{\max}(q))$ and $g(x;q) = 0$ for $x \in [x_{\max}(q), 1]$.

(A2) For each $q \in Q$, $\mu(\cdot;q) \in L^\infty_{\text{loc}}(0, x_{\max}(q))$ with $\mu(\cdot;q) \geq 0$ a.e. on $[0,1]$ and $\mu(\cdot;q) = 0$ on $[x_{\max}(q), 1]$. For any $\epsilon \in (0, x_{\max}(q))$, we define μ_ϵ by

$$\mu_\epsilon(x;q) = \begin{cases} \mu(x;q) & \text{for } x \in (0, x_{\max}(q)-\epsilon),\\ 0 & \text{otherwise,}\end{cases}$$

and assume that $\mu_\epsilon \in C(Q, L^\infty(0,1))$ for $0 < \epsilon < \min_{q\in Q} x_{\max}(q)$.

(A3) The function K is in $C(Q\times Q; L^\infty(0,1))$ with $K(q,\hat{q},\cdot) = 0$ a.e. on $[x_{\max}(q), 1]$.

We shall define an unbounded linear operator A and show that A is the infinitesimal generator of a C_0-semigroup of bounded linear operators on H. The domain $\mathcal{D}(A)$ of A is the set of all functions $u \in H$ satisfying the following conditions:

(i) For all $q \in Q$, $u(\cdot;q)$ is locally absolutely continuous on $[0, x_{\max}(q))$;

(ii) The function defined by $q \to \partial_x(g(x;q)u(x;q)) + \mu(x;q)u(x;q)$ for $x \in [0, x_{\max}(q))$, $q \in Q$, is an element in H;

(iii) For any $q \in Q$,

$$g(0;q)u(0;q) = \int_Q \int_0^{x_{\max}(\hat{q})} K(q,\hat{q},x)u(x;\hat{q})\,dx\,dm(\hat{q});$$

(iv) For any $q \in Q$, $\lim_{x \uparrow x_{\max}(q)} g(x;q)u(x;q) = 0$.

For $u \in \mathcal{D}(A)$, we define

$$(Au)(x;q) = -\partial_x\big(g(x;q)u(x;q)\big) - \mu(x;q)u(x;q), \quad 0 < x < x_{\max}(q),\ q \in Q.$$

The main result of this section is stated as follows.

Theorem 3.1. *Under assumptions* (A1) – (A3), *the operator A defined above is the infinitesimal generator of a C_0-semigroup $\{S(t),\ t \geq 0\}$ of bounded linear operators on H.*

The proof of this theorem is based on the use of the Lumer-Phillips theorem which requires that the infinitesimal generator A satisfies the following:

(1) The domain of A must be dense in H;
(2) There exists a real number ω such that $A - \omega I$ is m-dissipative, i.e., the range of $A - \omega I$ is H and $A - \omega I$ is dissipative.

3.1. Density of the domain $\mathcal{D}(A)$. Consider the following subset of H:

$$H_\epsilon = \{u \in H \mid u(x;q) = 0 \text{ for } x \in (x_{\max}(q) - \epsilon, 1) \text{ for all } q \in Q\}.$$

Let $H_0 = \cup_{\epsilon > 0} H_\epsilon$. From Lemma 2.1, we conclude that H_0 is dense in H. We first want to show that a subset of H_0 that contains functions differentiable in x is also dense in H. Consider a sequence $\{L_k\}$ of bounded linear operators $L_k : L^1(0,1) \to C^1(0,1)$ defined by

$$(L_k f)(x) = \int_0^1 f(y)\lambda_k(x-y)dy$$

where

$$\lambda_k(x) = \begin{cases} \theta_k e^{-(1-(kx)^2)^{-1}} & \text{for } x \in (-1/k, 1/k), \\ 0 & \text{elsewhere.} \end{cases}$$

The constant θ_k is chosen such that

$$\int_{-\infty}^{+\infty} \lambda_k(x)\,dx = \int_{-1/k}^{1/k} \lambda_k(x)\,dx = 1.$$

The following properties of L_k are useful.

Lemma 3.1. (i) *For any $f \in L^1(0,1)$ such that $f(x) = 0$ for all $x > \delta$ with a $\delta < 1$, we have $(L_k f)(x) = 0$ for all $x > \delta + 1/k$ whenever $\delta + 1/k < 1$.*

(ii) $\lim_{k \to \infty} \|f - L_k f\|_{L^1(0,1)} = 0$.

Proof. (i) is obvious. It is sufficient to show that (ii) holds for all continous functions. In fact,

$$\|f - L_k f\|_{L^1(0,1)} = \int_0^1 \Big| \int_0^1 \big(f(x) - f(y)\big)\lambda_k(x-y)\,dy\Big|\,dx$$
$$= \int_0^1 \Big| \int_{x-1/k}^{x+1/k} \big(f(x) - f(y)\big)\lambda_k(x-y)\,dy\Big|\,dx.$$

It is easy to see that

$$\Big|\int_{x-1/k}^{x+1/k} \big(f(x)-f(y)\big)\lambda_k(x-y)\,dy\Big| \leq \sup_{y\in(x-1/k,x+1/k)} |f(x)-f(y)| \leq 2 \sup_{x\in(0,1)} |f(x)|,$$

and therefore, since f is continuous,

$$\lim_{k\to\infty} \Big|\int_{x-1/k}^{x+1/k} \big(f(x) - f(y)\big)\lambda_k(x-y)\,dy\Big| = 0 \quad \text{for all } x \in (0,1).$$

By the dominated convergence theorem, we obtain the desired result (ii). □

Let V_ϵ be the subset of H_ϵ defined by

$$V_\epsilon = \{u \in H_\epsilon \mid u \in C(Q; C^1(0,1))\}$$

and let $V_0 = \cup_{\epsilon>0} V_\epsilon$. Then we have

Lemma 3.2. *V_0 is a dense subset of H.*

Proof. It is sufficent to show that every element $u \in H_0$ can be approximated by functions in V_0. Consider an element $u \in H_0$ and define u_k as follows,

$$u_k(\cdot;q) = L_k u(\cdot;q), \text{ for all } q \in Q.$$

We want to show that $u_k \in V_0$ for all k sufficiently large and $\lim_{k\to\infty} \|u_k - u\|_{\mathcal{H}} = 0$. Since $u \in H_0$, there exists $\epsilon > 0$ such that $u \in H_\epsilon$. For all $k > 2/\epsilon$, $u_k \in H_{\frac{\epsilon}{2}}$. On the other hand, for all $q \in Q$, $u_k(\cdot;q) \in C^1(0,1)$. Moreover, let C_k be a constant such that

$$\|L_k\|_{\mathcal{L}(L^1(0,1),C^1(0,1))} \leq C_k.$$

Then,

$$\|u_k(\cdot;q) - u_k(\cdot;\hat{q})\|_{C^1(0,1)} \leq C_k \|u(\cdot;q) - u(\cdot;\hat{q})\|_{L^1(0,1)}.$$

Since $u \in C(Q, L^1(0,1))$, we conclude that

$$\lim_{d(q,\hat{q})\to 0} \|u_k(\cdot;q) - u_k(\cdot;\hat{q})\|_{C^1(0,1)} = 0,$$

where d is the metric for Q.

As a result, for k sufficiently large, we have $u_k \in V_{\frac{\epsilon}{2}}$. Using Lemma 3.1, we find

$$\lim_{k\to\infty} \|u_k(\cdot;q) - u(\cdot;q)\|_{L^1(0,1)} = 0, \quad \text{for all } q \in Q.$$

On the other hand, by compactness of Q and the uniform boundedness of operators $L_k : L^1(0,1) \to L^1(0,1)$, we conclude that

$$\lim_{k\to\infty} \sup_{q\in Q} \|u_k(\cdot;q) - u(\cdot;q)\|_{L^1(0,1)} = 0,$$

or $u_k \to u$ in H, which establishes the density of V_0 in H. □

Finally, for a given element $u \in V_0$, we shall construct a sequence of functions u_ϵ in $\mathcal{D}(A)$ such that u_ϵ converges to u in H. The basic idea is to modify the function $u(\cdot;q)$ in the neighborhood of $x = 0$ so that the resulting function satisfies the boundary condition at $x = 0$. Let $\epsilon > 0$ be a small real value. We define the modified function u_ϵ by

$$u_\epsilon(x;q) = \begin{cases} (1 - x/\epsilon)v_\epsilon(q) + (x/\epsilon)u(\epsilon;q) & \text{for } 0 \le x \le \epsilon, \\ u(x;q) & \text{for } \epsilon < x \le 1, \end{cases}$$

where the function v_ϵ should be chosen such that $u_\epsilon \in \mathcal{D}(A)$. In particular, u_ϵ should satisfy

$$g(0;q)u_\epsilon(0;q) = \int_Q \int_0^{x_{\max}(\hat{q})} K(q,\hat{q},x)u_\epsilon(x)dx\, dm(\hat{q}).$$

Using the definition of u_ϵ, we must be able to choose the function $v = v_\epsilon$ to satisfy $\mathcal{T}_\epsilon v = v$ where the mapping $\mathcal{T}_\epsilon : C(Q) \to C(Q)$ is defined as

$$\begin{aligned}(\mathcal{T}_\epsilon v)(q) &= \int_Q \int_0^\epsilon (1 - \frac{x}{\epsilon}) \frac{K(q,\hat{q},x)}{g(0;q)} v(\hat{q})\, dx\, dm(\hat{q}) \\ &+ \int_Q \int_0^\epsilon \frac{K(q,\hat{q},x)}{g(0;q)} \frac{x}{\epsilon} u(\epsilon;\hat{q})\, dx\, dm(\hat{q}) \\ &+ \int_Q \int_\epsilon^{x_{\max}(\hat{q})} \frac{K(q,\hat{q},x)}{g(0;q)} u(x;\hat{q})\, dx\, dm(\hat{q}).\end{aligned}$$

To see that this is possible we argue:

Lemma 3.3. *For ϵ sufficiently small, the mapping $\mathcal{T}_\epsilon$ is a contraction from $C(Q)$ into $C(Q)$.*

Proof. The continuity of $\mathcal{T}_\epsilon v$ for $v \in C(Q)$ and $u \in V_0$ follows immediately from assumptions (A3) and (A1). On the other hand, for any $v,\ \hat{v} \in C(Q)$, we have

$$(\mathcal{T}_\epsilon v - \mathcal{T}_\epsilon \hat{v})(q) = \int_Q \int_0^\epsilon (1 - \frac{x}{\epsilon}) \frac{K(q,\hat{q},x)}{g(0,q)} (v(\hat{q}) - \hat{v}(\hat{q}))\, dx\, dm(\hat{q}).$$

As a result,

$$|(\mathcal{T}_\epsilon v - \mathcal{T}_\epsilon \hat{v})(q)| \le \sup_{\hat{q}\in Q} |v(\hat{q}) - \hat{v}(\hat{q})| \int_Q \int_0^\epsilon (1 - \frac{x}{\epsilon}) \frac{K(q,\hat{q},x)}{g(0,q)}\, dx\, dm(\hat{q}),$$

where

$$\lim_{\epsilon \to 0} \int_Q \int_0^\epsilon (1 - \frac{x}{\epsilon}) \frac{K(q, \hat{q}, x)}{g(0, q)} \, dx \, dm(\hat{q}) = 0.$$

Therefore, for all ϵ sufficiently small, there exists $\gamma_\epsilon < 1$ such that

$$\sup_{q \in Q} |(\mathcal{T}_\epsilon v - \mathcal{T}_\epsilon \hat{v})(q)| \leq \gamma_\epsilon \sup_{q \in Q} |(v - \hat{v})(q)|,$$

which proves the lemma. □

Using the contraction mapping theorem, there exists a unique fixed point v_ϵ for the mapping $\mathcal{T}_\epsilon$ in $C(Q)$. Since $u \in V_0$, it is not difficult to see that u_ϵ is piecewise continuously differentiable, and there exists $\delta > 0$ such that $u_\epsilon(x; q) = 0$ for $x \in [x_{\max}(q) - \delta, 1]$. Thus u_ϵ satisfies all the conditions required of an element of $\mathcal{D}(A)$. Moreover, we can also show that u_ϵ converges to u in the H-topology as $\epsilon \to 0$.

Lemma 3.4. *Let u_ϵ be defined as above. Then*

$$\lim_{\epsilon \to 0^+} \|u_\epsilon - u\|_{\mathcal{H}} = 0.$$

Proof. Considering the term $u_\epsilon(\cdot; q) - u(\cdot; q)$, we have the estimate

$$\begin{aligned} \|u_\epsilon(\cdot; q) - u(\cdot; q)\|_{L^1(0,1)} &\leq \int_0^\epsilon |u(x; q)| \, dx + \frac{1}{\epsilon} \int_0^\epsilon x |u(\epsilon; q)| \, dx + \int_0^\epsilon (1 - \frac{x}{\epsilon}) |v_\epsilon(q)| \, dx \\ &\leq \sup_{\hat{q} \in Q} \int_0^\epsilon |u(x; \hat{q})| dx + \frac{\epsilon}{2} \sup_{\hat{q} \in Q} \sup_{0 \leq x \leq 1} |u(x; \hat{q})| + \epsilon \sup_{\hat{q} \in Q} |v_\epsilon(q)|. \end{aligned}$$

From the definition of v_ϵ, we have

$$\sup_{\hat{q} \in Q} |v_\epsilon(q)| \leq \frac{C_0 + \epsilon C_2}{1 - \epsilon C_1}$$

where

$$C_1 = \sup_{q \in Q} \int_Q \frac{\|K(q, \hat{q}, \cdot)\|_{L^\infty(0,1)}}{g(0; q)} \, dm(\hat{q})$$

$$C_2 = \sup_{q \in Q} \int_Q \frac{\|K(q, \hat{q}, \cdot) u(\cdot; \hat{q})\|_{L^\infty(0,1)}}{g(0; q)} \, dm(\hat{q})$$

$$C_0 = \sup_{q \in Q} \int_Q \int_0^{x_{\max}(\hat{q})} \left| \frac{K(q, \hat{q}, x)}{g(0; q)} u(x; \hat{q}) \right| \, dx \, dm(\hat{q}).$$

Therefore

$$\sup_{\hat{q} \in Q} |v_\epsilon(q)| \leq C \quad \text{for all } \epsilon > 0 \text{ sufficiently small.}$$

As a result, we conclude that

$$\lim_{\epsilon\to 0}\sup_{q\in Q}\|u_\epsilon(\cdot;q)-u(\cdot;q)\|_{L^1(0,1)}=0.$$

□

Combining the above sequence of lemmas and results, we can conclude that $\mathcal{D}(A)$ is dense in H.

3.2. Dissipativity of the operator A. Let us recall that the operator $A-\lambda I$, $\lambda\in\mathbb{R}$, is dissipative if and only if

$$\tau_-(u,Au)\le\lambda\|u\|_{\mathcal{H}}\quad\text{for all } u\in\mathcal{D}(A).$$

Using Lemma 2.2, we have

$$\tau_-(u,Au)=\inf_{q\in\Omega(u)}\tau_-^{L^1}\big(u(\cdot;q),(Au)(\cdot;q)\big),\qquad u\in\mathcal{D}(A),$$

where $\Omega(u)=\{q\in Q\mid \|u(\cdot;q)\|_{L^1(0,1)}=\|u\|_{\mathcal{H}}\}$. By the same computations as presented in Lemma 2.1 of [3], we obtain immediately

$$\begin{aligned}\tau_-^{L^1}\big(u(\cdot;q),(Au)(\cdot;q)\big)\le g(0;q)|u(\cdot;q)| &= \Big|\int_Q\int_0^{x_{\max}(\hat q)}K(q,\hat q,x)u(x;\hat q)\,dx\,dm(\hat q)\Big|\\ &\le k_0\|u\|_{\mathcal{H}}\end{aligned}$$

for $q\in Q$, $u\in\mathcal{D}(A)$, where $k_0=\sup_{q,\hat q\in Q}\|K(q,\hat q,\cdot)\|_{L^\infty(0,1)}$. Thus $A-k_0I$ is dissipative on H.

3.3. The range of the operator $A-\lambda I$. Since the range of the operator $A-\lambda I$ and the range of a nonlinear perturbation of A can be found using similar arguments, we will try to establish a result that is useful in both cases. We first define a convex functional p on H by

$$p(u)=\|u\|_{\mathcal{H}}.$$

Furthermore, we define the nonlinear mapping $F:H\to H$ by

$$(Fu)(x;q)=-\mu_0(x;q)P(u)u(x;q),$$

where $\mu_0\in C(Q;L^\infty(0,1))$, $\mu_0(x;q)\ge 0$ for all $x\in(0,1)$ and $q\in Q$ and

$$P(u)=\int_Q\int_0^{x_{\max}(q)}|u(x;q)|\,dx\,dm(q).$$

We would like to show the following result.

Theorem 3.2. *Let the subset $C(r)$ of H be defined by $C(r)=\{u\in H\mid p(u)\le r\}$. Then under assumptions (A1) – (A3) and those above on μ_0, one can find a constant $a>0$ such that the following holds:*

Given $r\ge 0$, there exists a number $\lambda_0\equiv\lambda_0(r)\in(0,1/a)$ such that to each $u\in C(r)$ and each $\lambda\in(0,\lambda_0)$, there corresponds $u_\lambda\in\mathcal{D}(A)$ satisfying

(a) $u_\lambda - \lambda(A+F)u_\lambda = u$,
(b) $p(u_\lambda) \le (1-a\lambda)^{-1}p(u)$.

We observe that when $\mu_0 \equiv 0$, statement (a) implies that the range of the operator $A - \lambda I$ is H for all $\lambda > a$. On the other hand, (a) is equivalent to

$$\frac{1}{\lambda}u_\lambda(x;q) + \partial_x\big(g(x;q)u_\lambda(x;q)\big) + \big(\mu(x;q) + \mu_0(x;q)P(u_\lambda)\big)u_\lambda(x;q) = \frac{1}{\lambda}u(x;q)$$

for $x \in (0, x_{\max}(q))$, $q \in Q$. Combining this with the boundary condition (iii) in the definition of $\mathcal{D}(A)$, we find that u_λ must satisfy

$$\begin{aligned} u_\lambda(x;q) = {} & \frac{E(x,0;q,u_\lambda)}{g(x;q)} \int_Q \int_0^{x_{\max}(\hat{q})} K(q,\hat{q},y)u_\lambda(y;\hat{q})\, dy\, dm(\hat{q}) \\ & + \frac{1}{\lambda g(x;q)} \int_0^x E(x,z;q,u_\lambda)u(z;q)\, dz \end{aligned} \tag{3.1}$$

where

$$E(x,z;q,u_\lambda) = \exp\Big(-\int_z^x \frac{1/\lambda + \mu(\zeta;q) + \mu_0(\zeta;q)P(u_\lambda)}{g(\zeta;q)}\, d\zeta\Big).$$

The proof of the above theorem consists of establishing the existence of the solution of (3.1) with property (b) for a certain nonnegative constant a. Indeed, we will establish (b) with the constant a given in the following lemma.

Lemma 3.5. *If $u_\lambda \in H$ is a solution of the integral equation (3.1), then*

$$\|u_\lambda\|_{\mathcal{H}} \le (1-a\lambda)^{-1}\|u\|_{\mathcal{H}},$$

where

$$a = \sup_{\|u\|_{\mathcal{H}} \le 1, q \in Q} \Big|\int_Q \int_0^{x_{\max}} K(q,\hat{q},x)u(x;\hat{q})\, dx\, dm(\hat{q})\Big|.$$

Proof. From (3.1), we have

$$|u_\lambda(x;q)| \le \frac{E(x,0;q,\mu_\lambda)}{g(x;q)} a\|u_\lambda\|_{\mathcal{H}} + \Big|\frac{1}{\lambda g(x;q)} \int_0^x E(x,z;q,u_\lambda)u(z;q)\, dz\Big|.$$

If we integrate both sides of the above inequality, we obtain

$$\begin{aligned} \int_0^1 |u_\lambda(x;q)|\, dx \le {} & \int_0^1 \frac{E(x,0;q,\mu_\lambda)}{g(x;q)}\, dx\, a\|u_\lambda\|_{\mathcal{H}} \\ & + \int_0^1 \frac{1}{\lambda g(x;q)} \int_0^x E(x,z;q,u_\lambda)\, |u(z;q)|\, dz\, dx. \end{aligned}$$

But

$$\int_0^1 \frac{E(x,0;q,u_\lambda)}{g(x;q)}\, dx \le \int_0^1 \frac{1}{g(x;q)} \exp\Big(-\int_0^x \frac{d\zeta}{\lambda g(\zeta;q)}\Big)\, dx \le \lambda,$$

and

$$\int_0^1 \frac{1}{\lambda g(x;q)} \int_0^x E(x,z;q,u_\lambda)|u(z;q)|\,dz\,dx$$
$$= \int_0^1 |u(z;q)| \int_z^1 \frac{1}{\lambda g(x;q)} E(x,z;q,\mu_\lambda)\,dx\,dz.$$

Arguing as above, we find

$$\int_z^1 \frac{1}{\lambda g(x;q)} E(x,z;q,\mu_\lambda)\,dx \leq \int_z^1 \frac{1}{\lambda g(x;q)} \exp\Big(-\int_z^x \frac{d\zeta}{\lambda g(x;q)}\Big)\,dx \leq 1.$$

Therefore,

$$\int_0^1 \frac{1}{\lambda g(x;q)} \int_0^x E(x,z;q,u_\lambda)|u(z;q)|\,dz\,dx \leq \int_0^1 |u(x;q)|\,dx.$$

As a result, we have

$$\|u_\lambda\|_{\mathcal{H}} \leq \lambda a \|u_\lambda\|_{\mathcal{H}} + \|u\|_{\mathcal{H}},$$

or

$$\|u_\lambda\|_{\mathcal{H}} \leq \frac{\|u\|_{\mathcal{H}}}{1-\lambda a},$$

for $\lambda < 1/a$. □

The above lemma provides an *a priori* estimate for the solution of (3.1). To prove Theorem 3.2, it is sufficient to show that (a) has a solution in $\mathcal{D}(A)$. We define a mapping $T : H \to H$ by

$$(Tv)(x;q) = \frac{1}{g(x;q)} E(x,0;q,v) \int_Q \int_0^{x_{\max}(\hat{q})} K(q,\hat{q},\zeta)v(\zeta,\hat{q})\,d\zeta\,dm(\hat{q})$$
$$+ \frac{1}{\lambda g(x;q)} \int_0^x E(x,z;q,v)u(z;q)\,dz,$$

where

$$E(x,z;q,v) = \exp\Big(-\int_z^x \frac{1/\lambda + \mu(\zeta;q) + \mu_0(\zeta;q)P(v)}{g(\zeta;q)}\,d\zeta\Big),$$
$$P(v) = \int_Q \int_0^{x_{\max}(q)} |v(x;q)|\,dx\,dm(q).$$

Lemma 3.6. *For any given $r > 0$ and $u \in C(r)$, there exists $\lambda_0 \in (0, 1/a)$ such that, for all $\lambda \leq \lambda_0$, the mapping T is a contraction on $C(r/(1-\lambda_0 a))$.*

Proof. For any $v \in H$ from the definition of Tv, it is not difficult to see that $(Tv)(\cdot;q) \in L^1(0,1)$ for all $q \in Q$. Moreover, using the continuity properties of the functions v, K, μ and g, one can establish that $Tv \in H$.

Now consider two elements $v, \hat{v} \in H$. We assume without loss of generality that $P(v) \leq P(\hat{v})$. Then we have

$$\begin{aligned} &|(Tv)(x;q) - (T\hat{v})(x;q)| \\ &\qquad \leq \frac{1}{g(x;q)} E(x,0;q,v) \int\limits_Q \int\limits_0^{x_{\max}(\hat{q})} K(q,\hat{q},z)|v(z;\hat{q}) - \hat{v}(z;\hat{q})| \, dz \, dm(\hat{q}) \\ &\qquad + \frac{1}{g(x;q)} \big(E(x,0;q,v) - E(x,0;q,\hat{v})\big) \int\limits_Q \int\limits_0^{x_{\max}(\hat{q})} K(q,\hat{q},z)|\hat{v}(z;q)| \, dz \, dm(\hat{q}) \\ &\qquad + \frac{1}{\lambda g(x;q)} \int_0^x |u(z;q)| \big(E(x,z;q,v) - E(x,z;q,\hat{v})\big) \, dz. \end{aligned}$$

By integrating both sides we obtain

$$\begin{aligned} &\int\limits_0^{x_{\max}(q)} \big|(Tv)(x;q) - (T\hat{v})(x;q)\big| \, dx \\ &\quad \leq \int\limits_0^{x_{\max}(q)} \frac{1}{g(x;q)} E(x,0;q,v) \int\limits_Q \int\limits_0^{x_{\max}(\hat{q})} K(q,\hat{q},z)|v - \hat{v}| \, dz \, dm(\hat{q}) \, dx \\ &\quad + \int\limits_0^{x_{\max}(q)} \frac{1}{g(x;q)} \big(E(x,0;q,v) - E(x,0;q,\hat{v})\big) \int\limits_Q \int\limits_0^{x_{\max}(\hat{q})} K(q,\hat{q},z)|\hat{v}(z;q)| \, dz \, dm(\hat{q}) \, dx \\ &\quad + \int\limits_0^{x_{\max}(q)} \frac{1}{\lambda g(x;q)} \int_0^x |u(z;q)| \big(E(x,z;q,v) - E(x,z;q,\hat{v})\big) \, dz \, dx \\ &\quad =: I_1 + I_2 + I_3. \end{aligned}$$

The term I_1 can be estimated by

$$\begin{aligned} &\int\limits_0^{x_{\max}(q)} \frac{1}{g(x;q)} E(x,0;q,v) \int\limits_Q \int\limits_0^{x_{\max}(q)} K(q,\hat{q},z)|(v - \hat{v})(z;\hat{q})| \, dz \, dm(\hat{q}) \, dx \\ &\qquad \leq k_0 \|v - \hat{v}\|_{\mathcal{H}} \int\limits_0^{x_{\max}(q)} \frac{1}{g(x;q)} \exp\Big(-\int_0^x \frac{d\zeta}{\lambda g(\zeta;q)}\Big) \, dx = k_0 \lambda \|v - \hat{v}\|_{\mathcal{H}}. \end{aligned}$$

The term I_2 can be estimated by

$$\int_0^{x_{\max}(q)} \frac{1}{g(x;q)} \big(E(x,0;q,v) - E(x,0;q,\hat{v})\big) \int_Q \int_0^{x_{\max}(\hat{q})} K(q,\hat{q},z)|\hat{v}(z;q)|\,dz\,dm(\hat{q})\,dx$$

$$\begin{aligned}
&\le k_0\|\hat{v}\|_{\mathcal{H}}\Big(\int_0^{x_{\max}(q)} \frac{1}{g(x;q)} \exp\Big(-\int_0^x \frac{d\zeta}{\lambda g(\zeta;q)}\Big)\Big(\exp\Big(-\int_0^x \frac{\mu_0(\zeta;q)}{g(\zeta;q)} P(v)\,d\zeta\Big) \\
&\qquad\qquad - \exp\Big(-\int_0^x \frac{\mu_0(\zeta;q)}{g(\zeta;q)} P(\hat{v})\,d\zeta\Big)\Big)\,dx \\
&= -k_0\|\hat{v}\|_{\mathcal{H}}\lambda\Big(\exp\Big(-\int_0^x \frac{d\zeta}{\lambda g(\zeta;q)}\Big) \\
&\quad \times \Big(\exp\Big(-\int_0^x \frac{\mu_0(\zeta;q)}{g(\zeta;q)} P(v)\,d\zeta\Big) - \exp\Big(\Big(-\int_0^x \frac{\mu_0(\zeta;q)}{g(\zeta;q)} P(\hat{v})\,d\zeta\Big)\Big)\Big|_{x=0}^{x=x_{\max}(q)} \\
&\quad + \lambda \int_0^{x_{\max}(q)} \exp\Big(-\int_0^x \frac{d\zeta}{\lambda g(\zeta;q)}\Big) \frac{\mu_0(x;q)}{g(x;q)} \\
&\quad \times \Big(\big(P(v) - P(\hat{v})\big) \exp\Big(-\int_0^x \frac{\mu_0(\zeta;q)}{g(\zeta;q)} P(v)\,d\zeta\Big) \\
&\quad + P(\hat{v})\Big(\exp\Big(-\int_0^x \frac{\mu_0(\zeta;q)}{g(\zeta;q)} P(v)\,d\zeta\Big) - \exp\Big(-\int_0^x \frac{\mu_0(\zeta;q)}{g(\zeta;q)} P(\hat{v})\,d\zeta\Big)\Big)\Big)\,dx.
\end{aligned}$$

The first term in the above expression is equal to zero (assumption (A1) implies $\int_0^{x_{\max}(q)} \frac{d\zeta}{g(\zeta;q)} = \infty$). On the other hand, we have

$$\Big|\exp\Big(-\int_0^x \frac{\mu_0(\zeta;q)}{g(\zeta;q)} P(v)\,d\zeta\Big) - \exp\Big(-\int_0^x \frac{\mu_0(\zeta;q)}{g(\zeta;q)} P(\hat{v})\,d\zeta\Big)\Big| \le \theta|P(v) - P(\hat{v})|$$

for some constant θ. As a result, we have

$$\begin{aligned}
&\lambda \int_0^{x_{\max}(q)} \exp\Big(-\int_0^x \frac{d\zeta}{\lambda g(\zeta;q)}\Big) \frac{\mu_0(x;q)}{g(x;q)} \Big(\big(P(v) - P(\hat{v})\big) \exp\Big(-\int_0^x \frac{\mu_0(\zeta;q)}{g(\zeta;q)} P(v)\,d\zeta\Big) \\
&\qquad + P(\hat{v})\Big(\exp\Big(-\int_0^x \frac{\mu_0(\zeta;q)}{g(\zeta;q)} P(v)\,d\zeta\Big) - \exp\Big(-\int_0^x \frac{\mu_0(\zeta;q)}{g(\zeta;q)} P(\hat{v})\,d\zeta\Big)\Big)\Big)\,dx \\
&\le \lambda^2\|v - \hat{v}\|_{\mathcal{H}} \int_0^{x_{\max}(q)} \exp\Big(-\int_0^x \frac{d\zeta}{\lambda g(\zeta;q)}\Big) \frac{\mu_0(x;q)}{\lambda g(x;q)}\,dx \\
&\qquad + \lambda^2\theta\|v - \hat{v}\|_{\mathcal{H}} P(\hat{v}) \int_0^{x_{\max}(q)} \frac{\mu_0(x;q)}{\lambda g(x;q)} \exp\Big(-\int_0^x \frac{d\zeta}{\lambda g(\zeta;q)}\Big)\,dx.
\end{aligned}$$

Moreover, if $\mu_0 \in C(Q, L^\infty(0,1))$, we have

$$\int_0^{x_{\max}(q)} \frac{\mu_0(x;q)}{\lambda g(x;q)} \exp\Big(-\int_0^x \frac{d\zeta}{\lambda g(\zeta;q)}\Big)\,dx \le \|\mu_0\|_{C(Q,L^\infty(0,1))}.$$

The term I_3 can be estimated in a similar manner by

$$\begin{aligned}
&\int_0^{x_{\max}(q)} \frac{1}{\lambda g(x;q)} \int_0^x |u(z;q)|\big(E(x,z;q,v) - E(x,z;q,\hat{v})\big)\,dz\,dx \\
&\quad \le \int_0^{x_{\max}(q)} |u(z;q)| \int_z^{x_{\max}(q)} \frac{1}{\lambda g(x;q)} \exp\Big(-\int_z^x \frac{d\zeta}{\lambda g(\zeta;q)}\Big) \\
&\qquad \times \Big(\exp\Big(-\int_0^x \frac{\mu_0}{g} P(v)\,d\zeta\Big) - \exp\Big(-\int_z^x \frac{\mu_0}{g} P(\hat{v})\,d\zeta\Big)\Big)\,dx\,dz.
\end{aligned}$$

From the estimate of I_2, we have

$$\begin{aligned}
&\int_z^{x_{\max}(q)} \frac{1}{\lambda g(x;q)} \exp\Big(-\int_z^x \frac{d\zeta}{\lambda g(\zeta;q)}\Big) \\
&\qquad \times \Big(\exp\Big(-\int_0^x \frac{\mu_0(\zeta;q)}{g(\zeta;q)} P(v)\,d\zeta\Big) - \exp\Big(-\int_z^x \frac{\mu_0(\zeta;q)}{g(\zeta;q)} P(\hat{v})\,d\zeta\Big)\Big)\,dx \\
&\quad \le \lambda\|v-\hat{v}\|_{\mathcal{H}}(1+\theta) \int_z^{x_{\max}(q)} \frac{\mu_0(x;q)}{\lambda g(x;q)} \exp\Big(-\int_z^x \frac{d\zeta}{\lambda g(\zeta;q)}\Big)\,dx.
\end{aligned}$$

As a result, we have

$$\begin{aligned}
I_3 &\le \lambda\|v-\hat{v}\|(1+\theta)\|u\|_{\mathcal{H}} \int_0^{x_{\max}(q)} \frac{\mu_0(x;q)}{\lambda g(x;q)} \exp\Big(-\int_z^x \frac{d\zeta}{\lambda g(\zeta;q)}\Big)\,dx \\
&\le \lambda\|v-\hat{v}\|(1+\theta)\|u\|_{\mathcal{H}}\|\mu_0\|.
\end{aligned}$$

Combining these estimates, we find

$$\begin{aligned}
\|Tv - T\hat{v}\|_{\mathcal{H}} &\le \Big(k_0\lambda + \lambda^2\big(1+\theta P(\hat{v})\big)\|\mu_0\| + \lambda(1+\theta)\|u\|_{\mathcal{H}}\|\mu_0\|\Big)\|v-\hat{v}\|_{\mathcal{H}} \\
&= \lambda\Big(k_0 + \|\mu_0\|\big(\lambda(1+\theta P(\hat{v})\big) + (1+\theta)\|u\|_{\mathcal{H}}\Big)\|v-\hat{v}\|_{\mathcal{H}}.
\end{aligned}$$

Thus, we conclude that on every bounded subset M of H and for every $u \in H$, there exists λ_0 such that T is a contraction in M for all $\lambda \le \lambda_0$. Using the arguments and results of Lemma 3.5, we have that if $u \in C(r)$ and $v \in C(r/(1-\lambda_0 a))$, then

$$\|Tv\|_{\mathcal{H}} \leq \lambda a \frac{r}{1-\lambda_0 a} + r \leq \frac{r}{1-\lambda_0 a}$$

for all $\lambda \leq \lambda_0$. Therefore, there exists a λ_0, such that for all $\lambda \leq \lambda_0$, the mapping T is a contraction on $C(r/(1-\lambda_0 a))$. □

Proof of Theorem 3.2. Using Lemma 3.6, there exists a unique fixed point u_λ of T in H. It is not difficult to verify that $u_\lambda \in \mathcal{D}(A)$ and u_λ is a solution of (a). □

We return to Theorem 3.1. Combining the results of subsections 3.1, 3.2 and 3.3, we find that the operator A has the following properties: (i) $\mathcal{D}(A)$ is dense in H; (ii) $A-\lambda I$ is m-dissipative for all $\lambda \geq k_0$. Thus, by the Lumer-Phillips theorem (see [9]), A is the infinitesimal generator of a C_0-semigroup of bounded linear operators in H and we have established the results of Theorem 3.1.

4. A nonlinear perturbation of the model

A nonlinear version of the model (1.1) – (1.2) in Section 1 is given by

$$\begin{aligned}\partial_t u(t,x,q) = &-\partial_x\big(g(x;q)u(t,x;q)\big) - \mu(x;q)u(t,x;q)\\ &- \mu_0(x;q)P\big(u(t)\big)u(t,x;q),\ 0<x<x_{\max}(q),\end{aligned} \tag{4.1}$$

$$g(0;q)u(t,0;q) = \int_Q \int_0^{x_{\max}(\hat q)} K(q,\hat q,x)u(t,x;\hat q)\,dx\,dm(\hat q),\ t>0,\ q\in Q, \tag{4.2}$$

where $P(u(t)) = \int_Q \int_0^{x_{\max}(\hat q)} u(t,x;\dot q)\,dx\,dm(\hat q)$.

The new model can be viewed as a nonlinear perturbation of the linear model in the form

$$\frac{d}{dt}u(t) = (A+F)u(t)$$

where the operators A and F are defined in the previous section. Let p be the functional defined in Section 3 and the set $C(r)$ is defined as before. It is not difficult to see that the nonlinear operator $F+k_0 I : H \to H$ is locally Lipschiz continuous in the sense defined in [8] and the linear operator $A-k_0 I$ generates a C_0-semigroup of contractions. We may conclude the following.

Theorem 4.1. *There exists a continuous nonlinear semigroup of operators $\{S(t),\ t\geq 0\}$ on H satisfying*

(a) $S(t)u = e^{At}x + \int_0^t e^{A(t-\xi)}FS(\xi)u\,d\xi$,
(b) $\|S(t)u\|_{\mathcal{H}} \leq e^{at}\|u\|_{\mathcal{H}}$

for all $u \in H$. Moreover, $A+F$ is the infinitesimal generator of the semigroup $\{S(t),\ t\geq 0\}$.

Proof Using the results in Section 3 (more precisely Theorem 3 2), we may appeal to Theorems 1 and 2 of [8] with the convex functional $p(u) = \|u\|_{\mathcal{H}}$ □

5. Concluding remarks

In the above presentation, we have given a development of a linear semigroup formulation for rate distribution heterogeneous population models This presentation is structured so that extension to a nonlinear version of the model in which the mortality rate depends on the total population is readily obtained This is achieved by treating the nonlinear model as a perturbation of the linear model If we include a nonlinearity in the birth process, the system can no longer be viewed as one defined by the perturbation of a linear operator by an additive locally Lipschitz continuous nonlinear perturbation However, it may still be possible to adapt the ideas of [8] to obtain a nonlinear semigroup This would require one to consider the two sources of nonlinearity separately These and other related ideas are the subject of current efforts

References

1 H T BANKS L W BOTSFORD F KAPPEL C WANG *Modeling and estimation in size structured population models* Math Ecology T G Hallam L J Gross S A Levin (Eds) Singapore World Scientific 1988 pp 521 541

2 H T BANKS and B G FITZPATRICK *Estimation of growth rate distribution in size structure population models* Quart Appl Math **49** (1991) 215 235

3 H T BANKS and F KAPPEL *Transformation semigroups and L^1 approximation for size structured population models* Semigroup Forum **38** (1989) 141 155

4 N DUNFORD and J T SCHWARTZ *Linear Operators* Wiley Interscience New York 1971

5 B G FITZPATRICK *Vector valued measure approach for a size structured population model* J Math Anal Appl **172** (1993) 73 91

6 W HUYER *A size structured population model with dispersion* Institute fur Mathematik Technische Universitat Graz Universitat Graz Preprint No 212 1992 May 1992 J Math Analysis Appl to appear

7 A OKUBO *Diffusion and Ecological Problems Mathematical Models* Springer Verlag Berlin 1980

8 S OHARU and T TAKAHASHI *Locally Lipschiz continuous perterbations of linear dissipa tive operators and nonlinear semigroups* Proceedings of the American Mathematical Society **100** (1987) 187 194

9 A PAZY *Semigroups of Linear Operators and Applications to Partial Differential Equa tions* Springer New York 1983

10 G F WEBB *Theory of nonlinear age dependent population dynamics* Marcel Dekker Inc New York 1985

H T Banks
Center for Research in Scientific Computation
North Carolina State University
Raleigh, NC 27695-8205, USA
E-mail htbanks@crsc1 math ncsu edu

F Kappel
Institut fur Mathematik
Universitat Graz
Heinrichstrasse 36, A8010 Graz, Austria
E-mail kappel@edvz uni-graz ada at

C Wang
Department of Mathematics
University of Southern California
Los Angeles, CA 90089-1113, USA
E-mail cwang@mathp usc edu

DAMAGE DETECTION AND CHARACTERIZATION IN SMART MATERIAL STRUCTURES

H. T. BANKS AND Y. WANG

Center for Research in Scientific Computation
North Carolina State University

ABSTRACT. We present theoretical, computational and experimental findings in initial investigations related to methods for detection and geometric characterization of damage in piezoceramic based smart material structures. The feasibility of using self-exitation/self-sensing with piezoceramics in vibration nondestructive testing is demonstrated using a combination of experimental and simulated data computational tests.

1991 *Mathematics Subject Classification.* 35R30, 73D50, 73K05

Key words and phrases. Inverse problems, distributed parameter systems, smart materials, damage defection.

1. Introduction

It has been known for some time that damage such as cracks, corrosions, and delaminations in a structure produces changes in mass, stiffness, damping and other characteristics and material parameters in dynamic response models for the structure. Our focus in this paper is the development of vibration response ideas for piezoceramic based smart material structures in which self-testing nondestructive evaluation (NDE) techniques may be employed. There are several ways to formulate and analyze concepts and questions related to this quest. One could use the co-called "method-of-maps" as done in the thermal based tomography techniques of [BK1, BK2, BKW] wherein the damaged physical domain (the damaged beam, plate, elastic structure) is mapped into a regular domain on which dynamic equations with damage dependent coefficients must hold. An alternative technique involves direct estimation of the irregular damaged domain (i.e., structures with holes, cracks, corrosions, etc.) on which the usual dynamic equations hold.

Research supported in part by the Air Force Office of Scientific Research under grants AFOSR F49620-93-1-0198 and AFOSR F49620-93-1-0280. Part of this research was carried out while both authors were visiting scientists at the Institute for Computer Application in Science and Engineering (ICASE), NASA Langly Research Center, which is supported by NASA under Contract Nos. NAS1-18605 and NAS1-19480.

The approach we take here is, strictly speaking, neither of the above, but it is more in the spirit of the latter one; we attempt to estimate damage-related changes in actual physical parameters (stiffness, damping) based on changes in physical geometry and characteristics.

Most of the previous efforts in the substantial literature on vibration related damage detection are based on modal methods (e.g. [ABB, AC, ACPS, CR, IC, S]). The basis for such methods is that damage produces a decrease in dynamic stiffness EI. This decrease in turn produces decreases in natural frequencies for an undamped simple beam (recall the eigenvalues are given by $\lambda \sim \sqrt{EI/\rho A}$ where A is the cross sectional area of the beam and ρ is the mass density). While modal based methods may have certain advantages (e.g., they are simple if they do work, they can be used in development of rank-ordering of fractional eigenfrequency shift schemes that are reportedly insensitive to the magnitude of the damage[ABB]), modal based methods possess a number of major disadvantages. First of all, some of the modal based method investigations (e.g. [CR]) provide a strong argument for including geometry of the damage in any NDE testing scheme, something which is not easily done in frequency based methods. Indeed, mode and frequency charaterizations are not so simple in variable structure systems; there is ample evidence (e.g. see [S]) that one should not use modal methods based on uniform undamped simple beams or plates as is often done in the engineering literature in addressing damage assessment methodologies. Since material parameters are most properly considered as spatially dependent quantities with damage manifested by changes in geometry (and hence in the spatial dependence of these parameters), it is unlikely that any rigorous theoretical basis for modal based methods for variable material structures will emerge. But perhaps the most serious objection to modal based methods resides in the fact that modal based methods have been shown to be highly unreliable (indeed inadequate! - see [BI] and the references therein) for estimation of variable material parameters such as damping in composite material structures.

In light of the above comments, a question of rather great interest then is: can one develop analytically sound, non-modal based self-excitation/self-sensing methods for detection and characterization (geometrical and quantitative) of damage in smart material structures? Here we address this question in the context of embedded piezoceramic structures.

For an embedded piezoceramics smart material damage detection and characterization methodology, there are several distinct requirements. These include:

(a) One must be able to *estimate reliably* (repeatable across experiments) the *variable structure material parameters* of a piezoceramic loaded structure. This must be done using piezo actuation and sensing with accuracy comparable to that achievable with standard actuating (impulse hammers, solenoidal actuators) and sensing (accelerometers, strain gauges, laser vibrometers) devices in non-smart material testing schemes.

(b) One must be able to use the actuation and sensing properties of the piezoceramics to *excite the structure and analyze the response* (in a single experiment) for a reliable methodology that is the basis of self-excitation/self-sensing.

(c) One must be able to *detect and characterize damage via vibration self-excitation/ self-sensing* that relies only on the input/output signals for the piezoceramics.

In this paper we address each of these requirements in the context of a piezoceramic loaded beam. This particular structure is sufficiently representative to make a compelling case for feasibility of the ideas we propose. In Section 2 below, we summarize a rigorous theoretical well-posedness and approximation foundation for the distributed parameter identification methodology that is the focus of our recent efforts. This is followed by Section 3 in which we detail results from experiments and computations that provide a favorable comparison for piezo actuating/sensing with standard methods (requirement (a) above). We then present experimental/computational findings on self-excitation/self-sensing capabilities of piezoceramic loaded beams. These findings suggest that requirement (b) above can indeed be satisfied.

Finally, in Section 4 we offer preliminary simulation findings on the use of vibration experiments to detect and characterize (geometrically) the extent of several types of damage in a beam using piezo actuation and sensing. Experimental verification of these positive findings with regard to requirement (c) is currently underway.

2. Distributed parameter identification methodology

The system we consider here is a cantilevered beam with piezoelectric ceramic patches for actuation and sensing. Our choice of structure is motivated by its simplicity and its representative nature. This configuration has also been well studied by conventional approaches such as finite element methods and provides a standard test bed model for comparison. The structure and model reveal the difficulties and possibilities inherent in developing models and methods for more complex structures containing damages.

We consider a cantilevered Euler-Bernoulli beam of length ℓ fixed at $x = 0$ and free at $x = \ell$. The transverse vibrations $y = y(t, x)$ are described by the system

$$
\begin{aligned}
&\rho(x)\frac{\partial^2 y}{\partial t^2}(t,x) + \gamma\frac{\partial y}{\partial t}(t,x) + \frac{\partial^2 M}{\partial x^2}(t,x) = \tilde{f}(t,x), \quad 0 < x < \ell,\ t > 0, \\
&y(t,0) = \frac{\partial y}{\partial x}(t,0) = 0, \quad M(t,\ell) = \frac{\partial M}{\partial x}(t,\ell) = 0,
\end{aligned}
\tag{2.1}
$$

where $\rho(x)$ is the linear mass density, γ is the coefficient of viscous (air) damping, $M(t, x)$ is the internal moment and $\tilde{f}(t, x)$ represents the external loads. For a simple Euler-Bernoulli beam with Kelvin-Voigt or strain rate damping, the internal moment is composed of two components representing resistance to bending (with coefficient $EI(x)$) and damping (with coefficient $c_D I(x)$):

$$M(t,x) = EI(x)\frac{\partial^2 y}{\partial x^2}(t,x) + c_D I(x)\frac{\partial^3 y}{\partial x^2 \partial t}(t,x). \tag{2.2}$$

If piezoelectric elements are bonded to the beam in a configuration to produce (or sense) only bending, we have an actuator contribution $M_p(t,x)$ in the form of an input moment (or voltage output proportional to the accumulated strain in the element). For a pair of piezoelectric actuators located between x_1 and x_2 on opposite sides of the beam excited by a voltage $u(t)$ in an out-of-phase manner (see [BSW, CA, CFW, DFR, FC]), this moment term has the representation

$$M_p(t,x) = K_B\{H(x-x_1) - H(x-x_2)\}\, u(t) \tag{2.3}$$

where $H(x)$ is the Heaviside or unit step function and K_B is a piezoelectric material parameter depending on the material piezoelectric properties as well as geometry. When the moment in (2.3) is added to that of (2.2) and substituted into (2.1), we obtain the model

$$\begin{aligned} \rho\frac{\partial^2 y}{\partial t^2} + \gamma\frac{\partial y}{\partial t} + \frac{\partial^2}{\partial x^2}\left(EI\frac{\partial^2 y}{\partial x^2} + c_D I\frac{\partial^3 y}{\partial x^2 \partial t}\right) & \\ = K_B\left(\frac{d}{dx}\delta(x-x_2) - \frac{d}{dx}\delta(x-x_1)\right) u(t) + \tilde{f}(t,x) & \\ y(t,0) = \frac{\partial y}{\partial x}(t,0) = 0, \quad M(t,\ell) = \frac{\partial M}{\partial x}(t,\ell) = 0 & \end{aligned} \tag{2.4}$$

where δ is the Dirac delta function. This is formally equivalent to the equation in weak or variational form (we replace partial derivatives in time by subscript t and space by superscript $'$)

$$\begin{aligned} &\langle \rho y_{tt} + \gamma y_t, \phi\rangle + \langle EI\, y'' + c_D I\, y_t'' + K_B(H_1 - H_2)\, u(t), \phi''\rangle = \langle \tilde{f}(t,x), \phi\rangle \\ &y(t,0) = y'(t,0) = 0, \end{aligned} \tag{2.5}$$

for sufficiently smooth functions ϕ satisfying $\phi(0) = \phi'(0) = 0$. Here H_i is the shifted Heaviside function $H_i(x) = H(x - x_i)$, $i = 1,2$ and $\langle\cdot,\cdot\rangle$ is the usual L_2 inner product.

For the same configuration, when the beam is under deformation (bending), the generated charges in terms of voltage across the piezoelectric sensors has the expression (see [DIG])

$$K_s \int_{x_1}^{x_2} \frac{\partial^2 y}{\partial x^2}(t,x)dx = K_s\Big(\frac{\partial y}{\partial x}(t,x_2) - \frac{\partial y}{\partial x}(t,x_1)\Big)$$

where K_s is a sensor constant which is also a piezoelectric material properties and geometry related quantity.

The system (2.4) is a formal representation of the dynamics of a damped beam with piezoelectric actuators. To develop computational techniques (e.g., finite elements) based on rigorous convergence arguments, it is necessary to first have a precise formulation of this system.

We start with an abstract formulation. Let V and H be complex Hilbert spaces satisfying $V \hookrightarrow H = H^* \hookrightarrow V^*$ (see [W] for the construction of this Gelfand triple), where we denote their topological duals by V^* and H^*, respectively. Let Q be the admissible parameter set. The general second order system we consider here is given by

$$\text{(2.6)}\quad \begin{aligned} &\langle \ddot{y}(t), \psi\rangle_{V^*,V} + \sigma_1(q)(y(t),\psi) + \sigma_2(q)(\dot{y}(t),\psi) = \langle f(t),\psi\rangle_{V^*,V} \text{ for } \psi \in V,\\ &y(0) = y_0, \quad \dot{y}(0) = y_1. \end{aligned}$$

Here we use $\langle \cdot\,,\cdot\rangle_{V^*,V}$ to denote the usual [W] duality product. The term $\sigma_1(q)$ and $\sigma_2(q)$ are parameter dependent sesquilinear forms on V satisfying V-ellipticity and V-continuity conditions. That is, we assume that $\sigma_1(q)$ and $\sigma_2(q)$ satisfy

$$\text{(2.7)}\quad \operatorname{Re}\sigma_i(q)(\phi,\phi) \geq k_i|\phi|_V^2,$$

$$\text{(2.8)}\quad |\sigma_i(\phi,\psi)| \leq c_i|\phi|_V \cdot |\psi|_V, \qquad i = 1,2$$

for $k_i, c_i > 0$, $\phi, \psi \in V$. Under weak assumptions on f, the system (2.6) has a unique solution.

Theorem 2.1. *If the sesquilinear forms σ_1 and σ_2 satisfy conditions (2.7) and (2.8) with σ_1 symmetric and $f \in L_2\big((0,T),V^*\big)$, then, for each $w_0 = (y_0,y_1) \in \mathcal{H} = V \times H$, the initial value problem (2.6) has a unique solution $w(t) = \big(y(t),\dot{y}(t)\big) \subset L_2\big((0,T),V \times V\big)$. Moreover, this solution depends continuously on f and w_0 in the sense that the mapping $\{w_0, f\} \to w = (y,\dot{y})$ is continuous from $\mathcal{H} \times L_2\big((0,T),V^*\big)$ to $L_2\big((0,T),V \times V\big)$.*

For a detailed proof of this theorem see [BIW].

Returning to our beam problem, we define the sesquilinear forms by

$$\text{(2.9)}\quad \sigma_1(q)(y,\phi) = \langle EI\,y'', \phi''\rangle_{L_2}$$

$$\text{(2.10)}\quad \sigma_2(q)(\dot{y},\phi) = \langle c_D I\,\dot{y}'', \phi''\rangle_{L_2} + \langle \gamma\dot{y}, \phi\rangle_{L_2}$$

with the spaces defined by $V = H_L^2(0,\ell) = \{\phi \in H^2(0,\ell)\,|\,\phi(0) = \phi'(0) = 0\}$ and $H = L_2(0,\ell)$ with weighted inner product $\langle\cdot,\cdot\rangle_H = \langle\rho\cdot,\cdot\rangle_{L_2}$. We assume throughout that $\rho \in L_\infty(0,\ell)$ with $\rho(x) \geq \alpha$ for some $\alpha > 0$. The term $f(t) = f(t,x)$ is given by

$$\text{(2.11)}\quad f(t,x) = \frac{1}{\rho}K_B\big(H''(x-x_2) - H''(x-x_1)\big)u(t) + \frac{1}{\rho}\tilde{f}(t,x), \quad 0 \leq x, x_1, x_2 \leq \ell,$$

where $f(t,\cdot)$ belongs to the dual space (see [W]) $V^* = \left(H_L^2(0,\ell)\right)^*$ (we assume that $\tilde{f}(t,\cdot) \in V^*$ which is a weak assumption on $\tilde{f}$). If both $|EI(\cdot)|_{L_\infty} \geq \alpha$ and $|c_D I(\cdot)|_{L_\infty} \geq \alpha$ for some $\alpha > 0$, then σ_1 and σ_2 are V-elliptic and continuous with σ_1 symmetric; hence by Theorem 2.1 our beam equation is well posed for $f(t,x)$ given by (2.11).

The parameter estimation problems can be stated in terms of finding parameters which give the best fit of the parameter dependent solutions of the partial differential equations to the observation data for response of the system to various excitations. In our case, the parameters to be estimated include beam mass density $\rho(x)$, stiffness coefficient $EI(x)$ as well as damping parameters $c_D I(x)$, γ and piezoelectric material parameters K_B, K_s. Let the collection of unknown parameters be denoted by $q = (\rho(x), EI(x), c_D I(x), \gamma, K_B, K_s)$. For given observations $\{z_i\}$ corresponding to measurements at times t_i as obtained in most practical cases, we consider the least squares estimation problem of minimizing over $q \in Q$ the least squares functional

$$J(y,z;q) = \left|\tilde{C}_2\left(\tilde{C}_1\{y(t_i,\cdot;q)\} - \{z_i\}\right)\right|^2, \tag{2.12}$$

where $\{y(t_i,\cdot;q)\}$ are the parameter dependent weak solutions of (2.4) or (2.6) evaluated at each time t_i, $i = 1,2,\cdots,\bar{N}$ and $|\cdot|$ is an appropriately chosen Euclidian norm. The set Q is some admissible parameter set. The operator $\tilde{C}_1$ may have several forms depending on the type of sensors being used. When the collected data are displacement, velocity, or acceleration at a point $\bar{x}$ on the beam, we minimize

$$J_\nu(q) = \sum_{i=1}^{\bar{N}} \left|\frac{\partial^\nu y}{\partial t^\nu}(t_i,\bar{x};q) - z_i\right|^2, \tag{2.13}$$

for $\nu = 0,1,2$, respectively. In this case the operator $\tilde{C}_1$ involves differentiation (either $\nu = 0, 1$ or 2 times, respectively) with respect to time followed by pointwise evaluation in t and x. When a piezoelectric sensor is used, the functional to be minimized is

$$J_p(q) = \sum_{i=1}^{\bar{N}} \left|K_s\left(\frac{\partial y}{\partial x}(t_i,x_2;q) - \frac{\partial y}{\partial x}(t_i,x_1;q)\right) - z_i\right|^2, \tag{2.14}$$

for the piezoelectric elements being located on the beam between x_1 and x_2. Here $\{z_i\}$ are the measured voltages across the piezoelectric elements.

The operator $\tilde{C}_2$ may be the identity (corresponding to time domain identification procedures as in (2.13) and (2.14)) or the Fourier transform (corresponding to identification in the frequency domain). If the identification is carried out in the frequency domain and the operator $\tilde{C}_2$ is a Fourier transform operator, the corresponding cost functional is given by

$$(2.15)\quad \hat{J}(q)=\sum_{\ell=1}^{\bar{N}_M}\left(\epsilon_1\left|\,\mathrm{f}_{k_\ell^y}(q)-\mathrm{f}_{k_\ell^z}\right|^2+\epsilon_2\sum_{j=-n_\ell}^{N_\ell}\left|\|U(k_\ell^y+j;q)\|-\|Z(k_\ell^z+j)\|\,\right|^2\right),$$

where $U(k;q)$ and $Z(k)$ are the Fourier series coefficients of $\tilde{C}_1\{y(t_\imath,\bar{x};q)\}$ and $\{z_\imath\}$ respectively, $\mathrm{f}_{k_\ell^y}$ and $\mathrm{f}_{k_\ell^z}$ are the $(k_\ell^y)^{th}$ vibration frequency of the solution $U(k;q)$ and the $(k_\ell^z)^{th}$ frequency of the observation data $Z(k)$, ϵ_1, ϵ_2 are weighting constants, and n_ℓ, N_ℓ are certain lower and upper limits associated with the width (or the support) of the ℓth spikes. In obtaining (2.15), we have assumed that there are a finite and distinct number $\bar{N}_M$ ($< \bar{N}$) of "spikes", i.e. vibration frequencies, among the $Z(k)$ and the number of spikes of the solution $U(k;q)$ is the same as $\bar{N}_M$. We refer to [BW] for a detailed derivation and discussion of the cost functional (2.15).

The minimization in our parameter estimation problems involves an infinite dimensional state space and an infinite dimensional admissible parameter set (of functions). We thus consider Galerkin type approximations in the context of the variational formulation of (2.6). Let H^N be a sequence of finite dimensional subspaces of H, and Q^M be a sequence of finite dimensional sets approximating the parameter set Q. We define the orthogonal projections $P^N : H \to H^N$ of H onto H^N. Then a family of approximating estimation problems with finite dimensional state spaces and parameter sets can be formulated by seeking $q \in Q^M$ which minimizes

$$(2.16)\qquad J^N(y^N,z;q)=\left|\tilde{C}_2\Big(\tilde{C}_1\{y^N(t_\imath,\cdot;q)\}-\{z_\imath\}\Big)\right|^2,$$

where $y^N(t;q) \in H^N$ is the solution to the finite dimensional approximation of (2.6) given by

$$(2.17)\qquad \begin{aligned}&\langle \ddot{y}^N(t),\psi\rangle_{V^*,V}+\sigma_1(q)(y^N(t),\psi)+\sigma_2(q)(\dot{y}^N(t),\psi)=\langle f(t),\psi\rangle_{V^*,V}\\ &y^N(0)=P^N y_0,\ \ \dot{y}^N(0)=P^N y_1,\end{aligned}$$

for $\psi \in H^N$.

For the parameter sets Q and Q^M, and state spaces H^N, we make the following hypotheses:

(H1) The sets Q and Q^M lie in a metric space $\tilde{Q}$ with metric d with Q, Q^M compact in this metric and there is a mapping $i^M : Q \to Q^M$ so that $Q^M = i^M(Q)$. Furthermore, for each $q \in Q, i^M(q) \to q$ in $\tilde{Q}$ with the convergence uniform in $q \in Q$;

(H2) The finite dimensional subspaces H^N satisfy $H^N \subset V$ as well as the approximation properties: For each $\psi \in V$, $|\psi - P^N\psi|_V \to 0$ *as* $N \to \infty$.

In addition to (uniform in Q) ellipticity and continuity conditions (2.7) and (2.8), the sesquilinear forms $\sigma_1(q)$ and $\sigma_2(q)$ are assumed to be defined on Q and satisfy the continuity-with-respect-to-parameter conditions

$$|\,\sigma_i(q)(\phi,\psi) - \sigma_i(\tilde{q})(\phi,\psi)\,| \leq \gamma_i\, d(q,\tilde{q})\, |\phi|_V\, |\psi|_V, \qquad i = 1, 2, \tag{2.18}$$

for $\phi, \psi \in V$ and $q, \tilde{q} \in Q$ where the constants $\gamma_1,\ \gamma_2$ depend only on Q.

Solving the approximate estimation problems involving (2.16) and (2.17), we obtain a sequence of estimates $\{\bar{q}^{N,M}\}$. To obtain parameter estimate convergence and continuous dependence (with respect to the observations $\{z_i\}$) results, it has been shown in [B], [BK3] and [BWI] that it is suffices, under the assumption that Q is a compact set, to argue that for arbitrary $\{q^{N,M}\} \in Q$ with $q^{N,M} \to q \in Q$, we have

$$\tilde{C}_2\tilde{C}_1 y^N(t; q^{N,M}) \to \tilde{C}_2\tilde{C}_1 y(t;q) \tag{2.19}$$

for each t.

The following theorem and corollary establish the convergence (2.19) for all forms of $\tilde{C}_1,\ \tilde{C}_2$ that are of interest to us here.

Theorem 2.2. *Suppose that Q, Q^M satisfy (H1) and H^N satisfies (H2). Assume that the sesquilinear forms $\sigma_1(q)$ and $\sigma_2(q)$ satisfy (2.7), (2.8), (2.18). Furthermore, assume that*

$$q \to f(t;q) \text{ is continuous from } Q \text{ to } L_2\big((0,T), V^*\big).$$

Let $q^{N,M}$ be arbitrary in Q such that $q^{N,M} \to q$ in Q; then for $t > 0$ as $N, M \to \infty$ we have

$$\frac{\partial^\nu y^N}{\partial t^\nu}(t; q^{N,M}) \to \frac{\partial^\nu y}{\partial t^\nu}(t;q) \qquad \text{in } V\text{-norm for } \nu = 0, 1, 2,$$

where y^N, $\dot{y}^N$ are the solutions to (2.17) and y, $\dot{y}$ are the solutions to (2.6).

Furthermore we have

Corollary 2.3. *Under the assumptions of Theorem 2.2, we have as $N, M \to \infty$*

$$\frac{\partial y^N}{\partial x}(t; q^{N,M}) \to \frac{\partial y}{\partial x}(t;q) \qquad \text{in } H^1\text{-norm, for } t > 0,$$

$$\sum_{\ell=1}^{\bar{N}_M}\left(\epsilon_1\left|f_{k_\ell^{y^N}}(q^{N,M}) - f_{k_\ell^y}(q)\right|^2 + \epsilon_2 \sum_{j=-n_\ell}^{N_\ell}\left|\,\|U^N(k_\ell^{y^N}+j; q^{N,M})\| - \|U(k_\ell^y+j;q)\|\,\right|^2\right) \to 0,$$

where $U^N(k;q^{N,M})$ and $U(k;q)$ are the Fourier coefficients for $\tilde{C}_1\{y^N(t;q^{N,M})\}$ and $\tilde{C}_1\{y(t;q)\}$, respectively.

The corollary follows from Theorem 2.2. For a detailed proof when $\tilde{C}_2$ is the Fourier transform see [BWI]. For the proof of Theorem 2.2, we refer the reader to [BWIS] for the cases $\nu = 0, 1$ and [BR] for the case $\nu = 2$.

We have outlined a framework for methods to carry out parameter identification for distributed parameter systems in both the time domain and the frequency domain. The motivation for introducing a frequency domain technique is that the time domain techniques give poor results when the observations contain several vibration modes and the initial guesses for parameters are not close to the optimal ones ([WA]). In such a situation we use a hybrid of these two methods: we iterate in a frequency domain formulation to obtain a reasonable initial parameter estimate for the time domain formulation which is then used to complete the identification procedure.

In Theorem 2.2, we presented convergence results allowing approximation of the parameter set. This form of the theoretical foundation provides a sound necessary basis for the detection and quantification of damage in structures since damage affects the dynamic behavior of systems in terms of changes in the system parameters such as ρ and I.

In concluding this section we make further comments related to the various allowable forms of the operator $\tilde{C}_1$. The flexibility in the choice of $\tilde{C}_1$ and hence flexibility in the choice of sensors in the above theoretical framework makes it possible to carry out cross experiments to test the reliability of our methods. Even more important, this flexibility in our method enables us to use smart materials to carry out damage detection by employing only piezoelectric materials in actuating and sensing for the structure.

3. Quantitative and Experimental Results

In implementing the computational method outlined in Section 2, we chose cubic splines as the basis elements. The finite dimensional space H^N is defined as the span of the basis set for $S^N_{3,B}(0, \ell)$ (the cubic B-splines corresponding to mesh size ℓ/N - see [P]) with the basis set modified to satisfy the essential boundary conditions $\phi(0) = \phi'(0) = 0$. The finite dimensional space H^N constructed in this manner satisfies hypothesis (H2) (for details see [BWIS]). The dimension of the resulting approximation space is $N + 1$. For our calculations we chose $N = 10$, since the eigenvalues of this approximate finite dimensional system become stable at and after $N = 10$ in the sense that the eigenvalues did not change significantly as N increases beyond 10.

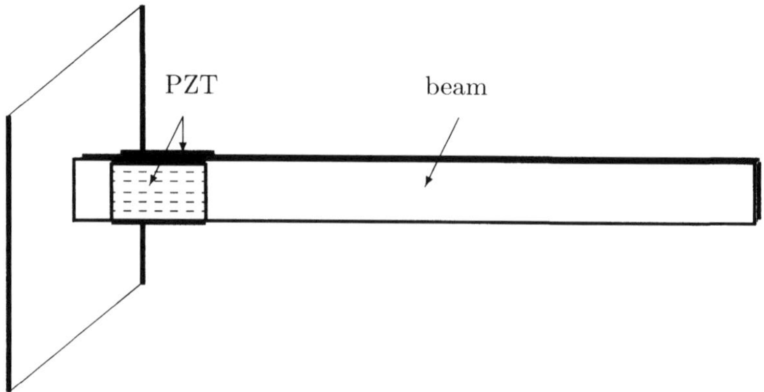

FIGURE 3.1. Test beam with piezoceramic patches.

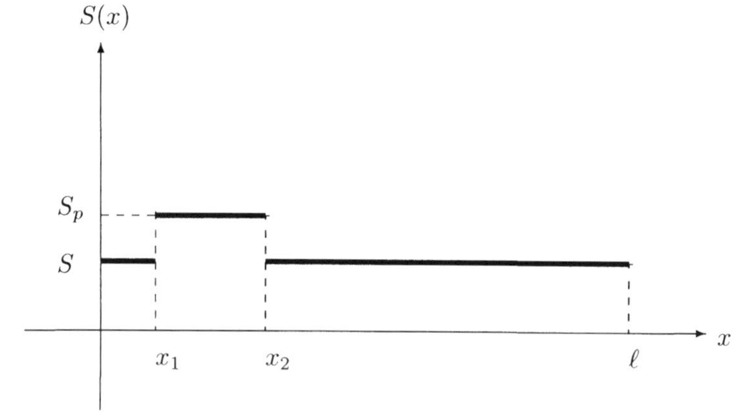

FIGURE 3.2. $\rho(x)$, $EI(x)$, $c_D I(x)$ function shape $S(x)$.

Unit	ℓ_b	w_b	t_b	ℓ_p	w_p	t_p
Metric	$45.73cm$	$2.03cm$	$0.16cm$	$6.37cm$	$2.03cm$	$0.0254cm$
English	$18''$	$0.8''$	$.0625''$	$2.5''$	$0.8''$	$0.01''$

TABLE 3.1. Dimension of the beam and piezoceramic patch.

Unit	E_b	ρ_b	E_p	ρ_p
Metric	$7.3 \times 10^6 N/cm^2$	$2.766g/cm^3$	$6.3 \times 10^6 N/cm^2$	$7.6g/cm^3$
English	$10.6 \times 10^6 psi$	$0.1\ell b/in^3$	$9.148 \times 10^6 psi$	$0.275\ell b/in^3$

TABLE 3.2. Characteristics of the beam and piezoceramic patch.

In the numerical examples we present next, an undamaged beam with two attached piezoceramic patches (PZT) as shown in Figure 3.1 was used as a test structure to demonstrate the consistency of identification results for our computational method across different excitation and sensing mechanisms. To accurately model the geometry of the structure, we assumed that the mass density, stiffness and Kelvin-Voigt damping coefficients are piecewise constant functions as depicted in Figure 3.2. These assumptions were subsequently verified to be most reasonable. For the choice of parameters $q = (\rho(x), EI(x), c_D I(x), \gamma, K_B, K_s)$, we take the parameter space $\tilde{Q}$ to be $[L_\infty(0,\ell)]^3 \times \mathbb{R}^3$, and the parameter set Q to be a uniformly bounded collection of piecewise constant functions each having jump discontinuities at most at x_1 and x_2. If we choose the mapping i^M as the identity, then hypothesis (H1) is satisfied.

Experiments with a test beam as depicted in Figure 3.1 were carried out at the Mechanical Systems Laboratory, then located at the State University of New York at Buffalo (now at Virginia Polytechnic Institute and State University). The measured dimensions of the actual structure are given in Table 3.1 and the handbook values for the physical characteristics of this aluminum (2024-T4) beam/piezo (G-1195 PZT ceramic) structure are given in Table 3.2. In the tables the subscripts indicate the materials: b for beam and p for piezoceramic, ℓ is length, w is width, t is thickness, E is the Young's modulus, and ρ is the mass density. The patches were bonded to the beam on the opposite sides of the beam at the same position.

The beam was clamped at $x = 0$. The piezoceramic patches were placed between $x_1 = 2.54cm$ and $x_2 = 8.89cm$ (or $1''$ and $3.5''$ respectively) on the beam. The time response data and input signal from the experimental beam were obtained using a Tektronix Analyzer (model 2600). For more detailed descriptions of the experimental setup and the parameter identification procedure, see [BWIS].

In examining the ability of our method to estimate parameters reliably, independent of the type of excitation or sensing device used, two separate experiments were performed. In the first experiment an impulse hammer was used as the input device and piezoceramic patches were used as a sensor. In the second experiment, the excitation was obtained via an input voltage to the patches and observation data was recorded from an accelerometer. At a sampling frequency of 256 Hz for 16 seconds, two modes (at 6.625 Hz and 38.375 Hz) were observed in the response in each experiment.

The results (graphs comparing the model response after ID of parameters with the experimental data) for the first identification experiment are given in Figure 3.3.

In this example, the beam was excited by an impulse force applied (via the impulse hammer) to the beam along the neutral axis at $2.54cm(1'')$ away from the clamped end. Then the forcing function in (2.11) becomes (since $u \equiv 0$)

$$f(t,x) = \tilde{f}(t,x)$$

with $\tilde{f}$ given by the impulse. The recorded signal exhibited a triangular shape. The voltages across the patches due to the beam vibrations were collected as observation data.

In the second experiment, a narrow triangular shaped voltages (approximating an impulse) was applied to the patches to excite the beam. The piezoceramic patches were excited out-of-phase so as to produce pure bending moments. Hence the input function is

$$f(t,x) = K_B\big(H''(x-x_2) - H''(x-x_1)\big)u(t).$$

An accelerometer weighing 0.5 gram, located at $\bar{x} = 2.14cm(0.844'')$ was used as the sensor. The parameter identification results are shown in Figure 3.4 where the data time history is compared to the model response corresponding to the best-fit parameters.

A summary of the estimated parameters is given in Table 3.3, where the units are for EI in $N \cdot m^2$, ρ in kg/m, c_DI in $s \cdot N \cdot m^2$, γ in $s \cdot N/m^2$, K_s in v, and K_B in $N \cdot m/v$. The measured and handbook quantities are listed in the table as "given" values; these were calculated according to the following equations

$$\rho(x) = \rho_b t_b w_b + 2\rho_p t_p w_p \chi_p(x) \tag{3.1}$$

$$EI(x) = \frac{1}{12} t_b^3 w_b E_b + \frac{2}{3}\left[\left(\frac{t_b}{2} + t_p\right)^3 - \left(\frac{t_b}{2}\right)^3\right] w_p E_p \chi_p(x) \tag{3.2}$$

where χ_p is a characteristic function given by

$$\chi_p(x) = \begin{cases} 1 & x_1 < x < x_2 \\ 0 & \text{otherwise.} \end{cases} \tag{3.3}$$

We next sought to demonstrate the capability of piezoelectric materials in smart structures (specifically in the area of self-evaluation of structures). We designed and performed an experiment on the same beam in which the piezoceramic

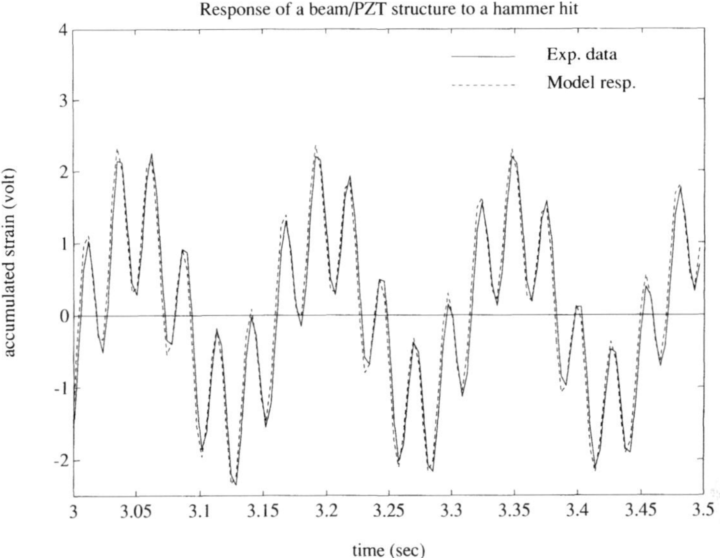

(a): Time history in [3, 3.5] seconds.

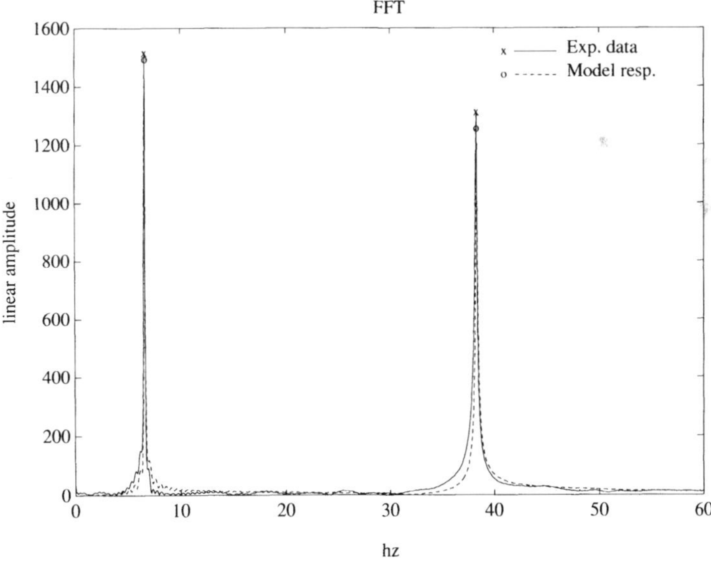

(b): Frequency content.

FIGURE 3.3. ID result for Experiment 1 (actuator–impulse hammer, sensor–PZT).

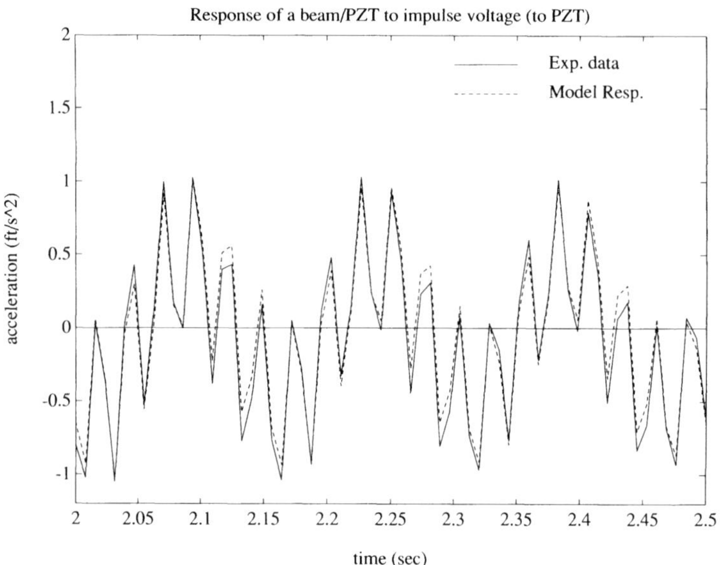

(a): Time history in [3, 3.5] seconds.

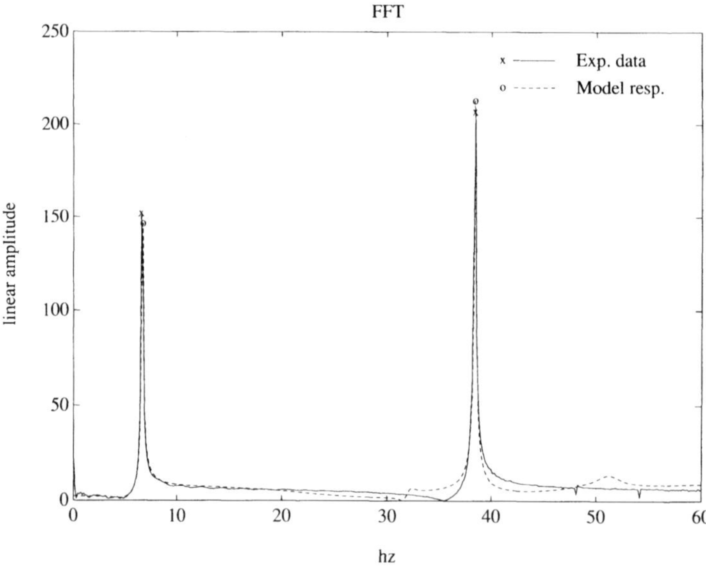

(b): Frequency content.

FIGURE 3.4. ID result for Experiment 2 (actuator–PZT; sensor–accelerometer).

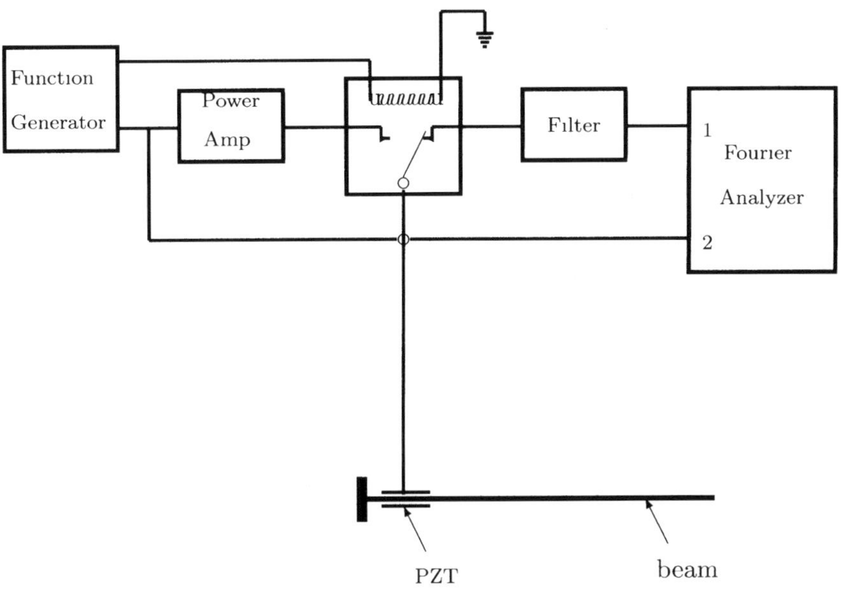

FIGURE 3.5. Circuit for the experiment using PZT as actuator/sensor.

patches were used as actuator and sensor in self-identification of material parameters. In this third experiment, the structure was excited by an input of sinusoidal voltage, out-of-phase, to the PZT patches for a time period of $[0, t_0)$ and was left to free vibrate for $t \geq t_0$. The circuit setup for this experiment is depicted in Figure 3 5

The accumulated strain in form of a voltage across the patches was recorded for approximately 14 seconds. The sample rate was 256 Hz and the excitation frequency was 38.25 Hz; this is approximately the second vibration mode of the composite structure (beam with bonded patches). The estimated parameters are reported in Table 3.4 in which the units are the same as those given for Table 3.3. Comparison of the model response (after parameter ID) to the experimental data is shown in Figure 3.6.

It can be observed that identification results for $\rho(x), x_1 < x < x_2$, in all three examples yield values that are substantially larger than the given one. One explanation for this is that equation (3.1) is simply a superposition of linear mass densities of the beam and the patches which ignores the mass of the bonding glue and the conducting copper foil.

Exp No	actuator / sensor	EI		ρ		$c_D I$ $(\times 10^{-5})$		γ	K_s	K_B
		beam	beam + PZT	beam	beam + PZT	beam	beam + PZT			
given		0 495	1 051	0 089	0 168	—	—	—	—	—
Exp 1	hammer /PZT	0 491	0 793	0 093	0 433	0 649	1 255	0 013	46 823	—
Exp 2	PZT/ accel	0 505	0 798	0 096	0 441	0 637	1 275	0 013	—	0 0175

TABLE 3.3. Given and estimated structural parameters for Experiments 1 and 2.

Exp No	actuator / sensor	EI		ρ		$c_D I$ $(\times 10^{-5})$		γ	K_s	K_B
		beam	beam + PZT	beam	beam + PZT	beam	beam + PZT			
given		0 495	1 051	0 089	0 168	—	—	—	—	—
Exp 3	PZT /PZT	0 505	0 796	0 095	0 483	0 697	1 275	0 013	46 823	0 0196

TABLE 3.4. Given and estimated structural parameters for Experiment 3.

In all our parameter estimation procedures, we adopted a so-called hybrid method (mentioned previously) , wherein we started with the cost function (2.15) (the cost function corresponding to frequency domain data) and then switched to (2.13) or (2.14) (time domain data) Our experience on these and numerous other inverse problems reveal that parameter identification in time domain is more sensitive and accurate if the initial guesses of parameters are close to the optimal ones. On the other hand, parameter identification in the frequency domain will yield quick and rough estimates with the resulting parameters in a neighborhood of the optimal values.

From the above reported results, we can conclude that the methodology outlined in the previous section yields theoretical, computational and experimental findings that are consistent. Moreover, it is feasible to use smart materials in nondestructive self-evaluation schemes such as presented in the next section.

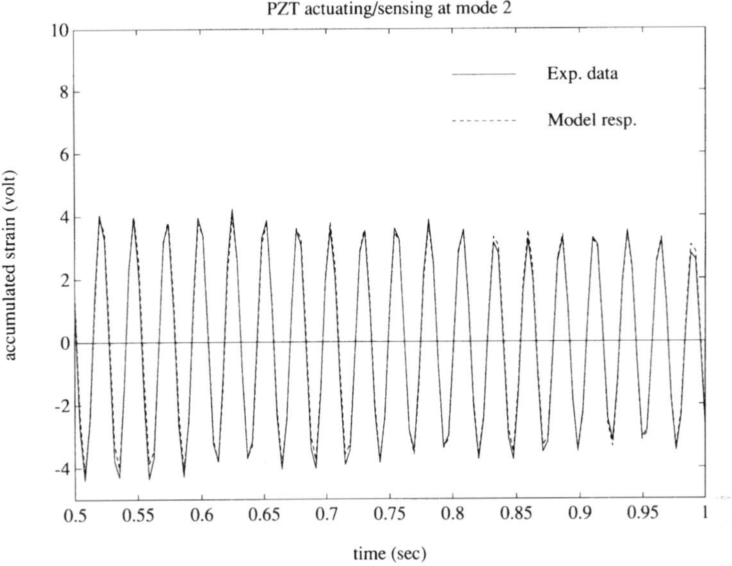

(a): Time history in [0.5, 1.0] seconds.

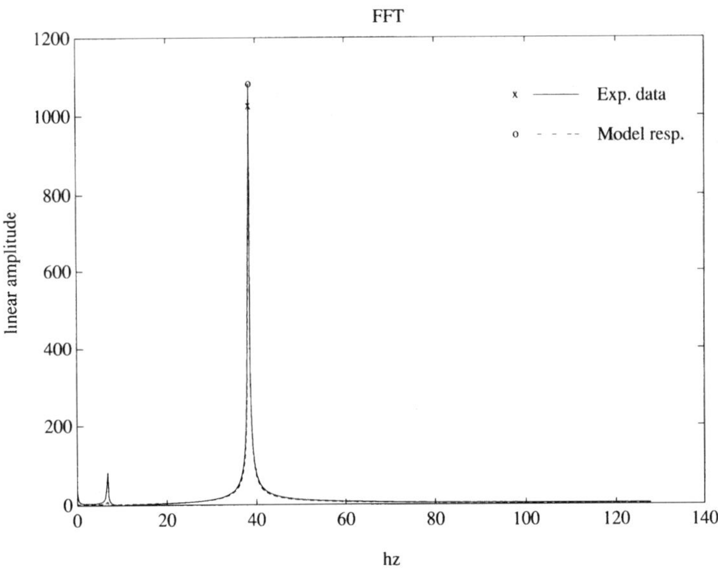

(b): Frequency content.

FIGURE 3.6. ID result for Experiment 3 (actuator–PZT; sensor–PZT).

4. Detection and Characterization of Damage

In our preliminary investigation of damage detection capabilities we have carried out simulation studies. The structure used in the simulation has the same dimensions and physical characteristics as the beam in Section 3. The difference is that the beam is assumed damaged by a hole. Since the beam model is based on Euler-Bernoulli theory, we assume that the center of the hole coincides with the neutral axis of the beam. Furthermore, we assume that the hole is either a circle centered at x_d with dimension r or a rectangle centered at x_d with dimension of $2r \times w_d$. With these assumptions, the functions $\rho(x)$ and $EI(x)$ can be expressed by

$$\rho(x) = \rho_b t_b w_b + 2\rho_p t_p w_p \chi_p(x) - \rho_b t_b (w_b - \tilde{w})\chi_d(x) \tag{4.1}$$

$$EI(x) = \frac{1}{12} t_b^3 w_b E_b + \frac{2}{3}\left[\left(\frac{t_b}{2} + t_p\right)^3 - \left(\frac{t_b}{2}\right)^3\right] w_p E_p \chi_p(x) - \frac{1}{12} t_b^3 E_b (w_b - \tilde{w})\chi_d(x) \tag{4.2}$$

where the characteristic function χ_p is the same as (3.3) and χ_d is given by

$$\chi_d(x) = \begin{cases} 1 & x_{d_1} < x < x_{d_2} \\ 0 & \text{otherwise.} \end{cases} \tag{4.3}$$

Here $x_{d_1} = x_d - r$ and $x_{d_2} = x_d + r$, and

$$\tilde{w} = \begin{cases} w_d & \text{for a rectangular hole,} \\ 2\sqrt{r^2 - (x - x_d)^2} & \text{for a circular hole.} \end{cases}$$

The Kelvin-Voigt damping function $c_D I(x)$ has the same form as (4.2) with E replaced by c_D. In both examples, the damage is parameterized by parameters r and x_d with w_d held fixed in the case of a rectangular hole at a value $w_d =$ 1.016cm(.4″).

In the simulations, we assumed that we knew the shape (rectangular or circular) of the damage. The damage detection (location and size of the hole) is carried out through parameter identification by fitting the solution of the model to the simulated observation data much in the same manner as described for the experimental data cases described in Section 3. Voltage measurements across the piezoceramic patches are used as observations.

The input data is a triangular (time function) voltage in a very short time period (see Figure 4.1) to simulate an impulse input. The choice of the triangular function was motivated by impulse hammer excitation records from actual experiments as reported in Section 3.

The simulated "observation data" $\{z_i\}$ were generated by solving the finite dimensional model (2.17). The number of finite elements was set at $N = 10$ with cubic spline basis elements. The time interval was $[0, 8]$ seconds with sampling rate

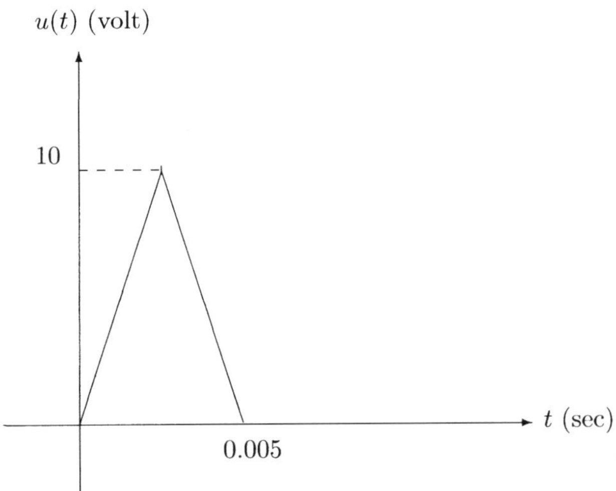

FIGURE 4.1. Input voltage.

512 Hz. Parameters estimated from experimental data and given in Table 3.4 were used for the simulated test structure.

In the following two examples, one with a rectangular hole and one with a circular hole, two cases were investigated. In Case I, the simulated observation data was generated with a damaged beam (a hole in the beam) and in Case II the data was generated with a undamaged beam (without holes). In these simple preliminary tests, these simulated data sets were used without additional noise. In the parameter estimation procedures, we also used 10 finite elements (cubic splines) in solving the inverse problems.

EXAMPLE 1. We assumed there was a rectangular hole in the beam. The parameters defining the hole in the simulated data were

$$r = 0.508cm(0.2''), \quad x_d = 10.414cm(4.1'').$$

The initial guesses and converged parameters are shown in Table 4.1.

EXAMPLE 2. In this example, the damage is characterized by a circular hole in the beam. The parameters for the hole were

$$r = 0.508cm(0.2''), \quad x_d = 10.414cm(4.1'').$$

The initial guesses and converged parameters are shown in Table 4.2.

As was noted, in these simulation examples we used noise free simulated observation data in the parameter estimation problems. We are currently studying the effects of noise in the data on the accuracy of the estimated parameters.

Case I.

Unit	Parameters	Initial	Final	Preassigned
Metric	$x_d(cm)$	3.048	10.414	10.414
	$r(cm)$	0.152	0.508	0.508
English	$x_d(in)$	1.2	4.1	4.1
	$r(in)$	0.06	0.2	0.2

Case II.

Unit	Parameters	Initial	Final	Preassigned
Metric	$x_d(cm)$	3.048	1.753	0
	$r(cm)$	0.152	1.592×10^{-6}	0
English	$x_d(in)$	1.2	0.69	0
	$r(in)$	0.06	6.27×10^{-7}	0

Case I. damage in shape of a rectangle; Case II. no damage

TABLE 4.1. Damage detection results

Case I.

Unit	Parameters	Initial	Final	Preassigned
Metric	$x_d(cm)$	3.048	10.414	10.414
	$r(cm)$	0.152	0.508	0.508
English	$x_d(in)$	1.2	4.1	4.1
	$r(in)$	0.06	0.2	0.2

Case II.

Unit	Parameters	Initial	Final	Preassigned
Metric	$x_d(cm)$	3.048	3.315	0
	$r(cm)$	0.152	4.55×10^{-3}	0
English	$x_d(in)$	1.2	1.305	0
	$r(in)$	0.06	1.791×10^{-3}	0

Case I. damage in shape of a circle; Case II. no damage

TABLE 4.2. Damage detection results

In many simulation studies found in the literature, the impulse function inputs are approximated by a smooth function, for example, $\sin^2(\alpha t)$. Our choice of a triangular shape for these input functions is in practice equivalent to the choice of a smooth function since discrete input data are used and hence smoothness of the time records does not play a role in the sampling process or the trajectories obtained.

The results reported in this section are very encouraging. These initial calculations suggest that our methods for parameter identification can reveal whether there is a damage or not. Our current efforts involve eliminating the assumed preknowledge of the shape of the damage. We also are investigating issues related to the detection and characterization of cracks in structures.

5. Concluding remarks

As we have noted in the introduction of this paper, the idea of using vibration testing as a basis for damage detection in structures is not new. However, most methods to date are based on modal techniques. In this paper we have presented a theoretical and computational non-modal framework for the identification of spatially dependent dynamic parameters in piezoceramic embedded structures using nondestructive vibration tests. This rigorous foundation permits one to use the piezoceramics to both excite and sense vibrations in a self-analysis framework that is a natural major feature of smart material structures.

Using data from beam experiments, we have demonstrated the feasibility of our approach in obtaining reliable physically meaningful dynamic parameters such as stiffness, damping, and mass density. In preliminary simulation studies on the possibility of detecting and geometrically characterizing damage such as holes, we have presented examples which suggest that such methods offer great promise. While much remains to be done in developing these ideas into general algorithms, we are most encouraged by our initial findings and are currently designing and carrying out physical experiments to support further efforts on geometric characterization of damage such as corrosions, cracks, and delaminations in composite material structures.

Acknowledgement

The authors would like to gratefully acknowledge continuing valuable discussions and encouragement from Dr. W. Winfree, NASA Langley Research Center, and Dr. D.J. Inman, Virginia Polytechnic Institute and State University. The authors are indebted to Dr. Inman and Dr. J. Slater (Wright State University) for their assistance and collaboration on the experimental efforts summarized in this paper and the references.

References

[ABB] D. Armon, Y. Ben-Hain, and S. Braun, *Crack detection in beams by rank-ordering of eigenfrequency shifts*, Mechanical Systems and Signal Processing, to appear.

[AC] R.D. Adams and P. Cawley, *The location of defects in structures from measurements of natural frequencies*, J. Strain Analysis 14 (1979), pp. 49-57.

[ACPS] R.D. Adams, P. Cauley, C.J. Pye and B.J. Stone, *A vibrational technique for nondestructively assessing the integrity of structures*, J. Mech. Engr. Sci. 20 (1979), pp. 93-100.

[B] H.T. Banks, *On a variational approach to some parameter estimation problems*, in Distributed Parameter Systems (ed. by F. Kappel, et. al.), Springer Lecture Notes in Control and Information Sciences, 75 (1985), pp. 1-23.

[BI] H.T. Banks and D.J. Inman, *On damping mechanisms in beams*, ICASE Rep. #89-64 August, 1989; ASME J. Applied Mechanics 58 (1991), pp. 716-723.

[BIW] H.T. Banks, K. Ito and Y. Wang, *Well posedness for damped second order systems with unbounded input operators*, CRSC-TR93-10, North Carolina State University, June, 1993; Differential and Integral Equations, to appear.

[BK1] H.T. Banks and F. Kojima, *Boundary shape identification problems in two dimensional domains related to thermal testing of materials*, Quart. Applied Math., 47 (1989), pp. 273-293.

[BK2] H.T. Banks and F. Kojima, *Boundary identification for 2-D parabolic systems arising in thermal testing materials*, Proceedings of 27th IEEE Cong. on Dec. and Control (1988), pp.1678-1683.

[BK3] H.T. Banks and K. Kunisch, *Estimation Techniques for Distributed Parameter Systems*, Birkhaüser, Boston, 1989.

[BKW] H.T. Banks, F. Kojima, and W.P. Winfree, *Boundary estimation problems arising in thermal tomography*, CAMS Tech. Rep 89-6, University of Southern California; Inverse Problems 6 (1990), pp. 897-921.

[BR] H.T. Banks and D.A. Rebnord, *Analytic semigroups: applications to inverse problems for flexible structures*, in Differential Equations with Applications (ed. by J. Goldstein, et. al.), Marcel Dekker, 1991, pp. 21-35.

[BSW] H.T. Banks, R. Smith and Y. Wang, *The modeling of piezoceramic patch interactions with shell, plates and beams*, Quarterly of Applied Mathematics, to appear.

[BW] H.T. Banks and Y. Wang, *Parameter identification in the frequency domain*, in Progress in Systems and Control Theory: Computation and Control III, (ed. by K Bowers and J. Lund) Birkhäuser, 1993, pp. 49-62.

[BWI] H.T. Banks, Y. Wang and D.J. Inman, *Bending and shear damping in beams: frequency domain estimation techniques*, CAMS Tech. Rep. 91-25, September, 1991, USC; ASME J. Vibration and Acoustics, to appear.

[BWIS] H.T. Banks, Y. Wang, D.J. Inman and J.C. Slater, *Approximation and parameter identification for damped second order systems with unbounded input operators*, CRSC-TR93-9, North Carolina State University, May, 1993; Control: Theory and Advanced Technology, to appear.

[CA] E.F. Crawley and E.H. Anderson, *Detailed models for piezoceramic actuation of beams*, AIAA Conf. paper 89-1388-CP, 1989, pp. 2000-2010.

[CFW] R.L. Clark, Jr., C.R. Fuller and A Wicks, *Characterization of multiple piezoelectric actuators for structural excitation*, J. Acoust. Soc. Amer., 1991, to appear.

[CR] P. Cawley and R.Ray, *A comparison of the natural frequency changes produced by cracks and slots*, ASME J. Vibration, Acoustics, Stress and Reliability in Design 110 (1988), pp. 366-370.

[DFR] E.K. Dimitriadis, C.R. Fuller and C.A. Rogers, *Piezoelectric actuators for distributed noise and vibration excitation of thin plates*, Proc. 8th ASME Conf. on Failure Prevention, Reliability, and Stress Analysis, Montreal, 1989, pp. 223-233.

[DIG] J Dosch, D J Inman, and E Garcia, *A self-sensing piezoelectric actuator for collocated control*, J Intell Material Systems and Structures, 3 1992, pp 166-185

[FC] J L Fanson and T K Caughey, *Positive position feedback control for large structures*, Structural Dynamics and Materials Conf , AIAA Conf Paper 87-0902, 1987, pp 588-598

[IC] A Isman and K Craig, *Damage detection in composite structures using piezoelectric materials*, 1st ARO Workshop on Smart Structures and Materials, U Texas-Arlington, Sept 22-24, 1993

[P] P M Prenter, *Splines and Variational Methods*, Wiley, New York, 1975

[S] H Sato, *Free vibration of beams with abrupt changes of cross-section*, J Sound and Vibration 89 (1983), pp 59-64

[WA] Y Wang, *Damping Modeling and Parameter Estimation in Timoshenko Beams*, Ph D Thesis, Brown University, May 1991

[W] J Wloka, *Partial Differential Equations*, Cambridge University Press, New York, 1987

H T Banks
Center for Research in Scientific Computation
North Carolina State University
Raleigh, NC 27695-8205, USA
E-mail htbanks@crsc1 math ncsu edu

Y Wang
Center for Research in Scientific Computation
North Carolina State University
Raleigh, NC 27695-8205, USA
E-mail ywang@crsc1 math ncsu edu

International Series of Numerical Mathematics, Vol. 118, © 1994 Birkhäuser Verlag Basel

OPTIMALITY CONDITIONS FOR NON-QUALIFIED PARABOLIC CONTROL PROBLEMS

M. BERGOUNIOUX, T. MÄNNIKKÖ, AND D. TIBA

Département de Mathématiques et d'Informatique
U.F.R. Sciences Université d'Orléans

Department of Mathematics
University of Jyväskylä

Institute of Mathematics of the Romanian Academy

ABSTRACT. We consider parabolic state constrained optimal control problems where the usual Slater condition is not necessarily satisfied. Instead, a weaker interiority property is assumed. Optimality conditions with a Lagrange multiplier are given. As an application we present an augmented Lagrangian algorithm. Numerical test results are included.

1991 *Mathematics Subject Classification.* 49K20, 49M29

Key words and phrases. Optimal control, optimality conditions, Lagrange multipliers, parabolic systems.

1. Introduction

Let us consider the following control problem with constraints:

$$(\mathcal{P}) \qquad \text{Minimize } \left\{ \frac{1}{2} \int_Q (y - z_d)^2 \, dx \, dt + \frac{N}{2} \int_Q u^2 \, dx \, dt \right\}$$

subject to

$$\frac{\partial y}{\partial t} - \Delta y = f + u \quad \text{in } Q, \tag{1.1}$$

$$y(x,t) = 0 \quad \text{on } \Sigma, \tag{1.2}$$

$$y(x,0) = y_0 \quad \text{in } \Omega, \tag{1.3}$$

$$u \in U_{ad}, \tag{1.4}$$

$$y \in K_0, \tag{1.5}$$

where $\Omega \subset \mathbb{R}^n$ is a bounded regular domain, $Q =]0,T[\times\Omega$, $T > 0$, $\Sigma = [0,T] \times \partial\Omega$ and $N > 0$. Moreover, $z_d, f \in L^2(Q)$, $y_0 \in H_0^1(\Omega) \cap H^2(\Omega)$, and U_{ad}, K_0 are closed convex subsets of $L^2(Q)$.

We shall examine the cases where both U_{ad} and K_0 have weak interiority properties of Slater type, or have even void interior in $L^\infty(Q)$. We give optimality conditions with a Lagrange multiplier which may be a function or a measure and which appears in a symmetric form with respect to the two constraints. As an application we present an augmented Lagrangian algorithm for the solution of $(\mathcal{P})$, and two numerical examples are included.

The approach of this paper is based on the penalization of the state system rather than of the constraints. An essential role is played by new estimates of the approximate Lagrange multiplier which can be mainly compared with the earlier works [1], [2], [3] of the first author. Another approach to such problems can be found in Bonnans and Casas [5], [6].

The main novelty of this paper consists in the treatment of the state constraints when K_0 has a void interior even in $L^\infty(Q)$, that is without satisfying the Slater condition. It is shown that a weak interiority property of U_{ad} is sufficient in order to ensure the existence of the Lagrange multiplier.

2. Optimality conditions

In what follows, we consider the case where the constraint (1.5) is of the form

$$\varphi(x,t) \le y(x,t) \le \psi(x,t) \quad \text{in } Q. \tag{2.1}$$

We assume that $\varphi, \phi \in C(\bar{Q})$, $\varphi \le \psi$ in Q, $\varphi \le 0 \le \psi$ on Σ and $\varphi(x,0) \le y_0(x) \le \psi(x,0)$ in Ω. Several more specific conditions on φ, ψ and U_{ad} with examples will be given in the sequel.

Let us define

$$\begin{aligned} K = \{\, y \in H^{2,1}(Q) \mid \varphi(x,t) \le y(x,t) \le \psi(x,t) \text{ a.e. in } Q, \\ y(x,0) = y_0(x) \text{ a.e. in } \Omega \,\}. \end{aligned} \tag{2.2}$$

We approximate $(\mathcal{P})$ by the penalized, "decoupled" optimization problem:

$$(\mathcal{P}_\varepsilon) \qquad \begin{aligned} \text{Minimize } \Big\{ \frac{1}{2}\int_Q (y - z_d)^2\,dx\,dt + \frac{N}{2}\int_Q u^2\,dx\,dt \\ + \frac{1}{2\varepsilon}\int_Q \left(\frac{\partial y}{\partial t} - \Delta y - f - u\right)^2 dx\,dt \Big\} \end{aligned}$$

subject to $u \in U_{ad}$, $y \in K$.

We denote by $[y^*, u^*]$ the unique optimal pair for $(\mathcal{P})$ (which exists under the standard admissibility condition) and by $[y_\varepsilon, u_\varepsilon]$ the optimal pair for $(\mathcal{P}_\varepsilon)$. We also denote

$$q_\varepsilon = \frac{1}{\varepsilon}\left(\frac{\partial y_\varepsilon}{\partial t} - \Delta y_\varepsilon - f - u_\varepsilon\right). \tag{2.3}$$

Moreover, let p_ε be the solution of the adjoint equation

$$-\frac{\partial p_\varepsilon}{\partial t} - \Delta p_\varepsilon = y_\varepsilon - z_d \quad \text{in } Q, \tag{2.4}$$

$$p_\varepsilon(x,t) = 0 \quad \text{on } \Sigma, \tag{2.5}$$

$$p_\varepsilon(x,T) = 0 \quad \text{in } \Omega, \tag{2.6}$$

and p^* the solution of (2.4)–(2.6) with y_ε replaced by y^*.

We may easily prove the following propositions (see Bergounioux [2], [3]):

Proposition 2.1. *When $\varepsilon \to 0$, we have*

$$y_\varepsilon \to y^* \quad \textit{strongly in } H^{2,1}(Q), \tag{2.7}$$

$$u_\varepsilon \to u^* \quad \textit{strongly in } L^2(Q), \tag{2.8}$$

$$p_\varepsilon \to p^* \quad \textit{strongly in } H^{2,1}(Q), \tag{2.9}$$

$$\{\sqrt{\varepsilon}\, q_\varepsilon\} \quad \textit{bounded in } L^2(Q). \tag{2.10}$$

Proposition 2.2. *The approximate optimality conditions may be written in the decoupled form*

$$\forall y \in K \quad \int_Q (p_\varepsilon + q_\varepsilon)\left(\frac{\partial}{\partial t}(y - y_\varepsilon) - \Delta(y - y_\varepsilon)\right) dx\, dt \geq 0, \tag{2.11}$$

$$\forall u \in U_{ad} \quad \int_Q (N u_\varepsilon - q_\varepsilon)(u - u_\varepsilon)\, dx\, dt \geq 0. \tag{2.12}$$

In order to obtain a good estimate also on $\{q_\varepsilon\}$ and to pass to the limit in (2.11), (2.12), we study several cases of interest.

Example 1

Let $\varphi \equiv a$, $\psi \equiv b$ in Q and $U_{ad} = \{\, u \in L^2(Q) \mid c \leq u(x,t) \leq d \text{ a.e. in } Q \,\}$. We strengthen the admissibility condition to the following Slater type assumption:

$$\begin{gathered} \exists \delta > 0,\ \exists \tilde{u} \in U_{ad} \ \text{ such that} \\ \tilde{y} = \mathcal{T}(\tilde{u}) \ \text{ satisfies } \ a + \delta \leq \tilde{y}(x,t) \leq b - \delta \ \text{ in } Q, \end{gathered} \tag{2.13}$$

where $\mathcal{T}$ is the affine bounded operator $u \mapsto y$ defined by (1.1)–(1.3). We notice that $U_{ad} \subset L^\infty(Q)$ in this example, and the relations (2.7)–(2.9) are valid in $L^p(Q)$, $W^{2,1,p}(Q)$ for any $p > 1$.

Let us consider the pair $[y_\kappa, \tilde{u}] = [\mathcal{T}(\tilde{u} - \rho\kappa), \tilde{u}]$, where $\tilde{u}$ is given by (2.13), $\rho > 0$ will be precised later and $\kappa \in L^p(Q)$, $|\kappa|_{L^p(Q)} = 1$ is arbitrary. By the continuity with respect to the data of $\mathcal{T}$, we get

$$|\tilde{y} - y_\kappa|_{W^{2,1,p}(Q)} \leq C\rho$$

with C depending only on p and Q. If $p > (n+2)/2$, the anisotropic Sobolev embedding theorem gives

$$|\tilde{y} - y_\kappa|_{C(\bar{Q})} \leq C_1 C\rho$$

with C_1 independent of κ, ρ. Taking $\rho = \delta/(2C_1 C)$, (2.13) gives $y_\kappa \in K$ for all κ.

Proposition 2.3. *Let p and ρ be as above. Then $\{q_\varepsilon\}$ is bounded in $L^{p'}(Q)$, where $1/p + 1/p' = 1$.*

Proof. We use the pair $[y_\kappa, \tilde{u}]$ in (2.11), (2.12) and we add the two relations to obtain

$$\int_Q (p_\varepsilon + q_\varepsilon)(\tilde{u} - \rho\kappa - \varepsilon q_\varepsilon - u_\varepsilon)\, dx\, dt + \int_Q (N u_\varepsilon - q_\varepsilon)(\tilde{u} - u_\varepsilon)\, dx\, dt \geq 0.$$

This gives

$$\int_Q p_\varepsilon(\tilde{u} - \rho\kappa - \varepsilon q_\varepsilon - u_\varepsilon)\, dx\, dt + \int_Q N u_\varepsilon(\tilde{u} - u_\varepsilon)\, dx\, dt - \varepsilon \int_Q q_\varepsilon^2\, dx\, dt \geq \rho \int_Q q_\varepsilon \kappa\, dx\, dt,$$

which implies that $\{q_\varepsilon\}$ is bounded in $L^{p'}(Q)$. □

So $q_\varepsilon \to q^*$ weakly in $L^{p'}(Q)$, and we can pass to the limit in the necessary conditions since the weak convergence in $L^{p'}(Q)$ will be coupled with the strong convergence in $L^p(Q)$. Thus we have

$$\text{(2.14)} \qquad \forall y \in K \quad \int_Q (p^* + q^*)\left(\frac{\partial}{\partial t}(y - y^*) - \Delta(y - y^*)\right) dx\, dt \geq 0,$$

$$\text{(2.15)} \qquad \forall u \in U_{ad} \quad \int_Q (N u^* - q^*)(u - u^*)\, dx\, dt \geq 0.$$

Remarks. (1) These conditions are also sufficient for optimality.

(2) We also notice the completely decoupled character of (2.14), (2.15) and the fact that the Lagrange multiplier q^* is in $L^{p'}(Q)$.

Example 2

We consider the more difficult case where the state constraint set has a void interior even in $L^\infty(Q)$. In the general problem $(\mathcal{P})$ we may take

$$\text{(2.16)} \qquad \varphi(x,t) = \psi(x,t) = 0 \quad \text{on } \Sigma,$$

which implies that K has a void interior in $C(\bar{Q})$.

We impose a very weak assumption on K and U_{ad}, which does not require any interiority conditions:

$$\exists f, g \in L^2(Q),\ \exists \rho > 0,\ \exists \tilde{u} \in U_{ad} \quad \text{such that}$$

$$f(x,t) + \rho \leq \tilde{u}(x,t) \leq g(x,t) - \rho \quad \text{a.e. in } Q, \tag{2.17i}$$

$$\varphi(x,t) \leq y_f(x,t) \leq y_g(x,t) \leq \psi(x,t) \quad \text{in } \bar{Q}, \tag{2.17ii}$$

where $y_f = \mathcal{T}(f)$ and $y_g = \mathcal{T}(g)$. We notice that the pair $[\tilde{y}, \tilde{u}] = [\mathcal{T}(\tilde{u}), \tilde{u}]$ is admissible for $(\mathcal{P})$, by comparison.

We define $y_\kappa = \mathcal{T}(\tilde{u} - \rho\kappa)$, $\forall \kappa \in L^\infty(Q)$, $|\kappa|_{L^\infty(Q)} = 1$. Since it holds

$$f(x,t) \leq \tilde{u}(x,t) - \rho\kappa(x,t) \leq g(x,t) \quad \text{a.e. in } Q$$

by (2.17i), we get again by comparison

$$y_f(x,t) \leq y_\kappa(x,t) \leq y_g(x,t) \quad \text{in } \bar{Q},$$

that is $y_\kappa \in K$ by (2.17ii). As a consequence, the pair $[y_\kappa, \tilde{u}]$ is admissible for the problem $(\mathcal{P}_\varepsilon)$.

Proposition 2.4. *The sequence $\{q_\varepsilon\}$ is bounded in $L^1(Q)$ and $q_\varepsilon \to q^*$ weakly in $L^\infty(Q)^* = \mathcal{M}(Q)$ on a generalized subsequence.*

The proof is identical with the proof of Proposition 2.3 by using the above admissible pairs. However, due to the weak convergence and regularity properties, the passage to the limit in the approximate optimality conditions is no more a direct one.

We observe that $\partial y_\varepsilon/\partial t - \Delta y_\varepsilon = \varepsilon q_\varepsilon + u_\varepsilon + f$. We use this in (2.11), and then we add (2.11), (2.12) to obtain

$$\int_Q (p_\varepsilon + q_\varepsilon)\left(\frac{\partial y}{\partial t} - \Delta y - \varepsilon q_\varepsilon - u_\varepsilon - f\right) dx\,dt + \int_Q (Nu_\varepsilon - q_\varepsilon)(u - u_\varepsilon)\,dx\,dt \geq 0.$$

Then we have

$$\begin{aligned} \int_Q (p_\varepsilon + q_\varepsilon)\left(\frac{\partial y}{\partial t} - \Delta y - f\right) dx\,dt - \int_Q p_\varepsilon(\varepsilon q_\varepsilon + u_\varepsilon)\,dx\,dt \\ + \int_Q Nu_\varepsilon(u - u_\varepsilon)\,dx\,dt - \int_Q q_\varepsilon u \geq 0 \quad \forall u \in U_{ad},\ \forall y \in K. \end{aligned} \tag{2.18}$$

Proposition 2.5. (i) *For any $u \in U_{ad} \cap L^\infty(Q)$, $y \in K$ such that $\partial y/\partial t - \Delta y - f \in L^\infty(Q)$, we have the first order necessary condition for $(\mathcal{P})$:*

$$\begin{aligned} \int_Q (p^* + q^*)\left(\frac{\partial y}{\partial t} - \Delta y - f\right) dx\,dt - \int_Q p^* u^*\,dx\,dt \\ + \int_Q Nu^*(u - u^*)\,dx\,dt - \int_Q q^* u \geq 0. \end{aligned} \tag{2.19}$$

(ii) *If the set of admissible regular pairs defined in* (i) *is dense in the set of all admissible pairs, then* (2.19) *is also a sufficient optimality condition.*

Proof. If the test functions y and u are regular, then (2.19) is a consequence of (2.18) and of Proposition 2.4.

Conversely, let $[y, u]$ be any admissible pair for $(\mathcal{P})$ and $[y_n, u_n]$ a regular admissible sequence for $(\mathcal{P})$ such that $y_n \to y$ in $H^{2,1}(Q)$, $u_n \to u$ in $L^2(Q)$. Then, by (2.19) and the fact that $u_n = \partial y_n/\partial t - \Delta y_n - f$, we have

$$\int_Q p^* \left(\frac{\partial y_n}{\partial t} - \Delta y_n - f \right) dx\, dt - \int_Q p^* u^*\, dx\, dt + \int_Q N u^* (u_n - u^*)\, dx\, dt \geq 0.$$

Passing to the limit $n \to \infty$, we get

$$\int_Q p^* \left(\frac{\partial y}{\partial t} - \Delta y - f \right) dx\, dt - \int_Q p^* u^*\, dx\, dt + \int_Q N u^* (u - u^*)\, dx\, dt \geq 0$$

for any admissible $[y, u]$. Then we have

$$\int_Q p^* \left(\frac{\partial}{\partial t}(y - y^*) - \Delta(y - y^*) \right) dx\, dt + \int_Q N u^* (u - u^*)\, dx\, dt \geq 0,$$

and integrating by parts, we obtain that the pair $[y^*, u^*]$ is optimal. □

Remarks. (1) The optimality conditions (2.19) are in a semidecoupled form since only the constraints are decoupled. If u^* (and consequently y^*) is regular, or q^* is in $L^2(Q)$, then the completely decoupled optimality system may be obtained.

(2) The above result cannot be inferred by the classical approach since the Slater assumption is not valid here.

Example 3

We give an alternative approach to Example 2. The same may be done in Example 1 as well.

Assume that X is any subspace of the control space U, possibly not dense. Suppose that there exist $\tilde{u} \in U$ (not necessarily in U_{ad}) and $\rho > 0$ such that $\tilde{y} = \mathcal{T}(\tilde{u}) \in K$ and $\tilde{u} + \rho v \in U_{ad}$ for any $v \in X$, $|v|_X = 1$.

The pairs $[\tilde{y}, \tilde{u} + \rho v]$ are admissible for $(\mathcal{P}_\varepsilon)$ and, by (2.11), (2.12), we infer

$$\int_Q (p_\varepsilon + q_\varepsilon)(\tilde{u} - \varepsilon q_\varepsilon - u_\varepsilon)\, dx\, dt + \int_Q (N u_\varepsilon - q_\varepsilon)(\tilde{u} + \rho v - u_\varepsilon)\, dx\, dt \geq 0.$$

Then

$$-\rho \int_Q q_\varepsilon v\, dx\, dt \geq - \int_Q N u_\varepsilon (\tilde{u} + \rho v - u_\varepsilon)\, dx\, dt + \int_Q p_\varepsilon (\tilde{u} - \varepsilon q_\varepsilon - u_\varepsilon)\, dx\, dt$$

for any $v \in X$, $|v|_X = 1$, that is $\{q_\varepsilon\}$ is bounded in X^*.

If X equals $L^p(Q)$, then we get that $\{q_\varepsilon\}$ is bounded in $L^{p'}(Q)$ even in the case where K has void interior in $L^\infty(Q)$. The argument may proceed as in the previous examples.

3. An augmented Lagrangian algorithm

As an application of the optimality conditions we present an augmented Lagrangian algorithm for state and control constrained control problems. We make the basic assumption

$$q^* \in L^2(Q), \tag{3.1}$$

which is satisfied in a variety of instances as discussed in Examples 1 and 3. Then the regularity conditions for y and u are no longer necessary and the optimality conditions are valid in the decoupled form (2.14), (2.15).

Let $J(y,u)$ denote the cost function in $(\mathcal{P})$. We consider the problem in the abstract form

$$\text{Min}\,\{\, J(y,u) \mid y \in K,\ u \in U_{ad},\ y = \mathcal{T}(u)\,\}, \tag{3.2}$$

where $\mathcal{T}$ is defined by (1.1)–(1.3). Let $[y^*, u^*]$ be the unique optimal pair for $(\mathcal{P})$ (which exists under the usual admissibility hypothesis). There is $q^* \in L^2(Q)$ such that the optimality conditions (2.14), (2.15) hold, and p^* is, as before, the solution of (2.4)–(2.6) with y_ε replaced by y^*. Let us define the augmented Lagrangian

$$\begin{aligned}\mathcal{L}_r(y,u,q) &= \frac{1}{2}|y - z_d|^2_{L^2(Q)} + \frac{N}{2}|u|^2_{L^2(Q)} \\ &\quad + \int_Q q\left(\frac{\partial y}{\partial t} - \Delta y - f - u\right) dx\,dt + \frac{r}{2}\left|\frac{\partial y}{\partial t} - \Delta y - f - u\right|^2_{L^2(Q)},\end{aligned} \tag{3.3}$$

where r is a positive penalty parameter. Our algorithm is then as follows:

Algorithm.

1. Initialize $q_0 \in L^2(Q)$ and $u_{-1} \in L^2(Q)$.

2. Iterates q_n and u_{n-1} being given,

2a. find $y_n \in K$, solution of

$$\begin{aligned}\text{Min}_{y\in K}\Big\{\frac{1}{2}\int_Q (y - z_d)^2\,dx\,dt &+ \int_Q q_n\left(\frac{\partial y}{\partial t} - \Delta y - f\right) dx\,dt \\ &+ \frac{r}{2}\int_Q \left(\frac{\partial y}{\partial t} - \Delta y - f - u_{n-1}\right)^2 dx\,dt\Big\};\end{aligned}$$

2b. find $u_n \in U_{ad}$, solution of

$$\begin{aligned}\text{Min}_{u\in U_{ad}}\Big\{\frac{N}{2}\int_Q u^2\,dx\,dt &- \int_Q q_n u\,dx\,dt \\ &+ \frac{r}{2}\int_Q \left(\frac{\partial y_n}{\partial t} - \Delta y_n - f - u\right)^2 dx\,dt\Big\}.\end{aligned}$$

3. If $\partial y_n/\partial t - \Delta y_n - f - u_n = 0$, then STOP, else

4. assign $q_{n+1} = q_n + \rho(\partial y_n/\partial t - \Delta y_n - f - u_n)$, where $0 < \rho_0 \le \rho \le r$, and go to step 2.

Theorem 3.1. *Under the above assumptions, we have*

$$u_n \to u^* \quad \text{strongly in } L^2(Q), \tag{3.4}$$

$$y_n \to y^* \quad \text{strongly in } H^{2,1}(Q), \tag{3.5}$$

$$q_{n+1} - q_n \to 0 \quad \text{strongly in } L^2(Q). \tag{3.6}$$

Proof. Integrating by parts in (2.14) and noticing that $\partial y^*/\partial t - \Delta y^* - f - u^* = 0$, we obtain

$$\begin{aligned}\int_Q (y^* - z_d)(y - y^*)\, dx\, dt + \int_Q q^* \left(\frac{\partial}{\partial t}(y - y^*) - \Delta(y - y^*)\right) dx\, dt \\ + r \int_Q \left(\frac{\partial y^*}{\partial t} - \Delta y^* - f - u^*\right)\left(\frac{\partial}{\partial t}(y - y^*) - \Delta(y - y^*)\right) dx\, dt \ge 0\end{aligned} \tag{3.7}$$

for every $y \in K$. Similarly, from (2.15) we get

$$\begin{aligned}N \int_Q u^*(u - u^*)\, dx\, dt - \int_Q q^*(u - u^*)\, dx\, dt \\ - r \int_Q \left(\frac{\partial y^*}{\partial t} - \Delta y^* - f - u^*\right)(u - u^*)\, dx\, dt \ge 0\end{aligned} \tag{3.8}$$

for every $u \in U_{ad}$. Steps 2a and 2b of the algorithm give

$$\begin{aligned}\int_Q (y_n - z_d)(y - y_n)\, dx\, dt + \int_Q q_n \left(\frac{\partial}{\partial t}(y - y_n) - \Delta(y - y_n)\right) dx\, dt \\ + r \int_Q \left(\frac{\partial y_n}{\partial t} - \Delta y_n - f - u_{n-1}\right)\left(\frac{\partial}{\partial t}(y - y_n) - \Delta(y - y_n)\right) dx\, dt \ge 0\end{aligned} \tag{3.9}$$

and

$$\begin{aligned}N \int_Q u_n(u - u_n)\, dx\, dt - \int_Q q_n(u - u_n)\, dx\, dt \\ - r \int_Q \left(\frac{\partial y_n}{\partial t} - \Delta y_n - f - u_n\right)(u - u_n)\, dx\, dt \ge 0\end{aligned} \tag{3.10}$$

for every $y \in K$ and $u \in U_{ad}$, respectively. Let us substitute $y = y_n$ to (3.7), $u = u_n$ to (3.8), $y = y^*$ to (3.9) and $u = u^*$ to (3.10), and then add the obtained inequalities. In this way we get

$$
\begin{aligned}
&|y_n - y^*|^2_{L^2(Q)} + N|u_n - u^*|^2_{L^2(Q)} \\
&+\int_Q (q_n - q^*)\left(\frac{\partial}{\partial t}(y_n - y^*) - \Delta(y_n - y^*) - (u_n - u^*)\right) dx\,dt \\
&+r\left|\frac{\partial}{\partial t}(y_n - y^*) - \Delta(y_n - y^*) - (u_n - u^*)\right|^2_{L^2(Q)} \\
&+r\int_Q \left(\frac{\partial}{\partial t}(y_n - y^*) - \Delta(y_n - y^*)\right)(u_n - u_{n-1})\,dx\,dt \le 0.
\end{aligned}
\tag{3.11}
$$

Let us denote for a while $\tilde{y}_n = y_n - y^*$ and $\tilde{u}_n = u_n - u^*$. Then we have

$$
\begin{aligned}
&\int_Q \left(\frac{\partial}{\partial t}(y_n - y^*) - \Delta(y_n - y^*)\right)(u_n - u_{n-1})\,dx\,dt \\
&\quad= \int_Q \left(\frac{\partial}{\partial t}(\tilde{y}_n - \tilde{y}_{n-1}) - \Delta(\tilde{y}_n - \tilde{y}_{n-1})\right)(\tilde{u}_n - \tilde{u}_{n-1})\,dx\,dt \\
&\quad+ \int_Q \left(\frac{\partial \tilde{y}_{n-1}}{\partial t} - \Delta\tilde{y}_{n-1} - \tilde{u}_{n-1}\right)(\tilde{u}_n - \tilde{u}_{n-1})\,dx\,dt \\
&\quad+\frac{1}{2}\left(|\tilde{u}_n|^2_{L^2(Q)} - |\tilde{u}_{n-1}|^2_{L^2(Q)} - |\tilde{u}_n - \tilde{u}_{n-1}|^2_{L^2(Q)}\right).
\end{aligned}
\tag{3.12}
$$

Let us write (3.10) first with $n-1$ and $u = u_n$, then with n and $u = u_{n-1}$, and add the two inequalities. Then, using the step 4 of the algorithm with $n-1$, we get

$$
\begin{aligned}
&r\int_Q \left(\frac{\partial}{\partial t}(\tilde{y}_n - \tilde{y}_{n-1}) - \Delta(\tilde{y}_n - \tilde{y}_{n-1})\right)(\tilde{u}_n - \tilde{u}_{n-1})\,dx\,dt \\
&\ge (N+r)|\tilde{u}_n - \tilde{u}_{n-1}|^2_{L^2(Q)} \\
&-\rho\int_Q \left(\frac{\partial y_{n-1}}{\partial t} - \Delta y_{n-1} - f - u_{n-1}\right)(\tilde{u}_n - \tilde{u}_{n-1})\,dx\,dt.
\end{aligned}
\tag{3.13}
$$

Combining (3.12) and (3.13), and noticing again that $\partial y^*/\partial t - \Delta y^* - f - u^* = 0$, we obtain

$$
\begin{aligned}
&r\int_Q \left(\frac{\partial}{\partial t}(y_n - y^*) - \Delta(y_n - y^*)\right)(u_n - u_{n-1})\,dx\,dt \\
&\quad\ge (N+\frac{r}{2})|u_n - u_{n-1}|^2_{L^2(Q)} + \frac{r}{2}\left(|u_n - u^*|^2_{L^2(Q)} - |u_{n-1} - u^*|^2_{L^2(Q)}\right) \\
&\quad+ (r-\rho)\int_Q \left(\frac{\partial y_{n-1}}{\partial t} - \Delta y_{n-1} - f - u_{n-1}\right)(u_n - u_{n-1})\,dx\,dt.
\end{aligned}
$$

Since $r - \rho \geq 0$, this is equivalent to

$$r \int_Q \left(\frac{\partial}{\partial t}(y_n - y^*) - \Delta(y_n - y^*)\right)(u_n - u_{n-1})\, dx\, dt$$
$$\geq (N + \frac{\rho}{2})|u_n - u_{n-1}|^2_{L^2(Q)} + \frac{r}{2}\left(|u_n - u^*|^2_{L^2(Q)} - |u_{n-1} - u^*|^2_{L^2(Q)}\right)$$
$$+ \frac{\rho - r}{2}\left|\frac{\partial y_{n-1}}{\partial t} - \Delta y_{n-1} - f - u_{n-1}\right|^2_{L^2(Q)},$$

and hence, (3.11) becomes

$$|y_n - y^*|^2_{L^2(Q)} + N|u_n - u^*|^2_{L^2(Q)}$$
$$+ \int_Q (q_n - q^*)\left(\frac{\partial y_n}{\partial t} - \Delta y_n - f - u_n\right) dx\, dt + r\left|\frac{\partial y_n}{\partial t} - \Delta y_n - f - u_n\right|^2_{L^2(Q)}$$
$$+ (N + \frac{\rho}{2})|u_n - u_{n-1}|^2_{L^2(Q)} + \frac{r}{2}\left(|u_n - u^*|^2_{L^2(Q)} - |u_{n-1} - u^*|^2_{L^2(Q)}\right)$$
$$+ \frac{\rho - r}{2}\left|\frac{\partial y_{n-1}}{\partial t} - \Delta y_{n-1} - f - u_{n-1}\right|^2_{L^2(Q)} \leq 0.$$

Let us denote

$$\lambda_n = \frac{r}{2}|u_{n-1} - u^*|^2_{L^2(Q)} + \frac{r - \rho}{2}\left|\frac{\partial y_{n-1}}{\partial t} - \Delta y_{n-1} - f - u_{n-1}\right|^2_{L^2(Q)},$$

so that we have

$$\begin{aligned} &|y_n - y^*|^2_{L^2(Q)} + N|u_n - u^*|^2_{L^2(Q)} \\ &\quad + \frac{r + \rho}{2}\left|\frac{\partial y_n}{\partial t} - \Delta y_n - f - u_n\right|^2_{L^2(Q)} + (N + \frac{\rho}{2})|u_n - u_{n-1}|^2_{L^2(Q)} \\ &\leq \lambda_n - \lambda_{n+1} - \int_Q (q_n - q^*)\left(\frac{\partial y_n}{\partial t} - \Delta y_n - f - u_n\right) dx\, dt. \end{aligned} \tag{3.14}$$

Moreover, let us denote

$$\mu_n = \lambda_n + \frac{1}{2\rho}|q_n|^2_{L^2(Q)} - \frac{1}{\rho}\int_Q q^* q_n\, dx\, dt$$

and use once again the step 4 of the algorithm. Then (3.14) becomes

$$|y_n - y^*|^2_{L^2(Q)} + N|u_n - u^*|^2_{L^2(Q)}$$
$$+ \frac{r + \rho}{2}\left|\frac{\partial y_n}{\partial t} - \Delta y_n - f - u_n\right|^2_{L^2(Q)} + (N + \frac{\rho}{2})|u_n - u_{n-1}|^2_{L^2(Q)}$$
$$\leq \mu_n - \mu_{n+1} + \frac{\rho}{2}\left|\frac{\partial y_n}{\partial t} - \Delta y_n - f - u_n\right|^2_{L^2(Q)}.$$

This implies

$$(3.15)\quad |y_n - y^*|^2_{L^2(Q)} + N|u_n - u^*|^2_{L^2(Q)} + \frac{r}{2}\left|\frac{\partial y_n}{\partial t} - \Delta y_n - f - u_n\right|^2_{L^2(Q)} \le \mu_n - \mu_{n+1},$$

so that $\{\mu_n\}$ is a decreasing sequence. Moreover, $\lambda_n \geq 0$ and hence

$$\mu_n \geq \frac{1}{2\rho}\left(|q_n|^2_{L^2(Q)} - 2\int_Q q^* q_n \, dx\, dt\right) \geq -\frac{1}{2\rho}|q^*|^2_{L^2(Q)},$$

that is, $\{\mu_n\}$ is bounded from below. Therefore, the sequence $\{\mu_n\}$ is convergent and $\mu_n - \mu_{n+1} \to 0$. The inequality (3.15) and the step 4 of the algorithm imply then the desired convergence results. □

4. Numerical experiments

We shall now present two example problems and give the numerical results obtained with the augmented Lagrangian algorithm. The discretization of the algorithm is performed by using the finite element method (FEM) with respect to space and the finite difference method (implicit Euler) with respect to time. We consider the problem $(\mathcal{P})$ with the state system (1.1)–(1.3) and the constraints

$$\begin{aligned} u \in U_{ad} &= \{\, u \in L^2(Q) \mid \alpha(x,t) \leq u(x,t) \leq \beta(x,t) \text{ a.e. in } Q \,\}, \\ y \in K &= \{\, y \in H^{2,1}(Q) \mid \varphi(x,t) \leq y(x,t) \leq \psi(x,t) \text{ a.e. in } Q, \\ &\qquad\qquad y(x,0) = y_0(x) \text{ a.e. in } \Omega \,\}. \end{aligned}$$

Example 1

In our first example we have $\Omega =]0,1[\times]0,1[\subset \mathbb{R}^2$, $T = 1$, $N = 1$ and

$$\begin{aligned} z_d(x,t) &= \sin(\pi x_1)\sin(\pi x_2)e^t, \\ f(x,t) &= (2\pi + 1)\sin(\pi x_1)\sin(\pi x_2)e^t - \sin(3\pi x_1)\sin(3\pi x_2)(1 - \frac{t}{2}), \\ y_0(x) &= \sin(\pi x_1)\sin(\pi x_2). \end{aligned}$$

Moreover, the constraint functions are

$$\begin{aligned} \alpha \equiv -1, &\quad \beta \equiv 1, \\ \varphi(x,t) &= 4(x_1 - x_1^2)(x_2 - x_2^2)e^t, \\ \psi(x,t) &= 16(x_1 - x_1^2)(x_2 - x_2^2)e^t. \end{aligned}$$

The "desired" state z_d belongs to K. Furthermore, z_d is very close to the upper bound ψ, and they coincide in the point $(x_1, x_2) = (0.5, 0.5)$. Functions φ and ψ at the time $t = 1$ are presented in Figure 4.1

FIGURE 4.1. Constraint functions φ and ψ at the time $t = 1$.

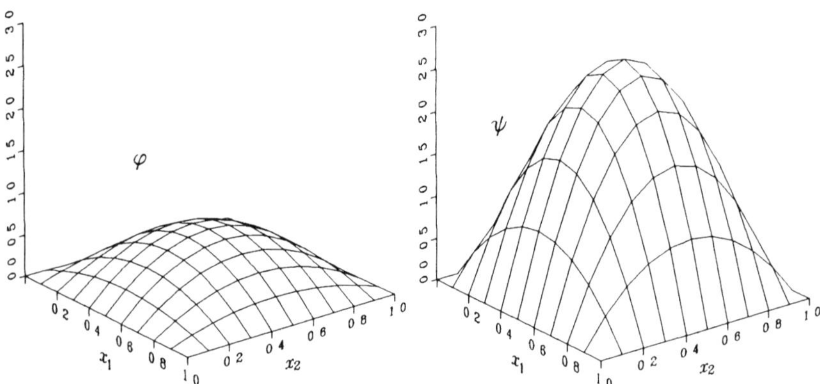

The data is chosen such that there is an admissible pair $[\hat{y}, \hat{u}]$, where $\hat{y} = z_d$ and $\hat{u}(x,t) = \sin(3\pi x_1)\sin(3\pi x_2)(1-t/2)$, which, however, is not the optimal pair.

In the space discretization we have $N_0 = 9 \times 9 = 81$ nodes ($h = 0.1$), and in the time discretization we have $K_t = 10$ time levels ($\Delta t = 0.1$). Thus, the dimension of the unknown functions is 810.

As the initial guesses we use $q_0 \equiv 0$ and $u_{-1} \equiv 0$. The minimizer used in the step 2a of the algorithm needs also an initial guess for y, for which we have chosen $y(x,t) = y_0(x)\ \forall t \in]0,1]$. The values of the cost function and the penalty term in this initial point are given in the upper part of Table 4.1

We have tested the algorithm using different values for the penalty parameter r. The parameter ρ in step 4 is chosen to be equal to r. The results are presented in the lower part of Table 4.1. In all four cases, the algorithm found a good solution already in one iteration. In the case $r = 1$, the minimizer used in step 2a failed to find a lower value for the cost function after two iterations, and in the other cases,

FIGURE 4.2. Control u at two time levels; Example 1.

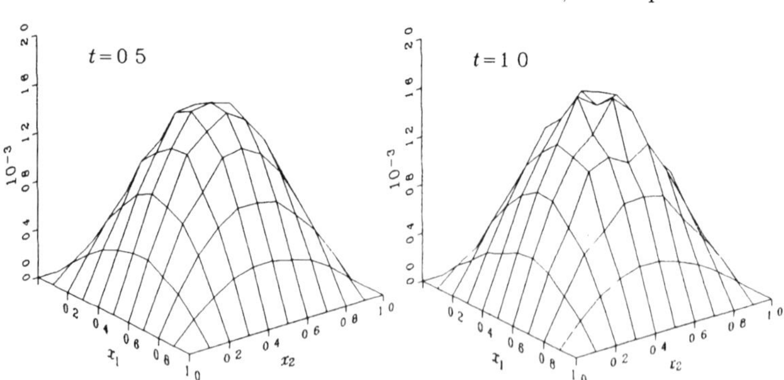

TABLE 4.1. Results of the Example 1

		$J(y,u)$	$\lvert\frac{\partial y}{\partial t}-\Delta y-f-u\rvert_{L^2(Q)}$
initial values		0.11049	9.9677

r	No. of iter.	$J(y,u)$	$\lvert\frac{\partial y}{\partial t}-\Delta y-f-u\rvert_{L^2(Q)}$
1	1	$0.18275 \cdot 10^{-3}$	$0.44888 \cdot 10^{-3}$
	2	$0.18237 \cdot 10^{-3}$	$0.77093 \cdot 10^{-3}$
10	1	$0.18337 \cdot 10^{-3}$	$0.90683 \cdot 10^{-5}$
100	1	$0.18344 \cdot 10^{-3}$	$0.28708 \cdot 10^{-6}$
1000	1	$0.18343 \cdot 10^{-3}$	$0.57937 \cdot 10^{-7}$

TABLE 4.2. Results of the Example 2

No. of	$r=1$	
iter.	$J(y,u)$	$\lvert\frac{\partial y}{\partial t}-\Delta y-f-u\rvert$
0	0.17204	0.58508
1	0.21115	0.27934
2	0.21037	0.27651
3	0.20906	0.27148
4	0.20854	0.26942
5	0.20808	0.26751
6	0.20844	0.26910
7	0.20838	0.26876
8	0.20802	0.26724
9	0.20796	0.26693
10	0.20828	0.26833

No. of	$r=10$	
iter.	$J(y,u)$	$\lvert\frac{\partial y}{\partial t}-\Delta y-f-u\rvert$
0	0.17204	0.58508
1	0.27574	0.11628
2	0.27666	0.11291
3	0.27875	0.10585
4	0.27687	0.11053
5	0.27873	0.10416
6	0.27755	0.10954
7	0.27932	0.10377
8	0.27953	0.10175
9	0.27851	0.10538
10	0.27893	0.10331

No. of	$r=100$	
iter.	$J(y,u)$	$\lvert\frac{\partial y}{\partial t}-\Delta y-f-u\rvert$
0	0.17204	0.58508
1	0.28451	0.10896
2	0.28618	0.10491
3	0.29196	0.09399
4	0.28772	0.10127
5	0.29129	0.09479
6	0.28837	0.10068
7	0.28726	0.10438
8	0.29046	0.09790
9	0.29063	0.09829
10	0.29101	0.09798

No. of	$r=1000$	
iter.	$J(y,u)$	$\lvert\frac{\partial y}{\partial t}-\Delta y-f-u\rvert$
0	0.17204	0.58508
1	0.28540	0.10855
2	0.28703	0.10453
3	0.28661	0.10539
4	0.28687	0.10575
5	0.28670	0.10754
6	0.28721	0.10687
7	0.28764	0.10709
8	0.28795	0.10760
9	0.28861	0.10730
10	0.28990	0.10645

this happened already after one iteration.

As could be expected, the larger the parameter r is, the better the constraint $\partial y/\partial t - \Delta y = f + u$ is satisfied. On the other hand, with smaller r the value of the cost function seems to get somewhat lower. In Figure 4.2 is presented the control u obtained in the case $r = 1$.

Example 2

FIGURE 4.3. Constraint functions φ and ψ when $x_2 = \frac{1}{2}$ and $t = 1$.

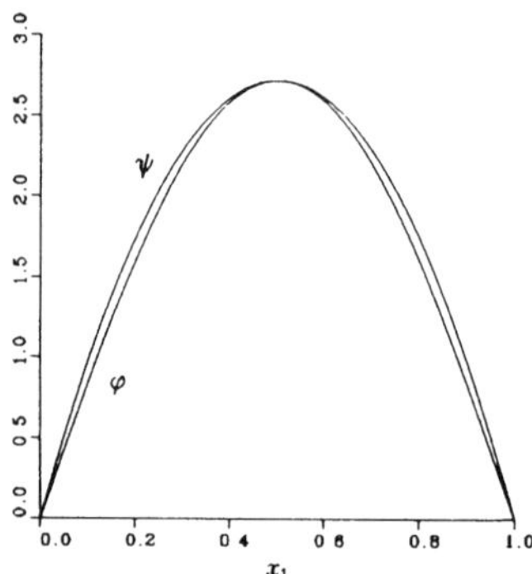

Our second example is a variant of the first one. The problem data is the same, except the following modifications have been made:

Now we have $z_d \equiv 1/2$, and

$$\varphi(x,t) = \sin(\pi x_1)\sin(\pi x_2)e^t,$$
$$\psi(x,t) = 16(x_1 - x_1^2)(x_2 - x_2^2)e^t.$$

FIGURE 4.4. Control u at two time levels; Example 2.

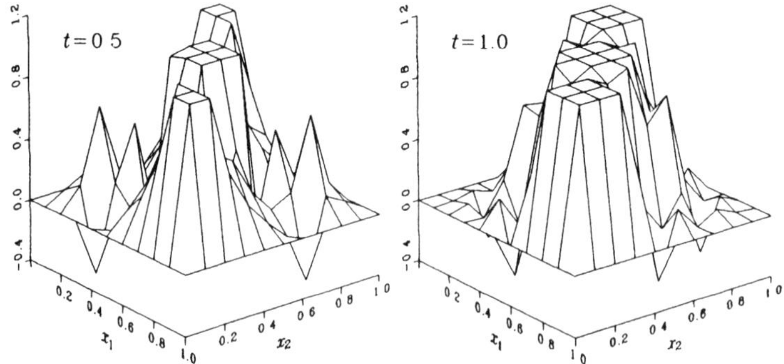

In this case, z_d does not belong to K. The constraint functions φ and ψ are chosen such that the set K is very "small"; see Figure 4.3. Moreover, there is no interior point at $(x_1, x_2) = (0.5, 0.5)$.

The pair $[\hat{y}, \hat{u}]$, where $\hat{y} = \varphi$ and $\hat{u}(x,t) = \sin(3\pi x_1)\sin(3\pi x_2)(1 - t/2)$ is admissible for this problem. The discretization is the same as before, as well as the initial guesses for q and u. The initial guess for y, however, is different, since $y(x,t) = y_0(x)$ $\forall t \in]0,1]$ does not belong to K; in this case we have used $y \equiv \varphi$.

Again, we have tested the algorithm using different values for r. The step parameter ρ is chosen to be equal to r. The results are presented in Table 4.2 (With the iteration '0' we mean the values at the initial point.)

In this example, the algorithm did not stop in the early iterations, as was the case in the previous example. However, due to the large computation time, only ten iterations have been calculated. In all four cases, the convergence is quite slow after the first iteration. If we look at the constraint $\partial y/\partial t - \Delta y = f + u$, the parameter $r = 100$ seems to give the "best" convergence. (The algorithm was tried also for $r = 10^4$ and $r = 10^5$; the results were similar to the case $r = 1000$.) In Figure 4.4 is presented the control u obtained in the case $r = 100$.

References

1 M Bergounioux, *Problèmes de contrôle avec contraintes sur l'état,* C R Acad Sci Paris, Sér I Math **310** (1990), 391–396

2 M Bergounioux, "Optimal control of parabolic problems with state constraints a penalization method for optimality conditions", Rapport de Recherche, Université d'Orléans, 1991

3 M Bergounioux, *A penalization method for optimal control of elliptic problems with state constraints,* SIAM J Control and Optimization **30** (1992), 305–323

4 M Bergounioux, T Mannikko, D Tiba, "On non-qualified parabolic control problems", Preprint 148, University of Jyvaskyla, Department of Mathematics, 1992

5 J F Bonnans, E Casas, *On the choice of the function spaces for some state constrained control problems,* Numer Funct Anal and Optim **7** (1984-85), 333–348

6 J F Bonnans, E Casas, *Optimal control of semilinear multistate systems with state constraints* SIAM J Control and Optimization **27** (1989), 446–455

M Bergounioux
Département de Mathématiques et d'Informatique
U F R Sciences Université d'Orléans
B P 6779
F-45067 Orléans Cedex 2, France

T Mannikko
Department of Mathematics
University of Jyvaskyla
P O Box 35
FIN-40351 Jyvaskyla, Finland

D. Tiba
Institute of Mathematics of the Romanian Academy
P.O. Box 1-764
RO-70700 Bucharest, Romania

CONVERGENCE OF TRAJECTORIES FOR A CONTROLLED VISCOUS BURGERS' EQUATION

CHRISTOPHER I BYRNES, DAVID S GILLIAM, AND VICTOR I SHUBOV

Department of Systems Science and Mathematics
Washington University

Department of Mathematics
Texas Tech University

ABSTRACT In this paper we consider a boundary control problem for Burgers' equation on a finite interval The controls enter as gain parameters in the boundary conditions as in [1, 2] The uncontrolled problem is obtained by equating the control parameters to zero while the zero dynamics system is obtained by equating the control parameters to infinity, or (intuitively) as the "high gain limit" of the system as the gains approach plus infinity The main result of the paper is a nonlinear enhancement of the classical root locus result which states that the trajectories of the closed loop system converge to the trajectories of the zero dynamics system as the gains are increased to infinity

1991 *Mathematics Subject Classification* 93B05, 93B28, 93B52

Key words and phrases Boundary control, nonlinear distributed parameter systems, zero dynamics, convergence of trajectories

1. Introduction

The need to control spatially distributed systems or systems in which rigid dynamics are coupled with the effects of fluid flow and flexible structural modes has sparked a tremendous effort aimed at the stabilization or control of distributed parameter systems. For example, the existence and analysis of boundary feedback control laws which stabilize classes of linear, distributed parameter systems has attracted the interest of many researchers, using a variety of techniques drawn from both state-space and frequency domain methods. While the differences between these two methodologies are certainly heightened in the distributed parameter case, one of the elegant aspects of linear geometric control theory is its capacity for giving intuitive state-space formulations of many frequency domain concepts and constructs. It is also worth noting that the extension of geometric methods to

This work was supported in part by grants from NSF, AFOSR and Texas Advanced Research Program

lumped nonlinear systems at the beginning of the last decade spearheaded a virtual renaissance in nonlinear control theory, yielding systematic methodologies for the design of feedback laws while retaining important intuitive aspects of classical state-space and frequency design methods.

One of the important feedback design methods of classical automatic control is root locus theory, based on the observation that in the frequency domain output the closed loop poles of a system vary from the open loop poles to the open loop zeros as the gain is increased from zero to infinity. Succesfully exploited for decades for finite dimensional systems, this fundamental method has been extended to the nonlinear finite dimensional case ([4]), where it is known that as certain gain parameters are tuned the closed-loop trajectories approach the trajectories of the zero dynamics. On the other hand, root locus methods have also recently been extended to the infinite dimensional case in [4] where a fairly complete analog of finite dimensional root locus theory is developed for a class of parabolic boundary control problems in which the inputs and outputs occur through certain boundary operators and a closed loop system is obtained by employing a proportional error feedback law, $u = -ky$. In this case, in [4] it is shown that the infinitely many closed loop poles vary from the open loop poles to the open loop zeros as the gain is increased from zero to plus or minus infinity, depending on the sign of the instantaneous gain. Defining the zero dynamics to be the system obtained by constraining the output y to zero, or equivalently, as the system obtained in the high gain limit, it is possible to enhance many of the classical results on stabilization of minimum phase systems (i.e., systems with exponentially stable zero dynamics). In particular, in [4] it is shown that the one parameter family of closed loop spatial operators A_k – the analog of $(A + BkC)$ in classical finite dimensional theory – form an analytic family, in the gain parameter k, of unbounded operators, in Hilbert state space, in the sense of norm resolvent convergence (cf. Kato [11]). This result together with a generalization the Trotter-Kato theorem provides a simple proof that the semigroups $S_k(t)$, with infinitesimal generators A_k, converge in the uniform operator topology to the semigroup generated by the zero dynamics, $S_\infty(t)$.

The main result of the present paper is to provide a nonlinear enhancement of the root locus results obtained for linear distributed parameter systems in [4] for a boundary controlled, viscous Burgers' equation. Explicitly, we show that the closed-loop trajectories approach the trajectories of the zero dynamics as certain gain parameters are tuned. Since the zero dynamics is exponentially stable, this design method also provides a class of stabilizing feedback laws. The most important result used in the proofs is that the linear part of the closed loop spatial Burgers' operator form a holomorphic family in the sense of norm resolvent convergence and that the nonlinear term is a Lipschitz operator with respect to the Hilbert scale norm generated by the square root of the linear part. This paper represents a special case of the results found in [6] where we consider the same system but with a nonzero external forcing term $f(x, t)$. In this case we do not obtain exactly the same result. Namely we do not obtain exponential stability. Further in [6] we show that for nonzero forcing terms which are independent of time, there exists, for every nonzero value of the gain parameter, a local attractor. It is also shown that as the gains tend to plus infinity the local attractors converge to the single global asymptotically stable equilibrium for the zero dynamics system. In [7] we consider the problem of convergence of trajectories using a priori estimates and Galerkin approximations for small L^2 initial data. This, of course, represents a marked improvemnt over the results for small H^1 initial data. We are not able to extend our semigroup methods used for H^1 initial data to this more general case. On the other hand we obtain stronger results in [6] which is to be expected for smoother initial data.

2. The Burgers' System and Zero Dynamics

Consider the uncontrolled viscous Burgers' system

$$\begin{aligned} &w_t(x,t) + \epsilon A_0 w(x,t) = F(w(x,t)), \qquad x \in (0,1),\ t > 0 \\ &0 = -w_x(0,t),\ 0 = w_x(1,t),\ w(x,0) = \phi(x) \end{aligned} \tag{2.1}$$

where A_0 is the unbounded operator $A = -\frac{d^2}{dx^2}$ with dense domain in $L^2(0,1)$

$$D(A_0) = \{f \in H^2 \mid f'(0) = f'(1) = 0\}$$

$(' = d/dx)$ and

$$F(w) = -\frac{dw}{dx}w.$$

The Controlled Burgers' system is defined by

$$\begin{aligned} &w_t(x,t) + \epsilon A_k w(x,t) = F(w(x,t)),\ \ x \in (0,1),\ t > 0 \\ &-w_x(0,t) + k_0 w(0,t) = 0,\ w_x(1,t) + k_1 w(1,t) = 0, \\ &w(x,0) = \phi(x). \end{aligned} \tag{2.2}$$

where $A_k = A$ with domain $\mathcal{D}(A_k) = \{f \in H^2(0,1) :\ f'(0) - k_0 f(0) = 0,\ f'(1) + k_1 f(1) = 0\}$ and $k = (k_0, k_1)$. Note that for $k_0 = k_1 = 0$, the uncontrolled system (2.1) is obtained and the parameters k_0 and k_1 with $k_0 + k_1 > 0$ are considered as boundary controls or gain parameters.

The zero dynamics for the controlled Burgers' system is the system obtained for infinite value of the gain parameters. This corresponds to the system

$$\begin{aligned} &w_t(x,t) + \epsilon A_\infty w(x,t) = F(w(x,t)),\ \ x \in (0,1),\ t > 0 \\ &w(1,t) = 0,\ w(0,t) = 0,\ w(x,0) = \phi(x) \end{aligned} \tag{2.3}$$

where $A_\infty = A$ with domain $\mathcal{D}(A_\infty) = \{f \in H^2(0,1) :\ f(0) = 0,\ f(1) = 0\}$.

3. The Linearization about Zero

The linearization about zero of (2.2) is the controlled heat equation

$$
\begin{aligned}
&w_t(x,t) + \epsilon A_k w(x,t), \quad x \in (0,1), \; t > 0 \\
&- w_x(0,t) + k_0 w(0,t) = 0 \\
&w_x(1,t) + k_1 w(1,t) = 0 \\
&w(x,0) = \phi(x).
\end{aligned} \tag{3.1}
$$

For $k_0 = k_1 = 0$, neither (2.2) nor the linearization (3.1) is asymptotically stable, but for $k_0 + k_1 > 0$ the linearization (3.1) is asymptotically stable. In fact, A_k is a strictly positive selfadjoint operator for $k_0 + k_1 > 0$, so that by the Lumer-Phillips Theorem $(-\epsilon A_k)$ generates a contraction semigroup $S_k(t)$ satisfying

$$||S_k(t)|| \leq e^{-\epsilon\lambda_1(k)t}$$

where $\lambda_1(k)$ is the first eigenvalue of A_k. In this case, it is easy to show that

$$0 > \lambda_1(k) \to \pi^2, \; k_0, k_1 \to \infty.$$

Thus the system (3.1) has solution $w(t)$ satisfying

$$\|w(t)\| = \|S_k(t)\phi\| \leq e^{-\epsilon\lambda_1(k)t}\|\phi\|$$

where $\|\cdot\|$ is the norm in $L^2(0,1)$ induced by the inner product

$$(\phi, \psi) = \int_0^1 \phi(x)\overline{\psi(x)}dx.$$

The spectrum of A_k is given by positive numbers $\lambda = \mu^2$ as the roots of the characteristic equation

$$(k_0 k_1 - \mu^2)\frac{\sin(\mu)}{\mu} + (k_0 + k_1)\cos(\mu) = 0.$$

This equation has infinitely many zeros $\{\mu_j\}_{j=1}^{\infty}$ satisfying

$$(j-1)\pi < \mu_j < j\pi, \; j = 1, 2, \cdots.$$

Corresponding to the eigenvalues $\lambda_j = \mu_j^2$, we have the complete orthonormal system of eigenfunctions

$$\psi_j(x) = \kappa_j\left(k_0\frac{\sin(\mu_j x)}{\mu_j} + \cos(\mu_j x)\right), \quad j = 1, 2, \cdots$$

in $L^2(0,1)$ where

$$\kappa_j = \sqrt{\frac{2\mu_j^2(k_1^2 + \mu_j^2)}{(k_0^2 + \mu_j^2)(k_1^2 + \mu_j^2) + (k_0 + k_1)(\mu_j^2 + k_0 k_1)}}.$$

is the normalization constant.

The operators A_k define an infinite scale of Hilbert spaces $\mathcal{H}^\alpha$ ($\alpha \in \mathcal{R}$). If $\alpha \geq 0$ then $\mathcal{H}^\alpha$ consists of vectors $\phi \in \mathcal{H}^0 = L^2(0,1)$ such that

$$\|\phi\|_\alpha = \left(\sum_{j=1}^{\infty} \lambda_j^\alpha (\phi, \psi_j)^2 \right)^{1/2} < \infty. \tag{3.2}$$

The space $\mathcal{H}^\alpha$ is the domain of the operator $A_k^{\alpha/2}$ and the inner product in this space is given by

$$(\phi, \psi)_\alpha = \left(A_k^{\alpha/2}\phi, A_k^{\alpha/2}\psi \right)$$

which is the same as (3.2) for $\psi = \phi$. The operator $A_k^{\alpha/2}$ is defined on $\mathcal{H}^\alpha$ by the formula

$$A_k^{\alpha/2}\phi = \sum_{j=1}^{\infty} \lambda_j^{\alpha/2} (\phi, \psi_j)\psi_j.$$

Using the fact that $\mathcal{D}(A_k)$ is a core in $\mathcal{D}(A_k^{1/2})$ it is easy to show that the norm in $\mathcal{H}^1$ can be written as

$$\|\varphi\|_1^2 = \int_0^1 |\varphi_x|^2 \, dx + k_0 |\varphi(0)|^2 + k_1 |\varphi(1)|^2. \tag{3.3}$$

The spaces $\mathcal{H}^\alpha$ have the following properties:

1. If $\beta > \alpha$ then $\mathcal{H}^\beta \subset \mathcal{H}^\alpha$ and $\|\phi\|_\alpha \leq \lambda_1(k)^{(\alpha-\beta)/2} \|\phi\|_\beta$ for all $\phi \in \mathcal{H}^\beta$; $\mathcal{H}^\beta$ is dense in $\mathcal{H}^\alpha$.
2. The imbedding $\mathcal{H}^\beta \subset \mathcal{H}^\alpha$ is compact.

The semigroup $S_k(t)$ is given by

$$S_k(t)f = \sum_{j=1}^{\infty} e^{-\epsilon \lambda_j(k)t} f_j \psi_j$$

from which it is easy to show that $\|S_k(t)\| \leq e^{-\epsilon \lambda_1(k)t}$.

4. Global Existence for Small Initial Data

The proof of the following result concerning global (in time) existence and exponential stability of solutions for small initial data in $H^1(0,1)$ follows exactly the proof given in Theorem 3.3.3, Theorem 6.3.3, page 199 of [10] and Lemma 5.6.7, page 159 of [12]. A more general situation is considered in the paper [6] where the controlled Burgers' equation has a additional forcing term $f(x,t)$. In this case we do not obtain the exponential convergence of the trajectories to zero as t goes to infinity. Nevertheless, even in this much more general situation we are able to obtain convergence of the trajectories for small forcing terms. Furthermore, if the forcing term is independent of t, then we establish the exist in [6] of local attractors in H^1 for each positive value of the gain parameters and prove that, in an appropriate sense, the local attractors converge to the single, global asymptotically stable equilibrium for the zero dynamics.

Theorem 4.1. *For $k = (k_0, k_1)$, k_0, $k_1 \geq 0$, $k_0 + k_1 > 0$ and any β satisfying $0 < \beta < \lambda_1(k) < \pi^2$, there exists $\rho > 0$ such that for $\phi \in H^1(0,1)$ with*

$$\|\phi\|_1 \leq \rho/2$$

there is a unique solution $z(t) \in H^1$ for all $t > 0$ of

$$z_t(t) + \epsilon A_k z(t) = F(z(t)), \quad z(0) = \phi$$

and, moreover, the solution satisfies

$$\|z(t)\|_1 \leq 2e^{-\epsilon\beta t}\|\phi\|_1, \quad t \geq 0. \tag{4.1}$$

With $c = \sqrt{2}(1 + \lambda_1(k)^{-1})^{1/2}$, ρ can be taken to be any number such that

$$\rho\frac{c}{\epsilon(\lambda_1(k) - \beta)}\left(\lambda_1(k)^{1/2} + \sqrt{\frac{\pi}{2}}(\lambda_1(k) - \beta)^{1/2}\right) < \frac{1}{2}. \tag{4.2}$$

The following simple estimates are used in establishing the theorem and also in the sequel. We state them without proof and refer to [6].

Lemma 4.1. *For each $\varphi \in \mathcal{H}^1$ we have*

$$\|\varphi\|^2 \leq \lambda_1(k)^{-1}\|\varphi\|_1^2,$$

where $0 < \lambda_1(k) < \pi^2$ is the first eigenvalue of A_k. Furthermore, the norm (3.3) is equivalent to the $H^1(0,1)$ Sobolev norm and, therefore $\mathcal{H}^1 = H^1(0,1)$.

Lemma 4.2. *For $z \in H^1(0,1)$ we have the estimates*

$$\|z\|_\infty \leq \sqrt{2}\|z\|_{H^1}, \quad \|z\|_\infty \leq c\|z\|_1$$

where

$$\|z\|_{H^1}^2 = \|z\|^2 + \|z_x\|^2, \ \|z\|_\infty = \operatorname*{ess\,sup}_{x\in[0,1]}|z(x)|$$

and

$$c = \sqrt{2}(1 + \lambda_1(k)^{-1})^{1/2}. \tag{4.3}$$

With these estimates it is easy to see (cf. [10]) that F is locally Lipschitz, i.e., for $\|z_1\|_1$, $\|z_2\|_1 < M$, there exists a constant, C depending only on M, and k but otherwise independent of z_1, z_2 such that

$$\|F(z_1) - F(z_2)\| \leq C\|z_1 - z_2\|_1.$$

This allows us to apply the basic local existence result found in [10] and [12].

Lemma 4.3. *For $z \in \mathcal{H}^1$ and $F(z) = -zz_x$ we have*

$$\|F(z)\| \leq c\left(\|z\|_1\right)^2 \tag{4.4}$$

where c is defined in (4.3).

Lemma 4.4. *The following estimates hold*

1. *If $g \in H^\alpha$ then for all $t > 0$, we have*
$$\|S_k(t)g\|_\alpha \leq e^{-\epsilon\lambda_1(k)t}\|g\|_\alpha.$$
2. *For $g \in L^2$ and all $t > 0$*

$$\|S_k(t)g\|_\alpha \leq \left(\left(\frac{\alpha}{2\epsilon t}\right)^{\alpha/2} + \lambda_1(k)^{\alpha/2}\right) e^{-\epsilon\lambda_1(k)t}\|g\|. \tag{4.5}$$

5. Convergence of Trajectories for Burgers' Equation

The main result of this paper is the following theorem where throughout the remainder of the paper we assume the hypotheses of Theorem 2.1.

Theorem 5.1. *Let z_k and z_∞ denote the solutions of (2.2) and (2.3), respectively for the same initial data $\phi \in \mathcal{H}^1$ with*

$$\|\phi\|_1 \leq \rho/2.$$

Then for arbitrary $\delta > 0$, there exists a $K > 0$ so that for $k_0, k_1 > K$ we have

$$\|z_k(\cdot, t) - z_\infty(\cdot, t)\|_\infty \leq \delta e^{-\beta t}$$

for all $t \geq 0$.

In order to establish both convergence in the spatial variable for increasing k_0 and k_1 as well as exponential convergence uniformly in time we introduce a shifted semigroup and infinitesimal generator. For $0 < \beta < \lambda_1(k)$ let

$$S_{k,\beta}(t) = e^{\epsilon\beta t}S_k(t) = e^{-\epsilon(A_k - \beta I)t}, \quad S_{\infty,\beta}(t) = e^{\epsilon\beta t}S_\infty(t) = e^{-\epsilon(A_\infty - \beta I)t}$$

with corresponding infinitesimal generators given by

$$-\epsilon A_{k,\beta} = \epsilon(-A_k + \beta I), \quad -\epsilon A_{\infty,\beta} = \epsilon(-A_\infty + \beta I).$$

The family of operators $A_{k,\beta}$ generate an infinite scale of Hilbert spaces which are related to, indeed equal as functions to, the Hilbert scale $\mathcal{H}^\alpha$ generated by A_k. Namely, with

$$C_\beta = \frac{\lambda_1(k) - \beta}{\lambda_1(k)}$$

we have

$$C_\beta^{\alpha/2}\|A_k^{\alpha/2}\phi\| \leq \|A_{k,\beta}^{\alpha/2}\phi\| \leq \|A_k^{\alpha/2}\phi\|.$$

The proof is straightforward: First note that

$$\|A_{k,\beta}^{\alpha/2}\phi\|^2 = \|(A_k - \beta I)^{\alpha/2}\phi\|^2 = \sum_{j=1}^{\infty}(\lambda_j - \beta)^{\alpha}|\phi_j|^2 \le \sum_{j=1}^{\infty}(\lambda_j)^{\alpha}|\phi_j|^2 = \|A_k^{\alpha/2}\phi\|^2$$

and with the definition of C_β above we have $(\lambda_j - \beta) > C_\beta \lambda_j$ for all j, so that

$$\begin{aligned} C_\beta^{\alpha}\|A_k^{\alpha/2}\phi\|^2 &= C_\beta^{\alpha}\sum_{j=1}^{\infty}(\lambda_j)^{\alpha}|\phi_j|^2 \\ &\le \sum_{j=1}^{\infty}(\lambda_j - \beta)^{\alpha}|\phi_j|^2 = \|(A_k - \beta I)^{\alpha/2}\phi\|^2 = \|A_{k,\beta}^{\alpha/2}\phi\|^2. \end{aligned}$$

Now recall that a classical solution of (2.2) satisfies the variation of parameters formula for $0 < k_0 + k_1 < \infty$

$$z_k(t) = S_k(t)\phi + \int_0^t S_k(t-\tau)F(z_k(\tau))\,d\tau,$$

as does the solution of the zero dynamics (2.3)

$$z_\infty(t) = S_\infty(t)\phi + \int_0^t S_\infty(t-\tau)F(z_\infty(\tau))\,d\tau.$$

Also from Lemma 4.2, we have the pointwise estimate

$$|z_k(x,t) - z_\infty(x,t)| \le c\|z_k(t) - z_\infty(t)\|_1$$

with c defined in (4.3).

So to compute a pointwise estimate for the difference of the solutions we need only consider the $\mathcal{H}^1$ norm of the difference.

$$e^{\epsilon\beta t}\|z_k(\cdot,t) - z_\infty(\cdot,t)\|_1 = e^{\epsilon\beta t}\|A_k^{1/2}(z_k - z_\infty)\| \le e^{\epsilon\beta t}\|A_k^{1/2}[S_k(t)\phi - S_\infty(t)\phi]\|$$

$$+ e^{\epsilon\beta t}\int_0^t \|A_k^{1/2}[S_k(t-\tau)F(z_k(\tau)) - S_\infty(t-\tau)F(z_\infty(\tau))]\|\,d\tau$$

$$= \|A_k^{1/2}[S_{k,\beta}(t)\phi - S_{\infty,\beta}(t)\phi]\| + \int_0^t \|A_k^{1/2}[S_{k,\beta}(t-\tau)e^{\epsilon\beta\tau}F(z_k(\tau))$$

$$- S_{\infty,\beta}(t-\tau)e^{\epsilon\beta\tau}F(z_\infty(\tau))]\|\,d\tau$$

$$\le C_\beta^{-1/2}\|A_{k,\beta}^{1/2}[S_{k,\beta}(t)\phi - S_{\infty,\beta}(t)\phi]\| + C_\beta^{-1/2}\|\int_0^t A_{k,\beta}^{1/2}S_{k,\beta}(t-\tau)e^{\epsilon\beta\tau}F(z_k(\tau))$$

$$- A_{\infty,\beta}^{1/2}S_{\infty,\beta}(t-\tau)e^{\epsilon\beta\tau}F(z_\infty(\tau))d\tau\|$$

$$+ C_\beta^{-1/2} \| \int_0^t [A_{k,\beta}^{1/2} - A_{\infty,\beta}^{1/2}] S_{\infty,\beta}(t-\tau) e^{\epsilon\beta\tau} F(z_\infty(\tau))\, d\tau \|$$

$$\equiv I + II + III.$$

For the term II we have the estimates

$$\begin{aligned} C_\beta^{1/2} II \;&\le\; \int_0^t \| A_{k,\beta}^{1/2} S_{k,\beta}(t-\tau) e^{\epsilon\beta\tau} (F(z_k(\tau)) - F(z_\infty(\tau))) \|\, d\tau \\ &\quad + \| \int_0^t [A_{k,\beta}^{1/2} S_{k,\beta}(t-\tau) - A_{\infty,\beta}^{1/2} S_{\infty,\beta}(t-\tau)] e^{\epsilon\beta\tau} F(z_\infty(\tau))\, d\tau \| \\ &\equiv\; IIa + IIb. \end{aligned}$$

For the term IIa we need the estimates

$$\begin{aligned} & e^{\epsilon\beta\tau} \|F(z_k) - F(z_\infty)\| = e^{\epsilon\beta\tau} \| z_k z_{kx} - z_\infty z_{\infty x} \| \\ & = e^{\epsilon\beta\tau} \| z_{kx}(z_k - z_\infty) + z_\infty (z_{kx} - z_{\infty x}) \| \\ & \le e^{\epsilon\beta\tau} \left(\|z_{kx}\| \|z_k - z_\infty\|_\infty + \|z_\infty\|_\infty \right) \|z_k - z_\infty\|_1) \\ & \le e^{\epsilon\beta\tau} \left(\|z_k\|_1 c \|z_k - z_\infty\|_1 + c\|z_\infty\|_1 \right) \|z_k - z_\infty\|_1) \\ & \le c e^{\epsilon\beta\tau} (\|z_k\|_1 + \|z_\infty\|_1) \|z_k - z_\infty\|_1 \\ & \le c e^{\epsilon\beta\tau} 2 e^{-\epsilon\beta\tau} \|\phi\|_1 \|z_k - z_\infty\|_1 \le c\rho e^{\epsilon\beta\tau} \|z_k - z_\infty\|_1 \end{aligned}$$

where we have used (4.1). Now using (4.5) we obtain

$$\begin{aligned} IIa \;&\le\; c\rho \int_0^t \left(\frac{1}{\sqrt{2\epsilon(t-\tau)}} + \lambda_1(k)^{1/2} \right) e^{-\epsilon\lambda_1(k)(t-\tau)} e^{\epsilon\beta\tau} \|z_k - z_\infty\|_1\, d\tau \\ &\le\; (1/2) \sup_{0<\tau<t} e^{\epsilon\beta\tau} \|z_k - z_\infty\|_1 \end{aligned}$$

where the last inequality follows from our choice of ρ in (4.2). Now defining

$$\omega(t) = \sup_{0<\tau<t} e^{\epsilon\beta\tau} \|z_k - z_\infty\|_1.$$

As in the proof of Theorem 4.1 it has been shown that

$$e^{\epsilon\beta t} \|z_k(\cdot,t) - z_\infty(\cdot,t)\|_1 \le I + IIb + III + \frac{1}{2}\omega(t) \tag{5.1}$$

and we need only show that the first three terms on the right can be made less that $\delta > 0$ for k sufficiently large, uniformly in t. Then we will have

$$\frac{1}{2}\omega(t) \leq \delta.$$

The term I is considered in detail in Lemmas 5.1 to 5.4. The term IIb is considered in Lemma 5.6 and the last term III is examined in Lemma 5.5.

As a consequence of the main theorem of [4] (see also [11], Theorem 1.14 and Example 1.15, page 374) we have that the negative selfadjoint operators $(-A_k)$ form a holomorphic family in $k_0, k_1 \in [0, \infty]$ with $k_0 + k_1 > 0$ in the sense of Kato ([11], Theorem 2.25, page 206 and Theorem 1.3 and Example 1.4, page 367). Therefore, defining

$$R_k(\lambda) = (\lambda I + A_k)^{-1}$$

for any $\lambda \notin (-\infty, 0)$ we have

$$\|R_k(\lambda) - R_\infty(\lambda)\| \to 0, \;\; k \to \infty$$

where $k \to \infty$ means $k_0, k_1 \to \infty$. In one form or another most of the following results repose on this strong statement concerning the fact that the resolvents converge in the uniform operator topology as k_0 and k_1 tends to infinity.

Let us now recall some facts concerning the resolvents.

Lemma 5.1. 1. *For every $f \in L^2$,*

$$\|sR_k(s)f - f\| \to 0, \;\; s \to +\infty$$

and

$$\|sR_k(s)\| \leq 1.$$

It also follows that

$$\|s^2 R_k(s)^2 f - f\| \to 0, \;\; s \to +\infty.$$

2. *For all $t \geq 0$ and $s > 0$*

$$\|R_k(s)\left(S_\infty(t) - S_k(t)\right)R_\infty(s)\| \leq te^{-\epsilon\lambda_1(k)t}\|R_k(s) - R_\infty(s)\|.$$

3. *For all $t \geq 0$ and $s > 0$*

$$\begin{aligned}[S_k(t) - S_\infty(t)]R_\infty(s)^2 &= R_k(s)[S_k(t) - S_\infty(t)]R_\infty(s) \\ &+ [R_k(s) - R_\infty(s)]S_\infty(t)R_\infty(s) - S_k(t)[R_k(s) - R_\infty(s)]R_\infty(s).\end{aligned}$$

4. *For all $t \geq 0$ and $s > 0$*

$$\|[S_k(t) - S_\infty(t)]R_\infty^2(s)\| \leq (t+2)e^{-\epsilon\lambda_1(k)t}\|R_k(s) - R_\infty(s)\|$$

where we recall that $\lambda_1(k) < \lambda_1(\infty)$.

Proof: Part 1 is a well known consequence of the fact that $-A_k$ generates a contraction semigroup (cf. [12]). Part 2 can be found in [11] (page 501, Theorem 2.14). Part 3 is simple algebra obtained by adding and subtracting certain terms. Part 4 follows from parts 1 through 3. □

Again we recall that the standing hypotheses are those of Theorem 2.1.

Lemma 5.2. *For every $\delta > 0$ we can find a K for which $k_0, k_1 > K$ implies*

$$\|S_k(t) - S_\infty(t)\| \le \delta e^{-\beta t}, \quad \forall t \ge 0.$$

Proof: Take $g \in L^2$ such that $\|g\| \le 1$, then

$$\begin{aligned}&\|[S_k(t) - S_\infty(t)]g\| \\ &\le \|[S_k(t) - S_\infty(t)](g - s^2 R_\infty(s)^2 g)\| + s^2\|[S_k(t) - S_\infty(t)]R_\infty^2 g\| \equiv \widetilde{I} + \widetilde{II}.\end{aligned}$$

Now from Lemma 5.1, part 1 there exists $s_0 > 0$ so that $s > s_0$ implies

$$\|g - s^2 R_\infty(s)^2 g\| \le \frac{\delta}{4}.$$

Thus by Lemma 5.1 part 2 we have

$$\widetilde{I} \le (\|S_k(t)\| + \|S_\infty(t)\|)\,\frac{\delta}{4} \le e^{-\epsilon\lambda_1(k)t}\frac{\delta}{2}.$$

Now by Lemma 5.1 part 4

$$\begin{aligned}\widetilde{II} &\le s^2(t+2)e^{-\epsilon\lambda_1(k)t}\|R_k(s) - R_\infty(s)\| \\ &\le s^2 C e^{-\epsilon\lambda_1(k)t/2}\|R_k(s) - R_\infty(s)\| \le \frac{\delta}{2}\end{aligned}$$

where

$$C = \max_{t\ge 0}(t+2)e^{-\epsilon\lambda_1(k)t/2} \le \frac{\max(2, 2e^{\epsilon\lambda_1(k)-1})}{\epsilon\lambda_1(k)}$$

and where for s fixed as above we have chosen K so that for $k_0, k_1 > K$ we have

$$\|R_k(s) - R_\infty(s)\| \le \frac{\delta}{2s^2 C}.$$

□

Lemma 5.3. *For every $\psi \in L^2(0,1)$, the bounded operators*

$$B_k = A_{k,\beta}^{1/2} A_{\infty,\beta}^{-1/2} \tag{5.2}$$

converge to the identity in L^2 in the strong operator topology. Further if $C \subset L^2(0,1)$ is a relatively compact set in L^2, then

$$\sup_{\psi\in C}\|[B_k - I]\psi\| \to 0, \quad k_0, k_1 \to \infty.$$

Proof: B_k is bounded by the closed graph theorem and the first part of the proof follows from norm resolvent convergence. See, for example, the proof of Theorem 3.13, page 459 of [11]. The second part is a simple general fact which can be found for example in [9], Theorem 3.2 page 124. □

Lemma 5.4. *For $\phi \in \mathcal{H}^1$ and $\delta > 0$ there exists a K such that for $k_0, k_1 > K$*

$$I = C_\beta^{1/2} \| A_{k,\beta}^{1/2} [S_{k,\beta}(t) - S_{\infty,\beta}(t)] \phi \| \leq \delta$$

for all $t \geq 0$.

Proof: For each fixed $t > 0$ a simple Banach algebra argument can be employed to establish this result (cf. [13] Theorem VIII.20, page 286). But to obtain the result uniformly in $t \geq 0$ a bit more work is required.

$$\|A_{k,\beta}^{1/2}[S_{k,\beta}(t) - S_{\infty,\beta}(t)]\phi\| \leq \|[A_{k,\beta}^{1/2} S_{k,\beta}(t) - A_{\infty,\beta}^{1/2} S_{\infty,\beta}(t)]\phi\|$$

$$+\|[A_{k,\beta}^{1/2} - A_{\infty,\beta}^{1/2}] S_{\infty,\beta}(t)]\phi\| = \|[S_{k,\beta}(t) A_{k,\beta}^{1/2} A_{\infty,\beta}^{-1/2} - S_{\infty,\beta}(t)] A_{\infty,\beta}^{1/2}\phi\|$$

$$+\|[A_{k,\beta}^{1/2} A_{\infty,\beta}^{-1/2} - I] A_{\infty,\beta}^{1/2} S_{\infty,\beta}(t)]\phi\| \leq \|[S_{k,\beta}(t)[A_{k,\beta}^{1/2} A_{\infty,\beta}^{-1/2} - I] A_{\infty,\beta}^{1/2}\phi\|$$

$$+\|[S_k(t) - S_{\infty,\beta}(t)] A_{\infty,\beta}^{1/2}\phi\|$$

$$+\|[A_{k,\beta}^{1/2} A_{\infty,\beta}^{-1/2} - I] A_{\infty,\beta}^{1/2} S_{\infty,\beta}(t)]\phi\|$$

$$\equiv \widetilde{I} + \widetilde{II} + \widetilde{III}.$$

Now define $\psi = A_{\infty,\beta}^{1/2}\phi$. For the first term we have

$$\widetilde{I} \leq e^{\epsilon\beta t} \|[A_{k,\beta}^{1/2} A_{\infty,\beta}^{-1/2} - I]\psi\| \leq C_\beta^{-1/2}\delta/3 \tag{5.3}$$

for $k_0, k_1 > K_1$. For the second term

$$\widetilde{II} = \|[S_{k,\beta}(t) - S_{\infty,\beta}(t)]\psi\| \leq C_\beta^{-1/2}\delta/3 \tag{5.4}$$

for $k_0, k_1 > K_2$ by Lemma 5.2. And finally for the third term we note that the set

$$\{S_{\infty,\beta}(t)\psi\}_{t \geq 0}$$

is a relatively compact set in $L^2(0,1)$ for $\psi \in L^2(0,1)$. Hence by Lemma 5.3, we have

$$\widetilde{III} \leq \sup_{t \geq 0} \|[A_{k,\beta}^{1/2} A_{\infty,\beta}^{-1/2} - I] S_{\infty,\beta}(t)\psi\| \leq C_\beta^{-1/2}\delta/3. \tag{5.5}$$

Combining (5.3), (5.4) and (5.5) the result follows. □

We have left to consider the two remaining terms IIb and III in (5.1).

$$IIb = \| \int_0^t [A_{k,\beta}^{1/2} S_{k,\beta}(t-\tau) - A_{\infty,\beta}^{1/2} S_{\infty,\beta}(t-\tau)] e^{\epsilon\beta\tau} F(z_\infty(\tau))\, d\tau \|,$$

$$C_\beta^{1/2} III = \| \int_0^t [A_{k,\beta}^{1/2} - A_{\infty,\beta}^{1/2}] S_{\infty,\beta}(t-\tau) e^{\epsilon\beta\tau} F(z_\infty(\tau))\, d\tau \|.$$

Lemma 5.5. *For any $\delta > 0$ there exists a $K > 0$ so that for $k_0, k_1 > K$ we have*

$$C_\beta^{1/2} III \leq \delta$$

for all $t \geq 0$.

Proof: First note that with B_k defined in (5.2)

$$C_\beta^{1/2} III = \| [B_k - I] \int_0^t A_{\infty,\beta}^{1/2} S_{\infty,\beta}(t-\tau) e^{\epsilon\beta\tau} F(z_\infty(\tau))\, d\tau \|.$$

Now we show that the set

$$\mathcal{S} = \bigcup_{t \geq 0} \left\{ \int_0^t A_{\infty,\beta}^{1/2} S_{\infty,\beta}(t-\tau) e^{\epsilon\beta\tau} F(z_\infty(\tau))\, d\tau \right\}$$

is a relatively compact set in $L^2(0,1)$ and hence the result will follow from Lemma 5.3. Take $\alpha = 1 + \gamma$ where $(1+\gamma)/2 < 1$. Now recall from (4.4) and (4.1) that

$$e^{\epsilon\beta\tau} \| F(z_\infty(\tau)) \| \leq c e^{-\epsilon\beta\tau} \|\phi\|^2.$$

We proceed to show that the set $\mathcal{S}$ is bounded in $\mathcal{H}^\gamma$ and hence it is a relatively compact set in $L^2(0,1)$. We have

$$\begin{aligned} &\left\| \int_0^t A_{\infty,\beta}^{1/2} S_{\infty,\beta}(t-\tau) e^{\epsilon\beta\tau} F(z_\infty(\tau))\, d\tau \right\|_\gamma \\ &= \left\| \int_0^t A_{\infty,\beta}^{\alpha} S_{\infty,\beta}(t-\tau) e^{\epsilon\beta\tau} F(z_\infty(\tau))\, d\tau \right\| \\ &\leq c\|\phi\|^2 \int_0^t \| A_{\infty,\beta}^{\alpha/2} S_{\infty,\beta}(t-\tau) \|\, d\tau \\ &\leq \widetilde{C} \int_0^\infty \left[\left(\frac{\alpha}{2\epsilon s} \right)^{\alpha/2} + \lambda_1(k)^{\alpha/2} \right] e^{-\epsilon(\lambda_1(k)-\beta)s}\, ds \\ &\leq \widetilde{\widetilde{C}} < \infty. \end{aligned}$$

□

For the term IIb we extend a result from Simon and Reed (cf. [13] Theorem VIII.20, page 286).

Lemma 5.6. *For any $\delta > 0$ there exists a $K > 0$ so that for $k_0, k_1 > K$ we have*

$$IIb \leq \delta$$

for all $t \geq 0$.

Proof: First we need to reduce our calculations to a compact time interval. Thus consider

$$\| \int_0^t [A_{k,\beta}^{1/2} S_{k,\beta}(t-\tau) - A_{\infty,\beta}^{1/2} S_{\infty,\beta}(t-\tau)] e^{\epsilon\beta\tau} F(z_\infty(\tau))\, d\tau \|$$

$$= \| \int_0^t [A_{k,\beta}^{1/2} S_{k,\beta}(\tau) - A_{\infty,\beta}^{1/2} S_{\infty,\beta}(\tau)] e^{\epsilon\beta(t-\tau)} F(z_\infty(t-\tau))\, d\tau \|$$

$$\leq c\|\phi\|^2 \int_0^t \|[A_{k,\beta}^{1/2} S_{k,\beta}(\tau) - A_{\infty,\beta}^{1/2} S_{\infty,\beta}(\tau)]\|\, d\tau$$

$$= c\|\phi\|^2 \int_0^{t_0} \|[A_{k,\beta}^{1/2} S_{k,\beta}(\tau) - A_{\infty,\beta}^{1/2} S_{\infty,\beta}(\tau)]\|\, d\tau$$

$$+ c\|\phi\|^2 \int_{t_0}^{t_1} \|[A_{k,\beta}^{1/2} S_{k,\beta}(\tau) - A_{\infty,\beta}^{1/2} S_{\infty,\beta}(\tau)]\|\, d\tau$$

$$+ c\|\phi\|^2 \int_{t_1}^{\infty} \|[A_{k,\beta}^{1/2} S_{k,\beta}(\tau) - A_{\infty,\beta}^{1/2} S_{\infty,\beta}(\tau)]\|\, d\tau.$$

For the first term after the last equality above we have

$$\int_0^{t_0} \|[A_{k,\beta}^{1/2} S_{k,\beta}(\tau) - A_{\infty,\beta}^{1/2} S_{\infty,\beta}(\tau)]\|\, d\tau$$

$$\leq \int_0^{t_0} \left(\|[A_{k,\beta}^{1/2} S_{k,\beta}(\tau)\| + \|A_{\infty,\beta}^{1/2} S_{\infty,\beta}(\tau)]\| \right) d\tau$$

$$\leq \widetilde{C} \int_0^{t_0} \frac{e^{-\epsilon(\lambda_1(k)-\beta)s}}{\sqrt{s}}\, ds$$

$$\leq \widetilde{C} \int_0^{t_0} s^{-1/2}\, ds \leq \widetilde{C}\sqrt{t_0} \leq \delta/3$$

for t_0 small enough.

For the last term above

$$\int_{t_1}^{\infty} \|[A_{k,\beta}^{1/2} S_{k,\beta}(\tau) - A_{\infty,\beta}^{1/2} S_{\infty,\beta}(\tau)]\| \, d\tau$$

$$\leq \int_{t_1}^{\infty} \left(\|[A_{k,\beta}^{1/2} S_{k,\beta}(\tau)\| + \|A_{\infty,\beta}^{1/2} S_{\infty,\beta}(\tau)]\| \right) \, d\tau$$

$$\leq \widetilde{C} \int_{t_1}^{\infty} \frac{e^{-\epsilon(\lambda_1(k)-\beta)s}}{\sqrt{s}} \, ds$$

$$\leq \frac{\widetilde{C}}{\sqrt{t_1}} \int_{0}^{\infty} e^{-\epsilon(\lambda_1(k)-\beta)s} \, ds \leq \frac{\widetilde{C}}{\sqrt{t_1}\epsilon(\lambda_1(k)-\beta)} \leq \delta/3$$

for t_1 large enough.

Finally, with t_0 and t_1 chosen above, we need only consider the integral over the fixed region $[t_0, t_1]$ which is compact. Thus we consider the term

$$\int_{t_0}^{t_1} \|[A_{k,\beta}^{1/2} S_{k,\beta}(\tau) - A_{\infty,\beta}^{1/2} S_{\infty,\beta}(\tau)]\| \, d\tau.$$

Let

$$g_\beta(\lambda, t) = \begin{cases} (\lambda-\beta)^{1/2} e^{-\epsilon(\lambda-\beta)t}, & \lambda > \beta, \\ 0, & \lambda \leq \beta. \end{cases}$$

The function g_β is a uniformly continuous function on $\mathcal{R} \times [t_0, t_1]$ and vanishes at infinity in λ. Furthermore by the spectral theorem

$$g_\beta(A_k, t) = A_{k,\beta}^{1/2} S_{k,\beta}(t), \; g_\beta(A_\infty, t) = A_{\infty,\beta}^{1/2} S_{\infty,\beta}(t),$$

and since $[t_0, t_1]$ is compact we can find $\{s_j\}_{j=1}^N$ and disjoint intervals $\{I_j\}$ such that

$$[t_0, t_1] = \bigcup_{j=1}^{N} \overline{I_j}$$

and

$$\sup_{\lambda \in \mathcal{R}, t \in I_j} |g_\beta(\lambda, t) - g_\beta(\lambda, s_j)| \leq \frac{\delta}{9(t_1 - t_0)}$$

for $j = 1, \cdots, N$ which implies by the spectral theorem

$$\sup_{\lambda \in \mathcal{R}, t \in I_j} \|A_{k,\beta}^{1/2} S_{k,\beta}(t) - A_{k,\beta}^{1/2} S_{k,\beta}(s_j)\| \leq \frac{\delta}{9(t_1 - t_0)}$$

for $j = 1, \cdots, N$. A similar expression holds for $k = \infty$.

Finally, and once again as a consequence of the spectral theorem (cf. [13] Theorem VIII 20, page 286) and the norm resolvent convergence as k_0, k_1 go to infinity, that there exists a $K > 0$ so that for $k_0, k_1 > K$ and for $j = 1, \cdots, N$

$$\sup_{t \in I_j} \|A_{k,\beta}^{1/2} S_{k,\beta}(s_j) - A_{\infty,\beta}^{1/2} S_{\infty\,\beta}(s_j)\| \leq \frac{\delta}{9(t_1 - t_0)}$$

Combining these results we can compute

$$\begin{aligned}
&\int_{t_0}^{t_1} \|[A_{k,\beta}^{1/2} S_{k,\beta}(\tau) - A_{\infty,\beta}^{1/2} S_{\infty,\beta}(\tau)]\| \, d\tau \\
&= \sum_{j=1}^{N} \int_{I_j} \|[A_{k,\beta}^{1/2} S_{k,\beta}(\tau) - A_{\infty,\beta}^{1/2} S_{\infty,\beta}(\tau)]\| \, d\tau. \\
&\leq \sum_{j=1}^{N} \left\{ \int_{I_j} \|[A_{k,\beta}^{1/2} S_{k,\beta}(\tau) - A_{k,\beta}^{1/2} S_{k,\beta}(s_j)]\| \, d\tau \right. \\
&\quad + \int_{I_j} \|[A_{k,\beta}^{1/2} S_{k,\beta}(s_j) - A_{\infty,\beta}^{1/2} S_{\infty,\beta}(s_j)]\| \, d\tau \\
&\quad \left. + \int_{I_j} \|[A_{\infty,\beta}^{1/2} S_{\infty,\beta}(\tau) - A_{\infty,\beta}^{1/2} S_{\infty,\beta}(s_j)]\| \, d\tau \right\} \\
&\leq \frac{\delta}{9(t_1 - t_0)} [3(t_1 - t_0)] \leq \frac{\delta}{3}
\end{aligned}$$

□

References

1 C I Byrnes, D S Gilliam, "Boundary feedback design for nonlinear distributed parameter systems," *Proc of IEEE Conf on Dec and Control*, Britton, England 1991

2 C I Byrnes and D S Gilliam, "Stability of certain distributed parameter systems by low dimensional controllers a root locus approach," *Proceedings 29th IEEE International Conference on Decision and Control*

3 C I Byrnes and D S Gilliam, "Boundary feedback stabilization of a viscous Burgers' equation," to appear in *Proceedings Third Conference on Computation and Control*, Bozeman, Montana

4 C I Byrnes and D S Gilliam, J He, "Root Locus and Boundary Feedback design for a Class of Distributed Parameter Systems," to appear in *SIAM J Control and Opt*

5 C I Byrnes, D S Gilliam and V I Shubov, "Boundary control, for a viscous burgers' equation," *Identification and control in systems governed by partial differential equations* H T Banks, R H Fabiano and K Ito (Eds), SIAM, 1993, 171-185

6 C I Byrnes, D S Gilliam and V I Shubov, "High gain limits of trajectories and attractors for a controlled viscous Burgers' equation," Submitted to *J Math Sys Est Control*

7 C I Byrnes, D S Gilliam and V I Shubov, "Zero and pole dynamics of a nonlinear distributed parameter system," Preprint

8 J Carr, "Applications of the center manifold theorem," Springer-Verlag, 1980

9 F Chatelin, "Spectral Approximation of Linear Operators," Academic Press, Series in Computer Science and Applied Mathematics, 1983
10 D Henry, "Geometric Theory of Semilinear Parabolic Equations," Springer-Verlag, Lec Notes in Math , Vol 840, 1981
11 T Kato, "Perturbation theory for linear operators," Springer-Verlag, New York, (1966)
12 A Pazy, "Semigroups of Linear Operators and Applications to Partial Differential Equations," Springer-Verlag, 1983
13 B Simon and M Reed, "Methods of Modern Mathematical Physics, Volume 1," Academic Press, New York and London, 1972

Christopher I Byrnes
Department of Systems Science and Mathematics
Washington University
St Louis, MO 63130, USA

David S Gilliam
Department of Mathematics
Texas Tech University
Lubbock, TX 79409, USA

Victor I Shubov
Department of Mathematics
Texas Tech University
Lubbock, TX 79409, USA

International Series of Numerical Mathematics Vol 118 © 1994 Birkhauser Verlag Basel

OPTIMALITY CONDITIONS FOR BOUNDARY CONTROL PROBLEMS OF PARABOLIC TYPE

PIERMARCO CANNARSA AND MARIA ELISABETTA TESSITORE

Dipartimento di Matematica
Universitá di Roma "Tor Vergata"

Dipartimento di Matematica
Universitá di Roma "La Sapienza"

ABSTRACT This paper is devoted to the study of finite horizon optimal control problems with boundary control We prove a sufficient condition for optimality of trajectory–control pairs, using a non–smooth analysis approach We formulate this condition in terms of an Hamiltonian system for which we show an existence and uniqueness result

1991 *Mathematics Subject Classification* 49K20, 49K27

Key words and phrases Optimality conditions, boundary control, parabolic equations, Neumann boundary conditions, analytic semigroups, backward uniqueness

1. Introduction

This paper is concerned with the problem of minimizing

$$\int_{t_0}^{T} L(s, x(s; t_0, x_0, u), u(s))ds + \phi(x(T; t_0, x_0, u)), \tag{1.1}$$

overall measurable functions $u : [t_0, T] \to U$ (such a function is usually called a *control*). Here $(t_0, x_0) \in [0, T] \times X$, and $x(s; t_0, x_0, u)$ is the mild solution of the infinite dimensional controlled system

$$\begin{cases} x'(t) = Ax(t) + F(x(t)) - ANu(t) \\ x(t_0) = x_0. \end{cases} \tag{1.2}$$

The state space X and the control space U are real Hilbert spaces and A is the infinitesimal generator of an analytic semigroup on X. Moreover, $F : X \to X$ is Lipschitz continuous and N is the so–called Neumann map.

Equation (1.2) is the abstract version of a parabolic equation that is controlled through a Neumann type boundary condition. Further details on this connection are recalled in section 2, where we also list all the assumptions on the various data. Here, we just wish to mention the fact that the running cost L is allowed to be unbounded, provided that a coercivity condition of the form

$$\exists \lambda_0 > 0, \lambda_1 \in \mathbb{R} \, : \, L(t,x,u) \geq \lambda_0 |u|^2 + \lambda_1, \forall t \in [0,T], u \in U.$$

is satisfied.

The literature on boundary control is huge. Clearly, the most investigated case is the Linear Quadratic problem, see for instance [16], [3], [15]. Among nonlinear boundary control problems one of the first examples to be studied was the convex case, see [1], [13], that requires a linear state equation and a convex payoff. As for general nonlinear boundary control problems, most of the results that are available in the literature are concerned with necessary optimality conditions, see e.g. [9], [10] and [20]. The Dynamic Programming approach to nonlinear boundary control problems is more recent and uses viscosity solutions, see [6] and [7].

In this paper we focus our attention on both necessary and sufficient conditions for optimality. Necessary conditions will be stated both in the well known traditional form of the Pontryagin Maximum Principle (see Theorem (3.1)) and in the Hamiltonian formulation

$$\left\{ \begin{array}{l} x'(t) = Ax(t) + F(x(t)) - (-A)^{\beta} D_p H(t, x(t), (-A)^{\beta} p(t)) \\ \\ -p'(t) = Ap(t) + [D_x F(x(t))]^* p(t) - D_x H(t, x(t), (-A)^{\beta} p(t)) \end{array} \right. \tag{1.3}$$

where H is the Hamiltonian associated with L. As usual, system (1.3) is assigned the initial–terminal condition

$$\left\{ \begin{array}{l} x(t_0) = x_0 \\ p(T) = -D\phi(x(T)) \, . \end{array} \right. \tag{1.4}$$

As it is done in [4] for distributed control problems, the formulation of the Pontryagin Maximum Principle is complemented with co–state inclusions of the type derived in [8] for finite dimensional control problems. Roughly speaking, such inclusions establish a connection between the co–state associated with some fixed optimal pair $\{x^*, u^*\}$, and some generalized gradient of the value function along the optimal trajectory x^*. We recall that the value function of problem (1.1), (1.2) is defined as

$$V(t_0, x_0) = \inf_{u \in U} \left\{ \int_{t_0}^{T} L(s, x(s; t_0, x_0, u), u(s)) ds + \phi(x(T; t_0, x_0, u)) \right\} . \tag{1.5}$$

The main object of this paper, however, is the generalization of a sufficient optimality condition obtained in [4] for distributed control problems. Such a result is based on the idea of replacing the initial–terminal condition (1.4) with an initial condition of the form

$$\left\{ \begin{array}{l} x(t_0) = x_0 \\ p(t_0) = p_0 \end{array} \right. \tag{1.6}$$

where $-p_0$ is a cluster point for the x–gradient of the value function V at (t_0, x_0). More precisely, extending the result of [4] to boundary control, we prove that, for any such p_0, problem (1.3)–(1.6) has a solution whose x–component is an optimal trajectory of (1.1)–(1.2). Other kinds of sufficient conditions for boundary control problems are discussed in [12].

Moreover, in this paper, we prove that the solution of system (1.3)–(1.6) is *unique*. This uniqueness result seems to be new even for distributed control problems. Since in system (1.3) one equation is forward and the other one is backward, a natural approach to the problem of uniqueness is to use backward uniqueness techniques. A detailed discussion of these techniques is contained in [14]. In this paper, however, we adapt (as in [5]) the method of [17].

We conclude with an outline of the paper. In §2 we recall basic material on boundary control problems, including some properties of the value function of problem (1.1)–(1.2). In §3 we state necessary conditions for optimality through the Pontryagin Maximum Principle (Theorem 3.1). We then formulate its Hamiltonian version and we show an existence and uniqueness result for this system, which is also a sufficient condition for optimality (Theorems 3.3 and 3.4).

2. Preliminaries

Let $T > 0, t_0 \in [0, T]$ and $\Omega \subset \mathbb{R}^n$ be open and bounded with smooth boundary. Consider the following boundary value problem with Neumann conditions

$$\begin{cases} \dfrac{\partial x}{\partial t}(t,\xi) = \Delta_\xi x(t,\xi) + f(x(t,\xi)) & \text{in } (t_0,T)\times\Omega \\ x(t_0,\xi) = x_0(\xi) & \text{on } \Omega \\ \dfrac{\partial x}{\partial \nu}(t,\xi) = u(t,\xi) & \text{on } (t_0,T)\times\Omega \end{cases} \tag{2.1}$$

where $x_0 \in L^2(\Omega)$, $u \in L^2(t_0, T; L^2(\partial\Omega))$, and $f : \mathbb{R} \to \mathbb{R}$ is Lipschitz continuous.

Given continuous functions $L : [0, T] \times L^2(\Omega) \times L^2(\partial\Omega) \to \mathbb{R}$ and $\phi : L^2(\Omega) \to \mathbb{R}$, let us consider the problem of minimizing the functional

$$J(t_0, x_0; u) = \int_{t_0}^{T} L(t, x(t,\cdot), u(t,\cdot))dt + \phi(x(T,\cdot)) \tag{2.2}$$

overall controls $u \in L^2(t_0, T; L^2(\partial\Omega))$, where x is the solution of (2.1).

Problem (2.1)–(2.2) may be rewritten in abstract form as follows.

Let $X = L^2(\Omega), U = L^2(\partial\Omega)$ and define the unbounded operator A in X by

$$D(A) = \left\{ x \in H^2(\Omega) \ : \ \frac{\partial x}{\partial \nu} = 0 \right\}$$

$$Ax = \Delta x - x.$$

Next, we define the Neumann map $N : U \to X$ as (see [9])

$$Nu = x \quad \Leftrightarrow \quad \begin{cases} \Delta x = x & \text{in } \Omega \\ \dfrac{\partial x}{\partial \nu} = u & \text{on } \partial\Omega \end{cases}$$

Formally, equation (2.1) may be written as

$$\text{(2.3)} \qquad \begin{cases} x'(t) = Ax(t) + F(x(t)) - ANu(t) \\ x(t_0) = x_0 \end{cases}$$

where

$$F(x)(\xi) = f(x(\xi)) + x(\xi).$$

The right–hand side of the equation (2.3) is not well defined as the range of N is not contained in $D(A)$. However, we note that N has some regularizing effect. Indeed, $N : L^2(\partial\Omega) \to H^{\frac{3}{2}}(\Omega)$, which may be expressed in abstract form using the fractional powers of $-A$. In fact,

$$D((-A)^\theta) = \begin{cases} H^{2\theta}(\Omega) & \text{if } 0 \le \theta < \frac{3}{4} \\ \left\{ x \in H^{2\theta}(\Omega) \;:\; \dfrac{\partial x}{\partial \nu} = 0 \right\} & \text{if } \frac{3}{4} < \theta \le 1. \end{cases}$$

The fact that operator A is the generator of an analytic semigroup in X implies that for every $\theta \in [0, 1]$ there exists a constant $M_\theta > 0$ such that

$$\text{(2.4)} \qquad |(-A)^\theta e^{tA} x| \le \frac{M_\theta}{t^\theta} |x|, \quad \forall t > 0, \forall x \in X.$$

Moreover let $\alpha \in]0, \frac{1}{2}[$. Then, a well known interpolation inequality, see e.g. [18], states that for every $\sigma > 0$ there exists $C_\sigma > 0$ such that

$$\text{(2.5)} \qquad |(-A)^\alpha x| \le \sigma |(-A)^{\frac{1}{2}} x| + C_\sigma |x|, \quad \forall x \in D((-A)^{\frac{1}{2}}).$$

By classical regularity results, $N : U \to D((-A)^\beta)$ for all $0 < \beta < \frac{3}{4}$. Therefore, equation (2.3) can be written as

$$\text{(2.6)} \qquad \begin{cases} x'(t) = Ax(t) + F(x(t)) + (-A)^\beta N_\beta u(t) \\ x(t_0) = x_0 \end{cases}$$

where $N_\beta = (-A)^{1-\beta} N$. Clearly, equation (2.6) has to be understood in mild form, that is

$$\text{(2.7)} \quad x(t) =$$

$$e^{(t-t_0)A} x_0 + \int_{t_0}^{t} e^{(t-s)A} F(x(s)) ds + (-A)^\beta \int_{t_0}^{t} e^{(t-s)A} N_\beta u(s) ds \;\; t_0 \le t \le T.$$

It is well known that (2.7) has a unique solution

$$x(\cdot;t_0,x_0,u)\in C([t_0,T]:X)$$

under the following assumptions:

$$(2.8)\quad \begin{cases} (i) & A:D(A)\subset X\to X \text{ is a densely defined closed linear operator;}\\ (ii) & \text{the inclusion } D(A)\subset X \text{ is compact and } A=A^*;\\ (iii) & \exists\omega>0 \text{ such that } \langle Ax,x\rangle\le-\omega|x|^2,\forall x\in D(A);\\ (iv) & F:X\to X \text{ is Lipschitz continuous;}\\ (v) & \beta\in]\frac{1}{4},\frac{1}{2}[. \end{cases}$$

The value function of problem (2.1)– (2.2) is defined as

$$(2.9)\quad V(t_0,x_0)=\inf_{u\in L^2(t_0,T;U)}\left\{\int_{t_0}^{T}L(s,x(s;t_0,x_0,u),u(s))ds+\phi(x(T;t_0,x_0,u))\right\}$$

where $L:[0,T]\times X\times U\to\mathbb{R}$ and $\phi:X\to\mathbb{R}$ are assumed to satisfy the following $\forall R>0$ and some constant $C_R>0$:

$$(2.10)\quad \begin{cases} (i) & L\in C([0,T]\times X\times U)\\ (ii) & |L(t,x,u)-L(t,y,u)|\\ & \qquad\le C_R|x-y|,\forall t\in[0,T],u\in U,|x|,|y|\le R\\ (iii) & L(t,x,\cdot) \text{ is strictly convex}\\ (iv) & \exists\lambda_0>0,\lambda_1\in\mathbb{R}\ :\ L(t,x,u)\ge\lambda_0|u|^2+\lambda_1,\forall t\in[0,T],u\in U\\ (v) & |L(t,x,u)-L(t,x,v)|\\ & \qquad\le C_R(1+|u|+|v|)|u-v|,\forall t\in[0,T],u,v\in U,|x|\le R\\ (vi) & |\phi(x)-\phi(y)|\le C_R|x-y|,\forall|x|,|y|\le R. \end{cases}$$

A control $u^*\in L^2(t_0,T;U)$ at which the infimum in (2.9) is attained, is said to be *optimal*. Equivalently, u^* is optimal if

$$V(t_0,x_0)=\int_{t_0}^{T}L(s,x(s;t_0,x_0,u^*),u^*((s))ds+\phi(x(T;t_0,x_0,u^*)).$$

The following Lemma will be useful in the sequel to prove the Pontryagin Maximum Principle. We denote by J the payoff functional associated to problem (2.9), i.e.

$$(2.11)\qquad J(t_0,x_0;u)=\int_{t_0}^{T}L(t,x(t;t_0,x_0,u),u(t))dt+\phi(x(T;t_0,x_0,u)).$$

Lemma 2.1. *Assume* (2.8), (2.10). *Then, for any $R > 0$ there exists a constant $M_R > 0$ such that, for any $x_0 \in X$, with $|x_0| \leq R$, and any control $u \in L^2(t_0, T; U)$, there exists $\overline{u} \in L^2(t_0, T; U)$ satisfying*

$$(i) \quad J(t_0, x_0; \overline{u}) \leq J(t_0, x_0; u)$$

$$(ii) \quad |\overline{u}(t)| \leq \frac{M_R}{(T-t)^\beta}, \forall t \in [t_0, T].$$

Proof – In the following, we will assume that hypothesis (2.10) holds true with $C_R = C$ independent of R. Then, the constant M_R in (ii) above is also independent of R, as we will show below. The proof of the general case is similar and is left to the reader.

Let $u \in L^2(t_0, T; U)$ and define, for any $n \in \mathbb{N}$,

$$I_n = \left\{ t \in [t_0, T] \ : \ |u(t)| > \frac{n}{(T-t)^\beta} \right\}$$

and

$$u_n(t) = \begin{cases} u(t) & \text{if } t \notin I_n \\ 0 & \text{if } t \in I_n. \end{cases}$$

Moreover, let us set

$$x(t) = x(t; t_0, x_0, u) \ , \ \ x_n(t) = x(t; t_0, x_0, u_n)$$

Then, denoting by $|I_n|$ the Lebesgue measure of I_n, we have

$$\begin{aligned} J(t_0, x_0; u_n) &= J(t_0, x_0; u) + \int_{t_0}^{T} [L(t, x_n(t), u_n(t)) - L(t, x_n(t), u(t))]dt \\ &+ \int_{t_0}^{T} [L(t, x_n(t), u(t)) - L(t, x(t), u(t))]dt + [\phi(x_n(T)) - \phi(x(T))] \\ &\leq J(t_0, x_0; u) + |\lambda_1| \, |I_n| - \lambda_0 \int_{I_n} |u(r)|^2 dr \\ &+ C \int_{t_0}^{T} |x_n(t) - x(t)|dt + C|x_n(T) - x(T)|. \end{aligned} \tag{2.12}$$

Now, recalling (2.7) and (2.4),

$$|x_n(s) - x(s)| \leq C_1 \int_{t_0}^{s} |x_n(r) - x(r)|dr + C_1 \int_{t_0}^{s} \frac{|u(r)|}{(s-r)^\beta} \chi_{I_n}(r) dr \tag{2.13}$$

where χ_{I_n} denotes the characteristic function of the set I_n. Let

$$\eta(t) = \int_{t_0}^{t} |x_n(s) - x(s)| ds.$$

Then, integrating (2.13),

$$\begin{aligned}\eta(t) &\leq C_1 \int_{t_0}^{t} \eta(s)ds + C_1 \int_{t_0}^{t} ds \int_{t_0}^{s} \frac{|u(r)|}{(s-r)^{\beta}} \chi_{I_n}(r)dr \\ &= C_1 \int_{t_0}^{t} \eta(s)ds + \frac{C_1}{1-\beta} \int_{t_0}^{t} |u(r)|(t-r)^{1-\beta} \chi_{I_n}(r)dr \\ &\leq C_1 \int_{t_0}^{t} \eta(s)ds + \frac{C_1 T^{1-\beta}}{1-\beta} \int_{I_n} |u(r)|dr.\end{aligned}$$

Hence, by Gronwall's Inequality,

$$\eta(t) \leq e^{C_1 T} \frac{C_1 T^{1-\beta}}{1-\beta} \int_{I_n} |u(r)|dr = C_2 \int_{I_n} |u(r)|dr.$$

From (2.13) and the estimate above, it follows that

$$|x_n(s) - x(s)| \leq C_3 \int_{I_n} |u(r)|dr + C_1 \int_{t_0}^{s} \frac{|u(r)|}{(s-r)^{\beta}} \chi_{I_n}(r)dr. \tag{2.14}$$

Using (2.14) in (2.12) we obtain, after some tedious computations,

$$\begin{aligned}&J(t_0, x_0; u_n) - J(t_0, x_0; u) \\ &\leq |\lambda_1|\, |I_n| - \lambda_0 \int_{I_n} |u(r)|^2 dr + C_4 \int_{I_n} |u(r)|dr + C_5 \int_{I_n} \frac{|u(r)|}{(T-r)^{\beta}} dr\end{aligned} \tag{2.15}$$

Finally, we claim that the right–hand side of (2.15) is negative for sufficiently large n, which will yield the conclusion of the Lemma. Indeed,

$$|\lambda_1|\, |I_n| - \frac{1}{3}\lambda_0 \int_{I_n} |u(r)|^2 dr \leq |\lambda_1|\, |I_n| - \frac{n^2 \lambda_0}{3T^{2\beta}} |I_n| < 0$$

provided n is large enough, say $n \geq n_1$. Similarly,

$$C_4 \int_{I_n} |u(r)|dr - \frac{1}{3}\lambda_0 \int_{I_n} |u(r)|^2 dr \leq C_4 \int_{I_n} |u(r)|dr - \frac{n\lambda_0}{3T^{\beta}} \int_{I_n} |u(r)|dr < 0$$

if $n \geq n_2$. Furthermore,

$$C_5 \int_{I_n} \frac{|u(r)|}{(T-r)^\beta} dr - \frac{1}{3}\lambda_0 \int_{I_n} |u(r)|^2 dr \leq C_5 \int_{I_n} \frac{|u(r)|}{(T-r)^\beta} dr - \frac{n\lambda_0}{3} \int_{I_n} \frac{|u(r)|}{(T-r)^\beta} dr < 0$$

if $n \geq n_3$. The claim follows and the proof is complete. ∎

We now recall the definitions of some generalized gradients that will be used in the sequel. Let D be an open subset of X. The superdifferential of a function $v : D \to \mathbb{R}$ at a point $x_0 \in D$ is the (possibly empty) set

$$D^+ v(x_0) = \left\{ p \in X \ : \ \limsup_{x \to x_0} \frac{v(x) - v(x_0) - \langle p, x - x_0 \rangle}{|x - x_0|} \leq 0 \right\}. \tag{2.16}$$

We denote by $D^* v(x_0)$ the of all vectors $p \in X$ for which there exists a sequence $\{x_n\}$ of points of D such that

$$\begin{cases} (i) & x_n \to x_0 \text{ as } n \to +\infty \\ (ii) & v \text{ is Fréchet differentiable at } x_n, \forall n \\ (iii) & Dv(x_n) \rightharpoonup p \text{ as } n \to +\infty \end{cases} \tag{2.17}$$

If v is Lipschitz in a neighborhood D_0 of x_0, then v is Fréchet differentiable on a dense subset of D_0 (see [19]). Hence, $D^* v(x_0) \neq \emptyset$.

Next, we recall a Lipschitz regularity result for the value function V defined in (2.9) proved in [7] (see also [6]).

Proposition 2.2. *Assume* (2.8), (2.10). *Then, the value function V defined in* (2.9) *is continuous on* $[0, T] \times X$. *Moreover, for any $R > \frac{1}{T}$ and $\theta \in [0, 1[$, there exists $C_{R,\theta} > 0$ such that*

$$|V(t, x) - V(t, y)| \leq C_{R,\theta} |(-A)^{-\theta}(x - y)|$$

for all $t \in \left[0, T - \frac{1}{R}\right]$ and all $x, y \in X$ satisfying $|x|, |y| \leq R$. In particular, for all such t and x,

$$D_x^+ V(t, x) \subset D((-A)^\theta). \tag{2.18}$$

3. Finite Horizon Problem

In this section we will derive necessary and sufficient conditions for the problem of

$$\text{minimizing } J(t_0, x_0; u) \text{ overall controls } u \in L^2(t_0, T; U) \tag{3.1}$$

where J is defined in (2.11). In addition to hypotheses (2.8),(2.10), we will assume that

$$\begin{matrix} (i) & F \text{ is Fréchet differentiable} \\ (ii) & L \text{ is Fréchet differentiable with respect to } x. \end{matrix} \tag{3.2}$$

Notice that the above assumptions and (2.8)–(2.10) imply that DF and D_xL are bounded on all bounded subsets of X. Moreover we assume that for suitable c_1, c_2

$$|D_xL(t,x,u)| \le c_1 L(t,x,u) + c_2 \tag{3.3}$$

Let $\overline{u} \in L^2(t_0,T;U), \overline{x}(\cdot) = x(\cdot;t_0,x_0,\overline{u})$ and $p_T \in D^+\phi(\overline{x}(T))$.

We recall that the *co-state* associated to the triplet $\{\overline{u},\overline{x},p_T\}$ is the mild solution to the problem

$$\begin{cases} -p'(t) = Ap(t) + [DF(\overline{x}(t)]^* p(t) + D_xL(t,\overline{x}(t),\overline{u}(t)) \,, & t \in [t_0,T[\\ p(T) = -p_T. \end{cases} \tag{3.4}$$

Theorem 3.1. *Assume* (2.8), (2.10), (3.2) *and* (3.3). *Let* $\{\overline{u},\overline{x}\}$ *be an optimal pair for problem* (3.1), $p_T \in D^+\phi(\overline{x}(T))$, *and let* $\overline{p}$ *be the corresponding co-state. Then,*

$$-\,\overline{p}(t) \in D_x^+V(t,\overline{x}(t)) \tag{3.5}$$

for all $t \in [t_0,T[$ *and*

$$-\,\langle N_\beta\overline{u}(t),(-A)^\beta\overline{p}(t)\rangle - L(t,\overline{x}(t),\overline{u}(t)) = H(t,\overline{x}(t),(-A)^\beta\overline{p}(t)) \tag{3.6}$$

for a.e. $t \in [t_0,T]$, *where*

$$H(t,x,p) = \sup_{u\in U}[-\langle N_\beta u,p\rangle - L(t,x,u)]. \tag{3.7}$$

Proof – In view of (3.5) and of Proposition 2.2 we conclude that $\overline{p}(t) \in D((-A)^\beta)$, $\forall t \in [t_0,T[$. Therefore the left–hand side of (3.6) is well defined. From the Dynamic Programming Principle and Lemma (2.1) it follows that for $|x_0| \le R$ and for $t_0 < T_0 < T$

$$(3.8)\quad V(t_0,x_0) =$$
$$\inf\left\{\int_{t_0}^{T_0} L(s,x(s;t_0,x_0,u),u(s))ds + V(T_0,x(T_0;t_0,x_0,u)) : |u| \le \frac{M_R}{(T-T_0)^\beta}\right\}$$

Then we can apply the Maximum Principle contained in [4], Theorem 6.1. So we conclude (3.5) and (3.6).

Remark 3.2. *From hypotheses* (2.10)(*iii*) *and* (3.2) *it follows that* H *is differentiable with respect to* (x,p). *So, by* (3.6) *and* (3.7)

$$\overline{u}(t) = -N_\beta^* D_pH(t,\overline{x}(t),(-A)^\beta\overline{p}(t)) \tag{3.9}$$

for a.e. $t \in [t_0,T]$. *The above equation and* (3.5) *yield the feedback law*

$$\overline{u}(t) \in -N_\beta^* D_pH(t,\overline{x}(t),(-A)^\beta D_x^+V(t,\overline{x}(t)) \tag{3.10}$$

for a.e. $t \in [t_0,T]$

The next result may be directly derived following the same reasonings contained in [4], Theorem 5.9.

Theorem 3.3. *Assume* (2.8), (2.10), (3.2) *and* (3.3). *Suppose that ϕ is Fréchet differentiable and, for all $R > 0$,*

$$|D_x H(t,x,p) - D_x H(t,y,q)| + |D_p H(t,x,p) - D_p H(t,y,q)| \leq C_R[|x-y| + |p-q|]$$

for some constant $C_R > 0$ and all $x, y, p, q \in X$ satisfying $|x|, |y| \leq R$. Let $(t_0, x_0) \in [0,T] \times X$ and $-p_0 \in D_x^* V(t_0, x_0)$. Then, the system

$$(3.11) \quad \begin{cases} x'(t) = Ax(t) + F(x(t)) - (-A)^\beta D_p H(t, x(t), (-A)^\beta p(t)) \\ \\ -p'(t) = Ap(t) + [D_x F(x(t))]^* p(t) - D_x H(t, x(t), (-A)^\beta p(t)) \end{cases}$$

with the initial-terminal condition

$$\begin{cases} x(t_0) = x_0 \\ p(T) = -D\phi(x(T)). \end{cases}$$

has a solution $(\overline{x}, \overline{p})$ such that $\overline{x}$ is an optimal trajectory for problem (3.1) *corresponding to some control $\overline{u}$. Moreover, $\overline{p}$ is the co-state associated to $\overline{u}$ and satisfies $\overline{p}(t_0) = p_0$.*

The above Theorem gives, in some sense, a sufficient condition for optimality. This condition would we more useful if one could garantee uniqueness of solutions for (3.11). Uniqueness results for problem (3.11) have been obtained in the linear case, see [11]. In the next Theorem we show an existence and uniqueness result for an Hamiltonian system of kind (3.11) replacing the terminal co-state datum with an initial one. By adapting the same technique of [17] we can derive the following:

Theorem 3.4. *Assume* (2.8), (2.10), (3.2) *and* (3.3). *Suppose that*

$$|D_x H(t,x,p) - D_x H(t,y,q)| + |D_p H(t,x,p) - D_p H(t,y,q)| \leq L_H[|x-y| + |p-q|]$$

for some constant $L_H > 0$ and all $x, y, p, q \in X$. Let $(t_0, x_0) \in [0,T] \times X$ and $-p_0 \in D_x^* V(t_0, x_0)$. Then the Hamiltonian system

$$(3.12) \quad \begin{cases} x'(t) = Ax(t) - (-A)^\beta D_p H(t, x(t), (-A)^\beta p(t)) \,, & x(t_0) = x_0 \\ \\ -p'(t) = Ap(t) - D_x H(t, x(t), (-A)^\beta p(t)) \,, & p(t_0) = p_0 \end{cases}$$

has a unique solution $(\overline{x}, \overline{p})$.

Proof – The existence part is a staightforward consequence of the above Theorem. Without loss of generality we set $t_0 = 0$. Let us make the following change of variables

$$y(t) = (-A)^{-\beta} x(t) \,.$$

Then system (3.12) becomes

$$(3.13)\quad \begin{cases} y'(t)=Ay(t)-D_pH(t,(-A)^\beta y(t),(-A)^\beta p(t)), & y(0)=(-A)^{-\beta}x_0 \\ p'(t)=-Ap(t)+D_xH(t,(-A)^\beta y(t),(-A)^\beta p(t)), & p(0)=p_0. \end{cases}$$

Let $(y_1(t),p_1(t))$ and $(y_2(t),p_2(t))$ be two distinct solutions to system (3.13) and consider $\tilde{y}(t)=y_1(t)-y_2(t)$ and $\tilde{p}(t)=p_1(t)-p_2(t)$. Then $\tilde{y}(t)$ and $\tilde{p}(t)$ satisfy the system

$$(3.14)\quad \begin{cases} \tilde{y}'(t)=A\tilde{y}(t)-D_p\tilde{H}(t,(-A)^\beta\tilde{y}(t),(-A)^\beta\tilde{p}(t))\,, & \tilde{y}(0)=0 \\ \tilde{p}'(t)=-A\tilde{p}(t)+D_x\tilde{H}(t,(-A)^\beta\tilde{y}(t),(-A)^\beta\tilde{p}(t))\,, & \tilde{p}(0)=0 \end{cases}$$

where

$$\begin{aligned} &D_p\tilde{H}(t,(-A)^\beta\tilde{y}(t),(-A)^\beta\tilde{p}(t)) \\ &\qquad = D_pH(t,(-A)^\beta y_1(t),(-A)^\beta p_1(t))-D_pH(t,(-A)^\beta y_2(t),(-A)^\beta p_2(t)) \end{aligned}$$

and

$$\begin{aligned} &D_x\tilde{H}(t,(-A)^\beta\tilde{y}(t),(-A)^\beta\tilde{p}(t)) \\ &\qquad = D_xH(t,(-A)^\beta y_1(t),(-A)^\beta p_1(t))-D_xH(t,(-A)^\beta y_2(t),(-A)^\beta p_2(t)). \end{aligned}$$

Let $\theta\in C^1(\mathbb{R})$ be a function such that

$$\theta(t)=\begin{cases} 1 & 0\le t\le \frac{T}{2} \\ \\ 0 & t=T \end{cases}$$

$$|\theta'(t)|\le\frac{4}{T}$$

We set

$$\overline{y}(t)=\theta(t)\tilde{y}(t) \text{ and } \overline{p}(t)=\theta(t)\tilde{p}(t)$$

then $\overline{y}(t)$ and $\overline{p}(t)$ satisfy the system

$$(3.15)\quad \begin{cases} \overline{y}'(t)=A\overline{y}(t)-D_p\overline{H}(t,(-A)^\beta\overline{y}(t),(-A)^\beta\overline{p}(t))+g_x(t)\,, & \overline{y}(0)=0 \\ \overline{p}'(t)=-A\overline{p}(t)+D_x\overline{H}(t,(-A)^\beta\overline{y}(t),(-A)^\beta\overline{p}(t))+g_p(t)\,, & \overline{p}(0)=0 \end{cases}$$

where

$$g_x(t)=\theta'(t)\tilde{y}(t) \quad\text{and}\quad g_p(t)=\theta'(t)\tilde{p}(t)$$

and

$$D_p\overline{H}(t,(-A)^\beta\overline{y}(t),(-A)^\beta\overline{p}(t))=\theta(t)D_p\tilde{H}(t,(-A)^\beta\tilde{y}(t),(-A)^\beta\tilde{p}(t))$$

$$D_x\overline{H}(t,(-A)^\beta\overline{y}(t),(-A)^\beta\overline{p}(t))=\theta(t)D_x\tilde{H}(t,(-A)^\beta\tilde{y}(t),(-A)^\beta\tilde{p}(t)).$$

Now we set

$$z(t)=e^{\frac{k(t-T)^2}{2}}\overline{y}(t) \quad\text{and}\quad q(t)=e^{\frac{k(t-T)^2}{2}}\overline{p}(t)$$

then $z(t)$ and $q(t)$ satisfy the system

$$
(3.16)\quad \begin{cases} z'(t) &= Az(t) + k(t-T)z(t) \\ & \qquad -D_z H(t,(-A)^\beta z(t),(-A)^\beta q(t)) + f_z(t)\,, \\ z(0) &= z(T) = 0 \\ q'(t) &= -Aq(t) + k(t-T)q(t) \\ & \qquad +D_q H(t,(-A)^\beta z(t),(-A)^\beta q(t)) + f_q(t)\,, \\ q(0) &= q(T) = 0 \end{cases}
$$

where

$$
f_z(t) = e^{\frac{k(t-T)^2}{2}} g_x(t) \ \text{ and } \ f_q(t) = e^{\frac{k(t-T)^2}{2}} g_p(t)
$$

and

$$
D_z H(t,(-A)^\beta z(t),(-A)^\beta q(t)) = e^{\frac{k(t-T)^2}{2}} D_p \overline{H}(t,(-A)^\beta \overline{y}(t),(-A)^\beta \overline{p}(t))
$$

$$
D_q H(t,(-A)^\beta z(t),(-A)^\beta q(t)) = e^{\frac{k(t-T)^2}{2}} D_x \overline{H}(t,(-A)^\beta \overline{y}(t),(-A)^\beta \overline{p}(t)).
$$

Then multiplying the first equation of system (3.16) by $z'(t)$ and the second equation by $q'(t)$ we get

$$
\begin{aligned} |z'(t)|^2 &= \langle Az(t), z'(t)\rangle + \langle k(t-T)z(t), z'(t)\rangle \\ & \quad -\langle D_z H(t,(-A)^\beta z(t),(-A)^\beta q(t)), z'(t)\rangle + \langle f_z(t), z'(t)\rangle \end{aligned}
$$

and

$$
\begin{aligned} |q'(t)|^2 &= \langle -Aq(t), q'(t)\rangle + \langle k(t-T)q(t), q'(t)\langle \\ & \quad +\langle D_q H(t,(-A)^\beta z(t),(-A)^\beta q(t)), q'(t)\rangle + \langle f_q(t), q'(t)\rangle\,. \end{aligned}
$$

The above equalities can be rewritten as

$$
\begin{aligned} |z'(t)|^2 &= \frac{1}{2}\frac{d}{dt}\{\langle Az(t), z(t)\rangle + k(t-T)|z(t)|^2\} - \frac{k}{2}|z(t)|^2 \\ & \quad -\langle D_z H(t,(-A)^\beta z(t),(-A)^\beta q(t)), z'(t)\rangle + \langle f_z(t), z'(t)\rangle \end{aligned}
$$

and

$$
\begin{aligned} |q'(t)|^2 &= \frac{1}{2}\frac{d}{dt}\{\langle -Aq(t), q(t)\rangle + k(t-T)|q(t)|^2\} - \frac{k}{2}|q(t)|^2 \\ & \quad +\langle D_q H(t,(-A)^\beta z(t),(-A)^\beta q(t)), q'(t)\rangle + \langle f_q(t), q'(t)\rangle\,. \end{aligned}
$$

Integrating on $[0,T]$, recalling that z and q vanish at initial and terminal points, we get

$$\begin{aligned}&\int_0^T (|z'(t)|^2+\frac{k}{2}|z(t)|^2)dt\\ &\le \int_0^T |z'(t)|^2dt+\frac{1}{2}\int_0^T(|D_zH(t,(-A)^\beta z(t),(-A)^\beta q(t))|^2+|f_z(t)|^2)dt\,,\end{aligned}\tag{3.17}$$

$$\begin{aligned}&\int_0^T (|q'(t)|^2+\frac{k}{2}|q(t)|^2)dt\\ &\le \int_0^T |q'(t)|^2dt+\frac{1}{2}\int_0^T(|D_qH(t,(-A)^\beta z(t),(-A)^\beta q(t))|^2+|f_q(t)|^2)dt\,.\end{aligned}$$

Therefore this yields

$$\begin{aligned}&k\int_0^T(|z(t)|^2+|q(t)|^2)dt\\ &\le \int_0^T(|f_z(t)|^2)+|f_q(t)|^2)dt+\int_0^T(|D_zH(t,(-A)^\beta z(t),(-A)^\beta q(t))|^2\\ &\qquad +|D_qH(t,(-A)^\beta z(t),(-A)^\beta q(t))|^2)dt\,.\end{aligned}\tag{3.18}$$

Moreover, from the interpolation inequality (2.5) it follows

$$\begin{aligned}&|D_zH(t,(-A)^\beta z(t),(-A)^\beta q(t))|\\ &=e^{\frac{k(t-T)^2}{2}}|D_p\overline{H}(t,(-A)^\beta\overline{y}(t),(-A)^\beta\overline{p}(t))|\\ &\le e^{\frac{k(t-T)^2}{2}}\theta(t)|D_pH(t,(-A)^\beta y_1(t),(-A)^\beta p_1(t))\\ &\quad -D_pH(t,(-A)^\beta y_2(t),(-A)^\beta p_2(t))|\\ &\le L_He^{\frac{k(t-T)^2}{2}}\theta(t)\left[|(-A)^\beta(y_1(t)-y_2(t))|+|(-A)^\beta(p_1(t)-p_2(t))|\right]\\ &=L_H[|(-A)^\beta z(t)|+|(-A)^\beta q(t)|]\\ &\le L_H\sigma[|(-A)^{\frac{1}{2}}z(t)|+|(-A)^{\frac{1}{2}}q(t)|]+L_HC_\sigma[|z(t)|+|q(t)|]\end{aligned}\tag{3.19}$$

Similarly we have

$$(3.20)\quad |D_qH(t,(-A)^\beta z(t),(-A)^\beta q(t))| \\ \leq L_H\sigma[|(-A)^{\frac{1}{2}}z(t)|+|(-A)^{\frac{1}{2}}q(t)|]+L_HC_\sigma[|z(t)|+|q(t)|].$$

Noticing that from estimates (3.19) and (3.20), for some positive constant C it follows

$$(3.21)\quad |D_zH(t,(-A)^\beta z(t),(-A)^\beta q(t))|^2 \\ \leq CL_H\sigma[|(-A)^{\frac{1}{2}}z(t)|^2+|(-A)^{\frac{1}{2}}q(t)|^2]+CL_HC_\sigma[|z(t)|^2+|q(t)|^2].$$

and

$$(3.22)\quad |D_qH(t,(-A)^\beta z(t),(-A)^\beta q(t))|^2 \\ \leq CL_H\sigma[|(-A)^{\frac{1}{2}}z(t)|^2+|(-A)^{\frac{1}{2}}q(t)|^2]+CL_HC_\sigma[|z(t)|^2+|q(t)|^2]\,.$$

Multiplying by $z(t)$ the first equation of (3.16) and integrating we obtain

$$\int_0^T |(-A)^{\frac{1}{2}}z(t)|^2dt \\ \leq (kT+1)\int_0^T |z(t)|^2dt+\frac{1}{2}\int_0^T(|D_zH(t,(-A)^\beta z(t),(-A)^\beta q(t))|^2+|f_z(t)|^2)dt.$$

Therefore from (3.21) it follows

$$\int_0^T |(-A)^{\frac{1}{2}}z(t)|^2dt \leq (kT+1)\int_0^T |z(t)|^2dt$$

$$(3.23)\qquad +\frac{CL_H\sigma}{2}\int_0^T(|(-A)^{\frac{1}{2}}z(t)|^2+|(-A)^{\frac{1}{2}}q(t)|^2)dt+\frac{1}{2}\int_0^T |f_z(t)|^2dt$$

$$+\frac{CL_HC_\sigma}{2}\int_0^T(|z(t)|^2+|q(t)|^2)dt.$$

Multiplying by $-q(t)$ the second equation of (3.16) and integrating we obtain

$$\int_0^T |(-A)^{\frac{1}{2}} q(t)|^2 dt \le (kT+1) \int_0^T |q(t)|^2 dt$$

$$(3.24) \qquad + \frac{CL_H\sigma}{2} \int_0^T (|(-A)^{\frac{1}{2}} z(t)|^2 + |(-A)^{\frac{1}{2}} q(t)|^2) dt + \frac{1}{2} \int_0^T |f_q(t)|^2 dt$$

$$+ \frac{CL_H C_\sigma}{2} \int_0^T (|z(t)|^2 + |q(t)|^2) dt.$$

Adding (3.23) and (3.24), choosing $0 < \sigma < \dfrac{1}{CL_H}$, we have

$$(3.25) \quad \int_0^T (|(-A)^{\frac{1}{2}} z(t)|^2 + |(-A)^{\frac{1}{2}} q(t)|^2) dt$$

$$\le C_1 \int_0^T (|z(t)|^2 + |q(t)|^2) dt + \frac{1}{2(1 - CL_H\sigma)} \int_0^T (|f_z(t)|^2 + |f_q(t)|^2) dt\,,$$

where

$$C_1 = \frac{kT + 1 + CL_H C_\sigma}{1 - CL_H\sigma}\,.$$

From estimates (3.18) and the above one, it follows

$$(3.26) \qquad \int_0^T (|z(t)|^2 + |q(t)|^2) dt \le \tilde{C}(k) \int_0^T (|f_z(t)|^2 + |f_q(t)|^2) dt\,,$$

where $\tilde{C}(k)$ is positive for k big enough and $0 < \sigma < \dfrac{1}{CL_H(1+2T)}$. It is defined as

$$\tilde{C}(k) = \frac{3 - 2CL_H\sigma}{2(1 - CL_H\sigma)(k - 2CL_H\sigma C_1 - 2CL_H C_\sigma)}\,,$$

notice that $\tilde{C} \to 0$ as $k \to \infty$.

From (3.26) directly follows

$$(3.27) \quad \int_0^T e^{k(t-T)^2} (|\overline{y}(t)|^2 + |\overline{p}(t)|^2) dt \le \tilde{C}(k) \int_0^T e^{k(t-T)^2} (|g_x(t)|^2 + |g_p(t)|^2) dt\,.$$

On the other hand

$$
(3.28)\quad \begin{aligned}
\int_0^T e^{k(t-T)^2}(|\overline{y}(t)|^2+|\overline{p}(t)|^2)dt &\geq \int_0^{\frac{T}{2}} e^{k(t-T)^2}(|\tilde{y}(t)|^2+|\tilde{p}(t)|^2)dt \\
&\geq e^{k\frac{T^2}{4}}\int_0^{\frac{T}{2}}(|\tilde{y}(t)|^2+|\tilde{p}(t)|^2)dt
\end{aligned}
$$

and the following holds

$$
(3.29)\quad \begin{aligned}
&\int_0^T e^{k(t-T)^2}(|g_x(t)|^2+|g_p(t)|^2)dt \\
&\leq \int_0^T e^{k(t-T)^2}|\theta'(t)|^2(|\tilde{y}(t)|^2+|\tilde{p}(t)|^2)dt \\
&\leq \left(\frac{4}{T}\right)^2\int_{\frac{T}{2}}^T e^{k(t-T)^2}(|\tilde{y}(t)|^2+|\tilde{p}(t)|^2)dt \\
&\leq \left(\frac{4}{T}\right)^2 e^{k\frac{T^2}{4}}\int_{\frac{T}{2}}^T(|\tilde{y}(t)|^2+|\tilde{p}(t)|^2)dt\,.
\end{aligned}
$$

In conclusion, from (3.27), (3.28) and (3.29) we get

$$
e^{k\frac{T^2}{4}}\int_0^{\frac{T}{2}}(|\tilde{y}(t)|^2+|\tilde{p}(t)|^2)dt \leq \tilde{C}(k)\left(\frac{4}{T}\right)^2 e^{k\frac{T^2}{4}}\int_{\frac{T}{2}}^T(|\tilde{y}(t)|^2+|\tilde{p}(t)|^2)dt\,.
$$

From the above inequality we obtain

$$
\int_0^{\frac{T}{2}}(|\tilde{y}(t)|^2+|\tilde{p}(t)|^2)dt \to 0 \ \text{ as } \ k\to\infty
$$

and we conclude that $|\tilde{y}(t)| = |\tilde{p}(t)| = 0$ on $[0,\frac{T}{2}]$. Iterating this procedure we obtain this result on [0,T]. ■

References

1 V Barbu, Boundary control problems with convex cost criterion, SIAM J Control and Optim , 18, 227–254, 1980

2 V Barbu and G Da Prato, Hamilton-Jacobi equations in Hilbert spaces, Pitman, Boston, 1982

3 A Bensoussan, G Da Prato, M C Delfour and S K Mitter, Representation and control of infinite dimensional systems, Birkhauser, Boston, 1992

4 P Cannarsa, H Frankowska, Value function and optimality conditions for semilinear control problems, Applied Math Optim , 26, 139–169, 1992

5 P Cannarsa and F Gozzi, On the smoothness of the value function along optimal trajectories, IFIP Workshop on Boundary Control and Boundary Variation (to appear in Lecture Notes in Control and Information Sciences, Springer–Verlag)

6 P Cannarsa, F Gozzi and H M Soner, A dynamic programming approach to nonlinear boundary control problems of parabolic type, J Funct Anal (to appear)

7 P Cannarsa and M E Tessitore, Cauchy problem for the dynamic programming equation of boundary control, Proceedings IFIP Workshop on Boundary Control and Boundary Variation (to appear)

8 F Clarke and R B Vinter, The relationship between the maximum principle and dynamic programming, SIAM J Control Optim , 25, 1291–1311, 1989

9 H O Fattorini, Boundary control system, SIAM J Control, 6, 349–385, 1968

10 H O Fattorini, A unified theory of necessary conditions for nonlinear nonconvex control systems, Appl Math Opt , 15, 141– 185, 1987

11 A Favini and A Venni, On a two–point problem for a system of abstract differential equations, Numer Funct Anal and Optimiz , 2(4), 301–322, 1980

12 H Goldberg and F Troltzsch, Second–order sufficient optimality conditions for a class of nonlinear parabolic boundary control problems, SIAM J Control and Optim , vol 31, 4, 1007–1025, 1993

13 V Iftode, Hamilton–Jacobi equations and boundary convex control problems, Rev Roumaine Math Pures Appl , 34, (1989), 2, 117–127

14 G E Ladas and V Lakshmikantham, Differential equations in abstract spaces, Academic Press, New York and London, 1972

15 I Lasiecka and R Triggiani, Differential and algebraic equations with application to boundary control problems continuous theory and approximation theory, Lecture Notes in Control and Information Sciences, Springer–Verlag, 164, 1991

16 J L Lions, Optimal control of systems described by partial differential equations, Springer–Verlag, Wiesbaden, 1972

17 J L Lions and B Malgrange, Sur l'unicité rétrograde dans les problemes mixtes paraboliques, Math Scand 8, 277–286, 1960

18 A Pazy, Semigroups of linear operators and applications to partial differential equations, Springer–Verlag, New York–Heidelberg– Berlin, 1983

19 D Preiss, Differentiability of Lipschitz functions on Banach spaces, J Funct Anal , 91, 312–345, 1990

20 F Troltzsch, On the semigroup approach for the optimal control of semilinear parabolic equations including distributed and boundary control, Zeitschrift fur Analysis und ihre Anwendungen Bd , 8, 431–443, 1989

Piermarco Cannarsa
Dipartimento di Matematica
Universitá di Roma "Tor Vergata"
Via O. Raimondo
I-00173 Roma, Italy

Maria Elisabetta Tessitore
Dipartimento di Matematica
Universitá di Roma "La Sapienza"
Piazzale A. Moro 2
I-00185 Roma, Italy

PONTRYAGIN'S PRINCIPLE FOR OPTIMAL CONTROL PROBLEMS GOVERNED BY SEMILINEAR ELLIPTIC EQUATIONS

EDUARDO CASAS

Departamento de Matemática Aplicada y Ciencias de la Computación
Universidad de Cantabria

ABSTRACT In this paper we prove an extension of Pontryagin's principle for optimal control problems governed by semilinear elliptic partial differential equations The control takes values in a bounded subset, not necessarily convex, of some Euclidean space, the cost functional is of Lagrange type, and some general equality and inequality state constraints are imposed To derive Pontryagin's principle we combine a suitable penalization of the state constraints with Ekeland's principle In the absence of equality state constraints we establish the optimality conditions in a qualified form for "almost all" problems In a first stage, the classical spike perturbations of the controls used to derive Pontryagin's principle are replaced by a new type of variations

1991 *Mathematics Subject Classification* 49K20, Secondary 35J65

Key words and phrases Pontryagin's principle, semilinear elliptic equations, state constraints, Ekeland's variational principle

1. Introduction

Many papers dealing with optimality conditions for control problems governed by partial differential equations have been written in the last twenty five years, but only a few of them have been devoted to Pontryagin's principle. This makes a big difference with the theory of optimal control for ordinary differential equations, where this principle is the usual way of formulating the conditions for optimality. The motive of this difference is that the methods used for finite dimensional systems can not be readily extended to infinite systems, arising some difficulties in the way of this extension. Nevertheless some new ideas have appeared leading to the first results; let us mention the papers by Fattorini and Frankowska [14], Li and Yao [15] and Li and Yong [16] for problems governed by parabolic equations; Bonnans and Casas [3], [4], [5], Bonnans and Tiba [6], Casas and Yong [9] and Yong [18] for the elliptic case.

This research was partially supported by Dirección General de Investigación Científica y Técnica (Madrid)

This paper deals with Pontryagin's principle for state-constrained control problems governed by semilinear elliptic equations. The principle proved here is stronger than any of the previous ones corresponding to elliptic equations because of the generality of the state constraints or the assumptions on the data defining the problem. In particular, less regularity of the domain and the coefficients of the equation is required, and no continuity with respect to the control of the functions involved in the problem is assumed. Moreover some proofs have been simplified by using different methods. In spite of these remarks, this paper could be viewed as a combination of the ideas introduced in [5], [9] and [18].

Following [5], we will prove two minimum principles, called weak and strong respectively. The term strong is due to the fact of providing optimality conditions in a qualified form. This strong version of Pontryagin's principle is proved only for problems without equality state constraints and it requires an additional hypothesis: a stability condition of the optimal cost with respect to small perturbations of the feasible state set. This condition was first used by Bonnans [2]; see also [5] and [9]. Unlike [5], no stability assumption is needed here to prove the weak principle, this is achieved by using a different penalization of the state constraints, previously utilized in [9] and [18].

As mentioned above, the proof of the minimum principle is based on a penalization procedure of the state constraints. In this process, Ekeland's principle plays an essential role. Another important feature of the proof is the use of some special variations of the controls different of the classical spike perturbations. This idea was first employed by Li and Yao [15] for evolution problems, and later it was adapted to elliptic problems in [18] and [9]. In this paper, we give a new proof of this adaptation, which is simpler and lets a better understanding of the nature of these perturbations.

The plan of this paper is as follows: in §2 we formulate the control problem and provide some examples; and in §3 and §4 we prove the weak and strong versions, respectively, of Pontryagin's principle.

2. Setting of the control problem

Let $\Omega \subset \mathbb{R}^n$, $n > 1$, be an open bounded set, with boundary Γ of class C^1 if $n > 3$ and Lipschitz if $n \leq 3$. In Ω we consider the differential operator

$$Ay = -\sum_{i,j=1}^{n} \partial_{x_j}[a_{ij}(x)\partial_{x_i} y],$$

where $a_{ij} \in C(\bar{\Omega})$, $1 \leq i, j \leq n$ and satisfy the usual ellipticity assumption: there exists $\Lambda > 0$ such that

$$\sum_{i,j=1}^{n} a_{ij}(x)\xi_i\xi_j \geq \Lambda|\xi|^2 \quad \forall x \in \Omega \text{ and } \forall \xi \in \mathbb{R}^n.$$

Let K be a compact subset of $\mathbb{R}^m$, $m \geq 1$, and $f : \Omega \times \mathbb{R} \times K \longrightarrow \mathbb{R}$ a Lebesgue measurable function of class C^1 with respect to the second variable satisfying

$$(1) \quad \begin{cases} \dfrac{\partial f}{\partial y}(x,y,u) \leq 0 \quad \forall (x,y,u) \in \Omega \times \mathbb{R} \times K \\ \forall M > 0\ \exists \psi_M \in L^s(\Omega),\ s > n/2, \text{ such that} \\ |f(x,0,u)| + \left|\dfrac{\partial f}{\partial y}(x,y,u)\right| \leq \psi_M(x) \ \ a.e.\ x \in \Omega,\ \forall |y| \leq M,\ \forall u \in K. \end{cases}$$

If Γ is not of class C^1, but it is only Lipschitz and $n = 2$, then we also assume that $s > 4/3$.

Given a function $u : \Omega \longrightarrow K$, we consider the state equation

$$(2) \quad \begin{cases} Ay = f(x, y(x), u(x)) & \text{in } \Omega \\ y = 0 & \text{on } \Gamma. \end{cases}$$

Theorem 1. *Under the previous assumptions, for every function $u : \Omega \longrightarrow K$, such that $x \in \Omega \longrightarrow f(x,y,u(x)) \in \mathbb{R}$ is measurable for each $y \in \mathbb{R}$, there exists a number $p > n$ such that (2) has a unique solution y_u in the Sobolev space $W_0^{1,p}(\Omega)$. Moreover there exists a constant $C_K > 0$, independent of u, such that*

$$(3) \quad \|y_u\|_{W_0^{1,p}(\Omega)} \leq C_K.$$

This theorem follows from the classical theory of partial differential equations. The first thing to remark is that f is a nonincreasing monotone function and it is enough to use a global or local cutting procedure of f (see, for instance, [3] or [7]) to obtain the existence of a solution. The $W_0^{1,p}(\Omega)$-regularity follows from the fact that $L^s(\Omega) \subset W^{-1,p}(\Omega)$ for every $p \in (n, ns/(n-s)]$ if $s < n$ and $p \in (n, \infty)$ if $s \geq n$, and that operator A is an isomorphism from $W_0^{1,p}(\Omega)$ onto $W^{-1,p}(\Omega)$; see Morrey [17]. When $n \leq 3$ and Γ is only Lipschitz, then the $W_0^{1,p}(\Omega)$-regularity is not always true, however there exists $\epsilon > 0$ depending on Ω such that for every

$$p \in \begin{cases} \left(\dfrac{4+\epsilon}{3+\epsilon}, 4+\epsilon\right) & \text{if } n = 2 \\ \left(\dfrac{3+\epsilon}{2+\epsilon}, 3+\epsilon\right) & \text{if } n = 3 \end{cases}$$

operator $A : W_0^{1,p}(\Omega) \longrightarrow W^{-1,p}(\Omega)$ is still an isomorphism; see Dahlberg [12].

An important consequence of the previous theorem is that $y_u \in W_0^{1,p}(\Omega) \subset C_0(\Omega)$, where $C_0(\Omega)$ denotes the space on continuous functions in $\bar{\Omega}$ vanishing on Γ. Moreover the previous inclusion is compact.

Established the fundamental result about the state equation, let us go to the formulation of the control problem. Firstly we consider a Lebesgue measurable function $L : \Omega \times \mathbb{R} \times K \longrightarrow \mathbb{R}$ of class C^1 with respect to the second variable satisfying

$$(4)\quad \begin{cases} \forall M > 0 \ \exists \phi_M \in L^1(\Omega) \quad \text{such that} \\ |L(x,0,u)| + \left|\dfrac{\partial L}{\partial y}(x,y,u)\right| \le \phi_M(x) \ \ a.e.\ x \in \Omega, \forall |y| \le M, \ \forall u \in K. \end{cases}$$

Let us denote by $\mathcal{U}$ the set of admissible controls: measurable functions $u : \Omega \to K$ such that

$$x \in \Omega \longrightarrow (f(x,y,u(x)), L(x,y,u(x))) \in \mathbb{R}^2 \ \text{ is measurable } \forall y \in \mathbb{R}.$$

Let Z be a separable Banach space and $Q \subset Z$ a closed convex subset with nonempty interior. Finally, let $G : W_0^{1,p}(\Omega) \longrightarrow Z$ and $F : W_0^{1,p}(\Omega) \longrightarrow \mathbb{R}^l$, $l \ge 1$, be two mappings of class C^1. Then we can state the optimal control problem as follows

$$(P) \begin{cases} \text{Minimize } J(u) = \displaystyle\int_\Omega L(x, y_u(x), u(x))dx \\ u \in \mathcal{U}, G(y_u) \in Q, F(y_u) = 0. \end{cases}$$

Let us show how the usual examples of state constraints can be handled with this formulation.

Example 1. Given a continuous function $g : \bar{\Omega} \times \mathbb{R} \longrightarrow \mathbb{R}$ of class C^1 respect to the second variable and a number $\delta > 0$, with $g(x,0) < \delta \ \ \forall x \in \Gamma$, the constraint $g(x, y_u(x)) \le \delta$ for all $x \in \Omega$ can be written in the above framework by putting $Z = C_0(\Omega)$, $G : W_0^{1,p}(\Omega) \longrightarrow C_0(\Omega)$, defined by $G(y) = g(\cdot, y(\cdot))$, and

$$Q = \{z \in C_0(\Omega) : z(x) \le \delta \ \ \forall x \in \Omega\}.$$

Example 2. The constraint

$$\int_\Omega |y_u(x)|dx \le \delta$$

is considered by taking $Z = L^1(\Omega)$, $G : W_0^{1,p}(\Omega) \longrightarrow L^1(\Omega)$, with $G(y) = y$, and Q the closed ball in $L^1(\Omega)$ of center at 0 and radius δ.

Example 3. For every $1 \le j \le k$ let $g_j : \Omega \times \mathbb{R} \longrightarrow \mathbb{R}$ be a measurable function of class C^1 with respect to the second variable such that for each $M > 0$ there exists a function $\eta_M^j \in L^1(\Omega)$ satisfying

$$|g_j(x,0)| + \left|\frac{\partial g_j}{\partial y}(x,y)\right| \le \eta_M^j(x) \ \ a.e.\ x \in \Omega, \ \forall |y| \le M.$$

Then the constraints

$$\int_\Omega g_j(x, y_u(x))dx \leq \delta_j, \quad 1 \leq j \leq k,$$

are included in the formulation of (P) by choosing $G = (G_1, \dots, G_k)^T$, with

$$G_j(y) = \int_\Omega g_j(x, y(x))dx,$$

$Z = \mathbb{R}^k$, and $Q = (-\infty, \delta_1] \times \cdots \times (-\infty, \delta_k]$.

Example 4. The equality constraints

$$\int_\Omega f_j(x, y_u(x))dx = 0, \quad 1 \leq j \leq l,$$

can also be included in problem (P) in the obvious way by assuming the same hypotheses as in Example 3.

Example 5. Let $\{x_j\}_{j=1}^l \subset \Omega$, then the equality constraints $y_u(x_j) = \delta_j$, $1 \leq j \leq l$, are well adapted to the above formulation. Indeed, since $W_0^{1,p}(\Omega) \subset C_0(\Omega)$, the functions $F_j : W_0^{1,p}(\Omega) \longrightarrow \mathbb{R}$ given by $F_j(y) = y(x_j) - \delta_j$ are well defined and they are of class C^1. Then it is enough to take $F = (F_1, \dots, F_l)^T$.

3. The weak Pontryagin's principle

The aim of this section is to prove the following result

Theorem 2. *Let $p > n$ be given by Theorem 1. If $\bar{u}$ is a solution of* (P), *then there exist $\bar{\alpha} \geq 0$, $\bar{y} \in W_0^{1,p}(\Omega)$, $\bar{\varphi} \in W_0^{1,p'}(\Omega)$, $p' = p/(p-1)$, $\bar{\mu} \in Z'$ and $\bar{\lambda} \in R^l$ such that*

$$\bar{\alpha} + \|\bar{\mu}\|_{Z'} + |\bar{\lambda}| > 0; \tag{5}$$

$$\begin{cases} A\bar{y} = f(x, \bar{y}(x), \bar{u}(x)) & \text{in } \Omega, \\ \bar{y} = 0 \quad \text{on } \Gamma; \end{cases} \tag{6}$$

$$\begin{cases} A^*\bar{\varphi} = \dfrac{\partial f}{\partial y}(x, \bar{y}(x), \bar{u}(x))\bar{\varphi} + \bar{\alpha}\dfrac{\partial L}{\partial y}(x, \bar{y}(x), \bar{u}(x)) \\ \qquad + [DG(\bar{y})]^*\bar{\mu} + [DF(\bar{y})]^*\bar{\lambda} \quad \text{in } \Omega, \\ \bar{\varphi} = 0 \quad \text{on } \Gamma; \end{cases} \tag{7}$$

$$\langle \bar{\mu}, z - G(\bar{y}) \rangle \leq 0 \quad \forall z \in Q; \tag{8}$$

$$\int_\Omega H_{\bar{\alpha}}(x, \bar{y}(x), \bar{u}(x), \bar{\varphi}(x))dx = \min_{v \in \mathcal{U}} \int_\Omega H_{\bar{\alpha}}(x, \bar{y}(x), v(x), \bar{\varphi}(x))dx; \tag{9}$$

where

$$H_\alpha(x, y, u, \varphi) = \alpha L(x, y, u) + \varphi f(x, y, u).$$

Moreover, if one of the following assumptions is satisfied:
(A1) *Functions f and L are continuous with respect to the third variable on K;*
(A2) *There exists a set $\Omega_0 \subset \Omega$, with $|\Omega_0| = |\Omega|$, such that the function*

$$x \in \Omega \longrightarrow (f(x, y, u), L(x, y, u)) \in \mathbb{R}^2$$

is continuous in Ω_0 for every $(y, u) \in \mathbb{R} \times K$;
then the following pointwise relation holds

$$H_{\bar\alpha}(x, \bar y(x), \bar u(x), \bar\varphi(x)) = \min_{v \in K} H_{\bar\alpha}(x, \bar y(x), v, \bar\varphi(x)) \quad a.e.\ x \in \Omega. \tag{10}$$

Let us see how the previous optimality conditions are written for some of the state constraints given in the examples of §2.

Example 6. Let us assume that the constraints associated with G and F are those given in the examples 1 and 4, respectively. Then, taking into account that $Z' = C_0(\Omega)'$ is identified with the space of real and regular Borel measures in Ω, denoted by $M(\Omega)$, Theorem 2 claims the existence of a measure $\bar\mu \in M(\Omega)$, a number $\bar\alpha \geq 0$, a vector $\bar\lambda \in R^l$ and two functions $\bar y \in W_0^{1,p}(\Omega)$ and $\bar\varphi \in W_0^{1,p'}(\Omega)$ such that (5), (6), (9) and possibly (10), and the following relations hold

$$\begin{cases} A^*\bar\varphi &= \dfrac{\partial f}{\partial y}(x, \bar y(x), \bar u(x))\bar\varphi + \bar\alpha \dfrac{\partial L}{\partial y}(x, \bar y(x), \bar u(x)) \\ & \quad + \dfrac{\partial g}{\partial y}(x, \bar y(x))\bar\mu + \displaystyle\sum_{j=1}^{l} \bar\lambda_j \dfrac{\partial f_j}{\partial y}(x, \bar y(x)) \ \text{ in } \Omega, \\ \bar\varphi = 0 \ \text{ on } \Gamma; \end{cases}$$

$$\int_\Omega (z(x) - g(x, \bar y(x)))d\bar\mu(x) \leq 0 \quad \forall z \in C_0(\Omega), \ \text{ with } z(x) \leq \delta \ \forall\, x \in \Omega.$$

Example 7. If the state constraints are given as in the examples 2 and 5, then $\bar\mu \in L^\infty(\Omega)$ and the adjoint state equation (7) and inequality (8) become now

$$\begin{cases} A^*\bar\varphi = \dfrac{\partial f}{\partial y}(x, \bar y(x), \bar u(x))\bar\varphi + \bar\alpha \dfrac{\partial L}{\partial y}(x, \bar y(x), \bar u(x)) + \bar\mu + \displaystyle\sum_{j=1}^{l} \bar\lambda_j \delta_{[x_j]} \ \text{ in } \Omega, \\ \bar\varphi = 0 \ \text{ on } \Gamma; \end{cases}$$

$$\int_\Omega \bar\mu(x)(z(x) - \bar y(x))dx \leq 0 \quad \forall z \in L^1(\Omega) \ \text{ with } \int_\Omega |z(x)|dx \leq \delta;$$

where $\delta_{[x_j]}$ denotes the Dirac measure centered at x_j.

Finally, let us remark that if the constraints associated to G are given as in Example 3, then $\bar{\mu} = \{\bar{\mu}_j\}_{j=1}^k \in \mathbb{R}^k$ and (8) states that $\bar{\mu}_j \geq 0$ for every $j = 1, \ldots, k$.

Now we go on to prove Theorem 2. First of all we need two lemmas, which describe the type of variations of the controls used in the proof.

Lemma 1. *Let $g \in L^1(\Omega)$ and $h \in L^s(\Omega)$, $1 \leq s \leq +\infty$. Given $\rho \in (0,1)$, let us denote*

$$\mathcal{E}_\rho = \{E \subset \Omega \textit{ measurable} : |E| = \rho|\Omega|\}.$$

Then

$$\inf_{E \in \mathcal{E}_\rho} \left\{ \left| \int_\Omega \left(1 - \frac{1}{\rho}\chi_E(x)\right) g(x)dx \right| + \left\| \left(1 - \frac{1}{\rho}\chi_E\right) h \right\|_{W^{-1,p}(\Omega)} \right\} = 0,$$

where χ_E is the characteristic function of E and p is any real number satisfying

$$\begin{cases} 1 < p \leq +\infty & \textit{if } s \geq n, \\ 1 < p \leq ns/(n-s) & \textit{if } s < n. \end{cases}$$

Proof. Given $\rho \in (0,1)$, let us take $\epsilon > 0$ arbitrary. Let us denote by B the closed unit ball of $W_0^{1,p'}(\Omega)$, where $p' = p/(p-1)$. From the assumptions on p and the Sobolev's imbeddings (see, for instance, Adams [1]) if follows

$$L^s(\Omega) \subset W^{-1,p}(\Omega) \quad \text{and} \quad W_0^{1,p'}(\Omega) \subset L^{s'}(\Omega),$$

the imbeddings being continuous. Above we denote $s' = s/(s-1)$. Let us assume that $s' < +\infty$, otherwise it is enough to change the integrals by the supremum norm to accomplish the proof.

We can take a partition $\{\Omega_j\}_{j=1}^r$ of Ω, with $|\Omega_j| > 0$, such that

$$\sum_{j=1}^r \int_{\Omega_j} \left| y(x) - \frac{1}{|\Omega_j|} \int_{\Omega_j} y(\xi)d\xi \right|^{s'} dx < \epsilon^{s'} \quad \forall y \in B. \tag{11}$$

This can be proved, for instance, by using the interpolation theory in Sobolev spaces; see Ciarlet [10].

For every $y \in B$ let us define $\tilde{y} : \Omega \longrightarrow \mathbb{R}$ by

$$\tilde{y}|_{\Omega_j}(x) = \frac{1}{|\Omega_j|} \int_{\Omega_j} y(\xi)d\xi \quad \forall x \in \Omega_j, \quad 1 \leq j \leq r.$$

Then (11) implies

$$\|y - \tilde{y}\|_{L^{s'}(\Omega)} < \epsilon \quad \forall y \in B. \tag{12}$$

Now we can approximate g and h in every Ω_j by simple functions

$$g_j(x) = \sum_{i=1}^{r_j} \alpha_{ij} \chi_{F^{ij}}(x), \quad h_j(x) = \sum_{i=1}^{r_j} \beta_{ij} \chi_{F^{ij}}(x),$$

with $\alpha_{ij}, \beta_{ij} \in \mathbb{R}$, $\{F^{ij}\}_{i=1}^{r_j}$ being a partition of Ω_j with $|F^{ij}| > 0$, and such that

$$\int_{\Omega_j} |g(x) - g_j(x)| dx + \int_{\Omega_j} |h(x) - h_j(x)| dx < \epsilon |\Omega_j|. \tag{13}$$

Let us take $E^{ij} \subset F^{ij}$ such that $|E^{ij}| = \rho |F^{ij}|$. Defining $E^j = \bigcup_{i=1}^{r_j} E^{ij}$, we have that

$$\int_{\Omega_j} g_j(x) dx = \int_{\Omega_j} \frac{1}{\rho} \chi_{E^j}(x) g_j(x) dx, \quad 1 \leq j \leq r, \tag{14}$$

and

$$\int_{\Omega_j} h_j(x) dx = \int_{\Omega_j} \frac{1}{\rho} \chi_{E^j}(x) h_j(x) dx, \quad 1 \leq j \leq r. \tag{15}$$

Finally we take $E = \bigcup_{j=1}^{r} E^j$. Then $|E| = \rho |\Omega|$ and for every $y \in B$ we have

$$\begin{aligned} \left| \int_\Omega \left(1 - \frac{1}{\rho} \chi_E(x)\right) h(x) y(x) dx \right| &\leq \left| \int_\Omega \left(1 - \frac{1}{\rho} \chi_E(x)\right) h(x) \tilde{y}(x) dx \right| \\ &\quad + \int_\Omega \left(1 + \frac{1}{\rho} \chi_E(x)\right) |h(x)| |y(x) - \tilde{y}(x)| dx \\ &\leq \left| \sum_{j=1}^{r} \int_{\Omega_j} \left(1 - \frac{1}{\rho} \chi_E(x)\right) h(x) \tilde{y}(x) dx \right| \\ &\quad + \left(1 + \frac{1}{\rho}\right) \|h\|_{L^s(\Omega)} \|y - \tilde{y}\|_{L^{s'}(\Omega)}. \end{aligned} \tag{16}$$

From (11) we deduce

$$\left(1 + \frac{1}{\rho}\right) \|h\|_{L^s(\Omega)} \|y - \tilde{y}\|_{L^{s'}(\Omega)} \leq \epsilon \left(1 + \frac{1}{\rho}\right) \|h\|_{L^s(\Omega)}. \tag{17}$$

On the other hand, from (13) and (15) we obtain for all $y \in B$

$$
\begin{aligned}
\left|\sum_{j=1}^{r}\int_{\Omega_j}\left(1-\frac{1}{\rho}\chi_E(x)\right)h(x)\tilde{y}(x)dx\right| &\le \left(1+\frac{1}{\rho}\right)\sum_{j=1}^{r}\left\{|\tilde{y}|_{\Omega_j}\int_{\Omega_j}|h(x)-h_j(x)|dx\right\}\\
&\quad+\left|\sum_{j=1}^{r}|\tilde{y}|_{\Omega_j}\int_{\Omega_j}\left(1-\frac{1}{\rho}\chi_E(x)\right)h_j(x)dx\right| \qquad (18)\\
&\le \left(1+\frac{1}{\rho}\right)\epsilon\|\tilde{y}\|_{L^1(\Omega)} \le C\left(1+\frac{1}{\rho}\right)\epsilon.
\end{aligned}
$$

Thus (16)–(18) imply that

$$
\left\|\left(1-\frac{1}{\rho}\chi_E\right)h\right\|_{W^{-1,p}(\Omega)} \le \epsilon\left(1+\frac{1}{\rho}\right)\left(C+\|h\|_{L^s(\Omega)}\right), \qquad (19)
$$

where we have used that $W^{-1,p}(\Omega) = \left(W_0^{1,p'}(\Omega)\right)'$.

On the other hand, using (13) and (14) we get

$$
\begin{aligned}
\left|\int_{\Omega}\left(1-\frac{1}{\rho}\chi_E\right)g(x)dx\right| &\le \sum_{j=1}^{r}\left|\int_{\Omega_j}\left(1-\frac{1}{\rho}\chi_E(x)\right)g(x)dx\right|\\
&\le \left(1+\frac{1}{\rho}\right)\sum_{j=1}^{r}\int_{\Omega_j}|g(x)-g_j(x)|dx \qquad (20)\\
&\quad+\sum_{j=1}^{r}\left|\int_{\Omega_j}\left(1-\frac{1}{\rho}\chi_E(x)\right)g_j(x)dx\right|\\
&\le \epsilon\left(1+\frac{1}{\rho}\right)|\Omega|.
\end{aligned}
$$

Then (19) and (20), together with the fact that $\epsilon > 0$ is arbitrary, prove the lemma. □

Lemma 2. *Suppose $p > n$ given by Theorem 1. Let $u \in \mathcal{U}$ and let y be its associated state. Then for every $\rho \in (0,1)$ we can find a measurable set $E_\rho \subset \Omega$, with $|E_\rho| = \rho|\Omega|$, such that if we define*

$$
u_\rho(x) = \begin{cases} u(x) & \text{if } x \in \Omega \setminus E_\rho \\ v(x) & \text{if } x \in E_\rho, \end{cases}
$$

with $v \in \mathcal{U}$, and if we denote by y_ρ the state corresponding to u_ρ, the following equalities hold

$$
y_\rho = y + \rho z + r_\rho, \quad \lim_{\rho\to 0}\frac{1}{\rho}\|r_\rho\|_{W_0^{1,p}(\Omega)} = 0, \qquad (21)
$$

and

$$J(u_\rho) = J(u) + \rho z^0 + r_\rho^0, \quad \lim_{\rho\to 0} \frac{1}{\rho}|r_\rho^0| = 0, \tag{22}$$

where $z \in W_0^{1,p}(\Omega)$ *satisfies*

$$Az = \frac{\partial f}{\partial y}(x,y,u)z + f(x,y,v) - f(x,y,u) \quad \text{in } \Omega \tag{23}$$

and

$$z^0 = \int_\Omega \left\{ \frac{\partial L}{\partial y}(x,y(x),u(x))z(x) + L(x,y(x),v(x)) - L(x,y(x),u(x)) \right\} dx. \tag{24}$$

Proof. Let us define

$$g(x) = L(x,y(x),v(x)) - L(x,y(x),u(x))$$

and

$$h(x) = f(x,y(x),v(x)) - f(x,y(x),u(x)).$$

Then, thanks to the hypotheses (1) and (4) we have that $h \in L^s(\Omega)$ and $g \in L^1(\Omega)$. Then we can apply Lemma 1 to obtain the existence of a measurable set $E_\rho \subset \Omega$, with $|E_\rho| = \rho|\Omega|$, satisfying

$$\left| \int_\Omega \left(1 - \frac{1}{\rho}\chi_{E_\rho}(x)\right) g(x)dx \right| + \left\| \left(1 - \frac{1}{\rho}\chi_{E_\rho}\right) h \right\|_{W^{-1,p}(\Omega)} \le \rho. \tag{25}$$

Putting

$$z_\rho = \frac{y_\rho - y}{\rho},$$

it is easy to obtain, by subtracting the equations satisfied by y_ρ and y, respectively, that

$$Az_\rho = a^\rho(x)z_\rho + \frac{1}{\rho}\chi_{E_\rho} h \quad \text{in } \Omega, \tag{26}$$

where

$$a^\rho(x) = \int_0^1 \frac{\partial f}{\partial y}(x, y(x) + \theta[y_\rho(x) - y(x)], u_\rho(x))d\theta.$$

It is not difficult to check with the help of the $W_0^{1,p}(\Omega)$–regularity of A, that

$$\|y_\rho - y\|_{W_0^{1,p}(\Omega)} \longrightarrow 0. \tag{27}$$

In particular, because of the imbedding $W_0^{1,p}(\Omega) \subset C(\bar\Omega)$, we deduce from (27) together with (1) that

$$a^\rho(x) \longrightarrow \frac{\partial f}{\partial y}(x,y(x),u(x)) \quad \text{in } L^s(\Omega). \tag{28}$$

This relation, (26) and the convergence $(1/\rho)\chi_{E_\rho}h \to h$ in $W^{-1,p}(\Omega)$ implies that $z_\rho \to z$ in $W_0^{1,p}(\Omega)$, which proves (21).

On the other hand

$$\frac{1}{\rho}|r_\rho^0| = \left|\frac{J(u_\rho) - J(u)}{\rho} - z^0\right|$$

$$\leq \int_\Omega \left|\int_0^1 \frac{\partial L}{\partial y}(x, y(x) + \theta[y_\rho(x) - y(x)], u_\rho(x))d\theta z_\rho(x) - \frac{\partial L}{\partial y}(x, y(x), u(x))z(x)\right| dx$$

$$+ \left|\int_\Omega \left(1 - \frac{1}{\rho}\chi_{E_\rho}(x)\right) g(x)dx\right| \longrightarrow 0 \quad \text{when } \rho \to 0,$$

which proves (22). □

Proof of Theorem 2.

Since Z is separable, we can take in Z a norm $\|\cdot\|_Z$ such that Z' endowed with the dual norm $\|\cdot\|_{Z'}$ is strictly convex. Then the function

$$d_Q : (Z, \|\cdot\|_Z) \longrightarrow \mathbb{R}$$

$$d_Q(z) = \inf_{y\in Q} \|y - z\|_Z$$

is convex, Lipschitz and Gâteaux differentiable at every point $z \notin Q$, with $\partial d_Q(z) = \{\nabla d_Q(z)\}$, where the Clarke's generalized gradient and the subdifferential in the sense of the convex analysis coincides for this function. Therefore, given $\xi \in \partial d_Q(y)$, we have that

$$\langle \xi, z - y\rangle + d_Q(y) \leq d_Q(z) \quad \forall z \in Z. \tag{29}$$

Moreover $\|\nabla d_Q(z)\|_{Z'} = 1$ for every $z \notin Q$; see Clarke [11] and Casas and Yong [9].

Let us take $J_\epsilon : \mathcal{U} \longrightarrow \mathbb{R}$ defined by

$$J_\epsilon(u) = \left\{[(J(u) - J(\bar{u}) + \epsilon)^+]^2 + d_Q(G(y_u))^2 + |F(y_u)|^2\right\}^{1/2}.$$

It is obvious that $J_\epsilon(u) > 0$ for every $u \in \mathcal{U}$ and $J_\epsilon(\bar{u}) = \epsilon$. On the other hand, if d denotes the Ekeland's distance in $\mathcal{U}$:

$$d(u, v) = |\{x \in \Omega : u(x) \neq v(x)\}|,$$

then J_ϵ is continuous in $(\mathcal{U}, d)$. Therefore we can apply Ekeland's variational principle [13] and deduce the existence of $u^\epsilon \in \mathcal{U}$ such that

$$d(u^\epsilon, \bar{u}) \leq \sqrt{\epsilon} \quad \text{and} \quad 0 < J_\epsilon(u^\epsilon) \leq J_\epsilon(u) + \sqrt{\epsilon}d(u^\epsilon, u) \quad \forall u \in \mathcal{U}. \tag{30}$$

Taking E_ρ and u_ρ^ϵ as in Lemma 2,

$$u_\rho^\epsilon(x) = \begin{cases} u^\epsilon(x) & \text{if } x \in \Omega \setminus E_\rho \\ v(x) & \text{if } x \in E_\rho, \end{cases}$$

with $v \in \mathcal{U}$ arbitrary, we get with the help of (21) and (22)

$$
\begin{aligned}
-\sqrt{\epsilon}|\Omega| &\le \frac{J_\epsilon(u_\rho^\epsilon) - J_\epsilon(u^\epsilon)}{\rho} \\
&= \frac{[(J(u_\rho^\epsilon) - J(\bar{u}) + \epsilon)^+]^2 - [(J(u^\epsilon) - J(\bar{u}) + \epsilon)^+]^2}{\rho[J_\epsilon(u_\rho^\epsilon) + J_\epsilon(u^\epsilon)]} \\
&\quad + \frac{d_Q(G(y_\rho^\epsilon))^2 - d_Q(G(y^\epsilon))^2 + |F(y_\rho^\epsilon)|^2 - |F(y^\epsilon)|^2}{\rho[J_\epsilon(u_\rho^\epsilon) + J_\epsilon(u^\epsilon)]} \qquad (31) \\
&\overset{\rho\to 0}{\longrightarrow} \frac{\{(J(u^\epsilon) - J(\bar{u}) + \epsilon)^+ z^{0,\epsilon} + \langle \xi^\epsilon, DG(y^\epsilon)z^\epsilon\rangle + \langle F(y^\epsilon), DF(y^\epsilon)z^\epsilon\rangle\}}{J_\epsilon(u^\epsilon)} \\
&= \alpha_\epsilon z^{0,\epsilon} + \langle [DG(y^\epsilon)]^* \mu^\epsilon, z^\epsilon\rangle + \langle [DF(y^\epsilon)]^*\lambda^\epsilon, z^\epsilon\rangle,
\end{aligned}
$$

where $z^\epsilon \in W_0^{1,p}(\Omega)$,

$$
\begin{aligned}
Az^\epsilon &= \frac{\partial f}{\partial y}(x, y^\epsilon, u^\epsilon)z^\epsilon + f(x, y^\epsilon, v) - f(x, y^\epsilon, u^\epsilon), \\
z^{0,\epsilon} &= \int_\Omega \left\{\frac{\partial L}{\partial y}(x, y^\epsilon, u^\epsilon)z^\epsilon + L(x, y^\epsilon, v) - L(x, y^\epsilon, u^\epsilon)\right\} dx, \\
\alpha_\epsilon &= \frac{(J(u^\epsilon) - J(\bar{u}) + \epsilon)^+}{J_\epsilon(u^\epsilon)}, \quad \mu^\epsilon = \frac{\xi_\epsilon}{J_\epsilon(u^\epsilon)}, \quad \lambda^\epsilon = \frac{F(y^\epsilon)}{J_\epsilon(u^\epsilon)}, \\
\xi^\epsilon &= \begin{cases} d_Q(G(y^\epsilon))\nabla d_Q G(y^\epsilon)) & \text{if } G(y^\epsilon) \notin Q \\ 0 & \text{otherwise.} \end{cases}
\end{aligned}
$$

Now let us remind that $A : W_0^{1,p}(\Omega) \longrightarrow W^{-1,p}(\Omega)$ is an isomorphism; see the comments after Theorem 1. Then $A^* : W_0^{1,p'}(\Omega) \longrightarrow W^{-1,p'}(\Omega)$ is also an isomorphism, therefore we can introduce the adjoint state $\varphi^\epsilon \in W_0^{1,p'}(\Omega)$ as solution of the equation

$$
A^*\varphi^\epsilon = \frac{\partial f}{\partial y}(x, y^\epsilon, u^\epsilon)\varphi^\epsilon + \alpha_\epsilon \frac{\partial L}{\partial y}(x, y^\epsilon, u^\epsilon) + [DG(y^\epsilon)]^*\mu^\epsilon + [DF(y^\epsilon)]^*\lambda^\epsilon. \tag{32}
$$

Thus, from (31) and (32), we obtain

$$
\int_\Omega [H_{\alpha_\epsilon}(x, y^\epsilon(x), u^\epsilon(x), \varphi^\epsilon(x)) - H_{\alpha_\epsilon}(x, y^\epsilon(x), v(x), \varphi^\epsilon(x))]dx \le \sqrt{\epsilon}|\Omega| \quad \forall v \in \mathcal{U}. \tag{33}
$$

Now we pass to the limit when $\epsilon \to 0$. To do this, let us remark that

$$
\alpha_\epsilon^2 + \|\mu^\epsilon\|_{Z'}^2 + |\lambda^\epsilon|^2 = 1. \tag{34}
$$

Then we take subsequences, denoted in the same way, satisfying

$$
\begin{cases} \alpha_\epsilon \to \bar{\alpha} \text{ in } \mathbb{R}, \quad \lambda^\epsilon \to \bar{\lambda} \text{ in } \mathbb{R}^n \\ \mu^\epsilon \to \bar{\mu} \text{ in the weak}^* \text{ topology of } Z'. \end{cases} \tag{35}
$$

On the other hand, the convergence $y^\epsilon \to \bar{y}$ in $W_0^{1,p}(\Omega) \subset C_0(\Omega)$ follows from (30). Then, using (35), it is easy to pass to the limit in (32) and (33) and to deduce (7) and (9). Now reminding the definition of μ^ϵ and ξ^ϵ and (29), we deduce

$$\langle \mu^\epsilon, z - G(y^\epsilon)\rangle \leq 0 \quad \forall z \in Q. \tag{36}$$

Passing to the limit in this expression we obtain (8). Let us prove (5). To do this, let us suppose that $\bar{\alpha} = |\bar{\lambda}| = 0$, then from (34) it follows $\|\mu^\epsilon\|_{Z'} \to 1$ as $\epsilon \to 0$. Let us take $z_0 \in \overset{o}{Q}$ and $r > 0$ such that $\bar{B}_r(z_0) \subset \overset{o}{Q}$. Then (36) implies that

$$\langle \mu^\epsilon, z + z_0 - G(y^\epsilon)\rangle \leq 0 \quad \forall z \in \bar{B}_r(z_0).$$

Hence

$$r\|\mu^\epsilon\|_{Z'} = \sup_{z \in \bar{B}_r(z_0)} \langle \mu^\epsilon, z\rangle \leq \langle \mu^\epsilon, G(y^\epsilon) - z_0\rangle.$$

Passing to the limit

$$0 < r \leq \lim_{\epsilon \to 0} \langle \mu^\epsilon, G(y^\epsilon) - z_0\rangle = \langle \bar{\mu}, G(\bar{y}) - z_0\rangle,$$

which proves that $\bar{\mu} \neq 0$

Finally let us deduce (10) from (9). Assume firstly that (A1) holds. Let us take a numerable dense subset $\{v_j\}_{j=1}^\infty$ of K. Let F and $\{F_j\}_{j=1}^\infty$ be measurable subsets of Ω, with $|F| = |\Omega| = |F_j|$ for every j, such that the Lebesgue point sets of functions $x \in \Omega \longrightarrow H(x, \bar{y}(x), \bar{u}(x), \bar{\varphi}(x))$ and $x \in \Omega \longrightarrow H(x, \bar{y}(x), v_j, \bar{\varphi}(x))$ are F and F_j, respectively. Let us set $F_0 = F \bigcap [\bigcap_{j=1}^\infty F_j]$. Then we have $|F_0| = |\Omega|$. Now given $x_0 \in F_0$ arbitrary, for every $\epsilon > 0$ and $j \geq 1$ we define the admissible controls

$$u_j^\epsilon(x) = \begin{cases} \bar{u}(x) & \text{if } x \notin B_\epsilon(x_0) \\ v_j & \text{otherwise.} \end{cases}$$

Then from (9) we deduce

$$\frac{1}{|B_\epsilon(x_0)|} \int_{B_\epsilon(x_0)} H_{\bar{\alpha}}(x, \bar{y}(x), \bar{u}(x), \bar{\varphi}(x))dx \leq \frac{1}{|B_\epsilon(x_0)|} \int_{B_\epsilon(x_0)} H_{\bar{\alpha}}(x, \bar{y}(x), v_j, \bar{\varphi}(x))dx, \quad 1 \leq j.$$

Passing to the limit when $\epsilon \to 0$ we get

$$H_{\bar{\alpha}}(x_0, \bar{y}(x_0), \bar{u}(x_0), \bar{\varphi}(x_0)) \leq H_{\bar{\alpha}}(x_0, \bar{y}(x_0), v_j, \bar{\varphi}(x_0)) \quad \forall x_0 \in F_0, \ \forall j \geq 1.$$

Taking into account that $v \longrightarrow H_{\bar{\alpha}}(x_0, \bar{y}(x_0), v, \bar{\varphi}(x_0))$ is continuous and that $\{v_j\}_{j=1}^\infty$ is dense in K, (10) follows from the above inequality.

Now let us suppose that assumption (A2) holds. Let $F_{\bar{\varphi}}$ be a measurable subset of Ω such that

$$\lim_{\epsilon \to 0} \frac{1}{|B_\epsilon(x_0)|} \int_{B_\epsilon(x_0)} |\bar{\varphi}(x) - \bar{\varphi}(x_0)|^{s'} dx = 0 \quad \forall x_0 \in F_{\bar{\varphi}}, \tag{37}$$

where $s' = s/(s-1)$, s given in (1). Then $|F_{\bar{\varphi}}| = |\Omega|$. Let $M = \|\bar{y}\|_\infty$ and $\psi_M \in L^s(\Omega)$ satisfying (1). Finally we put $F_0 = F_{\bar{\varphi}} \cap F_M \cap \Omega_0 \cap F$, where F is taken as above and F_M is the set of Lebesgue points of $|\psi_M|^s$. Thus we have that $|F_0| = |\Omega|$ and taking spike perturbations as before, we deduce

$$\frac{1}{|B_\epsilon(x_0)|} \int_{B_\epsilon(x_0)} H_{\bar{\alpha}}(x, \bar{y}(x), \bar{u}(x), \bar{\varphi}(x)) dx$$
$$\leq \frac{1}{|B_\epsilon(x_0)|} \int_{B_\epsilon(x_0)} H_{\bar{\alpha}}(x, \bar{y}(x), v, \bar{\varphi}(x)) dx \quad \forall x_0 \in F_0 \text{ and } \forall v \in K.$$

Since $x_0 \in F$, we can past to the limit on the left hand side of the inequality. Let us study the right hand side.

$$\frac{1}{|B_\epsilon(x_0)|} \int_{B_\epsilon(x_0)} H_{\bar{\alpha}}(x, \bar{y}(x), v, \bar{\varphi}(x)) dx$$
$$= \frac{1}{|B_\epsilon(x_0)|} \int_{B_\epsilon(x_0)} \bar{\alpha} L(x, \bar{y}(x), v) dx + \frac{1}{|B_\epsilon(x_0)|} \int_{B_\epsilon(x_0)} f(x, \bar{y}(x), v) dx \bar{\varphi}(x_0)$$
$$+ \frac{1}{|B_\epsilon(x_0)|} \int_{B_\epsilon(x_0)} [\bar{\varphi}(x) - \bar{\varphi}(x_0)] f(x, \bar{y}(x), v) dx.$$

The first two terms converge to $H_{\bar{\alpha}}(x_0, \bar{y}(x_0), v, \bar{\varphi}(x_0))$ because of the continuity of the integrands in x_0. Let us prove that the last term goes to zero.

$$\left| \frac{1}{|B_\epsilon(x_0)|} \int_{B_\epsilon(x_0)} [\bar{\varphi}(x) - \bar{\varphi}(x_0)] f(x, \bar{y}(x), v) dx \right|$$
$$\leq \left(\frac{1}{|B_\epsilon(x_0)|} \int_{B_\epsilon(x_0)} |\bar{\varphi}(x) - \bar{\varphi}(x_0)|^{s'} dx \right)^{1/s'} \left(\frac{1+M}{|B_\epsilon(x_0)|} \int_{B_\epsilon(x_0)} |\psi_M(x)|^s dx \right)^{1/s} \to 0,$$

thanks to (37) and the fact that $x_0 \in F_{\bar{\varphi}} \cap F_M$. □

4. The strong Pontryagin's principle

In this section we will prove that, in the absence of equality constraints, Theorem 2 holds with $\bar{\alpha} = 1$ for "almost all" control problems. We will precise this term later. The key to achieve this result is the introduction of a stability assumption of the optimal cost functional with respect to small perturbations of the set of feasible controls. This stability allows to accomplish an exact penalization of the state constraints. First of all let us formulate the control problem to be studied in this section

$$(\mathrm{P}_\delta)\left\{\begin{array}{l}\text{Minimize } J(u)\\ u\in\mathcal{U},\ G(y_u)\in Q_\delta\end{array}\right.$$

with the same notation and assumptions of §2 and setting $Q_\delta = Q + \bar{B}_\delta(0)$, for every $\delta > 0$.

Definition 1. We say that (P_δ) is strongly stable if there exist $\epsilon > 0$ and $C > 0$ such that

$$\inf(\mathrm{P}_\delta) - \inf(\mathrm{P}_{\delta'}) \leq C(\delta' - \delta) \quad \forall \delta' \in [\delta, \delta + \epsilon]. \tag{38}$$

This concept was first introduced in relation with optimal control problems by Bonnans [2]; see also Bonnans and Casas [5]. A weaker stability concept was used by Casas [8] to analyze the convergence of the numerical discretizations of optimal control problems. The following proposition states that almost all problems (P_δ) are strongly stable.

Proposition 1. *Let $\delta_0 \geq 0$ be a number such that (P_{δ_0}) has feasible controls. Then (P_δ) is strongly stable for all $\delta \geq \delta_0$ except at most a zero Lebesgue measure set.*

Proof. It is enough to consider the function $h : [\delta_0, +\infty) \longrightarrow \mathbb{R}$ defined by

$$h(\delta) = \inf(\mathrm{P}_\delta)$$

and remark that it is a nonincreasing monotone function and, consequently, differentiable at every point of $[\delta_0, +\infty)$ except at a zero measure set. Now it is obvious to check that (P_{δ_0}) is strongly stable at every point where h is differentiable. □

Now we carry out an exact penalization of the state constraint. To do this, we will use the distance function d_{Q_δ} associated to the set Q_δ, and define in the same way as in the proof of Theorem 2.

Proposition 2. *If (P_δ) is strongly stable and $\bar{u}$ is a solution of this problem, then there exists $q_0 > 0$ such that $\bar{u}$ is also a solution of*

$$\inf_{u\in\mathcal{U}} J_q(u) = J(u) + q d_{Q_\delta}(G(y_u)) \tag{39}$$

for every $q \geq q_0$.

Proof. Let us suppose that it is false. Then there exists a sequence $\{q_k\}_{k=1}^\infty$ of real numbers, with $q_k \to +\infty$, and elements $\{u_k\}_{k=1}^\infty \subset \mathcal{U}$ such that

$$J(u_k) + q_k d_{Q_\delta}(G(y_k)) < J(\bar{u}) \quad \forall k \geq 1,$$

where y_k is the state corresponding to u_k. From here we obtain that

$$d_{Q_\delta}(G(y_k)) < \frac{J(\bar{u}) - J(u_k)}{q_k} \longrightarrow 0 \quad \text{when } k \to +\infty$$

and $G(y_k) \notin Q_\delta$. Let $\delta_k > \delta$ be the smallest number such that $G(y_k) \in Q_{\delta_k}$. Since $\delta_k \to \delta$, we can use (38) to deduce

$$C(\delta_k - \delta) \geq \inf(\mathrm{P}_\delta) - \inf(\mathrm{P}_{\delta_k}) \geq J(\bar{u}) - J(u_k)$$

$$> q_k d_{Q_\delta}(G(y_k)) = q_k(\delta_k - \delta) \quad \forall k \geq k_\epsilon,$$

which is not possible. □

Since J_q is not Gâteaux differentiable on Q_δ, we are going to modify slightly this functional to attain the differentiability necessary for the proof.

Proposition 3. *Let us take $q \geq q_0$ and for every $\epsilon > 0$ let us consider the problem*

$$(\mathrm{P}_{\delta,\epsilon}) \ \inf_{u\in\mathcal{U}} J_{q,\epsilon}(u) = J(u) + q\left\{d_{Q_\delta}(G(y_u))^2 + \epsilon^2\right\}^{1/2}.$$

Then $\inf(\mathrm{P}_{\delta,\epsilon}) \to \inf(\mathrm{P}_\delta)$ *when* $\epsilon \to 0$.

Proof. It is an immediate consequence of the inequality

$$J_q(u) \leq J_{q,\epsilon}(u) \leq J_q(u) + q\epsilon \quad \forall u \in \mathcal{U}. \ \square$$

Finally we are ready to prove the strong Pontryagin's principle.

Theorem 3. *If (P_δ) is strongly stable and $\bar{u}$ is a solution of this problem, then Theorem 2 remains to be true with $\bar{\alpha} = 1$.*

Proof. Propositions 2 and 3 imply that $\bar{u}$ is a σ_ϵ^2-solution of $(\mathrm{P}_{\delta,\epsilon})$, with $\sigma_\epsilon \to 0$ when $\epsilon \to 0$, i.e.

$$J_{q,\epsilon}(\bar{u}) \leq \inf(\mathrm{P}_{\delta,\epsilon}) + \sigma_\epsilon^2.$$

Then we can apply again Ekeland's principle and deduce the existence of an element $u^\epsilon \in \mathcal{U}$ such that

$$d(u^\epsilon, \bar{u}) \leq \sigma_\epsilon, \quad J_{q,\epsilon}(u^\epsilon) \leq J_{q,\epsilon}(\bar{u}),$$

and

$$J_{q,\epsilon}(u^\epsilon) \leq J_{q,\epsilon}(u) + \sigma_\epsilon d(u^\epsilon, u) \quad \forall u \in \mathcal{U},$$

where once again d denotes the Ekeland's distance. Now we argue as in the proof of Theorem 2 and replace (31) by

$$-\sigma_\epsilon|\Omega| \leq \lim_{\rho\to 0} \frac{J_{q,\epsilon}(u_\rho^\epsilon) - J_{q,\epsilon}(u^\epsilon)}{\rho} = z^{0,\epsilon} + \langle \mu^\epsilon, DG(y^\epsilon)z^\epsilon \rangle,$$

where $\mu^\epsilon \in Z'$ is given by

$$\mu^\epsilon = \begin{cases} \dfrac{q d_{Q_\delta}(G(y^\epsilon))}{\{d_{Q_\delta}(G(y^\epsilon))^2 + \epsilon^2\}^{1/2}} \nabla d_{Q_\delta}(G(y^\epsilon)) & \text{if } G(y^\epsilon) \notin Q_\delta, \\ 0 & \text{otherwise.} \end{cases}$$

Therefore we have $\|\mu^\epsilon\|_{Z'} \leq q$ for every $\epsilon > 0$. Now we can take a subsequence that converges weakly* to an element $\bar{\mu} \in Z'$. The rest is as in the proof of Theorem 2, taking $\alpha_\epsilon = 1$. □

References

1. R.A. Adams, *Sobolev spaces*, Academic Press, New York, 1975.
2. J.F. Bonnans, *Pontryagin's principle for the optimal control of semilinear elliptic systems with state constraints*, 30th IEEE Conference on Control and Decision (Brighton, England), 1991, pp. 1976–1979.
3. J.F. Bonnans and E. Casas, *Un principe de Pontryagine pour le contrôle des systèmes elliptiques*, J. Differential Equations **90** (1991), no. 2, 288–303.
4. ______, *A boundary Pontryagin's principle for the optimal control of state-constrained elliptic equations*, Internat. Ser. Numer. Math. **107** (1992), 241–249.
5. ______, *An extension of Pontryagin's principle for state-constrained optimal control of semilinear elliptic equations and variational inequalities*, SIAM J. Control Optim. (To appear).
6. J.F. Bonnans and D. Tiba, *Pontryagin's principle in the control of semilinear elliptic variational equations*, Appl. Math. Optim. **23** (1991), 299–312.
7. E. Casas, *Boundary control of semilinear elliptic equations with pointwise state constraints*, SIAM J. Control Optim. **31** (1993), no. 4, 993–1006.
8. ______, *Finite element approximations for some state-constrained optimal control problems*, Mathematics of the Analysis and Design of Process Control (P. Borne, S.G. Tzafestas and N.E. Radhy, eds.), North-Holland, Amsterdam, 1992.
9. E. Casas and J. Yong, *Maximum principle for state-constrained optimal control problems governed by quasilinear elliptic equations*, Differential Integral Equations (To appear).
10. P.G. Ciarlet, *The finite element method for elliptic problems*, North-Holland, Amsterdam, 1978.
11. F.H. Clarke, *Optimization and nonsmooth analysis*, John Wiley & Sons, Toronto, 1983.
12. B.E.J. Dahlberg, *L^q-estimates for Green potentials in Lipschitz domains*, Math. Scand. **44** (1979), no. 1, 149–170.
13. I. Ekeland, *Nonconvex minimization problems*, Bull. Amer. Math. Soc. **1** (1979), no. 3, 76–91.
14. H.O. Fattorini and H. Frankowska, *Necessary conditions for infinite dimensional control problems*, Math. Control Signals Systems **4** (1991), 41–67.
15. X. Li and Y. Yao, *Maximum principle of distributed parameter systems with time lags*, Distributed Parameter Systems (New York), Springer–Verlag, 1985, Lecture Notes in Control and Information Sciences 75, pp. 410–427.
16. X. Li and J. Yong, *Necessary conditions of optimal control for distributed parameter systems*, SIAM J. Control Optim. **29** (1991), 1203–1219.
17. C.B. Morrey, *Multiple integrals in the calculus of variations*, Springer–Verlag, New York, 1966.
18. J. Yong, *Pontryagin maximum principle for semilinear second order elliptic partial differential equations and variational inequalities with state constraints*, Differential Integral Equations (To appear).

Eduardo Casas
Departamento de Matemática Aplicada y Ciencias de la Computación,
E.T.S.I. Caminos, C. y P.,
Universidad de Cantabria,
Av. Los Castros s/n.,
39071–Santander, Spain

E-mail: casas@masun1.unican.es

INVARIANCE OF THE HAMILTONIAN IN CONTROL PROBLEMS FOR SEMILINEAR PARABOLIC DISTRIBUTED PARAMETER SYSTEMS

H. O. FATTORINI

Department of Mathematics
University of California

ABSTRACT. We consider a general optimal control problem for a distributed parameter described by a semilinear parabolic equation. We show that the Hamiltonian is invariant over optimally controlled trajectories, and that it vanishes when the terminal time is free. The method is that of approximation of the original control system by a sequence of "smoothed" control systems

1991 *Mathematics Subject Classification.* 93E20, 93E25

Key words and phrases. Lagrange multiplier rule, Kuhn-Tucker conditions, maximum principle, optimal control.

1. Introduction

Let E be a Banach space, A the infinitesimal generator of a strongly continuous semigroup $S(t)$ in E. Consider the optimal control problem of minimizing the functional

$$y_0(t) = \int_0^t f_0(y(\tau), u(\tau))d\tau$$

among all trajectories of the autonomous semilinear differential system in E

$$y'(t) = Ay(t) + f(y(t), u(t)), \quad y(0) = \zeta,$$

subject to a control constraint $u(t) \in U (0 \leq t \leq \bar{t})$ and a target condition $y(\bar{t}) \in Y$. In situations where it can be proved, Pontryagin's maximum principle has two statements: one is that the Hamiltonian

$$H(\bar{y}(t), z_0, \bar{z}(t), v) = z_0 f_0(\bar{y}(t), v) + \langle \bar{z}(t), A\bar{y}(t) \rangle + \langle \bar{z}(t), f(\bar{y}(t), v) \rangle$$

This work was supported in part by the National Science Foundation under grant DMS-9221819.

is maximized over U by $v = \bar{u}(t)$ along the trajectory, where $\bar{u}(\cdot)$ is the optimal control, $\bar{y}(t)$ the optimal trajectory and the costate $\bar{z}(t)$ solves the adjoint variational equation

$$\bar{z}'(\bar{t}) = -\{A^* + \partial_y f(\bar{y}(t), \bar{u}(t))^*\}\bar{z}(t) - z_0 \partial_y f_0(\bar{y}(t), \bar{u}(t)), \ \bar{z}(\bar{t}) = z$$

in $0 \leq t \leq \bar{t}$; ∂_y indicates Fréchet derivative with respect to y and $(z_0, z) \neq 0$. The other statement is

$$H(\bar{y}(t), z_0, \bar{z}(t), \bar{u}(t)) = c \qquad (0 \leq t \leq \bar{t})$$

with $c = 0$ if the terminal time $\bar{t}$ is free. In both statements we face the problem that the Hamiltonian (precisely, the term $\langle \bar{z}(t), A\bar{y}(t) \rangle$) may not be defined; there are situations where $\bar{y}(t)$ (resp. $\bar{z}(t)$) does not belong to the domain of any fractional power of A (resp. of A^*). This is not important for the first statement, since $\langle \bar{z}(t), A\bar{y}(t) \rangle$ does not depend on v and can be dropped. However, this is not the case for the second statement, and the only results of this sort we know (coming from the maximum principle rather than from the dynamic programming approach) are those in [3]. They are very restrictive on nonlinearities; for instance the most general, if applied to a semilinear heat equation

$$y_t(t, x) = \Delta y(t, x) + f(y(t, x)) + u(t, x) \tag{1.1}$$

in $L^2(\Omega)$ (Ω a domain $\subseteq \mathbb{R}^m$) would require L^2-continuity of f, thus preempting (for instance) any powers of y.

We present in this paper a new approach to invariance of the Hamiltonian based on another of the results in [3] and on an approximation process. It is not more or less general than the ones in [3], but applies for instance to (1.1) in the space $C(\bar{\Omega})$ of continuous functions in $\bar{\Omega}$, and is very tolerant of nonlinearities; for instance, we may take as f in (1.1) a polynomial or arbitrarily high order or even a function of faster growth as long as certain dissipativity conditions that prevent finite time blowup of solutions are satisfied (see §2). See §6 for comparison with [3] and for possible generalizations.

2. The control system

Instead of the Laplacian, we work with

$$Ay = \sum_{j=1}^{m} \sum_{k=1}^{m} \partial^j (a_{jk}(x) \partial^k y) + \sum_{j=1}^{m} b_j(x) \partial^i y + c(x) y, \tag{2.1}$$

in a bounded domain $\Omega \subseteq \mathbb{R}^m$ with boundary Γ of class $C^{(2)}$. The leading coefficients a_{jk} are continuously differentiable in $\bar{\Omega}$, $a_{jk} = a_{kj}$ and $\Sigma\Sigma a_{jk}(x)\xi_j\xi_k \geq \kappa\|\xi\|^2$ ($x \in \Omega$, $\xi \in \mathbb{R}^m$); the b_j and c are continuous in $\bar{\Omega}$. We indicated by β either the Dirichlet boundary condition $y = 0$ ($x \in \Gamma$) or a variational boundary condition

$$\partial^\nu y(x) = \gamma(x) y(x) \qquad (x \in \Gamma),$$

where ∂^ν is the *conormal derivative*, i.e. the derivative in the direction of the conormal vector $\{\nu_j(x)\} = \{\Sigma a_{jk}(x)\nu_k(x)\}$, $\{\nu_k(x)\}$ the outer normal vector on Γ. The coefficient $\gamma(x)$ is assumed continuous in Γ. In general, $A(\beta)$ is A with domain restricted by a boundary condition β.

The *adjoint* A' of (2.1) is

$$A'y = \sum_{j=1}^{m} \sum_{k=1}^{m} \partial^j (a_{jk}(x)\partial^k y) - \sum_{j=1}^{m} \partial_j (b_j(x)y) + c(x)y;$$

and the *adjoint boundary condition* β' is $\beta' = \beta$ for the Dirichlet boundary condition; for a variational boundary condition β' is

$$\partial^\nu y(x) = (\gamma(x) + b(x))y(x)$$

with $b(x) = \Sigma\, b_j(x)\nu_j(x)$.

Denote by $C(\bar{\Omega})$ the space of continuous functions $y(\cdot)$ in $\bar{\Omega}$ endowed with the supremum norm; $C_0(\bar{\Omega})$ is the subspace defined by $y = 0$ on Γ. The dual $C(\bar{\Omega})^*$ of $C(\bar{\Omega})$ can be identified with the space $\Sigma(\bar{\Omega})$ of regular, bounded Borel measures μ in $\bar{\Omega}$ with the total variation norm; the duality pairing is $\langle \mu, f \rangle = \int f(x)\mu(dx)$. The dual $C_0(\bar{\Omega})$ can be identified with the subspace of $\Sigma(\bar{\Omega})$ whose elements vanish in Γ. We have the isometric imbeddings $C(\bar{\Omega}) \subseteq L^\infty(\Omega)$, $L^1(\bar{\Omega}) \subseteq \Sigma(\bar{\Omega})$.

Let β be a variational boundary condition, and let $A_c(\beta)$ be the operator in $C(\bar{\Omega})$ defined as follows:

$$D(A_c(\beta)) = \left\{ y \in \bigcap_{p \geq 1} W^{2,p}(\Omega)_\beta;\ Ay \in C(\bar{\Omega}) \right\}$$

and $A_c(\beta)y = Ay$, where $W^{2,p}(\Omega)_\beta$ denotes the subspace of the Sobolev space $W^{2,p}(\Omega)$ whose elements satisfy β on Γ. The operator $A_c(\beta)$ generates a compact analytic semigroup $S_c(t; A, \beta)$ in $C(\bar{\Omega})$. If β is the Dirichlet boundary condition the same is true in the space $C_0(\bar{\Omega})$; in the definition of the domain, we require $Ay \in C_0(\Omega)$.

The dual semigroup $S_c(t; A, \beta)^*$ in $\Sigma(\bar{\Omega})(\Sigma_0(\Omega)$ for the Dirichlet boundary condition) is compact and analysic in $t > 0$ but not strongly continuous at $t = 0$; however, $S_c(t; A, \beta)^*\mu \to \mu$ $C(\bar{\Omega})$-weakly in $\Sigma(\bar{\Omega})$ $(\Sigma_0(\bar{\Omega})$ for the Dirichlet boundary condition). The restriction of $S_c(t; A, \beta)^*$ to $L^1(\Omega)$ is a compact analytic semigroup $S_1(t; A', \beta')$ (A' the adjoint of A, β' the adjoint of β) whose infinitesimal generator $A_1'(\beta')$ is characterized as follows: $D(A_1'(\beta'))$ consists of all elements $y \in L^1(\Omega)$ such that there exists $z(= A_1'(\beta')y)$ in $L_1(\Omega)$ with

$$\int_\Omega y(x)(A(\beta)v)(x)dx = \int_\Omega z(x)v(x)dx$$

for every v in the subspace $C^{(2)}(\bar{\Omega})_\beta$ of all $y \in C^{(2)}(\bar{\Omega})$ that satisfy the boundary condition β. The dual semigroup $S_1(t; A', \beta')^*$ is compact and analytic in $t > 0$, but not strongly continuous at $t \to 0$, although has the same weak continuity properties as $S_c(t; A, \beta)^*$. Finally, $S_c(t; A, \beta)$ is the restriction of $S_1(t; A', \beta')^*$ to $C(\bar{\Omega})$ ($C_0(\bar{\Omega})$ for the Dirichlet boundary condition). Restriction relations also hold for the generators: $A_1'(\beta')$ is the restriction of $A_c(\beta)^*$ to the domain $D(A_1'(\beta')) = \{y \in L^1(\Omega); A_c(\beta)^* y \in L^1(\Omega)\}$ and $A_c(\beta)$ is the restriction of $A_1'(\beta')^*$ to $D(A_c(\beta)) = \{y \in C(\bar{\Omega});\ A_1'(\beta')^* y \in C(\bar{\Omega})\}$ ($C_0(\bar{\Omega})$ for the Dirichlet boundary condition).

We fix below the operator A and the boundary condition β and use the shorthand $S_c(t) = S_c(t; A, \beta)$, $S_\infty(t) = S_1(t; A', \beta')^*$, $S_1'(t) = S_1(t; A', \beta')$, $S_\Sigma'(t) = S_c(t; A, \beta)^*$, so that $S_c(t)$ (resp. $S_1'(t)$) is the restriction of $S_\infty(t)$ to $C(\bar{\Omega})$ (resp. of $S_\Sigma'(t)$ to $L^1(\Omega)$). Likewise, we write $A_c = A_c(\beta)$, $A_1' = A_1'(\beta')$. Also, we leave to the reader to add "$C_0(\bar{\Omega})$ for the Dirichlet boundary condition" or "$\Sigma_0(\bar{\Omega})$ for the Dirichlet boundary condition" whenever necessary.

The control system is

$$\mathbf{y}'(t) = A_c \mathbf{y}(t) + \mathbf{f}(\mathbf{y}(t)) + \mathbf{u}(t), \qquad \mathbf{y}(0) = \zeta \tag{2.2}$$

and is modelled on (1.1), so that the boldface letters mean: $\mathbf{y}(t)(x) = y(t, x)$, and $\mathbf{u}(t)(x) = u(t, x)$. Instantaneous values of $\mathbf{y}$ are in $C(\bar{\Omega})$. We want to admit $L^\infty((0, T) \times \Omega)$ as control space, thus we cannot take $\mathbf{u}(t) \in C(\bar{\Omega})$. To find the "right" control space we take an arbitrary separable Banach space X and call $L_w^\infty(0, T; X^*)$ the space of all X-weakly measurable X^*-valued functions $g(\cdot)$ endowed with the essential supremum norm (separability of X implies measurability of the norm). In this notation,

$$L^\infty((0, T) \times \Omega) = L_w^\infty(0, T; L^\infty(\Omega)) \tag{2.3}$$

thus we assume $\mathbf{u}(\cdot) \in L_w^\infty(0, T; L^\infty(\Omega))$ in (2.2).

By definition, solutions of (2.2) are $C(\bar{\Omega})$-valued continuous solutions of the associated integral equation

$$y(t) = S_c(t)\zeta + \int_0^t S_c(t - \tau)\mathbf{f}(\mathbf{y}(\tau))d\tau + \int_0^t S_\infty(t - \tau)\mathbf{u}(\tau)d\tau \tag{2.4}$$

where we use $S_\infty(t) \supseteq S_c(t)$ since $\mathbf{u}(\cdot)$ takes values in $L^\infty(\Omega)$. Although $\mathbf{u}(\cdot)$ is merely weakly measurable, $\tau \to S_\infty(t - \tau)\mathbf{u}(\tau)$ is strongly measurable, ([6, Lemma 4.1]) thus the last integral is a Bochner integral and produces a continuous $C(\bar{\Omega})$-valued function. The operator $\mathbf{f} : C(\bar{\Omega}) \to C(\bar{\Omega})$ in (2.2) and (2.4) is $\mathbf{f}(y)(x) = f(y(x))$. If $f(y)$ is locally Lipschitz continuous with respect to y then $\mathbf{f}$ is locally bounded and locally Lipschitz continuous and this is enough to insure local existence and uniqueness of solutions of (2.4); of course, we require $f(0) = 0$ for the Dirichlet boundary condition. We shall call $\mathbf{y}(t, \mathbf{u})$ (or $\mathbf{y}(t, \zeta, \mathbf{u})$ if ζ is not fixed) the solution of (2.4) for $\mathbf{u}(\cdot) \in L_w^\infty(0, T; L^\infty(\Omega))$.

We say that (2.2) (or any control system) has the *global existence property* (GEP) in an interval $0 \le t \le \bar{t}$ if for every $C,\ K > 0$ there exists $L = L(C, K) > 0$ such that if $\|\zeta\| \le C$ and $\|\mathbf{u}(\cdot)\| \le K$ then $\mathbf{y}(t, \zeta, \mathbf{u})$ exists in $0 \le t \le \bar{t}$ and $\|\mathbf{y}(t, \zeta, \mathbf{u})\| \le L$. One of the many conditions that produce the GEP is

Lemma 2.1. *Assume that*

$$\langle \mu, f(y)\rangle \le c(1 + \|y\|^2) \qquad (y \in C(\bar\Omega),\ \mu \in \Theta(y)), \tag{2.5}$$

where the duality set $\Theta(y) \subseteq \Sigma(\bar\Omega)$ *of an element* $y \in C(\bar\Omega)$ *is the set of all* $\mu \in \Sigma(\bar\Omega)$ *such that* $\langle \mu, y\rangle = \|\mu\|^2 = \|y\|^2$. *Then* (2.2) *has the GEP.*

To see what (2.5) means for, say, the Dirichlet boundary condition, note that if $y(\cdot) \in C_0(\bar\Omega)$ then $\Theta(y)$ consists of all $\mu \in \Sigma(\bar\Omega)$ supported by $e = \{x \in \bar\Omega; |y(x)| = \|y\|\}$ and such that $\|\mu\| = \|y\|$, $\mu \ge 0$ in $e_+ = \{x \in \bar\Omega;\ y(x) = \|y(x)\|\}$, $\mu \le 0$ in $e_- = \{x \in \bar\Omega;\ y(x) = -\|y(x)\|\}$. Accordingly, if

$$(sgn\ y) f(y) \le C(1 + |y|) \qquad (-\infty < y < \infty) \tag{2.6}$$

then

$$\begin{aligned}\langle \mu, f(y(\cdot))\rangle &= \int_e f(y(x))\mu(dx) = \int_e sgn\, y(x) f(y(x))|\mu|(dx) \\ &= c\int_e (1 + |y(x)|)|\mu|(dx) = c(\|y\| + \|y\|^2) \le 2c(1 + \|y\|^2),\end{aligned}$$

so that (2.5) holds. Condition (2.6) is easily seen to be satisfied by any polynomial containing only odd powers with negative coefficients, for instance $f(y) = -y^{2k+1}$.

3. Approximation of the state

The scheme uses

$$\mathbf{y}_n'(t) = A_c\mathbf{y}_n(t) + S_c(\varepsilon_n)\mathbf{f}(S_c(\varepsilon_n)\mathbf{y}_n(t)) + S_\infty(\varepsilon_n)\mathbf{u}(t),\ \mathbf{y}_n(0) = \zeta \tag{3.1.n}$$

with $\varepsilon_n > 0$, $\varepsilon_n \to 0$. As for (2.2), $\mathbf{y}_n(t, \zeta, \mathbf{u})$ indicates dependence on $\zeta, \mathbf{u}$. We assume that $\|S_c(t)\| \le Ce^{-\omega t}$ with $\omega > 0$ (which can always be achieved by a translation) so that fractional powers $(-A_c)^\alpha$ can be defined; since $S_c(t)$ is analytic, if $\alpha \ge 0$,

$$\|(-A_c)^\alpha S_c(t)\| \le C_\alpha t^{-\alpha} \qquad (t \ge 0). \tag{3.2}$$

Lemma 3.1. *Let* $0 \le \alpha < 1$. *Then the operator*

$$\Lambda_\alpha g(\cdot) = \int_0^t (-A_c)^\alpha S_\infty(t - \sigma)\mathbf{u}(\sigma)d\sigma, \tag{3.3}$$

is compact from $L^\infty_w(0, T; L^\infty(\Omega))$ *into* $C(0, T; C(\bar\Omega))$.

The proof is basically the same as that of its abstract counterpart, Theorem 3.1 (or, rather, Theorem 9.1) in [7].

We use below the fact that $L^\infty_w(0, T; L^\infty(\Omega)) = L^1(0, T; L^1(\Omega))^*$ (which is obvious by (2.3)). We recall that $L^1(0, T; L^1(\Omega))$-weak convergence of a sequence implies boundedness; also, a $L^1(0, T; L^1(\Omega))$-weakly compact set has to be bounded.

Corollary 3.2. *Let* $\mathbf{u}(\cdot) \in L^\infty_w(0,\bar{t}; L^\infty(\Omega))$ *be such that* $\mathbf{y}(t,\mathbf{u})$ *exists in* $0 \le t \le \bar{t}$, $\{\zeta_n\} \subset C(\bar{\Omega})$ *with* $\zeta_n \to \zeta$, $\{\mathbf{u}_n(\cdot)\} \subseteq L^\infty_w(0,\bar{t}; L^\infty(\Omega))$ *with* $\mathbf{u}_n(\cdot) \to \mathbf{u}(\cdot)$ $L^1(0,\bar{t}; L^1(\Omega))$*-weakly. Then, if* $\varepsilon > 0$ *there exists* $n = n_\varepsilon$ *such that* $\mathbf{y}_n(t,\zeta_n,\mathbf{u}_n)$ *exists in* $0 \le t \le \bar{t}$ *for* $n \ge n_\varepsilon$ *and*

$$\|\mathbf{y}_n(t,\zeta_n,\mathbf{u}_n) - \mathbf{y}(t,\zeta,\mathbf{u})\| \le \varepsilon \tag{3.4}$$

$$\|(-A_c)^\alpha \mathbf{y}_n(t,\zeta_n,\mathbf{u}_n)\|,\ \|(-A_c)^\alpha \mathbf{y}(t,\zeta,\mathbf{u})\| \le C_\alpha t^{-\alpha} \tag{3.5}$$

$$\|(-A_c)^\alpha \mathbf{y}_n(t,\zeta_n,\mathbf{u}_n) - (-A_c)^\alpha \mathbf{y}(t,\zeta,\mathbf{u})\| \to 0 \tag{3.6}$$

uniformly in $\delta \le t \le \bar{t}$ *for any* $\delta > 0$ *(in* $0 \le t \le \bar{t}$ *if* $\alpha = 0$*).*

Proof: Denote by $[0, t_n]$ the maximum subinterval of $[0,\bar{t}]$ where $\mathbf{y}_n(t,\zeta_n,\mathbf{u}_n)$ exists and satisfies (3.4). In this interval, we have

$$\begin{aligned}
(-A_c)^\alpha(\mathbf{y}_n(t,\zeta_n,\mathbf{u}_n) - \mathbf{y}(t,\zeta,\mathbf{u})) &= (-A_c)^\alpha S_c(t)(\zeta_n - \zeta) \\
&+ \int_0^t (-A_c)^\alpha S_c(t-\tau+\varepsilon_n)\{\mathbf{f}(S_c(\varepsilon_n)\mathbf{y}_n(\tau,\zeta_n,\mathbf{u}_n)) \\
&\quad - \mathbf{f}(S_c(\varepsilon_n)\mathbf{y}(\tau,\zeta,\mathbf{u}))\}d\tau \\
&+ \int_0^t (-A_c)^\alpha S_c(t-\tau+\varepsilon_n)\{\mathbf{f}(S_c(\varepsilon_n)\mathbf{y}(\tau,\zeta,\mathbf{u})) \\
&\quad - \mathbf{f}(\mathbf{y}(\tau,\zeta,\mathbf{u}))\}d\tau \\
&+ \int_0^t (-A_c)^\alpha (S_c(t-\tau+\varepsilon_n) - S_c(t-\tau))\mathbf{f}(\mathbf{y}(\tau,\zeta,\mathbf{u}))d\tau \\
&+ \int_0^t (-A_c)^\alpha (S_\infty(t-\tau+\varepsilon_n) - S_\infty(t-\tau))\mathbf{u}_n(\tau)d\tau \\
&+ \int_0^t (-A_c)^\alpha S_\infty(t-\tau)(\mathbf{u}_n(\tau) - \mathbf{u}(\tau))d\tau \\
&= \int_0^t (-A_c)^\alpha S_c(t-\tau+\varepsilon_n)\{\mathbf{f}(S_c(\varepsilon_n)\mathbf{y}_n(\tau,\zeta_n,\mathbf{u}_n)) \\
&\quad - \mathbf{f}(S_c(\varepsilon_n)\mathbf{y}(\tau,\zeta,\mathbf{u}))\}d\tau + \kappa_n(t)
\end{aligned} \tag{3.7}$$

which we first estimate for $\alpha = 0$. In the first three integrals making up $\kappa_n(t)$ we break up the interval of integration into $(0,\rho t)$ and $(\rho t, t)$ with $0 < \rho < 1$. In the first interval, we use uniform continuity of $S_c(t)$ and $S_\infty(t)$ in $t > 0$; in the second the bound for $\mathbf{u}_n(t)$. The three integrals are shown in this way to tend to zero uniformly in $0 \le t \le \bar{t}$. The same is true of the fourth using Lemma 3.1. We estimate by Gronwall's inequality and, taking $n \ge$ a certain n_ε we deduce that $\|\mathbf{y}_n(t,\zeta_n,\mathbf{u}_n) - \mathbf{y}(t,\zeta,\mathbf{u})\| \le \varepsilon/2$ and contradict the maximality of $[0,t_n]$ unless $t_n = \bar{t}$.

The estimations for $\alpha > 0$ use the *generalized Gronwall inequality* [11, p. 188 Lemma 7.1.1]; if $b \leq 0$, $\gamma > -1$ and $a(t)$, $u(t)$ are nonnegative and locally integrable in $0 \leq t \leq T$ with

$$u(t) \leq a(t) + b \int_0^t (t-\sigma)^\gamma u(\sigma) d\sigma \qquad (0 \leq t \leq T),$$

then there exists C depending only on γ, T such that

$$u(t) \leq a(t) + Cb \int_0^t (t-\sigma)^\gamma a(\sigma) d\sigma \qquad (0 \leq t \leq T).$$

The treatment of all integrals is the same and we omit the details. Note that the singular behavior at $t = 0$ comes only from the first term $(-A_c)^\alpha(\zeta_n - \zeta)$ of $\kappa_n(t)$, not from the integral terms.

Corollary 3.3. *Assume* (2.2) *has the GEP in* $0 \leq t \leq \bar{t}$. *Then* (3.1.n) *has the GEP in* $0 \leq t \leq \bar{t}$ *for* n *large enough, with constants* $C, K, L(C,K) + \varepsilon$, $\varepsilon > 0$.

Proof: Let $\|\zeta\| \leq C$, $\|\mathbf{u}(\cdot)\| \leq K$. Use Lemma 3.2 ($\alpha = 0$) with $\zeta_n = \zeta$, $\mathbf{u}_n(\cdot) = \mathbf{u}(\cdot)$. Notice that the estimation of the first three integrals making up $\kappa_n(t)$ is independent of $\mathbf{u}$ and that the fourth integral is absent. We then obtain that $\mathbf{y}_n(t, \zeta, \mathbf{u})$ exists in $0 \leq t \leq \bar{t}$ and satisfies $\|\mathbf{y}_n(t, \zeta, \mathbf{u})\| \leq L + \varepsilon$. Compactness of $S_c(t)$ (or Lemma 3.1) is not used here.

We assume from now on that (2.2) has the GEP. The proof of the following result is very similar to that of Corollary 3.3 and is thus omitted.

Corollary 3.4. *Let* $0 \leq \alpha < 1$, $\{\mathbf{u}_k(\cdot)\}$ *a* $L^1(0, \bar{t}; L^1(\Omega))$*-weakly convergent sequence in* $L^\infty_w(0, \bar{t}; L^\infty(\Omega))$. *Then*

$$\|(-A_c)^\alpha \mathbf{y}(t, \zeta, \mathbf{u}_k) - (-A_c)^\alpha \mathbf{y}(t, \zeta, \mathbf{u})\| \to 0 \quad \text{as} \quad k \to \infty \tag{3.8}$$

$$\|(-A_c)^\alpha \mathbf{y}_n(t, \zeta, \mathbf{u}_k) - (-A_c)^\alpha \mathbf{y}_n(t, \zeta, \mathbf{u})\| \to 0 \quad \text{as} \quad k \to \infty \tag{3.9}$$

uniformly in $0 \leq t \leq \bar{t}$.

A *cost functional* for (2.2) is a real valued function $y_0(t, \mathbf{y}, \mathbf{u})$ defined in $t > 0$ and for each t, for $\mathbf{y} = \mathbf{y}(\cdot) \in C(0, t; C(\bar{\Omega}))$ and $\mathbf{u} = \mathbf{u}(\cdot) \in L^\infty_w(0, t; L^\infty(\Omega))$. It is *weakly lower semicontinuous* (WLS) if

$$y_0(t, \bar{\mathbf{y}}, \bar{\mathbf{u}}) \leq \limsup_{k \to \infty} \; y_0(t_k, \mathbf{y}, \mathbf{u}_k) \tag{3.10}$$

for every sequence $\{t_k\} \subseteq \mathbb{R}$ with $t_k \to \bar{t}$, every sequence $\{\mathbf{y}_k(\cdot)\} \subseteq C(0, t_k; C(\bar{\Omega}))$ with $\mathbf{y}(\cdot) \to \mathbf{y}(\cdot)$ in $C(0, T; C(\bar{\Omega}))$ and every sequence $\{\mathbf{u}_k(\cdot)\} \subseteq L^\infty_w(0, T; L^\infty(\Omega))$ with $\mathbf{u}_k(\cdot) \to \bar{\mathbf{u}}(\cdot)$ $L^1(0, T; L^1(\Omega))$-weakly. (Here $T > \bar{t}$ and the passage from spaces in $[0, t_n]$ to spaces in $[0, T]$ involves obvious extensions.)

We equip the control system (2.2) with a WLS cost functional $y_0(t, \mathbf{y}(\zeta, \mathbf{u}), \mathbf{u})$ and a *control set* $U \subseteq L^\infty(\Omega)$; the *admissible control space* $C_{ad}(0, T; U)$ consists of all $\mathbf{u}(\cdot) \in L^\infty_w(0, t; L^\infty(\Omega))$ such that $\mathbf{u}(t) \in U$ a.e. The optimal control problem includes a *target condition* $\mathbf{y}(\bar{t}, \zeta, \mathbf{u}) = Y =$ *target set* $\subseteq C(\bar{\Omega})$, and we assume Y closed. Call m the minimum of $y_0(t, \mathbf{y}(\zeta, \mathbf{u}), \mathbf{u})$ in $C_{ad}(0, \bar{t}; U)$ subject to the target condition. Assuming that $-\infty < m < \infty$, a sequence $\{\mathbf{u}^k(\cdot)\}$, $\mathbf{u}^k(\cdot) \in C_{ad}(0, t_k; U)$ is a *minimizing sequence* if we have

$$
(3.11) \quad \limsup_{k\to\infty} y_0(t_k, \mathbf{y}(\zeta, \mathbf{u}^k), \mathbf{u}^k) \leq m, \quad \lim_{k\to\infty} \mathrm{dist}(\mathbf{y}(t_k, \zeta, \mathbf{u}^k), Y) \to 0.
$$

The definitions are the same for (3.1.n).

Theorem 3.5. *Let $C_{ad}(0, T; U)$ be $L^1(0, T; L^1(\Omega))$-weakly compact in $L^\infty_w(0, T; L^\infty(\Omega))$, and assume $y_0(t, \mathbf{y}, \mathbf{u})$ is weakly lower semicontinuous. Let $\{\mathbf{u}^k(\cdot)\}$, $\mathbf{u}^k(\cdot) \in C_{ad}(0, t_k; U)$ be a minimizing sequence with $t_k \to \bar{t} < \infty$. Then (a generalized sequence of) $\{\mathbf{u}^k(\cdot)\}$ converges $L^1(0, T; L^1(\Omega))$-weakly to an optimal control $\bar{\mathbf{u}}(\cdot)$ for* (2.2). *The same result holds for* (3.1.n).

Proof: By the GEP we know that $\{\mathbf{y}(t, \mathbf{u}^k)\}$ is bounded, so that using Alaoglu's theorem in $L^\infty_w(0,T;L^\infty(\Omega))$ we may select a subsequence (denoted equally $\{\mathbf{u}^k(\cdot)\}$) such that both $\{\mathbf{u}^k(\cdot)\}$ and $\{\mathbf{f}(\mathbf{y}(\cdot, \mathbf{u}^k))\}$ are $L^1(0, T; L^1(\Omega))$-weakly convergent. Then, using Corollary 3.4 we deduce that $\{\mathbf{y}(t, \mathbf{u}^k)\}$ is convergent in $C(0, T; C(\bar{\Omega}))$. Some acrobatics are necessary to take care of the moving t_n. For a very similar argument see for instance [9, Theorem 4.4]. The reasoning for (3.1.n) is of course the same.

We note that if $\mathbf{u}(\cdot) \in C_{ad}(0, \bar{t}; U)$ is such that $\mathbf{y}(\bar{t}, \zeta, \mathbf{u}) \in Y$ then a minimizing sequence for (2.2) automatically exists, although not necessarily with $\{t_k\}$ bounded. The same result holds for (3.1.n).

A popular cost functional for (1.2) is

$$
(3.12) \qquad y_0(t, \mathbf{y}, \mathbf{u}) = \int_0^t \int_\Omega f_0(y(\tau, x, \zeta, u),\ u(\tau, x)) dx d\tau
$$

where $u(t, x) = \mathbf{u}(t)(x)$, $y(t, x, \zeta, u) = \mathbf{y}(t, \zeta, \mathbf{u})(x)$. The proof of the following result is standard and we omit it.

Lemma 3.6. *Let $f_0(y, u)$ be continuous in $\bar{\Omega} \times \mathbb{R}$ and convex in u for y fixed. Then $y_0(t, \mathbf{u})$ is weakly lower semicontinuous.*

4. Approximation of the costate

We assume that f is continuously differentiable, so that $\mathbf{f}$ has a Fréchet derivative $\partial\mathbf{f}$ with respect to y in $C(\bar{\Omega})$ given by $(\partial\mathbf{f}(y)z)(x) = f'(y(x))z(x)$. The derivative is continuous with respect to y in the space $(C(\bar{\Omega}), C(\bar{\Omega}))$ of bounded operators from $C(\bar{\Omega})$ into itself. The *adjoint variational equation* corresponding to (2.2) is

$$\mathbf{z}'(s) = -\{A_1' + \partial\mathbf{f}(\mathbf{y}(s))^*\}\mathbf{z}(s) - \mathbf{g}(s), \qquad \mathbf{z}(\bar{t}) = z \tag{4.1}$$

in $L^1(\Omega)$, to be solved backwards in $0 \le t \le \bar{t}$; here $z \in \Sigma(\bar{\Omega})$, $\mathbf{g}(\cdot) \in L^\infty_w(0, T; \Sigma(\bar{\Omega}))$ and $\mathbf{y}(\cdot) \in C(0, \bar{t}; C(\bar{\Omega}))$. The operator $\partial\mathbf{f}(y(t))^*$ is multiplication by $f'(y(x))$ and is thus continuous in $(\Sigma(\bar{\Omega}),\ \Sigma(\bar{\Omega}))$ and in $(L^1(\Omega),\ L^1(\Omega))$. The treatment of (4.1) parallels that of (2.2) with $\Sigma(\bar{\Omega})$ taking the role of $L^\infty(\Omega)$ and $L^1(\Omega)$ taking the role of $C(\bar{\Omega})$. A solution of (4.1) is a continuous solution of the integral equation

$$\mathbf{z}(s) = S_\Sigma'(\bar{t} - s)z + \int_s^{\bar{t}} S_1'(\sigma - s)\partial\mathbf{f}(\mathbf{y}(\sigma))^*\mathbf{z}(s) + \int_s^{\bar{t}} S_\Sigma'(\sigma - s)\mathbf{g}(\sigma)d\sigma, \tag{4.2}$$

where the first integral is a Bochner integral. So is the second, on account of strong measurability of $\sigma \to S_\Sigma'(\sigma - s)\mathbf{g}(\sigma)$ [6, Lemma 4.1]. Note that we use S_1' in the first integral since solutions belong to $L^1(\Omega)$. Due to linearity, (4.2) enjoys global existence.

Since $\{S_c(\varepsilon_n)\partial\mathbf{f}(S_c(\varepsilon_n)\mathbf{y}(s))\}^* = S_c(\varepsilon_n)^*\partial\mathbf{f}(S_c(\varepsilon_n)\mathbf{y}(s))^*S_c(\varepsilon_n)^*$ and $S_1'(t)$ is the restriction to $L^1(\Omega)$ of $S_c(t)^*$, the adjoint variational equation corresponding to (3.1.n) is

$$\begin{aligned} &\mathbf{z}_n'(s) = -\{A_1' + S_1'(\varepsilon_n)\partial\mathbf{f}(S_c(\varepsilon_n)\mathbf{y}_n(s))^*S_1'(\varepsilon_n)\}\mathbf{z}_n(s) - \mathbf{g}_n(s), \\ &\mathbf{z}_n(\bar{t}) = z_n \end{aligned} \tag{4.3}$$

also to be solved backwards in $0 \le t \le \bar{t}$ with $z_n \in \Sigma(\bar{\Omega})$ and $\mathbf{g}_n(\cdot) \in L^\infty_w(0, T; \Sigma(\bar{\Omega}))$; as in (4.1), $\mathbf{y}_n(\cdot) \in C(0, \bar{t}; C(\bar{\Omega}))$. The associated integral equation is the corresponding replica of (4.2).

Lemma 4.1. *Let $0 \le \alpha < 1$. Then the operator*

$$\Lambda_\alpha g(\cdot) = \int_s^T (-A_1')^\alpha S_\Sigma'(\sigma - s)\mathbf{g}(\sigma)d\sigma, \tag{4.4}$$

is compact from $L^\infty_w(0, T; \Sigma(\bar{\Omega}))$ into $C(0, T; L^1(\Omega))$.

The proof follows from the same abstract scheme as does that of Lemma 3.1 (Theorems 3.1 and 9.1 in [7]).

Below, we use the equality $L^1(0, T; C(\bar{\Omega}))^* = L^\infty_w(0, T; \Sigma(\bar{\Omega}))$. This is a particular case of the formula $L^1(0, T; X)^* = L^\infty_w(0, T; X^*)$ valid for any Banach space X; see [7] for references.

Corollary 4.2. *Let $\{z_n\} \subseteq \Sigma(\bar{\Omega})$ be $C(\bar{\Omega})$-weakly convergent to $z \in \Sigma(\bar{\Omega})$ and let $\{\mathbf{g}_n(\cdot)\} \subseteq L^\infty_w(0, \bar{t}; \Sigma(\bar{\Omega}))$ be $L^1(0, \bar{t}; C(\bar{\Omega}))$-weakly convergent to $\mathbf{g}(\cdot) \in L^\infty_w(0, \bar{t}; \Sigma(\bar{\Omega}))$. Finally, let $\mathbf{y}_n(\cdot) \to \mathbf{y}(\cdot)$ in $C(0, \bar{t}; C(\bar{\Omega}))$, and $0 \le \alpha < 1$. Then*

$$\|(-A_1')^\alpha \mathbf{z}_n(\cdot)\|, \qquad \|(-A_1')^\alpha \mathbf{z}(t)\| \le C_\alpha t^{-\alpha} \tag{4.5}$$

$$\|(-A_1')^\alpha \mathbf{z}_n(\cdot) - (-A_1')^\alpha \mathbf{z}(t)\| \to 0 \tag{4.6}$$

uniformly in $0 \le t \le \bar{t} - \delta$ *for every* $\delta > 0$.

Proof: The proof is just the same as that of Corollary 3.2 with one minor difference; the nonintegral term on the right side of (3.5) is $(-A_c)^\alpha S_c(t)(\zeta_n - \zeta)$ with $\{\zeta_n\} \subseteq C(\bar{\Omega})$, $\zeta_n \to \zeta$ strongly, while here the corresponding term is $(-A_1')^\alpha S_\Sigma'(\bar{t}-s)(z_n - z)$ with $\{z_n\} \subseteq \Sigma(\bar{\Omega})$, $z_n \to z$ $C(\bar{\Omega})$-weakly; however compactness of $S_\Sigma'(\bar{t} - s)$ makes this a moot point.

5. The maximum principle and invariance of the Hamiltonian

We consider a cost functional

$$y_0(t, \mathbf{y}(\zeta, \mathbf{u}), \mathbf{u}) = \int_0^t \mathbf{f}_0(\mathbf{y}(\tau, \zeta, \mathbf{u}), \mathbf{u}(\tau))d\tau \tag{5.1}$$

where $f_0 : \bar{\Omega} \times L^\infty(\Omega) \to \mathbb{R}$ has a Fréchet differential $\partial_y \mathbf{f}_0(y, u)$ with respect to y. We assume that $\mathbf{f}_0(y, u)$ is $C(\bar{\Omega})$-continuous in y for u fixed and that $\partial_y \mathbf{f}_0(y, u)$ is $C(\bar{\Omega})$-continuous in y as a $\Sigma(\bar{\Omega})$-valued function for u fixed. Moreover, we require $t \to f_0(y, \mathbf{u}(t))$ measurable and bounded and $t \to \partial_y \mathbf{f}_0(y, \mathbf{u}(t))$ $C(\bar{\Omega})$-weakly measurable and bounded for every $\mathbf{u}(\cdot) \in L^\infty_w(0, T; L^\infty(\Omega))$. Finally, we assume that $y_0(t, \mathbf{y}, \mathbf{u})$ is WLS and that

$$\mathbf{f}_0(y, u) \ge \delta > 0 \quad (y \in C(\bar{\Omega}),\ u \in U). \tag{5.2}$$

For a functional (3.12), all of the above will hold if $\mathbf{f}_0(y, u)$ is continuous in $\bar{\Omega} \times \mathbb{R}$, continuously differentiable with respect to y in $\bar{\Omega} \times \mathbb{R}$ and positive. The system (2.2) is assumed to have the GEP, so that all the approximating systems (3.1.n) will also have the GEP for n large enough. We assume that the control set $C_{ad}(0, T; U)$ is $L^1(0, T; L^1(\Omega))$-weakly closed in $L^\infty_w(0, T; L^\infty(\Omega))$ and that the target set Y is closed.

Assume that a control $\mathbf{u}(\cdot) \in C_{ad}(0, T; U)$ exists with $\mathbf{y}(t, \zeta, \mathbf{u}) \in Y$ for some t. Then (see the remarks at the end of §3) a minimizing sequence for (2.2) exists. Condition (5.2) implies

$$y_0(t, \mathbf{y}(\zeta, \mathbf{u}), \mathbf{u}) \ge \delta t \tag{5.3}$$

thus for every minimizing sequence, $\{t_n\}$ is automatically bounded. Hence, by Theorem 3.5, an optimal control $\bar{\mathbf{v}}(\cdot)$ exists in an interval $0 \le t \le \tilde{t}$.

We consider the approximating problems (3.1.n) with the same admissible control space $C_{ad}(0, T; U)$ but a different initial condition ζ_n. The cost functional is

$$y_{on}(t, \mathbf{y}_n(\zeta, \mathbf{u}), \mathbf{u}) = \int_0^t \mathbf{f}_{on}(\mathbf{y}_n(\tau, \zeta_n, \mathbf{u}), \mathbf{u}(\tau))d\tau \tag{5.4}$$

where $\mathbf{f}_{on}(\mathbf{y}, \mathbf{u}) = \mathbf{f}_0(S_c(\varepsilon_n)y, u)$. Obviously, $\mathbf{f}_{on}(y, u)$ satisfies (5.2) and has a Fréchet differential $\partial_y \mathbf{f}_{on}(y, u) = S_c(\varepsilon_n)^* \partial_y \mathbf{f}_0(S_c(\varepsilon_n)y, u)$ sharing all the relevant properties of $\partial_y \mathbf{f}_0(y,u)$. We check easily that y_{on} is as well WLS. The target set Y_n is constructed as follows. As a particular case of Lemma 3.1,

$$\delta_n = \sup_{\mathbf{u}\in C_{ad}(0,\bar{t},U)} \|\mathbf{y}_n(\bar{t},\zeta,\mathbf{u}) - \mathbf{y}(\bar{t},\zeta,\mathbf{u})\| \to 0.$$

We set $Y_n = \{y; \mathrm{dist}(y,Y) \le 2\delta_n\}$ and choose $\zeta_n \in D(A_c)$ tending to ζ (in the norm of $C(\bar{\Omega})$) in such a way that $\mathbf{y}_n(\bar{t},\zeta_n,\bar{\mathbf{v}}) \in Y_n$. Accordingly, a minimizing sequence for (3.1.n) exists, thus an optimal control $\bar{\mathbf{u}}^n(\cdot)$ exists in some interval $[0,t_n]$ by Theorem 3.5. We have

$$\begin{aligned} &y_0(t_n,\mathbf{y}(\zeta,\bar{\mathbf{u}}^n),\bar{\mathbf{u}}^n) \\ &\quad = y_{on}(t_n,\mathbf{y}_n(\zeta_n,\bar{\mathbf{u}}^n),\bar{\mathbf{u}}^n) + \{y_0(t_n,\mathbf{y}(\zeta,\bar{\mathbf{u}}^n),\bar{\mathbf{u}}^n) - y_{on}(t_n,\mathbf{y}_n(\zeta_n,\bar{\mathbf{u}}^n),\bar{\mathbf{u}}^n)\} \\ &\quad \le y_{on}(\bar{t},\mathbf{y}_n(\zeta_n,\bar{\mathbf{v}}),\bar{\mathbf{v}}) + \{y_0(t_n,\mathbf{y}(\zeta,\bar{\mathbf{u}}^n),\bar{\mathbf{u}}^n) - y_{on}(t_n,\mathbf{y}_n(\zeta_n,\bar{\mathbf{u}}^n),\bar{\mathbf{u}}^n)\} \\ &\quad \le y_0(\bar{t},\mathbf{y}(\zeta,\bar{\mathbf{v}}),\bar{\mathbf{v}}) + \{y_{on}(\bar{t},\mathbf{y}_n(\zeta_n,\bar{\mathbf{v}}),\bar{\mathbf{v}}) - y_0(\bar{t},\mathbf{y}(\zeta,\bar{\mathbf{v}}),\bar{\mathbf{v}})\} \\ &\qquad + \{y_0(t_n,\mathbf{y}(\zeta,\bar{\mathbf{u}}^n),\bar{\mathbf{u}}^n) - y_{on}(t_n,\mathbf{y}_n(\zeta_n,\bar{\mathbf{u}}^n),\bar{\mathbf{u}}^n)\} \end{aligned}$$

with the two curly brackets tend to zero in view of Corollary 3.2. It then results that $\{\bar{\mathbf{u}}^n\}$ is a minimizing sequence for the original problem (2.2). Applying once again Theorem 3.5, we deduce that a subsequence (also denoted $\{\bar{\mathbf{u}}^n(\cdot)\}$) converges $L^1(0,T;L^1(\Omega))$-weakly in $L^\infty_w(0,T;L^\infty(\Omega))$ to a control $\bar{\mathbf{u}}(\cdot)$ which is optimal for (2.2) in an interval $0 \le t \le \bar{t}$.

Theorem 5.1. *There exists $(z_{on}, z_n) \in \mathbb{R} \times \Sigma(\bar{\Omega})$ such that*

$$z_{on} \le 0, \qquad z_n \in N_{Y_n}(\mathbf{y}_n(t_n,\zeta_n,\bar{\mathbf{u}}^n)), \qquad |z_{on}| + \|z_n\| = 1 \tag{5.5}$$

($N_{Y_n}(\mathbf{y}_n(t_n,\zeta_n,\bar{\mathbf{u}}^n))$ the normal cone to Y_n at $\mathbf{y}_n(t_n,\zeta_n,\bar{\mathbf{u}}^n)$) such that if $\mathbf{z}_n(t)$ is the solution of

$$\begin{aligned} \bar{\mathbf{z}}'_n(s) &= -\{A'_1 + S'_1(\varepsilon_n)\partial f(\mathbf{y}_n(s,\zeta_n,\bar{\mathbf{u}}^n))^* S'_1(\varepsilon_n)\}\bar{\mathbf{z}}_n(s) \\ &\quad - z_{on} S_c(\varepsilon_n)^* \partial_y \mathbf{f}_0(S_c(\varepsilon_n)\mathbf{y}_n(s,\zeta_n,\bar{\mathbf{u}}^n),\bar{\mathbf{u}}^n(s)), \ \bar{\mathbf{z}}_n(\bar{t}) = z_n \end{aligned} \tag{5.6}$$

and

$$\begin{aligned} H_n(\mathbf{y}(s,\zeta_n,\bar{\mathbf{u}}^n), z_{on}, \bar{\mathbf{z}}_n(s), \bar{\mathbf{u}}^n(s)) &= z_{on}\mathbf{f}_0(\mathbf{y}(s,\zeta_n,\bar{\mathbf{u}}^n),\bar{\mathbf{u}}^n(s)) \\ &+ \langle (-A_1)^{1/2}\bar{\mathbf{z}}_n(s), (-A_c)^{1/2}\mathbf{y}(s,\zeta_n,\bar{\mathbf{u}})\rangle \\ &+ \langle \bar{\mathbf{z}}_n(s), S_1(\varepsilon_n)\mathbf{f}(S_1(\varepsilon_n)\mathbf{y}(s,\zeta_n,\bar{\mathbf{u}}^n))\rangle + \langle S_1(\varepsilon_n)\bar{\mathbf{z}}_n(s), \bar{\mathbf{u}}^n\rangle, \end{aligned} \tag{5.7}$$

then

$$H_n(\mathbf{y}(s,\zeta_n,\bar{\mathbf{u}}^n), z_{on}, \bar{\mathbf{z}}_n(s), \bar{\mathbf{u}}^n(s)) = \max_{v\in U} H_n(\mathbf{y}(s,\zeta_n,\bar{\mathbf{u}}^n), z_{on}, \bar{\mathbf{z}}_n(s), v) \tag{5.8}$$

$$H_n(\mathbf{y}(s,\zeta_n,\bar{\mathbf{u}}^n), z_{on}, \bar{\mathbf{z}}_n(s), \bar{\mathbf{u}}^n(s)) = c_n \tag{5.9}$$

a.e. in $0 \le t \le \bar{t}$, with $c_n = 0$ if the terminal time is free.

Theorem 5.1 is a particular case of [6, Theorem 6.3] except for (5.9); note that the target set contains interior points, which guarantees the third condition (5.5) (see the comments in [6] after Theorem 6.4).

We sketch a proof of (5.9). We note first that (3.1.n) can be written in the form $\mathbf{y}'_n(t, \zeta_n, \bar{\mathbf{u}}^n) = A_c\mathbf{y}_n(t, \zeta_n, \bar{\mathbf{u}}^n) + \boldsymbol{\phi}(t)$ where $\mathbf{y}'_n(0, \zeta_n, \bar{\mathbf{u}}^n) = \zeta_n \in D(A_c)$ and where $\boldsymbol{\phi}(t) = S_c(\varepsilon_n)\mathbf{f}(S_c(\varepsilon_n)\mathbf{y}_n(t, \zeta_n, \bar{\mathbf{u}}^n)) + S_\infty(\varepsilon_n)\mathbf{u}(t)$ and $A_c\boldsymbol{\phi}(t)$ are strongly measurable and bounded. It follows then from the theory of linear abstract parabolic equations that $\mathbf{y}_n(t, \zeta_n, \bar{\mathbf{u}}^n) \in D(A_c)$ a.e. and is Lipschitz continuous and differentiable a.e. with $\mathbf{y}'_n(t, \zeta_n, \bar{\mathbf{u}}^n) = A_c\mathbf{y}_n(t, \zeta_n, \bar{\mathbf{u}}^n) + S_c(\varepsilon_n)\mathbf{f}(S_c(\varepsilon_n)\mathbf{y}_n(t, \zeta_n, \bar{\mathbf{u}}^n)) + S_\infty(\varepsilon_n)\mathbf{u}(t)$. Exactly the same considerations apply to $\bar{\mathbf{z}}_n(s)$ but, since the final condition z_n is in $\Sigma(\bar{\Omega})$, only in intervals $0 \le s \le \bar{t} - \delta$, $\delta > 0$. At any rate the function $\langle \mathbf{y}_n(s, \zeta_n, \bar{\mathbf{u}}^n), \bar{\mathbf{z}}_n(s)\rangle$ is locally Lipschitz continuous in $0 \le s \le \bar{t}$ and its derivative can be directly calculated at any s where $\mathbf{y}_n(s, \zeta_n, \bar{\mathbf{u}}^n)$ and $\mathbf{z}_n(s)$ can be differentiated; likewise, the derivative of $\mathbf{f}_0(\mathbf{y}_n(s, \zeta_n, \mathbf{u}), \mathbf{u}(t))$ (as a function of s) can be calculated a.e. Doing this and using the differential equations for $\mathbf{y}_n(s, \zeta_n, \bar{\mathbf{u}}^n)$ and $\bar{\mathbf{z}}_n(s)$,

$$(\partial/\partial s)|_{s=t}\phi(s,t) = (\partial/\partial s)|_{s=t}H_n(\mathbf{y}(s, \zeta_n, \bar{\mathbf{u}}^n), z_{on}, \mathbf{z}_n(s), \bar{\mathbf{u}}^n(t)) = 0$$

in a set e of full measure in $[0, \bar{t}]$ independent of t. Since $\phi(s,t)$ is s-Lipschitz continuous independently of t then the function $\psi(s) = \inf_{0 \le t \le \bar{t}} \phi(s,t)$ is Lipschitz continuous, hence absolutely continuous, so that there exists a subset $e_0 \subseteq e$, still of full measure in $[0, \bar{t}]$ where $\psi'(s)$ exists. In this set we have

$$\begin{aligned}\frac{\psi(s+h) - \psi(s)}{h} &= \frac{\phi(s+h, s+h) - \phi(s,s)}{h} \\ &= \frac{\phi(s+h, s+h) - \phi(s+h, s)}{h} + \frac{\phi(s+h, s) - \phi(s,s)}{h}.\end{aligned}$$

If $s \in e_0$ the left side and the second term on the right have a limit as $h \to 0$, thus so does the first term on the right. Since $\phi(s+h, s+h) \le \phi(s+h, s)$ for h of any sign, this limit must be zero. It follows that $\psi'(s) = 0$ almost everywhere, thus $\psi(s)$ is constant, proving (5.9). The statement for free terminal time problems involves using an auxiliary system and is essentially the same used in [3 pp. 130-32], so we omit the proof.

Theorem 5.2. *Let the target set Y be convex with nonempty interior. Then there exists $(z_0, z) \in \mathbb{R} \times \Sigma(\bar{\Omega})$ such that*

$$z_0 \le 0, \qquad |z_0| + \|z\| = 1 \tag{5.10}$$

and such that, if $\mathbf{z}(s)$ solves

$$\begin{aligned}\bar{\mathbf{z}}'(s) &= -\{A'_1 + \partial\mathbf{f}(\mathbf{y}(s, \zeta, \bar{\mathbf{u}}))^*\}\bar{\mathbf{z}}(s) \\ &\qquad - z_0\partial_y\mathbf{f}_0(\mathbf{y}(s, \zeta, \bar{\mathbf{u}}), \bar{\mathbf{u}}(s)) \\ \bar{\mathbf{z}}(\bar{t}) &= z\end{aligned} \tag{5.11}$$

and

$$H(\mathbf{y}(s,\zeta,\bar{\mathbf{u}}),z_0,\bar{\mathbf{z}}(s),\bar{\mathbf{u}}(s)) = z_0\mathbf{f}_0(\mathbf{y}(s,\zeta,\bar{\mathbf{u}}),\bar{\mathbf{u}}(s)) \tag{5.12}$$
$$+ \langle(-A_1)^{1/2}\bar{\mathbf{z}}(s),(-A_c)^{1/2}\mathbf{y}(s,\zeta,\bar{\mathbf{u}})\rangle$$
$$+ \langle\bar{\mathbf{z}}(s),\mathbf{f}(\mathbf{y}(s,\zeta,\bar{\mathbf{u}}))\rangle + \langle\bar{\mathbf{z}}(s),\bar{\mathbf{u}}\rangle$$

then

$$H(\mathbf{y}(s,\zeta,\bar{\mathbf{u}}),z_0,\bar{\mathbf{z}}(s),\bar{\mathbf{u}}(s)) = \max_{v\in U} H(\mathbf{y}(s,\zeta,\bar{\mathbf{u}}),z_0,\bar{\mathbf{z}}(s),v) \tag{5.13}$$

$$H(\mathbf{y}(s,\zeta,\bar{\mathbf{u}}),z_0,\bar{\mathbf{z}}(s),\bar{\mathbf{u}}(s)) = c \tag{5.14}$$

with $c = 0$ if the problem is free terminal time.

Theorem 5.2 will be proved (of course) taking limits in Theorem 5.1. Note first that the sequences $\{z_{on}\}$, $\{z_n\}$ coming out of Theorem 5.1 are bounded; moreover, everything in H_n is bounded, (the second term by Corollary 3.4 and Corollary 4.2) so that c_n must be bounded. Hence, we may take a subsequence if necessary and assume that

$$c_n \to c, \qquad z_{on} \to z_0, \qquad z_n \to z, \tag{5.15}$$

the last $\to$ understood $C(\bar{\Omega})$-weakly in $\Sigma(\bar{\Omega})$. To take limits directly in the Hamiltonian would not work, since $\{\bar{\mathbf{u}}^n(\cdot)\}$ is just weakly convergent; we use the integrated version of (5.8) and (5.9). The first is

$$\int_c H_n(\mathbf{y}_n(s,\zeta_n,\bar{\mathbf{u}}^n),z_{on},\bar{\mathbf{z}}_n(s),\mathbf{v}(s))ds \tag{5.16}$$
$$\leq \int_e H_n(\mathbf{y}_n(s,\zeta_n,\bar{\mathbf{u}}^n),z_{on},\bar{\mathbf{z}}_n(s),\bar{\mathbf{u}}^n(s))ds$$

for measurable sets $e \subseteq [0,\bar{t}]$ and every $\mathbf{v}(\cdot) \in C_{ad}(0,\bar{t};U)$; the second

$$\int_e H_n(\mathbf{y}_n(s,\zeta_n,\bar{\mathbf{u}}),z_0,\mathbf{z}_n(s),\bar{\mathbf{u}}^n(s))ds = c_n\mathrm{meas}(e). \tag{5.17}$$

Clearly, we can go from (5.16) and (5.17) to (the integrated versions of) (5.13) and (5.14) if we can show that

$$\int_e H(\mathbf{y}(s,\zeta,\bar{\mathbf{u}}),z_0,\bar{\mathbf{z}}(s),\mathbf{v}(s))ds \tag{5.18}$$
$$= \lim_{n\to\infty}\int_e H_n(\mathbf{y}(s,\zeta_n,\bar{\mathbf{u}}^n),z_{on},\bar{\mathbf{z}}_n(s),\bar{\mathbf{v}}^n(s))ds$$

in just two cases: $\mathbf{v}^n = \mathbf{v}(\cdot) \in C_{ad}(0,\bar{t};U)$, and $\mathbf{v}^n(\cdot) = \bar{\mathbf{u}}^n(\cdot)$. We concentrate on the second case. To take limits in the second term of the Hamiltonian (5.7) we use Corollary 3.4 for $\mathbf{y}_n(s,\zeta_n,\bar{\mathbf{u}}^n)$, Corollary 4.2 for $\bar{\mathbf{z}}^n(t)$ and the dominated convergence theorem. The second term is handled in the same way, using Corollary 3.4 and continuity of $\mathbf{f}$. Finally, the third term is disposed of using Corollary 4.2 (which implies that $\mathbf{z}_n(\sigma, \bar{\mathbf{u}}^n)$, thus $S_1(\varepsilon_n)\mathbf{z}_n(\sigma, \bar{\mathbf{u}}^n)$, is convergent in $L^1(0,\bar{t}; L^1(\Omega)))$ and the $L^1(0,\bar{t}; L^1(\Omega))$-weak convergence of $\bar{\mathbf{u}}^n(\cdot)$ in $L^\infty_w(0,T; L^\infty(\Omega))$. At this point, everything is over if $z_{on} \to 0$. Otherwise, we still have to prove that

$$\int_e \mathbf{f}_{on}(\mathbf{y}_n(s,\zeta_n,\bar{\mathbf{u}}^n), \bar{\mathbf{u}}^n(s))ds = \lim_{n\to\infty} \int_e \mathbf{f}_0(\mathbf{y}(s,\zeta,\bar{\mathbf{u}}), \bar{\mathbf{u}}(s))ds. \tag{5.19}$$

This is obvious if we write

$$\mathbf{f}_{on}(\mathbf{y}(s,\zeta_n,\bar{\mathbf{u}}^n), \bar{\mathbf{u}}^n(s)) = \{c_n s - H_{n1}(\mathbf{y}_n(s,\zeta_n,\bar{\mathbf{u}}^n), z_{on}, \bar{\mathbf{z}}_n(s), \bar{\mathbf{u}}(s))\}/z_{on}$$

where H_{n1} comprises the terms in the Hamiltonian with whose convergence we can count on. This ends with (5.13) and (5.14). We must still prove the second condition (5.10) or, equivalently, that $(z_0, z) \neq 0$. If $z_0 = \lim z_{on} \neq 0$ we are through; otherwise, $\|z_n\| \to 1$. We use the second condition (5.5), written in the form

$$\langle z_n, y\rangle \leq 0 \quad (y \in T_{Y_n}(\mathbf{y}_n(t_n,\zeta_n,\bar{\mathbf{u}}^n)))\,,$$

Y_n the tangent cone to Y_n at $\mathbf{y}_n(t_n,\zeta_n,\bar{\mathbf{u}}^n)$. Since Y is convex and has interior points all the $T_{Y_n}(\mathbf{y}_n(t_n,\zeta_n,\bar{\mathbf{u}}^n))$ contain a common ball and $z \neq 0$ by [6, Corollary 2.9], thus ending the proof of Theorem 5.2. We note that the convexity condition on Y is probably excessive, and it is likely that assumptions on the tangent cones of Y of the type used in [6, §2] suffice, although we have not been able to prove this.

6. Final comments

The Hilbert space setting in [3] is due to the use of "maximal regularity" results [1] for $y'(t) = Ay(t) + f(t)$. These have been extended to Banach spaces including L^p spaces in [2], but L^p extensions would still be very restrictive on nonlinearities. On the other hand, the results in [3] are not strictly less general than the ones here; for instance, the GEP is not needed.

Relaxed controls ((that is, measure-valued controls $\mu(t)$) can be installed in any semilinear system $y'(t) = Ay(t) + f(t, y(t), u(t))$ under very general assumptions [4]; with them, the system becomes $y'(t) = Ay(t) + F(t, y(t))\mu(t)$, linear in the control, and all the theory in this paper can be developed with minor changes: the "two-space" set up $C(\bar{\Omega}), L^\infty(\Omega)$ and its dual $L^1(\Omega), \Sigma(\bar{\Omega})$ is handled as in [6] or using the theory of Phillips adjoints (see [7]). However, it is essential that the semigroup $S(t)$ generated by A be compact, as the arguments in this paper make clear. It seems possible, however, to extend at least some of the results to the case where the control u itself enters in the equation through a compact control term.

Another interesting line of generalization (requiring an analytic semigroup) would be to relax assumptions on the nonlinearity by means of "front-and-back smoothing with fractional powers", a trick well known for the Navier-Stokes equations; we assume here that $((-A_1')^{-\beta})^*\mathbf{f}((-A_c)^{-\alpha}(\cdot))$, not $f(\cdot)$ is locally bounded and locally Lipschitz continuous. $D((-A_c)^\alpha)$-valued solutions of the system (2.2) are constructed through the integral equation

$$(-A_c)^\alpha \mathbf{y}(t) = (-A_c)^\alpha S_c(t)\zeta + \int_0^t (-A_c)^{\alpha+\beta} S_c(t-\tau)((-A_1')^{-\beta})^* \mathbf{f}(\mathbf{y}(\tau))d\tau + \int_0^t (-A_c)^{-\alpha} S_\infty(t-\tau)\mathbf{u}(\tau)d\tau$$

which can be solved in the customary way when $\alpha + \beta < 1$. The adjoint equation can be dealt with in the same way. However, convergence of $(-A_1)^{1/2}\mathbf{z}_n(t, \bar{\mathbf{u}}^n)$ and of $(-A_c)^{1/2}\mathbf{y}(t, \bar{\mathbf{u}}^n)$, essential in the arguments in §4, seems out of reach unless $\alpha + \beta < 1/2$. This is excessive for dealing with realistic nonlinearities involving grad $y(x)$.

References

1 L DE SIMON, *Un'applicazione della teoria degli integrali singolari allo studio delle equazioni differenziali astratte del primo ordine*, Rend Sem Mat Univ Padova **34** (1964), 205–22

2 G DORE and A VENNI, *On the closedness of the sum of two closed operators*, Math Z **196** (1987), 189–201

3 H O FATTORINI, *Constancy of the Hamiltonian in infinite dimensional systems*, in Control and Estimation of Distributed Parameter Systems, (F Kappel, K Kunisch, W Schappacher, Eds), Int Series Numerical Math , vol 91, Birkhauser, Basel, 1989, pp 123–133

4 H O FATTORINI, *Relaxed controls in infinite dimensional control systems*, in Estimation and Control of Distributed Parameter Systems, (W Desch, F Kappel, K Kunisch, Eds), International Series of Numerical Mathematics, vol 100, Birkhauser, Basel, 1991, pp 115–128

5 H O FATTORINI, *Optimal control problems for distributed parameter systems governed by semilinear parabolic equations in L^1 and L^∞ spaces*, in Optimal Control of Partial Differential Equations, (K -H Hoffmann, W Krabs, Eds), Springer Lecture Notes in Control and Information Sciences, vol 149, 1991, pp 68–80

6 H O FATTORINI, *Optimal control problems in Banach spaces*, Appl Math Optim **28** (1993), 225–257

7 H O FATTORINI, *Relaxation in semilinear infinite dimensional control systems*, in L Markus Festschrift, (K D Elworthy, W N Everitt, E B Lee, Eds), Marcel Dekker, New York, 1993, pp 505–522

8 H O FATTORINI and H FRANKOWSKA, *Necessary conditions for infinite dimensional control problems*, Math Control, Signals and Systems **4** (1991), 41–67

9 H O FATTORINI and T MURPHY, *Optimal problems for nonlinear parabolic boundary control systems*, to appear in SIAM J Control & Optimization

10 H FRANKOWSKA, *Some inverse mapping theorems*, Ann Inst Henri Poincaré **7** (1990), 183–234

11 D HENRY, " Geometric Theory of Semilinear Parabolic Equations", Springer, Berlin 1981

12 E HILLE and R S PHILLIPS, "Functional Analysis and Semi-Groups", AMS Colloquium Pubs vol 31, Providence, 1957

H. O. Fattorini
Department of Mathematics
University of California
Los Angeles, California 90024, USA

RATE DISTRIBUTION MODELING FOR STRUCTURED HETEROGENEOUS POPULATIONS

BEN G FITZPATRICK

Center for Research in Scientific Computation
Department of Mathematics
North Carolina State University

ABSTRACT We propose a modeling strategy for structured populations, in which individuals are not necessarily identical The heterogeneity is obtained by modeling the population as comprising homogeneous subpopulations By using a vector measure, we combine the subpopulations with an abstract integral to obtain the density of the population We show that this approach leads to a semigroup formulation of the dynamics in a space of vector measures, and we develop some estimation methods for determining the initial structure from observed data

1991 *Mathematics Subject Classification* 46G10, 92D25, 92D40

Key words and phrases Rate distributions, vector measures, semigroups, estimation

1. Introduction

Many biological applications of dynamical systems and control theory involve populations that are structured in terms of size, age, or spatial distribution. From predicting future populations to developing optimal harvesting strategies, models provide important information for biologist. Thus, it is crucial to develop models that predict population behavior in an accurate manner. This paper focuses on a particular aspect of model improvement, that of incorporating individual-based information into aggregate, population-wide models.

The original motivation for this work involved certain behavior exhibited by observed data for a size structured population of mosquitofish, behavior which is inconsistent with the commonly used Sinko-Streifer model:

$$v_t + (gv)_x = -\mu v, \quad x_0 < x < x_1, t > 0.$$

Here $v = v(t,x)$ denotes the density of indiviuals of size x at time t, $g = g(t,x)$ denotes individual growth rate, and $\mu = \mu(t,x)$ is the mortality rate. The observed data, as discussed in [BBKW, BF], exhibits a dispersion in size as time increases,

Research supported in part under AFOSR contract F49620-93-1-0153 and in some cases unimodal initial densities develop into bimodal densities at later time: the Sinko-Streifer model does not predict these phenomena without biologically unrealistic assumptions on the parameters.

In order to improve model predictions, various researchers (see [BBKW, BF]) have attempted to examine the model on the individual level. In general, individuals do not have identical growth rates (a basic assumption for the Sinko-Streifer model). To introduce individual variations, we model the population as being composed of homogeneous subpopulations, with each subpopulation obeying a Sinko-Striefer law having different parameters. The approach in [BBKW, BF] involves subpoulation densities $v(t, x; g)$, with the growth rate parameter g differentiating the subpopulations. The subpopulations are then be combined by integrating the densities with respect to a measure on the space of parameters:

$$u(t, x) = \int_G v(t, x; g)\, dP(g).$$

The measure P, called the growth rate distribution, represents the proportion of individuals having a given growth rate.

In this paper, we consider the more general problem of a population whose dynamics are modeled with a parameter-dependent C_0 semigroup $T(t; q)$ on a Hilbert space X. These operators typically arise as solution operators for a differential equation such as the Sinko-Streifer equation given above. This general approach allows us to consider several different kinds of models from many differential equations – age, size, spatial structure models, or a combination. Within the context of population models, the parameter q typically comprises several individual rate parameters, such as growth, mortality, and fecundity. The rate distribution idea then is to model the population as a combination of subpopulations which are modeled with the original semigroup dynamics $T(t; q)$. The question then becomes how to build the whole population from the subpopulations.

Suppose for the moment that the number of subpopulations is finite. Then, we denote the densities by $v(t, x; q_i) = T(t, q_i)\varphi_i(x)$, for $1 \leq i \leq n$. Then the density $u(t, x)$ of the whole population is given by

$$u(t, x) = \sum_{i=1}^{n} T(t, q_i)\varphi_i(x). \tag{1.1}$$

In [BF], the initial densities were assumed to satisfy

$$\varphi_i = p_i \varphi, \tag{1.2}$$

for some $\varphi \in X$, and some collection $\{p_i\}$ with $p_i \geq 0$, and $\sum p_i = 1$. Then the population density can be given in the form

$$u(t, x) = \sum_{i=1}^{n} T(t, q_i)\varphi(x)p_i = \int_Q T(t; q)\varphi(x)\, dP(q),$$

where P is the discrete probability measure on Q with support $\{q_i\}$ and weights $\{p_i\}$. The integral form is generalized naturally to the case mentioned above in which P is any probability measure on Q, and the estimation theory of [BF] generalizes in a straightforward manner for this problem.

To generalize this model further, we relax (1.2), so that the subpopulations may have different initial structure as well as distinct rate parameters. Thus we are led to a generalization of the form

$$u(t) = \int_Q T(t;q)\varphi(q)\,dP(q), \tag{1.3}$$

where the X valued function $\varphi(q)$ gives the parameter dependent initial structure. A further step is to write

$$u(t) = \int_Q T(t;q)\,d\mathbf{m}(q), \tag{1.4}$$

where $\mathbf{m}$ is a vector-valued measure on Q (taking values in X). Intuitively, we have $\mathbf{m}(dq) = \varphi(q)\,P(dq)$. Also, when $X = L^2$, say, $\int_a^b \mathbf{m}(A)(x)\,dx$ denotes the number of individuals whose structure variable x is between a and b, and whose parameter q lies in the set $A \subset Q$. Note that formula (1.4) above involves integrating an operator valued function with respect to a vector measure. There are some subtle questions of measurability in both (1.3) and (1.4) that must be resolved from properties of the original semigroup model (see [F]).

The advantage to using the form (1.4) over (1.3) is not in a greater level of generality, for under rather general conditions a vector measure can be expressed in terms of a vector valued density and a scalar measure. However, as we shall see below, the vector measure approach leads directly to a semigroup formulation of the dynamics in a vector measure space, and hence we may view the state as possessing not only the original (size, age, or spatial) structure but also the individual (rate) structure.

We shall consider the population state at time t to be $\mathbf{m}_t$, which can be given in mild form at

$$\mathbf{m}_t = \mathcal{T}(t)\mathbf{m}_0 + \int_0^t \mathcal{T}(t-s)\mathbf{f}(s)\,ds, \tag{1.5}$$

where $\mathbf{f}$ is a "forcing function" modeling changes in parameter structure due to external environmental changes. Below we shall derive the semigroup $\mathcal{T}$ from $T(t;q)$: the idea is that $\mathcal{T}(t)\mathbf{m}_0(dq) = T(t;q)\varphi(q)\,\mu(dq)$, where we have $\mathbf{m}_0(dq) = \varphi(q)\,\mu(dq)$.

The inverse problem is to estimate the measure $\mathbf{m}_0$ and the forcing function $\mathbf{f}$ from observations of the population. In size structured populations, one typically obtains "histogram" data which represents numbers of individuals whose size lies in given "size bins." We assume that the observations are of the form $\{\widehat{u}(t_i)\colon 1 \leq$

$i \le n\} \subset Z$, where the data space Z is a Hilbert space, and we use the least squares criterion

$$J(\mathbf{m}_0, \mathbf{f}) = \sum_{k=1}^{n} \|\widehat{u}(t_k) - \mathcal{C}\mathbf{m}_t(Q)\|_Z^2,$$

where $\|\cdot\|_Z$ denotes the norm in Z, and where $\mathcal{C}\colon X \to Z$ is the observation mapping (e.g., a projection onto the step functions in the histogram case). Computation of solutions to this inverse problem requires several approximations – for the semigroup, the vector integral in (1.5), and the vector measure itself. It does, however, provide an advantage over identification of the parameter q: the least squares problem for the measure is a linear least squares problem. A very similar idea, relaxed control, has been used in control theory for quite some time (see, e.g. [W]).

A computational advantage in this approach is that the problem contains a high level of parallelism. Different subpopulations can be simulated independently; hence, parallel computing platforms can be used very efficiently.

This paper is organized as follows. In Section 2, we discuss the functional analytic tools necessary to put (1.5) on a rigorous foundation. Section 3 contains the approximation methods for inverse problems, and some remarks on future work are contained in Section 4.

2. Semigroup Formulation of the Model

Our goal in this section is to extend the dynamics of a population comprising identical individuals to a heterogeneous population, distributed not only in its original structure but also in its rate structure. We begin with the original dynamics,

$$\dot{v}(t;q) = A(q)v(t;q), \tag{2.1}$$

where A is the infinitestimal generator of a C_0 semigroup $T(t;q)$ on a Hilbert space X. In the extension we are seeking, the solutions $v(t;q)$ represent the subpopulation density having parameter q. Of course, when we refer to this function of q as a density, we need a measure with respect to which v is a density. We also need an extended state space which incorporates the additional structure.

One natural approach to a state space formulation is to take the initial population to be an X-valued measure. The space $\mathcal{M}$ of regular, countably additive X-valued Borel measures of bounded variation on Q forms a Banach space which can be identified with $C(Q, X)^*$. We include here some definitions and notation which will be very useful below. The interested reader may consult [DU] or [D] for the details of vector measures. First, we recall that a *countably additive vector measure* (or more concisely, a *vector measure*) is a function from a σ-algebra $\mathcal{F}$ of sets in Q to X satisfying

$$\mathbf{m}\Big(\bigcup_{i=1}^{\infty} E_i\Big) = \sum_{i=1}^{\infty} \mathbf{m}(E_i)$$

for every collection $\{E_\imath\} \subset \mathcal{F}$ of disjoint sets. The limit in the infinite sum holds in the X norm sense. We define the *variation of* $\mathbf{m}$, $|\mathbf{m}|$, by

$$|\mathbf{m}|(E) = \sup_{\pi \in \Pi} \sum_{E_\imath \in \pi} \|\mathbf{m}(E_\imath)\|_X,$$

where Π is the collection of finite partitions of E. If $|\mathbf{m}|(Q) < \infty$, we say that $\mathbf{m}$ is *of bounded variation*. The *semivariation* is defined by

$$\|\mathbf{m}\|(E) = \sup\Big\{|\langle x, \mathbf{m}(E)\rangle| : x \in X, \|x\|_X = 1\Big\},$$

where $\langle \cdot, \cdot \rangle$ denotes the inner product in X. We say that $\mathbf{m}$ is *of bounded semivariation* if $\|\mathbf{m}\|(Q) < \infty$. A vector measure is of bounded semivariation if and only if its range is bounded in X (p. 5 of [DU]) so that we refer to measures of bounded semivariation simply as *bounded measures*. Unless otherwise stated, integrals of vector valued functions are taken to be Bochner integrals (see [DU, Chapter 2]).

Measures of bounded variation can also be expressed as $\mathbf{m}(dq) = \varphi(q)\,\mu(dq)$, where $\varphi \in L^1(Q, \mu; X)$ and μ is a positive real-valued measure on Q, such that $0 \le \mu(A) \le \|\mathbf{m}\|(A)$, for each measurable A. This fact follows from a theorem of Bartle, Dunford, and Schwarz and the fact that Hilbert spaces have the Radon-Nikodym property (see [DU, pp. 14, 61, and 100]).

Another state space possibility is given by $C(Q; X) \times M$, where $M(= C(Q)^*)$ denotes the finite real-valued Borel measures on Q, for Q a compact separable metric space. This approach has been successfully employed in [BKW] within the context of model development.

Here we shall focus on the vector measure approach, which has proven effective in inverse problems (see [F] and the following sections). We begin by extending the original semigroup to the vector measure space. The following result is the first necessary step.

Lemma 2.1. *Suppose that X is a Hilbert space and that $T(t;q)$ is a semigroup on X that is strongly jointly continuous on $[0,\infty) \times Q$, where Q is a compact, separable metric space. Suppose that μ is a finite regular nonnegative measure on (the Borel sets of) Q and that $\varphi \in L^1(Q, \mu; X)$. Then, for each t, the map $q \mapsto T(t,q)\varphi(q)$ is in $L^1(Q, \mu; X)$, and the vector measure $\mathbf{n}$ given by $\mathbf{n}(A) = \int_A T(t,q)\varphi(q)\,\mu(dq)$ is in $\mathcal{M}$.*

Proof. First, we note that if the above map is measurable, then we have that

$$\begin{aligned}\int_Q \|T(t;q)\varphi(q)\|_X\,\mu(dq) &\le \int_Q \|T(t;q)\|\,\|\varphi(q)\|_X\,\mu(dq)\\ &\le Me^{\omega t}\int_Q \|\varphi(q)\|_X\,\mu(dq) < \infty,\end{aligned}$$

since T is jointly strongly continous (see [DU, F]). Thus, to prove that the above map is L^1, it remains to show measurability.

We recall that an X-valued function is measurable if and only if it is the (μ-a.e.) pointwise strong limit of simple functions (see [DU]). Given that $\varphi \in L^1(Q, \mu; X)$, we may choose a sequence of simple functions $\varphi^n \to \varphi$ strongly, for μ-a.e. $q \in Q$. Next, for each n, we choose a finite collection of balls of radius $1/n$, $B(q_i^n, 1/n)$, for $1 \leq i \leq K_n$ that cover Q. We redefine these balls in order to make them disjoint, by setting $B_1^n = B(q_1^n, 1/n)$, and letting

$$B_k^n = B(q_k^n, 1/n) \cap (B_1^n)^c \cap \dots (B_{K_n}^n)^c.$$

We then set

$$\psi^n(q) = \sum_{k=1}^{K_n} \sum_{i=1}^{N_n} T(t, q_k^n) \phi_i^n \chi_{A_i^n \cap B_k^n}, \tag{2.2}$$

where $\varphi = \sum_{i=1}^{N_n} \phi_i^n \chi_{A_i^n}$. Clearly, ψ^n is a simple function. Moreover, if q_0 is a point for which $\varphi^n(q_0) \to \varphi(q)$, then there is a sequence of sets $A_{i_n}^n \cap B_{k_n}^n$ that contain q_0 and whose radii are tending to 0. By the strong continuity of T, together with the convergence of φ^n, we have that $T(t; q_{k_n}^n)\varphi^n(q^0) \to T(t; q_0)\varphi(q_0)$, as desired.

For the last claim, we have that $\mathbf{n}(A) = \int_A T(t, q)\varphi(q)\, \mu(dq)$ is a countably additive, bounded variation vector measure (from [DU, p. 46]) and that in fact

$$|\mathbf{n}|(A) = \int_A \|T(t, q)\varphi(q)\|_X \, \mu(dq).$$

To prove regularity of $\mathbf{n}$, we must show that for every measurable set A and for every $\varepsilon > 0$, there exist sets K, compact, and O, open satisfying $K \subset A \subset O$, and $|\mathbf{n}|(O - K) < \varepsilon$.

Note that if $\varepsilon > 0$, since $q \mapsto T(t, q)\varphi(q)$ is in $L^1(Q, \mu; X)$, there exists $\delta > 0$ such that if B is measurable and $\mu(B) < \delta$, then $|\mathbf{n}|(B) < \varepsilon$. The regularity of μ then provides the result.

Using this lemma, we define the operators $\mathcal{T}(t)$ on $\mathcal{M}$ by setting

$$\mathcal{T}(t)\mathbf{m}(dq) = T(t; q)\varphi(q)\mu(dq). \tag{2.3}$$

To prove that this definition gives us a semigroup on $\mathcal{M}$, we must first show that $\mathcal{T}(t)$ is well defined. That is, suppose $d\mathbf{m} = \varphi \, d\mu = \psi \, d\nu$. (In fact, many such distinct representations exist: see [DU, p. 269, Corollary 3]). We must show that both representations lead to the same linear operator; i.e.,

$$T(t; q)\varphi(q)\mu(dq) = T(t; q)\psi(q)\nu(dq).$$

Toward that end, we take B_k^n and q_k^N as in the above proof, and we define a sequence of operators

$$T^n(t; q)x = \sum_{k=1}^{K_n} \chi_{B_k^n}(q) T(t; q_k^N)x,$$

which converges strongly to $T(t;q)$. In a manner similar to (2.2), we set

$$\mathbf{n}_1^n(A) = \sum_{k=1}^{K_n} \int_{A\cap B_k^n} T(t;q_k^N)\varphi\,\mu(dq) = \sum_{k=1}^{K_n} T(t;q_k^N) \int_{A\cap B_k^n} \varphi\,\mu(dq),$$

and

$$\mathbf{n}_2^n(A) = \sum_{k=1}^{K_n} \int_{A\cap B_k^n} T(t;q_k^N)\psi\,\nu(dq) = \sum_{k=1}^{K_n} T(t;q_k^N) \int_{A\cap B_k^n} \psi\,\nu(dq).$$

From the hypothesis on φ, ψ, μ, and ν, we see that $\mathbf{n}_1^n = \mathbf{n}_2^n$. Moreover, by the dominated convergence theorem, we have that $\mathbf{n}_i^n(A) \to \mathbf{n}_i(A)$ for $i = 1, 2$, where

$$\mathbf{n}_1(A) = \int_A T(t,q)\varphi(q)\,\mu(dq)$$

and $\mathbf{n}_2$ is similarly defined. Since $\mathbf{n}_1^n = \mathbf{n}_2^n$, we must have $\mathbf{n}_1 = \mathbf{n}_2$, which implies that $\mathcal{T}(t)$ is well defined.

We also note that since

$$|\mathbf{n}|(A) = \int_A \|T(t,q)\varphi(q)\|_X\,\mu(dq) \le Me^{\omega t}|\mathbf{m}|(A),$$

$\mathcal{T}(t)$ is a bounded linear operator on $\mathcal{M}$. Using these facts we have the following.

Theorem 2.2. *Under the hypotheses of Lemma 2.1, we have that* $\mathcal{T}(t), t \ge 0$ *is a strongly continuous semigroup of linear operators on* $\mathcal{M}$.

Proof. The semigroup property is easily seen to follow from (2.3), as is the fact that $\mathcal{T}(0) = I$. We must now argue strong continuity. Note that

$$|\mathcal{T}(t)\mathbf{m} - \mathbf{m}|(Q) = \int_Q \|T(t,q)\varphi(q) - \varphi(q)\|_X\,\mu(dq),$$

which goes to 0 as t goes to 0, by the dominated convergence theorem. Thus strong continuity is proved.

From this theorem we obtain a model for dynamics in the vector measure state space, $\mathcal{M}$, through the mild form given in (1.5):

$$\mathbf{m}_t = \mathcal{T}(t)\mathbf{m}_0 + \int_0^t \mathcal{T}(t-s)\mathbf{f}(s)\,ds,$$

where $\mathbf{f} \in L^1([0,t_f];\mathcal{M})$. The state vector now is $\mathbf{m}_t$ which contains structure in both the parameter space and the original space X. Moreover, the mild form also contains a forcing function which can be used to model externally induced parameter changes. Having developed a semigroup formulation for the extended dynamics, we now turn to inverse problems.

3. Inverse Problems for the Rate Distribution Model

It is typically the case in a population biology application that the individual rate constants cannot be measured directly. Thus, in population wide observations, one would be unlikely to have access to measurements of parameter variation in the population. On the other hand, size and spatial structure are observable in many cases. Thus, we are confronted with an inverse problem in which the observations $\hat{u}(t_i) \in Z$ correspond to $\mathcal{C}\big(\int_Q T(t_i;q)\varphi(q)\,\mu(dq)\big)$, where $\mathcal{C}\colon X \to Z$ is an observation operator and the data space Z is a separable Hilbert space. In many population biology examples, the data are histograms giving the number of individuals whose size (age, position) lies in particular intervals. In such a case Z is finite dimensional and $\mathcal{C}$ is an integral "averaging" operator. The integral in q denotes the summing of the population over the parameter space: we observe the total population, structured only in X. We thus pose the inverse problem as the determination of $\mathbf{m}$ from such measurements. We shall use the least squares approach, in which we seek to minimize

$$J(\mathbf{m}) = \sum_{i=1}^{n} \|\mathcal{C}\mathcal{T}(t_i)\mathbf{m}(Q) - \hat{u}(t_i)\|_Z, \tag{3.1}$$

which, we note, is an infinite dimensional linear least squares problem.

A possible approach to the minimization is through standard linear least squares theory and the normal equations $B^*B\mathbf{m} = \hat{u}$, with $B = (\mathcal{T}(t_1), \dots, \mathcal{T}(t_n))^T \in \mathcal{L}(X^n)$, and $\hat{u} = (\hat{u}(t_1), \dots, \hat{u}(t_n))^T \in X^n$. We do not have at this point any general conditions on the original semigroup under which the normal equations have a unique solution. Hence, we take here a different approach.

We treat the problem as a constrained minimization of J over a set $\mathcal{M}_{ad}$ of admissible parameters. Since the unit ball in $C(Q,X)^*$ is weak-* compact (by Alaoglu's theorem), we take $\mathcal{M}_{ad} \subset \mathcal{M}$ to be the measures with variation bounded by M, which is a fixed positive number. Under the assumption that the adjoint semigroup, $T^*(t;q)$, is strongly continuous in t and q, we have that $\mathbf{m}^N \to \mathbf{m}$ weak-* implies that $\mathcal{T}(t)\mathbf{m}^N(Q) \to \mathcal{T}(t)\mathbf{m}(Q)$ weakly in X, for

$$\begin{aligned}\langle x, \int_Q T(t;q)\varphi(q)\,\mu(dq)\rangle &= \int_Q \langle x, T(t;q)\varphi(q)\rangle\,\mu(dq) \\ &= \int_Q \langle T^*(t;q)x, \varphi(q)\rangle\,\mu(dq).\end{aligned}$$

Since T^* is strongly continuous, $q \mapsto T^*(t,q)x \in C(Q,X)$, and the weak convergence becomes apparent. From this convergence we obtain the following.

Theorem 3.1. *Suppose that $\mathcal{C} \in \mathcal{L}(X,Z)$ is compact. Under the assumptions of Theorem 2.2, together with strong continuity of the adjoint semigroup $T^*(t;q)$, the functional J is continuous with respect to the weak-* topology on $\mathcal{M}_{ad}$; hence, J attains a minimum over $\mathcal{M}_{ad}$.*

Proof. First, note that the weak-* topology on $\mathcal{M}_{ad}$ and the weak topology of $\mathcal{T}(t)(\mathcal{M}_{ad}) \subset X$ are metrizable (see, e.g., [R, p. 203]). Thus, the above argument shows that $\mathbf{m} \mapsto \mathcal{T}(t)\mathbf{m}$ is continuous with respect to these topologies. Since $\mathcal{C}$ is compact, it maps weakly convergent sequences to strongly convergent sequences, and we have the desired result.

The optimization problem posed is an infinite dimensional one, so we need some approximation methods for compuational purposes. We assume that we have for the original problem a numerical scheme that produces a sequence of semigroups $T^N(t;q)$ defined on finite dimensional subspaces X^N of X, taken to be of the form $X^N = span\{\psi_1^N, \ldots, \psi_N^N\}$, with P^N being the orthogonal projection from X to X^N. From these semigroups we construct cost functionals

$$J^N(\mathbf{m}) = \sum_{k=1}^{n} \|\hat{u}(t_k) - \mathcal{C}\mathcal{T}^N(t)\mathbf{m}(Q)\|_Z^2,$$

where $\mathcal{T}^N$ is defined in the obvious manner. This cost functional must be minimized over a finite dimensional set. Let $\{q^1, q^2, \ldots\}$ be a countable, dense subset of Q. We set

$$\mathcal{M}^N = \{\sum_{j,k=1}^{N} a_{jk}\psi_j^N \delta_{q^k}(dq) \; : \; |\sum_{j,k=1}^{N} a_{jk}\psi_j^N \delta_{q^k}|(Q) \le M\},$$

which is a finite dimensional subset of $\mathcal{M}_{ad}$. Furthermore (as in [F]), any element of $\mathcal{M}_{ad}$ can be approximated in the weak-* sense by a sequence from these sets. With these observations, we may obtain the following approximation result.

Theorem 3.2. *Assume that the hypotheses of Theorem 3.2 hold. Furthermore let the sequence $T^N(t;q)$ satisfy $T^N(t;q)P^N x \to T(t;q)x$, and $(T^N)^*(t;q)P^N x \to T^*(t;q)x$ strongly in X and uniformly on $Q \times [0,\tau]$. Then minimizers of J^N over $\mathcal{M}^N$ converge subsequentially in the weak-* topology of $\mathcal{M}_{ad}$ to minimizers of J.*

Proof. In view of the abstract least squares theory of Banks (see, e.g., [BK, pp. 144-5]) it only remains to show that $J^N(\mathbf{m}^N) \to J(\mathbf{m})$, whenever $\mathbf{m}^N \to \mathbf{m}$ in the weak-* sense. Toward that end, we note that

$$\begin{aligned}\langle x, \int_Q T^N(t;q)P^N\varphi^N(q)\,\mu^N(dq)\rangle &= \int_Q \langle x, T^N(t;q)P^N\varphi^N(q)\rangle\,\mu^N(dq) \\ &= \int_Q \langle P^N x, T^N(t;q)P^N\varphi^N(q)\rangle\,\mu^N(dq) \\ &\quad - \int_Q \langle x, T(t;q)\varphi^N(q)\rangle\,\mu^N(dq) \\ &\quad + \int_Q \langle x, T(t;q)\varphi^N(q)\rangle\,\mu^N(dq)\end{aligned}$$

$$= \int_Q \langle (T^N(t,q))^* P^N x, \varphi^N(q)\rangle \, \mu^N(dq)$$
$$- \int_Q \langle T^*(t,q)x, \varphi^N(q)\rangle \, \mu^N(dq)$$
$$+ \langle x, \int_Q T(t,q)\varphi^N(q) \, \mu^N(dq)\rangle$$
$$\to \langle x, \int_Q T(t,q)\varphi(q) \, \mu(dq)\rangle,$$

which implies $\mathcal{T}^N(t)\mathbf{m}^N \to \mathcal{T}(t)\mathbf{m}$ weakly in X As in Theorem 3 1, we then obtain the desired convergence for J^N

4. Conclusions

We have examined here a semigroup formulation of a rate distribution model for structured populations Being based merely on an original semigroup formulation for subpopulations of identical individuals, this model is quite general Using basic properties of vector measures, we studied the basic well-posed question for the semigroup formulation of the distributed model, and developed an estimation framework for fitting the model to data

We have performed a variety of computational examples based on the Sinko-Streifer problem mentioned above The implementation involves solving the Sinko-Streifer PDE for many growth rate functions, and we have successfully used the 128-processor Intel iPSC/860 at Oak Ridge National Laboratory for the computations These results, reported in [BFZ], demonstrate not only the computational feasibility of the techniques we have described, but also the capability of obtaining high quality fits to real observed size structured data that were not previously treatable with inverse techniques

Future studies will focus on vector measure formulations for nonlinear problems (such as [H]) as well as models which incorporate subpopulation mixing (as in [BKW])

References

[BBKW] Banks, H T , L Botsford F Kappel and C Wang Modeling and Estimation in Size Structured Population Models *Proceedings 2nd Course on Mathematical Ecology,* (Trieste December 8 12, 1986), World Press Singapore (1988), 521-541

[BF] Banks, H T , and B G Fitzpatrick Estimation of Growth Rate Distributions in Size Structured Population Models *Q Applied Math,* **49**, no 2, 1991, pp 215 235

[BFZ] Banks, H T , B G Fitzpatrick and Y X Zhang 'A Parallel Algorithm for Rate Distribution Estimation in Size Structured Population Models," to appear

[BKW] Banks, H T , F Kappel, and C Wang "Semigroup formulation of a heterogeneous population model and its nonlinear perturbations," to appear

[BK] Banks, H T , and K Kunisch "Estimation Techniques for Distributed Parameter Systems " Birkhauser, Boston, 1989

[DU] Diestel, J , and J J Uhl, Jr "Vector Measures," Mathematical Surveys **15**, American Mathematical Society, Providence, 1977

[D] Dinculeanu, N "Vector Measures," Permagon Press, Oxford, 1967

[DS] Dunford, N , and J T Schwartz "Linear Operators, Part I General Theory," Wiley Interscience, New York, 1966

[F] Fitzpatrick, B G Modeling and Estimation Problems for Structured Heterogeneous Populations, *J Math Anal Appl*, **172** no 1, 1993, pp 73-91

[H] Huyer, W "A Size Structured Population Model with Dispersion," *J Math Anal Appl*, to appear

[W] Warga, J "Optimal Control of Differential and Functional Equations," Academic Press, New York, 1972

Ben G Fitzpatrick
Center for Research in Scientific Computation
Department of Mathematics
North Carolina State University
Raleigh, NC 27695-8205, USA

A MODEL FOR A TWO-LAYERED PLATE WITH INTERFACIAL SLIP

SCOTT W HANSEN

Department of Mathematics
Iowa State University

ABSTRACT In this paper we derive a model for a two-layered plate in which slip can occur at the interface We assume that a "glue" layer of negligible thickness bonds the two adjoining surfaces in such a way that the restoring force created by the glue is proportional to the amount of slippage Within each plate the assumptions of Timoshenko beam theory (namely, that the straight filaments orthogonal to each center sheet at equilibrium remain straight during deformation) are applied and the equations of motion are derived through the principle of virtual work We relate the resulting system to the Mindlin-Timoshenko-Reissner plate system and also to the Kirchhoff plate system by singular perturbations involving passing the shear stiffness parameter and the glue strength parameter to infinity

1991 *Mathematics Subject Classification* 73K10, 73K20

Key words and phrases Multi-layer plate, Mindlin plate, Reissner plate

1. Introduction

Over the past decades, composite materials have found increasingly broad application in many areas from design of sporting goods to space structures. These materials can be designed to produce either favorable damping characteristics or high strength to mass ratios, and often for a combination of each. Consequently there has been a considerable amount of effort in modelling these structures and analyzing the dynamical properties thereof.

Models for multilayered beams and plates have existed in the literature since at least the late fifties [Yu], [Ke]. Since then, numerous more elaborate models have been developed (see [He], [Di], [Me], [YD] and the numerous references therein). For those models which were constructed with the idea of modelling damping the usual approach has been to assume the material consists of three separate layers which are bonded together (no-slip at the interfaces). The outer layers are modelled by plates which allow little or no shear while the inner layer is modelled as a material with a "complex shear modulus" but usually without other material assumptions. Kinematic restrictions are then artificially imposed which couple the strains or stresses in each layer with one another and a plate model involving only the transverse displacement is produced.

In this paper we derive a model for a two-layered plate in which slip can occur at the interface. We imagine that a "glue" layer of negligible thickness bonds the two adjoining surfaces in such a way that the restoring force created by the glue is proportional to the amount of slippage. Within each plate the assumptions of Timoshenko beam theory [Ti] (namely, that the straight filaments orthogonal to each center sheet at equilibrium remain straight during deformation) are applied and the equations of motion are derived through the principle of virtual work.

In the approach used here, we impose no *a-priori* kinematic restrictions coupling the motion of the upper plate to the lower one. (Although we do assume within each plate standard assumptions of Timoshenko beam theory). However, at least for the present article we will assume a geometric symmetry in that we consider only the case where the upper and lower plates are identical. This allows for a decoupling of the in-plane motions from the bending motions.

Initially we obtain a system of nine equations which describe the dynamics of the composite plate. The part describing the bending has five equations in the transverse displacement, the *effective* rotation angles and two for the components of the slip. We show that this system is an analog of the Mindlin-Timoshenko-Reissner plate system (called the Mindlin-Timoshenko plate in [LL], [La]) in the sense that 1) when there is no glue our system reduces to an MTR system, 2) when the strength of the glue tends to infinity, solutions of our system converge in a weak sense to solutions of an MTR system.

By letting the shear-stiffiness parameter tend to infinity we obtain a limiting system of three equations which is analogous to the Kirchhoff system in the same sense that our original system is analogous to the MTR system.

A major reason for modelling slip (as opposed to including a center layer for the glue) is that one can then easily introduce frictional forces into the dynamics of the slip. If one models this force as simple sliding friction (frictional force proportional the velocity) one obtains two new *structurally damped* plate models. It is known that much of the dissipation in composite materials is in fact due to friction between adjacent fibers. Thus a goal of this research is to characterize the damping one obtains in plates due to internal friction. However in the present paper we concern ourselves with the case in which no frictional forces are included.

2. Basic assumptions

Much of our development and notation will follow that of Lagnese and Lions [LL, chapters 1,2].

Our plate consists of two identical nonhomogeneous (homogeneous in the transverse direction), isotropic thin plates which are in contact with each other on a *middle surface* which occupies the region $\Omega \subset \mathbb{R}^2$ at equilibrium. Each plate is assumed to have uniform thickness $h/2$.

Let us use the rectangular coordinates $\underline{x} = (x_1, x_2)$ to denote points in Ω and $x = (\underline{x}, x_3) = (x_1, x_2, x_3)$ to denote points in $Q^- = \Omega \times (-h/2, 0)$ or $Q^+ = \Omega \times (0, h/2)$. For $x \in Q^+ \cup Q^-$ let $U(x) = (U_1, U_2, U_3)(x)$ denote the displacement vector of the point which, when the plate in equilibrium has coordinates $x = (x_1, x_2, x_3)$. (In the dynamic case these variables also depend upon time, but we will suppress all time dependence in this section.) In addition let us define u^+ and u^- by

$$u^{\pm}(\underline{x}) = \lim_{x_3 \to 0^{\pm}} U(\underline{x}, x_3) \quad \forall \underline{x} \in \Omega.$$

2.1. Stress-strain relations. Let $\sigma_{ij}, \epsilon_{ij}$ $(i, j = 1, 2, 3)$ denote the stress and strain tensors, respectively. For a small displacement theory we assume

$$\epsilon_{ij} = \frac{1}{2}\left(\frac{\partial U_i}{\partial x_j} + \frac{\partial U_j}{\partial x_i}\right). \tag{2.1}$$

Let $E = E(\underline{x}) > E_0 > 0$ denote Young's modulus and $\mu(\underline{x})$ denote Poisson's ratio, $0 < \mu_0 < \mu < \mu_1 < 1/2$. Since the plates are assumed to be isotropic we have the following stress-strain relations:

$$\sigma_{ij} = \frac{E}{1+\mu}\left(\epsilon_{ij} + \frac{\mu}{1-2\mu}\epsilon_{kk}\delta_{ij}\right), \tag{2.2}$$

where we have used the summation convention for repeated indices.

As in [LL], we assume that σ_{33} is negligible, hence to highest order one can solve for ϵ_{33} in terms of the other principle strains:

$$\epsilon_{33} = \frac{-\mu}{1-\mu}(\epsilon_{11} + \epsilon_{22}). \tag{2.3}$$

From (2.2) and (2.3) we have

$$\begin{cases} \sigma_{11} &= \frac{E}{1-\mu^2}(\epsilon_{11} + \mu\epsilon_{22}) \\ \sigma_{22} &= \frac{E}{1-\mu^2}(\mu\epsilon_{11} + \epsilon_{22}) \\ \sigma_{33} &= 0 \\ \sigma_{ij} &= \frac{E}{1+\mu}\epsilon_{ij}, i \neq j. \end{cases} \tag{2.4}$$

However, in anticipation of the fact that we will model each plate according to the assumptions of Timoshenko beam theory, a *shear correction* coefficient k is incorporated into (2.4) to account for the fact that (physically) the transverse shear strains do not remain constant throughout the thickness of each plate. (In the Timoshenko theory the shear strains are assumed to be constant throughout the thickness.) Thus the shear stresses in (2.4) are modified:

$$\sigma_{13} = \frac{kE}{1+\mu}\epsilon_{13}, \quad \sigma_{23} = \frac{kE}{1+\mu}\epsilon_{23}. \tag{2.5}$$

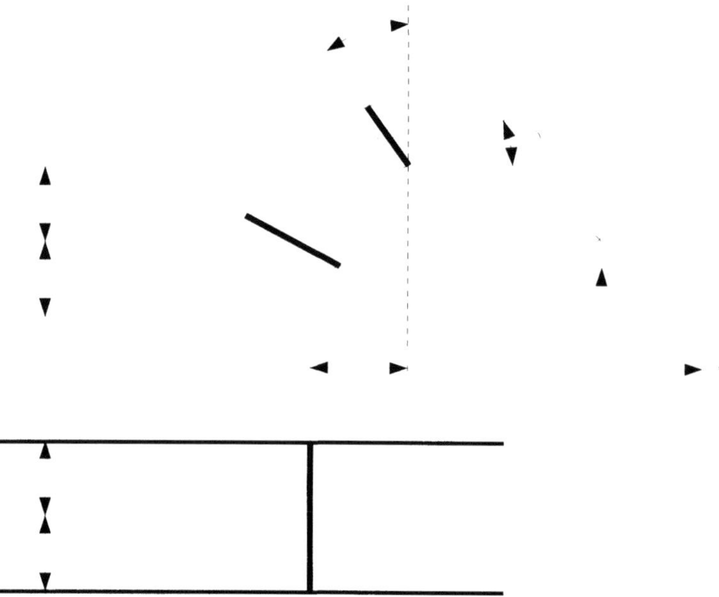

FIG. 1. A schematic of the x_i-x_3 cross section of the undeformed filament A-B-C-D (bottom) located at $(\underline{x}, \cdot)$ and the corresponding deformed filaments A'-B', C'-D' (top) whose positions are described by $\psi_i^{\pm}(\underline{x}), \theta_i^{\pm}(\underline{x}), u_i^{\pm}(\underline{x})$.

2.2. Plate assumptions. The MTR plate model [LL, p.14] is the simplest plate model which incorporates the effects of shear as an independent quantity. In this model straight filaments originally orthogonal to a "middle surface" remain straight after deformation, however are free to rotate relative to the deformed "middle surface". Thus the MTR system is described by three equations; one for the dynamics of the transverse deflection and two for the shear angles. If in addition, one allows for in-plane translations, one obtains an additional two equations for the dynamics of these motions.

Both the upper and lower plates of our composite plate will be modelled by MTR plates which include the in-plane translations. Thus within each plate, the straight filaments are allowed both rotate and translate so that each filament in each plate has (for the moment) five degrees of freedom; three translational and two rotational. Thus the displacements of the filament $\mathcal{F} = \mathcal{F}^+ \cup \mathcal{F}^-$ which occupies $\{(x_1, x_2, x_3) : -h/2 < x_3 < h/2\}$ at equilibrium is completely determined by specifying the translational displacements $u^{\pm}(\underline{x})$ and a rotation angle for $\mathcal{F}^+$ and $\mathcal{F}^-$. ($\mathcal{F}^+$ belongs to the "upper plate" and $\mathcal{F}^-$ is in the "lower plate".)

Rather than attempt to define shear angles relative to a "middle surface" of each plate, it will be more convenient to define the angles relative to the surfaces $\mathcal{S}^{\pm} = \{(\underline{x}, 0) + u^{\pm}(\underline{x}) : \underline{x} \in \Omega\}$. Let $\varphi^{\pm}(\underline{x}) = (\varphi_1^{\pm}, \varphi_2^{\pm})(\underline{x})$ denote shear angles; that is, $\varphi_i^\pm(\underline{x})$ is the angle, measured with positive orientation that the filament $\mathcal{F}^\pm$ has rotated relative to the normal to $\mathcal{S}^\pm$ at $(\underline{x},0)+u^\pm(\underline{x})$. Let $\theta^\pm(\underline{x}) = (\theta_1^\pm,\theta_2^\pm)(\underline{x})$ denote the rotation angles from the equilibrium position, with positive orientation, of $\mathcal{S}^\pm$ at $(\underline{x},0)+u^\pm(\underline{x})$. Let $\psi^\pm(\underline{x}) = (\psi_1^\pm,\psi_2^\pm)(\underline{x})$ denote the total rotation angle the filament $\mathcal{F}^\pm$ has rotated with respect to the x_3 axis, again, with positive orientation. (See Fig. 1.) The angles $\theta^\pm$, $\psi^\pm$ and $\varphi^\pm$ are thus related by

$$\psi^\pm(\underline{x}) = \theta^\pm(\underline{x}) + \varphi^\pm(\underline{x}) \quad \forall x \in \Omega. \tag{2.6}$$

Under the assumption that the rotation angles of the middle surface are small we may use the approximation

$$\theta^\pm(\underline{x}) \simeq \nabla u_3^\pm(\underline{x}) \quad \forall \underline{x} \in \Omega. \tag{2.7}$$

Then the displacements can be written in terms of the translations $u^\pm$ and total rotation angles $\psi^\pm$ as

$$\begin{cases} U_1(x_1,x_2,x_3) &= \begin{cases} u_1^+(x_1,x_2) - x_3\psi_1^+(x_1,x_2) & x_3>0 \\ u_1^-(x_1,x_2) - x_3\psi_1^-(x_1,x_2) & x_3<0 \end{cases} \\ U_2(x_1,x_2,x_3) &= \begin{cases} u_2^+(x_1,x_2) - x_3\psi_2^+(x_1,x_2) & x_3>0 \\ u_2^-(x_1,x_2) - x_3\psi_2^-(x_1,x_2) & x_3>0 \end{cases} \\ U_3(x_1,x_2,x_3) &= \begin{cases} u_3^+(x_1,x_2) & x_3>0 \\ u_3^-(x_1,x_2) & x_3<0. \end{cases} \end{cases} \tag{2.8}$$

Substituting (2.8) into (2.1), and using (2.6), (2.7) gives

$$\begin{cases} \epsilon_{11} &= \dfrac{\partial u_1^\pm}{\partial x_1} - x_3\dfrac{\partial \psi^\pm}{\partial x_1} \qquad \epsilon_{22} = \dfrac{\partial u_2^\pm}{\partial x_2} - x_3\dfrac{\partial \psi_2^\pm}{\partial x_2} \\ \epsilon_{12} &= \frac{1}{2}\left[\dfrac{\partial u_1^\pm}{\partial x_2} + \dfrac{\partial u_2^\pm}{\partial x_1} - x_3\left(\dfrac{\partial \psi_1^\pm}{\partial x_2} + \dfrac{\partial \psi_2^\pm}{\partial x_1}\right)\right] \\ \epsilon_{13} &= \frac{1}{2}(-\varphi_1^\pm) \qquad \epsilon_{23} = \frac{1}{2}(-\varphi_2^\pm), \end{cases} \tag{2.9}$$

where the + superscript (respectively, − subscript) applies to $x_3>0$ (respectively, $x_3<0$). Instead of using (2.3) however, we set (in accordance with [LL, p. 9])

$$\epsilon_{33} = 0. \tag{2.10}$$

One can check that (2.8)-(2.9) is the exact analogue of the strain-displacement equations which one obtains for the MTR system as in [LL, p.14]. However up to this point no assumptions have been made regarding the amount of slip between the two plates.

2.3. Interface assumptions. For $\underline{x} \in \Omega$ let us define a dimensionless slip variable $s(\underline{x}) = (s_1, s_2)(\underline{x})$ by

$$s_i(\underline{x}) = \frac{u_i^+(\underline{x}) - u_i^-(\underline{x})}{2h} \quad i = 1, 2. \tag{2.11}$$

For the moment consider the glue layer to have a uniform thickness d and that shearing motions obey Hooke's law. In this case the slip $s(\underline{x})$ is proportional to the amount of shear at $\underline{x}$ and thus integrating over Ω the product of the stress and the strain gives the following expression for the strain energy within the glue:

$$\mathcal{P}_g = \frac{h^2}{2} \int_{\Omega} \gamma(s_1^2(\underline{x}) + s_2^2(\underline{x}))d\underline{x}, \tag{2.12}$$

where $\gamma = \gamma(\underline{x}) \geq 0$ is proportional to d and the shear modulus of the glue.

Henceforth we shall regard (2.12) as the only (non-negligible) contribution by the glue to energy for the composite plate, where γ is an arbitrary non-negative essentially bounded function defined on $\bar{\Omega}$. Thus implicitly we assume that the mass and thickness of the "glue" which bonds the surfaces is negligible (compared to those of the plates) and can be ignored. While the argument used to motivate the assumption (2.12) is heuristic, (2.12) provides the simplest possible means by which the slip can be penalized.

Let

$$w(\underline{x}) = \frac{u_3^+(\underline{x}) + u_3^-(\underline{x})}{2}, \quad \theta(\underline{x}) = \frac{\theta^+(\underline{x}) + \theta^-(\underline{x})}{2}, \quad \underline{x} \epsilon \Omega.$$

To highest order, Taylor's formula gives (we will use the usual "dot product" notation):

$$u_3^{\pm}(\underline{x}) - w(\underline{x}) = \pm h \nabla w(\underline{x}) \cdot s(\underline{x}) \quad \underline{x} \epsilon \Omega. \tag{2.13}$$

To obtain a linear theory, we assume the time and spatial derivatives of the right hand side of (2.13) are negligible in comparison to those of w. Thus let us assume

$$w(\underline{x}) = u_3^+(\underline{x}) = u_3^-(\underline{x}). \tag{2.14}$$

In (2.14) we are actually making the same assumption that is made in the MTR theory, namely that the transverse displacements on the deformed filaments $\mathcal{F}$ (described earlier in this section) are independent of x_3.

2.4. Strain and kinetic energy. The strain energy for the composite plate consists of the strain energy for the glue (2.12) and the strain energy for the plates

$$\mathcal{P}^{\pm} = \frac{1}{2} \int_{Q^{\pm}} \epsilon_{ij} \sigma_{ij} d\underline{x} dx_3. \tag{2.15}$$

From (2.4) and (2.5) we can write $\mathcal{P}_{\pm}$ in terms of the strains:

$$\mathcal{P}^- + \mathcal{P}^+ = \frac{1}{2} \int\limits_{Q^+ \cup Q^-} \frac{E}{(1-\mu^2)} \Big(\epsilon_{11}^2 + 2\mu\epsilon_{11}\epsilon_{22} + \epsilon_{22}^2 + 2(1-\mu)(k\epsilon_{12}^2 + k\epsilon_{13}^2 + \epsilon_{23}^2)\Big)\, d\underline{x}dx_3. \tag{2.16}$$

For $\underline{x} \in \Omega$ define $u = (u_1, u_2)$, $\psi = (\psi_1, \psi_2)$, $\eta = (\eta_1, \eta_2)$ by

$$\begin{cases} u_i(\underline{x}) & = \frac{1}{2}(u_i^+(\underline{x}) + u_i^-(\underline{x})) \quad i = 1,2 \\ \psi(\underline{x}) & = \frac{1}{2}(\psi^+(\underline{x}) + \psi^-(\underline{x})) \\ \eta(\underline{x}) & = \frac{1}{2}(\psi^+(\underline{x}) - \psi^-(\underline{x})). \end{cases} \tag{2.17}$$

Then we have

$$\begin{cases} u_i^+(\underline{x}) & = u_i(\underline{x}) + hs_i(\underline{x}) \quad i = 1,2 \\ u_i^-(\underline{x}) & = u_i(\underline{x}) - hs_i(\underline{x}) \quad i = 1,2 \\ \psi^+(\underline{x}) & = \psi(\underline{x}) + \eta(\underline{x}) \\ \psi^-(\underline{x}) & = \psi(\underline{x}) - \eta(\underline{x}). \end{cases} \tag{2.18}$$

Substitution of (2.18), (2.9), (2.10), (2.14), into (2.16) gives an expression for $\mathcal{P}^+ + \mathcal{P}^+ -$ in terms of s, u, ψ and η. For example,

$$\begin{aligned} \int\limits_{-h/2}^{h/2} \epsilon_{11}^2 dx_3 = 2 \int\limits_0^{h/2} & \left(\frac{\partial u_1}{\partial x_1}\right)^2 + h^2 \left(\frac{\partial s_1}{\partial x_1}\right)^2 + x_3^2 \left(\left(\frac{\partial \psi_1}{\partial x_1}\right)^2 + \left(\frac{\partial \eta_1}{\partial x_1}\right)^2\right) \\ & - 2x_3 \left(\frac{\partial u_1}{\partial x_1}\frac{\partial \eta_1}{\partial x_1} + \frac{\partial \psi_1}{\partial x_1} h \frac{\partial s_1}{\partial x_1}\right) dx_3 \\ = \frac{h^3}{12} & \left[\left(\frac{\partial}{\partial x_1}(\psi_1 - 3s_1)\right)^2 + 3\left(\frac{\partial s_1}{\partial x_1}\right)^2\right] \\ & + \frac{h}{12}\left[\left(\frac{\partial}{\partial x_1}(h\eta_1 - 3u_1)\right)^2 + 3\left(\frac{\partial u_1}{\partial x_1}\right)^2\right]. \end{aligned}$$

Let

$$\xi(\underline{x}) = \psi(\underline{x}) - 3s(\underline{x}), \quad \underline{x}\epsilon\Omega \tag{2.19}$$

$$\nu(\underline{x}) = h\eta(\underline{x}) - 3u(\underline{x}), \quad \underline{x}\epsilon\Omega. \tag{2.20}$$

Also let us define an average shear angle $\varphi = (\varphi_1, \varphi_2)$ by

$$\varphi(\underline{x}) = \psi(\underline{x}) - \nabla w(\underline{x}) = \xi(\underline{x}) + 3s(\underline{x}) - \nabla w(\underline{x}). \tag{2.21}$$

When each of the terms in (2.16) are computed one finds that the potential energy of the plates splits into a *bending potential energy* and a *stretching potential energy*:

$$\mathcal{P}^- + \mathcal{P}^+ = \mathcal{P}_b + \mathcal{P}_s, \tag{2.22}$$

where

$$\begin{aligned} \mathcal{P}_b = \frac{1}{2}\int_\Omega D\Bigg[&\left(\frac{\partial \xi_i}{\partial x_i}\right)^2 + 3\left(\frac{\partial s_i}{\partial x_i}\right)^2 + 2\mu\left(\frac{\partial \xi_2}{\partial x_2}\frac{\partial \xi_1}{\partial x_1} + 3\frac{\partial s_1}{\partial x_1}\frac{\partial s_2}{\partial x_2}\right) \\ &+ \left(\frac{1-\mu}{2}\right)\left[\left(\frac{\partial \xi_1}{\partial x_2} + \frac{\partial \xi_2}{\partial x_1}\right)^2 + 3\left(\frac{\partial s_1}{\partial x_2} + \frac{\partial s_2}{\partial x_1}\right)^2\right]\Bigg] + K\varphi\cdot\varphi\, d\underline{x} \end{aligned} \tag{2.23}$$

and

$$\begin{aligned} \mathcal{P}_s = \frac{1}{2}\int_\Omega \frac{Eh}{12(1-\mu^2)}\Bigg[&\left(\frac{\partial \nu_i}{\partial x_i}\right)^2 + 3\left(\frac{\partial u_i}{\partial x_i}\right)^2 + 2\mu\left(\frac{\partial \nu_2}{\partial x_2}\frac{\partial \nu_1}{\partial x_1} + 3\frac{\partial u_1}{\partial x_1}\frac{\partial u_2}{\partial x_2}\right) \\ &+ \left(\frac{1-\mu}{2}\right)\left[\left(\frac{\partial \nu_1}{\partial x_2} + \frac{\partial \nu_2}{\partial x_1}\right)^2 + 3\left(\frac{\partial u_1}{\partial x_2} + \frac{\partial u_2}{\partial x_1}\right)^2\right]\Bigg]\, d\underline{x} \end{aligned}$$

where $D = Eh^3/12(1-\mu^2)$ is the *modulus of flexural rigidity* (see [LL, p. 10]) for a plate of thickness h, with the same material properties as the two plates (which each have thickness $h/2$), and $K = kEh/2(1+\mu)$ is the *modulus of elasticity in shear* [LL, p 14], also for the same plate, but with thickness h.

The kinetic energy $\mathcal{K} = \mathcal{K}^+ + \mathcal{K}^-$ is defined by

$$\mathcal{K}_\pm = \frac{1}{2}\int_{Q^\pm} \rho(\dot{U}_1^2 + \dot{U}_2^2 + \dot{U}_3^2)\, d\underline{x}\, dx_3,$$

where $\dot{} = d/dt$ and $\rho = \rho(\underline{x}) > \rho_0 > 0$ denotes the mass density per unit volume.

As with the potential energy, the kinetic energy decouples into a bending part and a stretching part:

$$\mathcal{K} = \mathcal{K}_b + \mathcal{K}_s. \tag{2.24}$$

One finds that

$$\mathcal{K}_b = \frac{1}{2}\int_\Omega \rho h(\dot{w})^2 + I_\rho\left(\dot{\xi}\cdot\dot{\xi} + 3\dot{s}\cdot\dot{s}\right) d\underline{x}, \tag{2.25}$$

$$\mathcal{K}_s = \frac{1}{2}\int_\Omega \frac{\rho h}{12}\left(\dot{\nu}\cdot\dot{\nu} + 3\dot{u}\cdot\dot{u}\right) d\underline{x}, \tag{2.26}$$

where $I_\rho = \rho h^3/12$ is the density function for the mass-moment of inertia about the interface.

2.5. Work. Assume that Ω has smooth boundary Γ. To set ideas we will assume that the plate is clamped on a portion of its edge $\Gamma_0 \subset \Gamma$. Furthermore denote $\Gamma_1 = \Gamma - \Gamma_0$.

Now assume the composite plate is subject to a volume distribution of forces $(\tilde{f}_1, \tilde{f}_2, \tilde{f}_3)$ and a distribution of forces $(\tilde{g}_1, \tilde{g}_2, \tilde{g}_3)$ along Γ_1. The work done on the plate by these forces is

$$\mathcal{W} = \int_{-h/2}^{h/2} \int_{\Omega} \tilde{f}_i U_i d\underline{x} dx_3 + \int_{-h/2}^{h/2} \int_{\Gamma_1} \tilde{g}_i U_i d\Gamma dx_3.$$

It will not be necessary to consider the pointwise forces, but only the resultants

$$(2.27)\quad \begin{cases} f_i(\underline{x}) = \int_{-h/2}^{h/2} \tilde{f}_i dx_3 & g_i(\underline{x}) = \int_{-h/2}^{h/2} \tilde{g}_i dx_3, \quad i = 1,2,3 \\ M_i(\underline{x}) = \int_{-h/2}^{h/2} \tilde{f}_i x_3 dx_3 & m_i(\underline{x}) = \int_{-h/2}^{h/2} \tilde{g}_i x_3 dx_3 \quad i = 1,2 \\ P_i(\underline{x}) = \int_{-h/2}^{h/2} \tilde{f}_i \text{ sgn } (x_3) dx_3 & p_i(\underline{x}) = \int_{-h/2}^{h/2} \tilde{g}_i \text{ sgn } (x_3) dx_3 \; i = 1,2 \\ Q_i(\underline{x}) = \int_{-h/2}^{h/2} \tilde{f}_i |x_3| dx_3 & q_i(x) = \int_{-h/2}^{h/2} \tilde{g}_i |x_3| dx_3 \quad i = 1,2. \end{cases}$$

As with the energy expressions, $\mathcal{W}$ decouples into two parts

$$(2.28)\qquad \mathcal{W} = \mathcal{W}_s + \mathcal{W}_b;$$

$$(2.29)\qquad \begin{aligned} \mathcal{W}_b = &\int_{\Omega} hs_1 P_1 + hs_2 P_2 - \psi_1 M_1 - \psi_2 M_2 + w f_3 d\underline{x} \\ &+ \int_{\Gamma_1} hs_1 p_1 + hs_2 p_2 - \psi_1 m_1 - \psi_2 m_2 + w g_3 d\Gamma, \end{aligned}$$

$$(2.30)\qquad \begin{aligned} \mathcal{W}_s = &\int_{\Omega} u_1 f_1 + u_2 f_2 - \eta_1 Q_1 - \eta_2 Q_2 d\underline{x} \\ &+ \int_{\Gamma_1} u_1 g_1 + u_2 g_2 - \eta_1 q_1 - \eta_2 q_2 d\Gamma. \end{aligned}$$

3. Equations of motion

The Lagrangian $\mathcal{L}$ on $(0, T)$ is defined by

$$\mathcal{L} = \int_0^T \mathcal{K}(t) + \mathcal{W}(t) - \mathcal{P}(t) dt.$$

Because of the uncoupling in (2.22), (2.24) and (2.28) of the variables (w, s, ψ) from (u, η) in the expressions for $\mathcal{K}$, $\mathcal{W}$ and P, the Lagrangian also decouples as $\mathcal{L} = \mathcal{L}_b + \mathcal{L}_s$, where

$$\mathcal{L}_b = \int_0^T \mathcal{K}_b(t) + \mathcal{W}_b(t) - \mathcal{P}_b(t) - \mathcal{P}_g(t)dt \tag{3.1}$$

$$\mathcal{L}_s = \int_0^T \mathcal{K}_s(t) + \mathcal{W}_s(t) - \mathcal{P}_s(t)dt. \tag{3.2}$$

3.1. Weak form of equations. According to the principle of virtual work, the solution trajectory is the trajectory which renders stationary the Lagrangian under all kinematically admissible displacements. Due to the decoupling of (3.1), (3.2) this will hold if and only if the variations of $\mathcal{L}_b$ and $\mathcal{L}_s$ vanish separately. Thus the equations of motion one obtains decouples into two *completely independent parts*. By setting the variation of $\mathcal{L}_s = 0$ one obtains a system of equations in the variables u and η which describe the in-plane or *stretching* motions. We mention what these are in Remark 3.3.

On the other hand, the equations one obtains from $\mathcal{L}_b$ describe *bending* motions. Since our primary motivation is to study the effect that slip has upon bending we are mainly interested in the equations which one obtains from $\mathcal{L}_b$.

It thus follows that for the purpose of calculating (ξ, s, w) we may assume *with no loss of generality* that

(H1) Solutions satisfy
   (i) $\psi^+ = \psi^- = \psi$
   (ii) $u^+ + u^- = 0$.
(H2) All terms in $\mathcal{W}_s$ vanish, i.e.
   (i) $f_1 = f_2 = g_1 = g_2 = 0$
   (ii) $Q_1 = Q_2 = q_1 = q_2 = 0$.

Let $(\hat{\xi}, \hat{s}, \hat{w})$ denote a test function on $\Omega \times (0, T)$ for which

$$(\hat{\xi}, \hat{s}, \hat{w}) = (\frac{\partial \hat{\xi}}{\partial n}, \frac{\partial \hat{s}}{\partial n}, \frac{\partial \hat{w}}{\partial n}) = 0 \quad \text{on } \Gamma_0 \times (0, T)$$

$$(\hat{\xi}, \hat{s}, \hat{w})|_{t=0} = \frac{\partial}{\partial t}(\hat{\xi}, \hat{s}, \hat{w})|_{t=0} = (\hat{\xi}, \hat{s}, \hat{w})|_{t=T} = \frac{\partial}{\partial t}(\hat{\xi}, \hat{s}, \hat{w})|_{t=T} = 0 \quad \text{in } \Omega$$

where n is the outward unit normal to Γ. We set

$$0 = \langle \mathcal{L}_b'(\xi, s, w), (\hat{\xi}, \hat{s}, \hat{w}) \rangle = \lim_{\epsilon \to 0} \frac{\mathcal{L}_b\Big((\xi, s, w) + \epsilon(\hat{\xi}, \hat{s}, \hat{w})\Big) - \mathcal{L}_b(\xi, s, w)}{\epsilon}$$

to obtain the equations of motion in weak form:

$$\int_0^T c\left(\dot{\xi}, \dot{s}, \dot{w}; \dot{\hat{\xi}}, \dot{\hat{s}}, \dot{\hat{w}}\right) - a\left(\xi, s, w; \hat{\xi}, \hat{s}, \hat{w}\right) + (f_3, \hat{w})_\Omega - (M, \hat{\psi})_\Omega + h(P, \hat{s})_\Omega dt + \int_0^T \int_{\Gamma_1} hp \cdot \hat{s} - m \cdot \hat{\psi} + \hat{w} g_3 d\Gamma dt = 0 \tag{3.3}$$

where $(f, g)_\Omega = \int_\Omega f g d\Omega$, $\hat{\psi} = \hat{\xi} + 3\hat{s}$ and

$$c(\xi, s, w; \hat{\xi}, \hat{s}, \hat{w}) = h(\rho w, \hat{w})_\Omega + (I_\rho \xi, \hat{\xi})_\Omega + (3 I_\rho s, \hat{s})_\Omega \tag{3.4}$$

$$a(\xi, s, w; \hat{\xi}, \hat{s}, \hat{w}) = a_0(\xi; \hat{\xi}) + 3a_0(s; \hat{s}) + a_1(\xi, s, w; \hat{\xi}, \hat{s}, \hat{w}) + a_2(s; \hat{s}) \tag{3.5}$$

$$\begin{aligned} a_0(\xi, \hat{\xi}) &= \left(D\frac{\partial \xi_1}{\partial x_1}, \frac{\partial \hat{\xi}_1}{\partial x_1}\right)_\Omega + \left(D\frac{\partial \xi_2}{\partial x_2}, \frac{\partial \hat{\xi}_2}{\partial x_2}\right)_\Omega \\ &+ \left(\mu D\frac{\partial \xi_2}{\partial x_2}, \frac{\partial \hat{\xi}_1}{\partial x_1}\right)_\Omega + \left(\mu D\frac{\partial \xi_1}{\partial x_1}, \frac{\partial \hat{\xi}_2}{\partial x_2}\right)_\Omega \\ &+ \left(\left(\frac{1-\mu}{2}\right) D \left(\frac{\partial \xi_1}{\partial x_2} + \frac{\partial \xi_2}{\partial x_1}\right), \left(\frac{\partial \hat{\xi}_1}{\partial x_2} + \frac{\partial \hat{\xi}_2}{\partial x_1}\right)\right)_\Omega \end{aligned} \tag{3.6}$$

$$a_1(\xi, s, w; \hat{\xi}, \hat{s}, \hat{w}) = (K(\xi + 3s - \nabla w), \hat{\xi} + 3\hat{s} - \nabla \hat{w})_\Omega \tag{3.7}$$

$$a_2(s, \hat{s}) = 4h^2(\gamma s, \hat{s})_\Omega. \tag{3.8}$$

The force terms in (3.3) can be rewritten

$$(hP, \hat{s})_\Omega - (M, \hat{\psi})_\Omega = -(M, \hat{\xi})_\Omega + (\mathcal{K}, \hat{s})_\Omega; \quad \mathcal{K} = hP - 3M \tag{3.9}$$

and

$$\int_{\Gamma_1} hp \cdot \hat{s} - m \cdot \hat{\psi} d\mu = \int_{\Gamma_1} -m \cdot \hat{\xi} + \kappa \cdot \hat{s}; \quad \kappa = hp - 3m. \tag{3.10}$$

3.2. Associated boundary value problem. The boundary value problem associated with (3.3) can be obtained through the usual integration by parts procedure. Before we state the result let us first introduce some simplifying notation.

If $\varepsilon = (\varepsilon)_{ij}$ denotes a symmetric 2×2 matrix then $M[\varepsilon]$ defined by

$$M[\varepsilon] = D \begin{pmatrix} \varepsilon_{11} + \mu\varepsilon_{22} & (1-\mu)\varepsilon_{12} \\ (1-\mu)\varepsilon_{12} & \mu\varepsilon_{11} + \varepsilon_{22} \end{pmatrix}, \tag{3.11}$$

defines another symmetric matrix. For differentiable $\psi = (\psi_1, \psi_2)$ define $\epsilon(\psi)$ by

$$\epsilon(\psi) = \frac{1}{2}(\nabla\psi + (\nabla\psi)^T) = \frac{1}{2}\left(\frac{\partial\psi_i}{\partial x_j} + \frac{\partial\psi_j}{\partial x_i}\right)_{ij}. \tag{3.12}$$

Then (3.11)-(3.12) defines a symmetric 2×2 matrix $\mathcal{M}[z]$ by

$$\mathcal{M}[z] = M[\epsilon(z)]. \tag{3.13}$$

We define the divergence of a symmetric matrix to be the divergence of each row:

$$div \begin{bmatrix} a_{11} & a_{12} \\ a_{12} & a_{22} \end{bmatrix} = (\operatorname{div}(a_{11}, a_{12}), \operatorname{div}(a_{12}, a_{22}))\,.$$

Then $L\psi = (L_1\psi, L_2\psi) = div\ \mathcal{M}[\psi]$ defines a second order operator which is given explicitly by

$$\begin{aligned} L_i\psi = {} & \frac{\partial}{\partial x_i}\left(D\frac{\partial\psi_i}{\partial x_i}\right) + \frac{\partial}{\partial x_j}\left(\frac{(1-\mu)}{2}D\frac{\partial\psi_i}{\partial x_j}\right) + \frac{\partial}{\partial x_j}\left(\frac{(1-\mu)}{2}D\frac{\partial\psi_j}{\partial x_i}\right) \\ & + \frac{\partial}{\partial x_i}\left(\mu D\frac{\partial\psi_j}{\partial x_j}\right), \quad (i,j) = (1,2),\ (2,1). \end{aligned} \tag{3.14}$$

Let us also define the boundary operator $\mathcal{B}\psi = (\mathcal{B}_1(\psi_1, \psi_2), (\mathcal{B}_2(\psi_1, \psi_2))$ by

$$\mathcal{B}\psi = \mathcal{M}[\psi]n, \tag{3.15}$$

where $n = (n_1, n_2)$ denotes the outward unit normal to Γ. Explicitly one has

$$\mathcal{B}_1(\psi_1, \psi_2) = D\left[\left(\frac{\partial\psi_1}{\partial x_1}n_1 + \mu\frac{\partial\psi_2}{\partial x_2}n_1\right) + \left(\frac{1-\mu}{2}\right)\left(\frac{\partial\psi_1}{\partial x_2} + \frac{\partial\psi_2}{\partial x_1}\right)n_2\right]$$

$$\mathcal{B}_2(\psi_1, \psi_2) = D\left[\left(\frac{\partial\psi_2}{\partial x_2}n_2 + \mu\frac{\partial\psi_1}{\partial x_1}n_2\right) + \left(\frac{1-\mu}{2}\right)\left(\frac{\partial\psi_2}{\partial x_1} + \frac{\partial\psi_1}{\partial x_2}\right)n_1\right].$$

An integration by parts in t of (3.3) followed by an application of Green's theorem leads to the associated boundary value problem:

$$\begin{cases} \text{(i)} & \rho h\ddot{w} + \operatorname{div}(K\varphi) = f_3 & \text{in } \Omega \times \mathbb{R}^+ \\ \text{(ii)} & I_\rho\ddot{\xi} + K\varphi - L\xi = -M & \text{in } \Omega \times \mathbb{R}^+ \\ \text{(iii)} & 3I_\rho\ddot{s} + 3K\varphi + 4\gamma h^2 s - 3Ls = \mathcal{K} & \text{in } \Omega \times \mathbb{R}^+ \\ \text{(iv)} & \varphi = \xi + 3s - \nabla w & \text{in } \Omega \times \mathbb{R}^+ \end{cases} \tag{3.16}$$

with the boundary conditions

$$\begin{cases} \text{(i)} & (w, \xi, s) = 0 & \text{on } \Gamma_0 \times \mathbb{R}^+ \\ \text{(ii)} & K\varphi \cdot n = g_3 & \text{on } \Gamma_1 \times \mathbb{R}^+ \\ \text{(iii)} & \mathcal{B}\xi = -m & \text{on } \Gamma_1 \times \mathbb{R}^+ \\ \text{(iv)} & 3\mathcal{B}s = \kappa & \text{on } \Gamma_1 \times \mathbb{R}^+. \end{cases} \tag{3.17}$$

Initial conditions can be given as

$$\text{(3.18)} \qquad \begin{cases} \xi_i^0(\underline{x}) = \xi_i(\underline{x}, 0); \;\; \xi_i^1(\underline{x}) = \dot{\xi}_i(\underline{x}, 0) & i = 1, 2 \\ s_i^0(\underline{x}) = s_i(\underline{x}, 0); \;\; s_i^1(\underline{x}) = \dot{s}_i(\underline{x}, 0) & i = 1, 2 \\ w^0(\underline{x}) = w(\underline{x}, 0); \;\; w^1(\underline{x}) = \dot{w}(\underline{x}, 0) & \underline{x} \in \Omega. \end{cases}$$

Of course, depending upon the regularity of the initial data the solution of (3.16)-(3.17) will have different regularity properties.

Many other types of boundary conditions are possible, and easily determined by inspection of (3.17). In the boundary conditions we consider, (i) is precisely the usual "clamped" b.c.'s together with the extra condition that the slip on Γ_0 vanishes. (ii) specifies the transverse shear on Γ_1 while (iii) and (iv) are moment conditions. It's worth noting that the moments $\mathcal{K}$ and κ are the unique ones (up to multiplicity) for which $M \cdot \mathcal{K} = 0$ at each $\underline{x} \in \Omega$ and $m \cdot \kappa = 0$ at each $t \in \Gamma_1$.

Remark 3.1. The natural substitution $\xi = \psi - 3s$ used in (2.19) suggests that s has an interpretation as an angle. This is justified by noting that to highest order ξ_i is the rotation angle (projected onto the x_i-x_3 plane with positive orientation as measured from the x_3 axis) of the ray from $(0,0,0)$ to $(U_1(\cdot,\cdot,h/3)$, $U_2(\cdot,\cdot,h/3)$, $U_3(\cdot,\cdot,h/3))$. Thus ξ can be thought of as an *effective angle of rotation.*

Remark 3.2. The fact that the Lagrangian decouples into a bending part and a stretching part relied upon finding the change of variables (2.11), (2.17). If however, we did not assume that the top and bottom plates were identical, then except in very special cases, there does not exist a change of variables for which the Lagrangian decouples. Said another way, without the symmetry in the upper and lower plates, the in-plane motions *do not decouple* from the transverse motions.

Remark 3.3. In the same way that (3.16)-(3.17) were obtained, one may obtain the equations for the in-plane motions. In the following, the u variables represent the coordinates of the dislpacement of the center of mass while the ν variables are related to the difference in the shear angles of the top and bottom plate (see (2.17),(2.20)).

$$\begin{cases} \text{(i)} & I_\rho \ddot{\nu} - L\nu = -Qh & \text{in } \Omega \times \mathbb{R}^+ \\ \text{(ii)} & I_\rho \ddot{u} - Lu = h^2(f_1, f_2) - 3Qh & \text{in } \Omega \times \mathbb{R}^+ \end{cases}$$

with the boundary conditions

$$\begin{cases} \text{(i)} & \nu = u = 0 & \text{on } \Gamma_0 \times \mathbb{R}^+ \\ \text{(ii)} & \mathcal{B}\nu = -qh & \text{on } \Gamma_1 \times \mathbb{R}^+ \\ \text{(iii)} & \mathcal{B}u = h^2(g_1, g_2) - 3hq & \text{on } \Gamma_1 \times \mathbb{R}^+. \end{cases}$$

4. Existence, uniqueness, regularity

Denote

$$H^1_{\Gamma_0} = \{\varphi : \varphi \in H^1(\Omega), \varphi = 0 \text{ on } \Gamma_0\}.$$

Recall Γ_0 could be part or all of Γ. In the latter case $H^1_{\Gamma_0} = H^1_0(\Omega)$. Also define

$$\mathcal{V} = (H^1_{\Gamma_0})^5 \quad \mathcal{H} = (L^2(\Omega))^5.$$

4.1. Variational formulation. Let us consider our problem in the absence of external forces. In this case the variational formulation of (3.16)-(3.18) is: Find functions ξ, s, w such that

$$(\xi, s, w) \in C([0,T]; \mathcal{V}) \cap C^1([0,T]; \mathcal{H}), \tag{4.1}$$

$$\frac{d}{dt} c(\dot{\xi}, \dot{s}, \dot{w}; \hat{\xi}, \hat{s}, \hat{w}) + a(\xi, s, w; \hat{\xi}, \hat{s}, \hat{w}) = 0 \qquad \forall (\hat{\xi}, \hat{s}, \hat{w}) \in \mathcal{V} \tag{4.2}$$

in the sense of distributions on $(0,T)$ and

$$\begin{cases} (\xi(0), s(0), w(0)) = (\xi^0, s^0, w^0) & \text{given in } \mathcal{V} \\ (\dot{\xi}(0), \dot{s}(0), \dot{w}(0)) = (\xi^1, s^1, w^1) & \text{given in } \mathcal{H}. \end{cases} \tag{4.3}$$

The forms $c(\cdot\,;\cdot)$, $a(\cdot\,;\cdot)$, $a_0(\cdot\,;\cdot)$, $a_1(\cdot\,;\cdot)$ and $a_2(\cdot\,;\cdot)$ are each symmetric bilinear functions of their arguments. Let us denote by $c(\cdot)$, $a(\cdot)$ and so forth the corresponding (non-negative) quadratic functions, e.g., $a(\xi, s, w) = a(\xi, s, w; \xi, s, w)$.

Theorem 4.1. *The variational problem (4.1)-(4.3) admits a unique solution. Moreover the mapping* $\{\xi^0, s^0, w^0\} \times \{\xi^1, s^1, w^1\} \to \{\xi, s, w\}$ *is continuous from* $\mathcal{V} \times \mathcal{H} \to C([0,T]; \mathcal{V}) \cap C^1([0,T]; \mathcal{H})$.

Proof. The proof is almost identical to the proof of Theorems 2.1 and 2.2, p. 44 and 47, respectively of [LL]. Thus we mention only the main points. Clearly $c(\cdot)$ is equivalent to $\|\cdot\|_{\mathcal{H}}$. In particular there exists $\delta > 0$ for which

$$c(\xi, s, w) \geq \delta \|\{\xi, s, w\}\|^2_{\mathcal{H}}.$$

What remains is to obtain a coercive estimate on $a(\cdot)$ over $\mathcal{V}$ of the type:

$$a(\xi, s, w) + \lambda \|\{\xi, s, w\}\|^2_{\mathcal{H}} \geq \delta \|\{\xi, s, w\}\|^2_{\mathcal{V}} \tag{4.4}$$

for some $\lambda > 0$, $\delta > 0$. Indeed, if (4.4) holds then by the Lax-Milgram theorem and the usual variational theory (J.L. Lions [Li]) one concludes that (4.1)-(4.3) is well-posed.

In the case $\Gamma_0 = \emptyset$ (4.4) follows by [La, p.29]. However let us discuss only the case $\Gamma_0 \neq 0$. By [LL, p.43] $a_0(\cdot)$ is coercive over $H^1(\Omega)$. Therefore there exists $\delta_0 > 0$ for which

$$a_0(\xi) + 3a_0(s) \geq \delta_0 [\|\xi\|^2_{(H^1(\Omega))^2} + \|s\|^2_{(H^1(\Omega))^2}]. \tag{4.5}$$

Next, due to Poincáre's inequality, for some $\delta_1 > 0$

$$\|\nabla w\|^2_{(L^2(\Omega))^2} \geq \delta_1 \|w\|^2_{H^1(\Omega)}. \tag{4.6}$$

Combining (4.5) and (4.6) it follows that for some $\delta > 0$

$$a_0(\xi) + 3a_0(s) + a_1(\xi, s, w) > \delta \|\{\xi, s, w\}\|_{\mathcal{V}}.$$

In particular, (4.4) holds with $\lambda = 0$ when $\Gamma_0 \neq \emptyset$. □

Remark 4.1. The system (3.16)-(3.18) reduces exactly to the MTR system in the following two situations:

(I) **If $\gamma = 0$** there is no glue to cause interaction between the top and bottom plate. Hence one would expect to recover the MTR system for a plate of thickness $h/2$. Let us see that this is so.
Theorem 4.1, in particular, holds when $\gamma = 0$. Thus consider (3.16)-(3.17) in the homogeneous case. Let $\psi = \xi + 3s$ so that $\varphi = \psi - \nabla w$. Adding (ii) and (iii) of (3.16) and dividing the sum by 8 we obtain

$$\frac{1}{8} I_\rho \ddot{\psi} + \frac{1}{2} K\varphi - \frac{1}{8} L\psi = 0. \tag{4.7}$$

Since D and I_ρ are cubic in h while $K = kEh/2(1 + \mu)$ is linear, if we double h in (4.7) and (3.16-i) we obtain

$$\begin{cases} \text{(i)} & \rho h \ddot{w} + \operatorname{div}(K\varphi) = 0 & \text{in } \Omega \times \mathbb{R}^+ \\ \text{(ii)} & I_\rho \ddot{\psi} + K\varphi - L\psi = 0 & \text{in } \Omega \times \mathbb{R}^+ \\ \text{(iii)} & \varphi = \psi - \nabla w & \text{in } \Omega \times \mathbb{R}^+. \end{cases} \tag{4.8}$$

In the same way we obtain the boundary conditions

$$\begin{cases} \text{(i)} & (w, \psi) = 0 & \text{on } \Gamma_0 \times \mathbb{R}^+ \\ \text{(ii)} & K\varphi \cdot n = 0 & \text{on } \Gamma_1 \times \mathbb{R}^+ \\ \text{(iii)} & \mathcal{B}\psi = 0 & \text{on } \Gamma_1 \times \mathbb{R}^+. \end{cases} \tag{4.9}$$

Equations (4.8)-(4.9) are precisely the MTR equations for a plate of thickness h. (We defined ψ with the opposite sign as in [LL] hence there is a sign difference between [LL, (3.7) p.15] and (4.8)-(4.9).)

(II) **If $\gamma \longrightarrow \infty$** the glue becomes infinitely stiff and one expects to see no slip in the limit. Hence one would expect the limiting system to reduce to the MTR system for a plate of thickness h since the individual plates are themselves MTR plates and by (H1) the shear angles are the same in each layer. In fact one can show that as $\gamma \to \infty$ solutions of (3.16)-(3.18) tend in a certain weak sense to solutions of the MTR system (4.8)-(4.9). We make this precise in section 6.

4.2. Semigroup formulation. Solutions of (3.16)-(3.18) can also be defined in terms of semigroups on a range of spaces which depend upon the regularities of the body and boundary forces. We limit our discussion here however, for the sake of brevity, to the homogenous case.

First assume $\Gamma_0 \neq \emptyset$.

We identify $\mathcal{H}$ with its dual $\mathcal{H}'$ and have the dense and continuous embeddings

$$\mathcal{V} \subset \mathcal{H} = \mathcal{H}' \subset \mathcal{V}'.$$

Let $\langle \cdot , \cdot \rangle$ denote the duality pairing between $\mathcal{V}$ and $\mathcal{V}'$ which coincides with $(\cdot , \cdot)_{\mathcal{H}}$ when both arguments are in $\mathcal{H}$. The forms $c(\cdot\,;\cdot)$ and $a(\cdot\,;\cdot)$ define scalar products on $\mathcal{H}$ and $\mathcal{V}$, respectively, which are equivalent to the natural ones. Thus we may define operators $C \in \mathcal{L}(\mathcal{H})$ and $A \in \mathcal{L}(\mathcal{V}, \mathcal{V}')$ by

$$\begin{aligned}\langle C\mathbf{u} , \mathbf{v}\rangle &= c(\mathbf{u}; \mathbf{v}) \quad \forall\ \mathbf{u}, \mathbf{v} \in \mathcal{H}\\ \langle A\mathbf{u} , \mathbf{v}\rangle &= a(\mathbf{u}; \mathbf{v}) \quad \forall\ \mathbf{u}, \mathbf{v} \in \mathcal{V}.\end{aligned}$$

Let us set

$$\mathbf{u} = (\xi, s, w) \quad \mathbf{u}^0 = (\xi^0, s^0, w^0) \quad \mathbf{u}^1 = (\xi^1, s^1, w^1)$$

and rewrite (4.2) as

$$C\ddot{\mathbf{u}} + A\mathbf{u} = 0 \quad \text{in } \mathcal{V}'. \tag{4.10}$$

In first order form (4.10) becomes

$$\frac{d}{dt}\begin{pmatrix} A & 0 \\ 0 & C \end{pmatrix}\begin{pmatrix} \mathbf{u} \\ \dot{\mathbf{u}} \end{pmatrix} + \begin{pmatrix} 0 & -A \\ A & 0 \end{pmatrix}\begin{pmatrix} \mathbf{u} \\ \dot{\mathbf{u}} \end{pmatrix} = 0,$$

or

$$\mathcal{C}\dot{\mathbf{U}} + \mathcal{A}\mathbf{U} = 0; \quad \mathcal{A} = \begin{pmatrix} 0 & -A \\ A & 0 \end{pmatrix}, \quad \mathcal{C} = \begin{pmatrix} A & 0 \\ 0 & C \end{pmatrix}, \mathbf{U} = \begin{pmatrix} \mathbf{u} \\ \dot{\mathbf{u}} \end{pmatrix}. \tag{4.11}$$

Assume we wish to solve (4.11) in $\mathcal{V} \times \mathcal{H}$. Define $\mathcal{D}(A)$ and $\mathcal{D}(\mathcal{A})$ by

$$\mathcal{D}(A) = \{\mathbf{u} \in \mathcal{H} : A\mathbf{u} \in \mathcal{H}\}, \quad \mathcal{D}(\mathcal{A}) = \mathcal{D}(A) \times \mathcal{V}.$$

Then $\mathcal{A} : \mathcal{D}(\mathcal{A}) \to \mathcal{V}' \times \mathcal{H}$ and $\mathcal{C}^{-1} : \mathcal{V}' \times \mathcal{H} \to \mathcal{V} \times \mathcal{H}$. Hence we can rewrite (4.11) as

$$\dot{\mathbf{U}} + \mathcal{C}^{-1}\mathcal{A}\mathbf{U} = 0 \quad \text{in } \mathcal{V} \times \mathcal{H} \tag{4.12}$$

One can easily verify that $\mathcal{C}^{-1}\mathcal{A}$ is densely defined and furthermore for all $(\mathbf{u}, \mathbf{v}) \in \mathcal{D}(\mathcal{A})$ we have

$$(\mathcal{C}^{-1}\mathcal{A}(\mathbf{u}, \mathbf{v}), (\mathbf{u}, \mathbf{v}))_{\mathcal{V}\times\mathcal{H}} = 0. \tag{4.13}$$

Thus by e.g., Pazy [Pa, p.41], $-\mathcal{C}^{-1}\mathcal{A}$ is the generator of a unitary group on $\mathcal{V}\times\mathcal{H}$.

If $\Gamma_0 = \emptyset$ then $\mathcal{A}$ is no longer an isomorphism. However $\mathcal{C}$ remains an isomorphism and by (4.4) $A + \lambda I$ remains an isomorphism $\mathcal{V} \to \mathcal{V}'$. This is enough:

By a simple argument involving the change of variables $\mathbf{u} = e^{\sigma t}\mathbf{w}, \sigma > 0$ it can be shown that $-\mathcal{C}^{-1}\mathcal{A}$ remains the generator of a C_0-semigroup when $\Gamma_0 = \emptyset$. (See [La; p.32].) Furthermore one can show directly that (4.13) remains valid when $\Gamma_0 = \emptyset$. Thus again we reach the conclusion that $-\mathcal{C}^{-1}\mathcal{A}$ is the generator of a unitary group on $\mathcal{V} \times \mathcal{H}$.

By considering appropriate extensions or restrictions of the generator we are lead to the following:

Theorem 4.2. *The operator $\mathcal{C}^{-1}\mathcal{A}$ is the generator of a unitary group on each of the spaces: $\mathcal{V} \times \mathcal{H}$, $\mathcal{D}(A) \times \mathcal{V}$, $\mathcal{H} \times \mathcal{V}'$, $\mathcal{V}' \times (\mathcal{D}(A))'$. We thus have the following regularity results for (4.12):*

$$\begin{aligned}
&(i) \quad (\mathbf{u}^0, \mathbf{u}^1) \in \mathcal{D}(A) \times \dot{\mathcal{V}} \Rightarrow \quad && (\mathbf{u}, \dot{\mathbf{u}}) \in C([0,T]; \mathcal{D}(A) \times \mathcal{V})\\
&(ii) \quad (\mathbf{u}^0, \mathbf{u}^1) \in \mathcal{V} \times \mathcal{H} \Rightarrow && (\mathbf{u}, \dot{\mathbf{u}}) \in C([0,T]; \mathcal{V} \times \mathcal{H})\\
&(iii) \quad (\mathbf{u}^0, \mathbf{u}^1) \in \mathcal{H} \times \mathcal{V}' \Rightarrow && (\mathbf{u}, \dot{\mathbf{u}}) \in C([0,T]; \mathcal{H} \times \mathcal{V}')\\
&(iv) \quad (\mathbf{u}^0, \mathbf{u}^1) \in \mathcal{V}' \times (\mathcal{D}(A)) \Rightarrow && (\mathbf{u}, \dot{\mathbf{u}}) \in C([0,T]; \mathcal{V}' \times (\mathcal{D}(A))').
\end{aligned}$$

5. Limiting behavior of (3.12)-(3.13) as $K \to \infty$

When one lets $K \to \infty$ in the MTR plate one finds that the solutions converge in a weak sense to solutions of a Kirchhoff plate equation [LL]. In this section we show that as $K \to \infty$ solutions of the system (4.1)-(4.3) converge in a weak sense to solutions of a new plate equation which we show is an analog of the Kirchhoff plate.

This limit creates a singular perturbation in which a family of eigenvalues for the shear modes tends to infinity along the imaginary axis. In application, K is generally several orders of magnitude larger than the other parameters (in particular D) which appear in (3.16)-(3.17). Hence in principle, the limiting PDE should provide a good approximation to the low-frequency characteristics of the original system.

Let us recall the variational problem (4.1)-(4.3): Find functions $\xi = (\xi_1, \xi_2)$, $s = (s_1, s_2), w$ such that

$$(5.1) \quad \begin{cases} (a) & (\xi, s, w) \in C([0,T]; \mathcal{V}) \cap C^1([0,T]; \mathcal{H}), \\ (b) & c(\ddot{\xi}, \ddot{s}, \ddot{w}; \hat{\xi}, \hat{s}, \hat{w}) + a(\xi, s, w; \hat{\xi}, \hat{s}, \hat{w}) = 0 \quad \forall (\hat{\xi}, \hat{s}, \hat{w}) \in \mathcal{V}, \\ (c) & (\xi(0), s(0), w(0)) = (\xi^0, s^0, w^0) \quad \text{given in } \mathcal{V}, \\ (d) & (\dot{\xi}(0), \dot{s}(0), \dot{w}(0)) = (\xi^1, s^1, w^1) \quad \text{given in } \mathcal{H}, \end{cases}$$

where

$$(5.2) \quad a(\xi, s, w; \hat{\xi}, \hat{s}, \hat{w}) = a_0(\xi; \hat{\xi}) + 3a_0(s; \hat{s}) + a_1(\xi, s, w; \hat{\xi}, \hat{s}, \hat{w}) + a_2(s; \hat{s})$$

and c, a_0, a_1, a_2 were defined in (3.4)-(3.8).

5.1. The limiting variational problem as $K \to \infty$. Let

$$H^2_{\Gamma_0} = \{w \in H^2(\Omega) \cap H^1_{\Gamma_0} : \frac{\partial w}{\partial v} = 0 \text{ on } \Gamma_0\}$$

and define

$$W = \{(\xi, s, w) \in (H^1_{\Gamma_0})^4 \times H^2_{\Gamma_0} : \xi + 3s - \nabla w = 0 \text{ in } \Omega\}$$
$$V = \{(\xi, s, w) \in (L^2(\Omega))^4 \times H^1_{\Gamma_0} : \xi + 3s - \nabla w = 0 \text{ in } \Omega\}.$$

We will show in Theorem 5.2 that solutions of (5.1) with initial data in $W \times V$ converge in a weak sense to solutions of the following variational problem:

Find functions (ξ, s, w) for which

$$(\xi, s, w) \in C([0,T]; W) \cap C^1([0,T]; V), \tag{5.3}$$

$$\frac{d}{dt} c(\dot{\xi}, \dot{s}, \dot{w}; \hat{\xi}, \hat{s}, \hat{w}) + b(\xi, s, w; \hat{\xi}, \hat{s}, \hat{w}) = 0 \quad \forall (\hat{\xi}, \hat{s}, \hat{w}) \in W \tag{5.4}$$

$$\begin{cases} (\xi(0), s(0), w(0)) = (\xi^0, s^0, w^0) \text{ given in } W, \\ (\dot{\xi}(0), \dot{s}(0), \dot{w}(0)) = (\xi^1, s^1, w^1) \text{ given in } V, \end{cases} \tag{5.5}$$

where (5.4) is understood in the sense of distributions on $(0, T)$ and

$$b(\xi, s, w; \hat{\xi}, \hat{s}, \hat{w}) = a_0(\xi; \hat{\xi}) + 3a_0(s; \hat{s}) + a_2(s; \hat{s}). \tag{5.6}$$

Theorem 5.1. *The variational problem (5.3)-(5.5) admits a unique solution. Moreover the mapping $(\xi^0, s^0, w^0) \times (\xi^1, s^1, w^1) \to (\xi, s, w)$ is continuous from $W \times V \to C([0,T]; W) \cap C^1([0,T]; V)$.*

Proof. First assume $\Gamma_0 \neq \emptyset$.

We may eliminate ξ from (5.3)–(5.6) with $\xi = \nabla w - 3s$. By (4.5)

$$\begin{aligned} a_0(\nabla w - 3s) + 3a_0(s) &\geq \delta_0 \left(\|\nabla w - 3s\|^2_{(H^1(\Omega))^2} + \|s\|^2_{H^1(\Omega))^2} \right) \\ &\geq \delta_1 \left(\|\nabla w\|^2_{(H^1(\Omega))^2} + \|s\|^2_{(H^1(\Omega))^2} \right), \\ &\geq \delta_2 \left(\|w\|^2_{(H^2(\Omega))} + \|s\|^2_{(H^1(\Omega))^2} \right), \end{aligned} \tag{5.7}$$

where the last inequality holds since $\Gamma_0 \neq \emptyset$. It follows that $b(\cdot, \cdot)$ is coercive over W. Since $c(\cdot; \cdot)$ is coercive on $\mathcal{H}$, we have (by similar inequalities as in (5.7))

$$c(\nabla w - 3s, s, w) \geq \delta_3 \left(\|w\|^2_{H^1(\Omega)} + \|s\|^2_{L^2(\Omega)} \right). \tag{5.8}$$

It follows that $c(\cdot; \cdot)$ is coercive over V and hence by the standard variational theory (5.3)-(5.5) is well-set. If $\Gamma_0 = \emptyset$ then (5.7) doesn't hold on W, however it is enough (see (4.4)) to note that (5.7) holds when $\lambda((w, w)_\Omega + (s, s)_\Omega)$ are added to the left-hand side. (See [LL, p.46].) □

Theorem 5.2. *Let (ξ^0, s^0, w^0), (ξ^1, s^1, w^1) be given in $W \times V$. Then as $K \to \infty$, the solution (ξ_K, s_K, w_K) to (5.1) tends to (ξ, s, w) in the sense that*

$$(5.9) \qquad \begin{cases} (\xi_K, s_K, w_K) \to (\xi, s, w) \text{ in } L^\infty(0,T;\mathcal{V}) \text{ weak star} \\ (\dot\xi_K, \dot s_K, \dot w_K) \to (\dot\xi, \dot s, \dot w) \text{ in } L^\infty(0,T;\mathcal{H}) \text{ weak star,} \end{cases}$$

where ξ, s, w are the solutions of (5.3)-(5.5).

Proof. This result is similar to [LL, p. 50] where it is shown that the solutions of the Kirchhoff plate equation may be obtained as similar weak limits of solutions of the MTR system. The idea of that proof remains valid here, with the exception that we impose the (additional) compatibility condition $(\xi^1, s^1, w^1) \in V$ and obtain a stronger result which includes (5.19).

We will assume $\Gamma_0 \neq \emptyset$. The modifications necessary to handle the general case are the same as those in [LL, p. 51].

Theorem 4.1 shows that (5.1) is well-set and Theorem 4.2 implies the energy

$$(5.10) \qquad \mathcal{E}(t) = \frac{1}{2}a(\xi, s, w) + \frac{1}{2}c(\dot\xi, \dot s, \dot w)$$

remains constant along trajectories. By (5.5) we have $a_1(\xi^0, s^0, w^0) = 0$. Then by conservation of the energy (5.10) as $K \to \infty$,

$$(5.11) \qquad \begin{cases} \{\xi_K, s_K\} & \text{bounded in } L^\infty(0,T;(H^1_{\Gamma_0})^4) \\ \{\dot\xi_K, \dot s_K, \dot w_K\} & \text{bounded in } L^\infty(0,T;\mathcal{H}) \end{cases}$$

$$(5.12) \qquad \{K^{1/2}(\xi_K + 3s_K - \nabla w_K)\} \quad \text{bounded in } L^\infty(0,T;(L^2(\Omega))^2).$$

Therefore, in particular, since $\Gamma_0 \neq \emptyset$

$$(5.13) \qquad \{w_K\} \text{ remains bounded in } L^\infty(0,T;H^1_{\Gamma_0}(\Omega)).$$

Thus by extracting a subsequence of $\{\xi_K, s_K, w_K\}$, (5.9) follows from (5.11) and (5.13). (Of course, we have yet to identify this limit as a solution of (5.3)-(5.5).) From (5.12) we know $\{\xi_K + 3s_K - \nabla w_K\} \to 0$ strongly in $L^\infty(0,T;(L^2(\Omega))^2$ and hence also $\{\dot\xi_K + 3\dot s_K - \nabla \dot w_K\} \to 0$ in the sense of distributions. Therefore

$$(5.14) \qquad \begin{cases} \xi + 3s - \nabla w = 0 & 0 \le t \le T \\ \dot\xi + 3\dot s - \nabla \dot w = 0 & \text{almost everywhere } 0 < t < T. \end{cases}$$

Thus from (5.14) and (5.9) one sees that

$$(5.15) \qquad (\xi, s, w) \in L^\infty(0,T;W), \qquad (\dot\xi, \dot s, \dot w) \in L^\infty(0,T;V).$$

To see that (ξ, s, w) satisfy (5.4), we take in (5.1) test functions $(\hat\xi, \hat s, \hat w) \in W$ so that (5.1-b) (with $\xi = \xi_K$, $s = s_K$, $w = w_K$) reduces to

$$(5.16) \qquad \frac{d}{dt}c(\dot\xi_K, \dot s_K, \dot w_K; \hat\xi, \hat s, \hat w) + a_0(\xi_K; \hat\xi) + 3a_0(s_K, \hat s) + a_2(s_K, \hat s) = 0.$$

Using (5.9) we pass to the limit (in the sense of distributions) and obtain (5.4).

Let us now check that $(\xi(0), s(0), w(0)) = (\xi^0, s^0, w^0)$. By (5.11), (5.13) we know that $\{(\xi_K, s_K, w_K)\}$ is bounded in $L^\infty(0,T;\mathcal{V}) \cap W^{1,\infty}(0,T;\mathcal{H})$, and thus by compactness, in addition to (5.9) we have

$$(\xi_K, s_K, w_K) \to (\xi, s, w) \quad \text{in} \quad C([0,T];\mathcal{H}).$$

Consequently by passing limits we obtain

$$(\xi(0), s(0), w(0)) = (\xi^0, s^0, w^0). \tag{5.17}$$

As is well-known, the solution to (5.1) is the same as the semigroup solution in *(ii)* of Theorem 4.2. When this solution is differentiated with respect to time it is easy to see from (4.12) that $(\dot{\mathbf{u}}, \ddot{\mathbf{u}}) \in C([0,T];\mathcal{H} \times \mathcal{V}')$. Thus for each K $(\ddot{\xi}_K, \ddot{s}_K, \ddot{w}_K)\} \in C([0,T];\mathcal{V}')$ and if we use the hypothesis that $(\xi^0, s^0, w^0) \in W$, $(\xi^1, s^1, w^1) \in V$, then (4.12) implies $\ddot{w}_K(0) = 0$ and $\{\ddot{s}_K(0), \ddot{\xi}_K(0)\} \in \left((H^1_{\Gamma_0})'\right)^4$ independent of K. Furthermore by Theorem 4.2 we also know that conservation of (norm) energy holds in that space. Thus we are in a position to repeat with $\{\dot{\xi}_K, \dot{s}_K, \dot{w}_K\}$ the same argument that was applied to $\{\xi_K, s_K, w_K\}$. We obtain

$$\{\ddot{\xi}_K, \ddot{s}_K, \ddot{w}_K\} \quad \text{remain bounded in} \quad L^\infty(0,T;\mathcal{V}'). \tag{5.18}$$

Consequently

$$\{\dot{\xi}_K, \dot{s}_K, \dot{w}_K\} \to (\dot{\xi}, \dot{s}, \dot{w}) \quad \text{in} \quad C([0,T];(\mathcal{V}')),$$

and hence

$$(\dot{\xi}(0), \dot{s}(0), \dot{w}(0)) = (\xi^1, s^1, w^1). \tag{5.19}$$

Since the initial data were given in $W \times V$, by Theorem 5.1 there is a unique solution $(\tilde{\xi}, \tilde{s}, \tilde{w})$ to (5.4)-(5.5) within the space $C([0,T];W) \cap C^1([0,T];V)$. Furthermore by a regularity improvement theorem in Lions and Magenes [LM, p.275], (5.15) implies that after modification on a set of measure zero,

$$(\xi, s, w) \in C([0,T];W) \cap C^1([0,T];V). \tag{5.20}$$

Hence $(\xi, s, w) = (\tilde{\xi}, \tilde{s}, \tilde{w})$ is uniquely determined and satisfies (5.3)-(5.5). □

Remark 5.1. If the hypothesis that (ξ^1, s^1, w^1) is weakened to

$$(\xi^1, s^1, w^1) \in \mathcal{H} : w^1 \in H^1_{\Gamma_0}$$

then (5.18) will not in general hold and consequently one can not infer (5.19). However as a consequence of (5.9) and (5.15) there will exist limit points satisfying (5.17), (5.20).

5.2. Semigroup formulation of limiting system. The semigroup formulation of (5.3)-(5.5) is analogous to that of (5.1) and a result similar to Theorem 4.2 also holds here. Thus for simplicity we assume for this discussion that $\Gamma_0 \neq \emptyset$.

We introduce the forms

$$c_1(s, w; \hat{s}, w) = c(\nabla w - 3s, s, w; \nabla\hat{w} - 3\hat{s}, \hat{s}, \hat{w}) \tag{5.21}$$

$$b_0(s, w; \hat{s}, \hat{w}) = a_0(\nabla w - 3s; \nabla\hat{w} - 3\hat{s}) + 3a_0(s; \hat{s}), \tag{5.22}$$

and continue to use the notation $c_1(s, w) = c_1(s, w; s, w)$ and likewise for the form b_0.

When ξ and $\hat{\xi}$ are eliminated from (5.3)-(5.5) the variational equation may be written

$$\frac{d}{dt}[c_1(\dot{s}, \dot{w}; \hat{s}, \hat{w})] + b_0(s, w; \hat{s}, \hat{w}) + a_2(s, \hat{s}) = 0, \ \forall(\hat{w}, \hat{s}) \in H^2_{\Gamma_0} \times (H^1_{\Gamma_0})^2, \tag{5.23}$$

with initial conditions

$$\begin{cases} \{s(0), w(0)\} = \{s^0, w^0\} & \text{given in } W_0 \equiv (H^1_{\Gamma_0})^2 \times H^2_{\Gamma_0} \\ \{\dot{s}(0), \dot{w}(0)\} = \{s^1, w^1\} & \text{given in } V_0 = (L^2(\Omega))^2 \times H^1_{\Gamma_0}. \end{cases} \tag{5.24}$$

It is easy to see that $c_1(\cdot\,;\cdot)$ defines a scalar product on V_0 equivalent to the natural one and since $\Gamma_0 \neq \emptyset, b_0(\cdot\,;\cdot)$ defines a scalar product on W_0 equivalent to the natural one. Let $\langle\cdot\,;\cdot\rangle$ defined on $W_0' \times W_0$ be the duality pairing of W_0' and W_0 with respect to $H' = H = (L^2(\Omega))^3$.

We thus have the continuous dense embeddings

$$W_0 \subset V_0 \subset H \subset V_0' \subset W_0'$$

Define $A \in \mathcal{L}(W_0, W_0')$ and $C \in \mathcal{L}(V_0, V_0')$ by

$$\langle A\{s, w\}, \{\hat{s}, \hat{w}\}\rangle = b_0(s, w; \hat{s}, \hat{w}) + a_2(s; \hat{s}),$$
$$\langle C\{s, w\}, \{\hat{s}, \hat{w}\}\rangle = c_1(s, w; \hat{s}, \hat{w}).$$

Following the approach of section 4.2 we write (5.23) as

$$\dot{\mathbf{U}} + \mathcal{C}^{-1}\mathcal{A}\mathbf{U} = 0; \quad \mathcal{A} = \begin{pmatrix} 0 & -A \\ A & 0 \end{pmatrix}, \quad \mathcal{C} = \begin{pmatrix} A & 0 \\ 0 & C \end{pmatrix}, \tag{5.25}$$

where $\mathbf{U} = \{\mathbf{u}, \mathbf{v}\}^T$; $\mathbf{u} = (s, w)$, $\mathbf{v} = (\dot{s}, \dot{w})$. (Here T denotes transposition.) The domains of A and $\mathcal{A}$ are defined by

$$\mathcal{D}(A) = \{\mathbf{u} \in : A\mathbf{u} \in V_0'\}, \quad \mathcal{D}(\mathcal{A}) = \{\{\mathbf{u}, \mathbf{v}\} : \mathbf{u} \in \mathcal{D}(A), \ \mathbf{v} \in W_0\}.$$

Then just as in section 4.2, $\mathcal{C}^{-1}\mathcal{A} : \mathcal{D}(\mathcal{A}) \to W_0 \times V_0$ is densely defined and also satisfies (4.13). Thus by a proof similar to the one Theorems 4.2 we have the following.

Theorem 5.3. *The operator $-\mathcal{C}^{-1}\mathcal{A}$ is the generator of a unitary group on any of the spaces $\mathcal{D}(A) \times W_0$, $W_0 \times V_0$, $V_0' \times W_0'$, $W_0' \times (\mathcal{D}(A))'$. Furthermore one has the following regularity results*

$$\begin{array}{lll}
(i) & (\mathbf{u}^0, \mathbf{u}^1) \in \mathcal{D}(A) \times W_0 \Rightarrow & (\mathbf{u}, \dot{\mathbf{u}}) \in C([0,T]; \mathcal{D}(A) \times W_0) \\
(ii) & (\mathbf{u}^0, \mathbf{u}^1) \in W_0 \times V_0 \Rightarrow & (\mathbf{u}, \dot{\mathbf{u}}) \in C([0,T]; W_0 \times V_0) \\
(iii) & (\mathbf{u}^0, \mathbf{u}^1) \in V_0' \times W_0' \Rightarrow & (\mathbf{u}, \dot{\mathbf{u}}) \in C([0,T]; V_0' \times W_0') \\
(iv) & (\mathbf{u}^0, \mathbf{u}^1) \in W_0' \times (\mathcal{D}(A))' \Rightarrow & (\mathbf{u}, \dot{\mathbf{u}}) \in C([0,T]; W_0' \times (\mathcal{D}(A))').
\end{array}$$

5.3. Boundary value problem for limiting system as $K \to \infty$. Assume the plate is subject to body forces $(f_3, M, \mathcal{K})$ and boundary forces (g_3, m, κ) as in (3.10), (3.16), (3.17). (More precisely, $M, m, \mathcal{K}, \kappa$ are force moments.) For consistency with no shear we impose

$$\int_{-h/2}^{0} \tilde{f}_i(x_3 + h/4)dx_3 = \int_{0}^{h/2} f_i(x_3 - h/4)dx_3 = 0 \quad i = 1,2. \tag{5.26}$$

Adding the two equations in (5.26) gives $M_i(\underline{x}) - \frac{h}{4}P_i(\underline{x}) = 0 \quad \forall \underline{x} \in \Omega$. Consequently from (3.9) we find $M(\underline{x}) = \mathcal{K}(\underline{x}) \quad \forall \underline{x} \in \Omega$. Likewise from (3.10) the same condition holds on Γ_1. Thus for compatibility with lack of shear notions, we impose the conditions

$$\begin{cases} M_i = \mathcal{K}_i & i = 1,2 \quad \text{in } \Omega \times (0,T) \\ m_i = \kappa_i & i = 1,2 \quad \text{on } \Gamma_1 \times (0,T) \end{cases} \tag{5.27}$$

When the force terms are included with (4.2) (from (3.2)) we have

$$\begin{aligned}0 = c(\ddot{\xi}, \ddot{s}, \ddot{w}; \hat{\xi}, \hat{s}, \hat{w}) + a_0(\xi; \hat{\xi}) + 3a_0(s; \hat{s}) + +a_2(s; \hat{s}) \\ - \int_\Omega f_3\hat{w} - M \cdot \hat{\xi} + \mathcal{K} \cdot \hat{s}\, d\underline{x} - \int_{\Gamma_1} g_3\hat{w} - m \cdot \hat{\xi} + \kappa \cdot \hat{s}\, d\Gamma.\end{aligned}$$

We then use $\hat{\xi} = \nabla\hat{w} - 3s$, and (5.27) to obtain

$$\begin{aligned}0 = c(\ddot{\xi}, \ddot{s}, \ddot{w}; \nabla\ddot{w} - 3\ddot{s}, \hat{s}, \hat{w}) + a_0(\xi; \nabla\hat{w} - 3\hat{s}) + 3a_0(s; \hat{s}) + a_2(s; \hat{s}) \\ - \int_\Omega f_3\hat{w} - M \cdot \nabla\hat{w} + 4M \cdot \hat{s}\, d\underline{x} - \int_{\Gamma_1} g_3\hat{w} - m \cdot \nabla\hat{w} + 4m \cdot \hat{s}\, d\Gamma.\end{aligned}$$

By Green's theorem one finds

$$0 = \int_\Omega \hat{w}(\rho h\ddot{w} - \operatorname{div}(I_\rho(\ddot{\xi}) - L\xi)) + 3\hat{s}\cdot(I_\rho\ddot{s} - Ls + \frac{4}{3}h^2\gamma s - I_\rho(\ddot{\xi}) + L\xi)d\underline{x}$$
$$+ \int_{\Gamma_1}(\nabla\hat{w} - 3\hat{s})\cdot\mathcal{B}\xi + 3\hat{s}\cdot\mathcal{B}(s) + \hat{w}\left[I_\rho\left(\frac{\partial\ddot{w}}{\partial n} - 3\ddot{s}\cdot n\right) - (L\xi)\cdot n\right]d\Gamma$$
$$- \int_\Omega f_3\hat{w} + (\operatorname{div} M)\hat{w} + 4M\cdot\hat{s}\,d\underline{x} - \int_{\Gamma_1} g_3 - M\cdot n\hat{w} - m\cdot\nabla\hat{w} + 4m\cdot\hat{s}\,d\Gamma.$$

Either $\Gamma_0 = \emptyset$ or w vanishes on Γ_0. Either way we may use the identity

$$\int_{\Gamma_1}\nabla\hat{w}\cdot\psi d\Gamma = \int_{\Gamma_1}\frac{\partial\hat{w}}{\partial n}\psi\cdot n - \hat{w}\frac{\partial}{\partial\tau}(\psi\cdot\tau)d\Gamma \tag{5.28}$$

where $n = (n_1, n_2)$ is the unit normal vector, $\tau = (-n_2, n_1)$ is the unit tangent vector to Γ and $\psi = (\psi_1, \psi_2)$. By collecting terms and applying (5.28) we obtain the boundary value problem:

$$\begin{cases}\rho h\ddot{w} - \operatorname{div}(I_\rho\ddot{\xi}) + \operatorname{div} L(\xi) = f_3 + \operatorname{div} M \\ 4I_\rho\ddot{s} - I_\rho\nabla\ddot{w} + L(\xi - s) + \frac{4}{3}h^2\gamma s = \frac{4}{3}M \\ \xi = \nabla w - 3s & \text{in } \Omega\times(0,T)\end{cases} \tag{5.29}$$

$$w = \frac{\partial w}{\partial n} = s = 0 \qquad \text{on } \Gamma_0\times(0,T) \tag{5.30}$$

$$\begin{cases}I_\rho\left(\frac{\partial\ddot{w}}{\partial n} - 3\ddot{s}\cdot n\right) + J_1\xi = g_3 - M\cdot n + \frac{\partial}{\partial\tau}(m\cdot\tau) \\ J_2\xi = -m\cdot n \\ \mathcal{B}(4s - \nabla w) = \frac{4}{3}m & \text{on } \Gamma_1\times(0,T)\end{cases} \tag{5.31}$$

where $J_1\psi = -(L\psi)\cdot n - \frac{\partial}{\partial\tau}(\mathcal{B}\psi\cdot\tau)$ and $J_2\psi = \mathcal{B}\psi\cdot n$. Initial conditions may be given as in (5.5).

Remark 5.2. One sees that the boundary value problem (5.29)-(5.31) is an analog of the Kirchhoff plate equations [LL, p. 13]. Indeed, in the following two situations, the system (5.29)-(5.31) reduces exactly to a Kirchhoff system.

(I) **If $\gamma = 0$**, then there is no interaction between the two plates. Thus one expects to obtain a Kirchhoff plate with thickness $h/2$. For simplicity, assume there are no external forces and $\Gamma_0 \neq \emptyset$. Furthermore assume the initial conditions satisfy $(s^0, s^1) = (\nabla w^0/4, \nabla w^1/4)$. We know by Theorem 5.1 that that the system is well posed when $\gamma = 0$. Let $y = 4s - \nabla w$. Then from (5.29),

$$I_\rho\ddot{y} - Ly = 0 \quad \text{in } \Omega\times(0,T) \tag{5.32}$$

with boundary conditions

$$(5.33) \qquad y = 0 \text{ on } \Gamma_0 \times (0,T), \quad \mathcal{B}y = 0 \text{ on } \Gamma_1 \times (0,T).$$

Since $-L$ is a positive second order operator (corresponding to the positive quadratic form a_0) and $\mathcal{B}$ is the associated conormal derivative operator (see (3.15)) there exists a unique solution to (5.32)-(5.33) with initial conditions

$$y(0) = \dot{y}(0) = 0.$$

Thus $y = 0$ in $\Omega \times (0,T)$ and we obtain

$$s = \frac{\nabla w}{4} \text{ in } \Omega \times (0,T).$$

What remains is

$$(5.34) \begin{cases} \rho\frac{h}{2}\ddot{w} - \operatorname{div}\left(\frac{I_\rho}{8}\nabla\ddot{w}\right) + \frac{1}{8}\operatorname{div} L(\nabla w) = 0 & \text{in } \Omega \times (0,T) \\ w = \frac{\partial w}{\partial n} = 0 & \text{on } \Gamma_0 \times (0,T) \\ \frac{I_\rho}{8}\left(\frac{\partial \ddot{w}}{\partial n}\right) + \frac{1}{8}J_1\nabla w = 0 & \text{on } \Gamma_1 \times (0,T) \\ \frac{1}{8}J_2\nabla w = 0 & \text{on } \Gamma_1 \times (0,T) \\ (w(0), \dot{w}(0)) \quad \text{given} & \text{in } H^2_{\Gamma_0} \times H^1_{\Gamma_0}. \end{cases}$$

One easily sees that (5.34) is precisely the Kirchoff system for a plate of thickness $h/2$. (See Remark 4.1 for the effect of the scaling.)

(II) **If $\gamma \to \infty$**, then one expects the system (5.29)-(5.31) to reduce to a Kirchhoff system for a plate of thickness h since infinitely strong glue will prevent any slip from occurring. As $\gamma \to \infty$ we expect $s \to 0$. Putting $s \equiv 0$ in (5.29)-(5.31) gives precisely the Kirchhoff system for a plate of thickness h:

$$(5.35) \begin{cases} \rho h\ddot{w} - \operatorname{div}(I_\rho\nabla\ddot{w}) + \operatorname{div} L(\nabla w) = 0 & \text{in } \Omega \times (0,T) \\ w = \frac{\partial w}{\partial n} = 0 & \text{on } \Gamma_0 \times (0,T) \\ I_\rho\left(\frac{\partial \ddot{w}}{\partial n}\right) + J_1\nabla w = 0 & \text{on } \Gamma_1 \times (0,T) \\ J_2\nabla w = 0 & \text{on } \Gamma_1 \times (0,T) \\ (w(0), \dot{w}(0)) \quad \text{given} & \text{in } H^2_{\Gamma_0} \times H^1_{\Gamma_0}. \end{cases}$$

We discuss this situation in more detail in the next section.

6. Behavior of (5.1) and (5.3)-(5.5) as $\gamma \to \infty$

If $\gamma \to \infty$ then one expects the system (5.1) to reduce to the MTR system for a plate of thickness h since in the limit there will be no slip. Likewise, since one obtains the Kirchhoff system as a weak limit as $K \to \infty$ of solutions of the MTR system [LL], one would expect to obtain a Kirchhoff plate of thickness h as a weak limit as $\gamma \to \infty$ of solutions of (5.3)-(5.5). We make precise statements to this effect in this section.

6.1. Behavior of (5.1) as $\gamma \to \infty$. The MTR system (4.8)-(4.9) may be written in variational form as

$$(6.1)\ \frac{d}{dt}c(\dot\psi, 0, \dot w; \hat\psi, 0, \hat w) + a_0(\psi; \hat\psi) + a_1(\psi, 0, w; \hat\psi, 0, \hat w) = 0,\ \forall(\hat\psi, \hat w) \in (H^1_{\Gamma_0})^3.$$

For each initial data $\{(\psi^0, w^0), (\psi^1, w^1)\} \in (H^1_{\Gamma_0})^3 \times (L^2(\Omega))^3$ there is a unique solution (see [LL]) with

$$(6.2)\qquad (\psi, w) \in C([0,T]; (H^1_{\Gamma_0})^3) \cap C^1([0,T]; (L^2(\Omega))^3).$$

Theorem 6.1. *Let (ψ^0, s^0, w^0), (ψ^1, s^1, w^1) be given in $\mathcal{H} \times \mathcal{V}$ with $s^0 = s^1 = 0$. Then as $\gamma \to \infty$, the solution $(\psi_\gamma, s_\gamma, w_\gamma)$ to (5.1) tends to $(\psi, 0, w)$ in the sense that $s \to 0$ in $L^\infty(0,T; L^2(\Omega))$ and*

$$(6.3)\qquad \begin{cases} (\psi_\gamma, w_\gamma) \to (\psi, w) & \text{in } L^\infty(0,T; (H^1_{\Gamma_0})^3))\ \text{weak star} \\ (\dot\psi_\gamma, \dot w_\gamma) \to (\dot\psi, \dot w) & \text{in } L^\infty(0,T; (L^2(\Omega))^3\ \text{weak star,} \end{cases}$$

where (ψ, w) are the solutions of (6.1).

Sketch of Proof. Since the proof involves the same steps as the proof of Theorem 5.2 we only mention the main points. For simplicity we assume $\Gamma_0 \neq \emptyset$.

We use conservation of the energy (5.10) and find that

$$(6.4)\qquad \begin{cases} \{\psi_\gamma, s_\gamma\} & \text{bounded in } L^\infty(0,T; (H^1_{\Gamma_0})^4) \\ \{\dot\psi_\gamma, \dot s_\gamma, \dot w_\gamma\} & \text{bounded in } L^\infty(0,T; \mathcal{H}) \end{cases}$$

$$(6.5)\qquad \{\psi_\gamma + 3s_\gamma - \nabla w_\gamma\} \qquad \text{bounded in } L^\infty(0,T; (L^2(\Omega))^2)$$

$$(6.6)\qquad \{\gamma^{1/2} s\} \qquad \text{bounded in } L^\infty(0,T; (L^2(\Omega))^2).$$

Thus $s \to 0$ strongly as $\gamma \to \infty$. Since $\Gamma_0 \neq \emptyset$ (6.5) implies that w_γ remains bounded in $L^\infty(0,T; H^1_{\Gamma_0})$. Thus by taking a subsequence (6.3) holds.

It remains to show that (ψ, w) are defined by (6.1)-(6.2). We take in (5.2) test functions $\hat\psi, \hat s, \hat w$ with $\hat s = 0$ so (5.2) (with $\psi = \psi_\gamma$, $s = s_\gamma$, $w = w_\gamma$) reduces to

$$\frac{d}{dt}c(\dot\psi_\gamma, 0, \dot w_\gamma; \hat\psi, 0, \hat w) + a_0(\psi_\gamma; \hat\psi) + a_1(\psi_\gamma, 0, w_\gamma; \hat\psi, 0, \hat w) = 0 \quad \forall(\hat\psi, \hat w) \in (H^1_{\Gamma_0})^3.$$

Using (6.3) we pass to the limit and obtain (6.1). Finally we need to check that $(\psi(0), w(0)) = (\psi^0, w^0)$ and $(\dot\psi(0), \dot w(0)) = (\psi^1, w^1)$. The first of these follows from the boundedness of the time derivitaves in (6.4). To obtain the second we differentiate (4.12) and note that by Theorem 4.2 the resulting equation in the variables $(\dot{\mathbf{u}}_\gamma, \ddot{\mathbf{u}}_\gamma)$ is well-posed on $\mathcal{H} \times \mathcal{V}'$ and the norm of the solution at time t is conserved. Since $s^1 = 0$ we may repeat the same argument to obtain

$$\{\ddot\psi_\gamma, \ddot s_\gamma, \ddot w_\gamma\} \qquad \text{remain bounded in } L^\infty(0,T; \mathcal{V}').$$

Thus we obtain convergence of the initial values in $\mathcal{H} \times \mathcal{V}'$. However since the data was given in $\mathcal{V} \times \mathcal{H}$ and the MTR system is well-posed on that space we actually have the regularity in (6.2). $\square$

6.2. Behavior of (5.3)-(5.5) as $\gamma \to \infty$. The Kirchhoff system (5.36) can be written in variational form as

$$\frac{d}{dt}[c(\nabla \dot{w}, 0, \dot{w}; \nabla \hat{w}, 0, \hat{w})] + a_0(\nabla w, \nabla \hat{w}) = 0 \qquad \forall \hat{w} \in H^2_{\Gamma_0} \tag{6.7}$$

$$w \in C([0,T]; H^2_{\Gamma_0}) \cap C^1([0,T]; H^1_{\Gamma_0}) \tag{6.8}$$

$$(w(0), \dot{w}(0)) = (w^0, w^1) \qquad \text{in } H^2_{\Gamma_0} \times H^1_{\Gamma_0}. \tag{6.9}$$

By [LL, p.48], (6.7)-(6.9) has a unique solution.

The variational problem (5.3)-(5.5) was shown to have a unique solution in Theorem 5.1. We can eliminate the variable ξ and rewrite (5.3)-(5.5) in the form (5.23)-(5.24) and in this form, the problem has a unique solution with

$$(s, w) \in C([0,T]; W_0) \cap C^1([0,T]; V_0]), \tag{6.10}$$

where the spaces W_0 and V_0 are defined by (5.24).

Theorem 6.2. *Let (s^0, w^0), (s^1, w^1) be given in $W_0 \times V_0$ with $s^0 = s^1 = 0$. Then as $\gamma \to \infty$, the solution (s_γ, w_γ) to (5.23)-(5.24) tends to $(0, w)$ in the sense that $s_\gamma \to 0$ in $L^\infty(0,T; L^2(\Omega))$ and*

$$\begin{cases} \{w_\gamma\} \to w & \text{in } L^\infty(0,T; H^2_{\Gamma_0}) \text{ weak star} \\ \{\dot{w}_\gamma\} \to \dot{w} & \text{in } L^\infty(0,T; H^1_{\Gamma_0}) \text{ weak star,} \end{cases} \tag{6.11}$$

where w is the solution to (6.7)-(6.9).

Sketch of Proof. Again, let us mention only the main points. The details can be filled in following the proof of Theorem 5.2. For simplicity assume $\Gamma_0 \neq \emptyset$.

From Theorem 5.3 the energy

$$\mathcal{E}(t) = \frac{1}{2}c_1(s, w) + \frac{1}{2}b_0(s, w) + \frac{1}{2}a_2(s)$$

is conserved along all solution trajectories of (5.23)-(5.24). From this we know

$$\{\nabla w_\gamma - 3s_\gamma, s_\gamma\} \quad \text{bounded in } L^\infty(0,T; (H^1_{\Gamma_0})^4) \tag{6.12}$$

$$\{\nabla \dot{w}_\gamma - 3\dot{s}_\gamma, \dot{s}_\gamma, \dot{w}_\gamma\} \text{ bounded in } L^\infty(0,T; (L^2(\Omega))^5) \tag{6.13}$$

$$\{\gamma^{1/2} s\} \quad \text{bounded in } L^\infty(0,T; (L^2(\Omega))^2). \tag{6.14}$$

Thus $s \to 0$ strongly in $L^\infty(0,T; L^2(\Omega)^2)$ as $\gamma \to \infty$. Since $\Gamma_0 \neq \emptyset$ (6.12) implies that w_γ remains bounded in $L^\infty(0,T; H^2_{\Gamma_0})$. Thus by taking a subsequence (6.11) holds.

To see that w satisfies (6.7) we take in (5.23) test functions $\hat{s}, \hat{w}$ with $\hat{s} = 0$ so (5.23) (with $s = s_\gamma$, $w = w_\gamma$) reduces to

$$(6.15)\ \frac{d}{dt}c(\nabla \dot{w}_\gamma - 3s_\gamma, 0, \dot{w}_\gamma; \nabla\hat{w}, 0, \hat{w}) + a_0(\nabla w_\gamma - 3s_\gamma; \nabla\hat{w}) = 0\ \forall \hat{w} \in H^2_{\Gamma_0}$$

Since $s_\gamma \to 0$ strongly in $L^2(\Omega)$ and by (6.12) s_γ converges weakly to a limit function in $L^\infty(0,T; H^1_{\Gamma_0})$ this limit must also be zero. Thus passing the limit in (6.15) gives (6.7). What remains is to obtain the convergence of the initial values (6.9). This is done exactly in the same way as in the proof of Theorem 5.2, i.e., we use Theorem 5.3 to infer essential boundedness of $\ddot{w}$ in a weaker space and use this to infer (6.9), but in the weaker space. Since the initial data were given in $H^2_{\Gamma_0} \times H^1_{\Gamma_0}$ we then by well posedness of (6.7)-(6.9) obtain the regularity in (6.9). □

Acknowledgements. The author is indebted to J.E. Lagnese for many very helpful suggestions and also to Jiongmin Yong and Enrique Zuazua for their help on some technical points. Most of this research was conducted while the author was supported by Institute for Mathematics and its Applications (IMA) of the University of Minnesota during the "Year in Control Theory" 1992-93. The author is grateful for their hospitality.

References

[Di] DiTaranto, R A *Theory of vibratory bending for elastic and viscoelastic layered finite-length beams*, J Appl Mech **32** (1965), 881–886

[He] Heuer, R *Static and dynamic analysis of transversely isotropic, moderately thick sandwich beams by analogy*, Acta Mechanica **91** (1992), 1–9

[Ke] Kerwin, E M *Damping of flexural waves by constrained visco-elastic layer*, J Acoustical Soc of Am **31** (1959), 952–962

[La] Lagnese, J E *Boundary Stabilization of Thin Plates*, series "SIAM Studies in Applied Mathematics" SIAM, Philadelphia 1989

[LL] Lagnese, J E and Lions, J -L *Modelling Analysis and Control of Thin Plates*, collection "Recherches en Mathématiques Appliquées", RMA 6, Springer-Verlag, New York, 1989

[Li] Lions, J -L *Equations Différentielles Opérationnelles et Problemes aux Limites*, Grundlehren B III, Springer-Verlag, Berlin 1961

[LM] Lions, J -L and Magenes, E *Non-Homogeneous Boundary Value Problems and Applications*, Vol 1, Springer-Verlag, Berlin 1972

[Me] Mead, D J *A comparison of some equations for the flexural vibration of damped sandwich beams*, J Sound Vib **83** (1982), 363–377

[Pa] Pazy, A *Semigroups of linear operators and applications to partial differential equations*, Springer-Verlag series "Applied Mathematical Sciences " #44 Springer-Verlag, New York, 1983

[Ti] Timoshenko, S , Young, D H , Weaver, W *Vibration problems in engineering*, New York, Wiley, 1974

[YD] Yan, M -J , Dowell, E H *Governing equations for vibrating constrained-layer damping sandwich plates and beams*, J Appl Mech **39** (1972) 1041–1046

[Yu] Yu, Y -Y *Simple thickness-shear modes of vibration of infinite sandwich plates*, J Appl Mech **26** (1959), 679–681

Scott W. Hansen
Department of Mathematics
Iowa State University,
Ames, Iowa 50011, USA

NUMERICAL SOLUTION OF A CONSTRAINED CONTROL PROBLEM FOR A PHASE FIELD MODEL

M HEINKENSCHLOSS AND E W SACHS*

Interdisciplinary Center for Applied Mathematics
Department of Mathematics
Virginia Polytechnic Institute and State University

FB IV – Mathematik
Universitat Trier

ABSTRACT A phase field model is considered for an evolution process which describes the phase change from solid to liquid The process is controlled by inputs which are uniformly bounded The objective is that the phase and the temperature follow certain desired functions as close as possible with a cost term for the control The necessary optimality conditions of first order are formulated for this control problem These conditions include projections which make the application of Newton's method difficult We present an approach to circumvent this difficulty and obtain an algorithm which is very efficient numerically The numerical examples indicate that the required time to solve the problem with control constraints is of the same magnitude as for the unconstrained problem

1991 *Mathematics Subject Classification* 49M15

Key words and phrases Optimal control, bound constraints, projected Newton method, phase field model

1. Introduction

Phase field models describe the solid–liquid phase transition in a pure material They involve the temperature of the material and the phase function which indicates the liquid or solid state of the material These models consist of a system of two nonlinear parabolic differential equations, [4], [7], [8], [13], [15] We consider the phase field model in [4], [7], [8]

The corresponding control problem is to follow a prescribed temperature and/or phase by controlling the temperature through a system of diffusion equations Problems of this type have been considered in [5], [6], [11] For these problems

*This work was supported in part by AFOSR under Grant F49620–93 1 0280 while the author was a visiting scientist at the Interdisciplinary Center for Applied Mathematics, Virginia Polytechnic Institute and State University

necessary optimality conditions were obtained. Computations in [6] were done using a gradient method combined with a discretization scheme of the differential equation. In [10] a different approach was presented which exploits the necessary optimality conditions, a system of four coupled nonlinear partial differential equations. The problem was reformulated and Newton's method was applied. Due to the reformulation a decoupled system needed to be solved at each step. This was carried out for a discretized version using GMRES with a hierarchical basis preconditioner.

In this paper we extend the model and add simple bounds as constraints on the controls. Existence of optimal controls and necessary optimality conditions have been shown in [11]. We rewrite these and obtain a fixed point problem for the optimal control, temperature, phase, and their adjoints. A classical approach for the solution of these box constrained problems is the gradient projection method. However, this method is a first order method and converges rather slowly. Moreover, it requires the solution of the nonlinear state equation in each step. Newton's method, on the other hand, cannot be applied directly to the optimality system because the projection involved for the controls is nondifferentiable.

It is the goal of this paper to obtain a projected Newton algorithm with favorable convergence properties. For minimization problems in finite dimensions the gradient projection method has been adapted in [2] to successfully utilize Newton steps. In [12] an extension of this approach is used to solve the necessary optimality conditions in infinite dimensions obtained from a control problem with ordinary differential equations and control constraints. The advantage over the approach in [2] is that at each step only the linearized state equation needs to be solved instead of the original nonlinear one. A convergence proof in infinite dimensions is given for an ordinary control problem with a control in one dimension.

For the phase field control problem we can also formulate an algorithm which leads to a projected Newton method. To do this we extend the approach in [12] to control problems governed by systems of parabolic equations and combine this with the multilevel Newton method used in [10] for the unconstrained problem. Each step requires the solution of a system of linear boundary value problems where the decoupling strategy [10] from the unconstrained case can be used again. This yields an efficient solver for the linearized state and adjoint equations. As in the unconstrained case this allows to eliminate the control in the linear system and we obtain an algorithm which requires hardly any more work than for the unconstrained problem. The numerical results underscore this observation and show that the projected Newton method is an efficient solver for the phase field control problem with control constraints.

2. The Control Problem

Solid–liquid phase transitions in pure materials can be modeled by so–called phase field models. These models can be viewed as extensions of the classical Stefan problem. Instead of using an explicit condition for the interface boundary between solid and liquid phase, phase field models consist of systems of two nonlinear parabolic equations. The interface can be constructed from the so-called phase function. The advantage of phase field models is that the explicit use of the free boundary is avoided and that phenomena associated with surface tension and supercooling are incorporated into the model. In this paper we consider the phase field model proposed in [4], [7], [8]. Let Ω be a subset of $\mathbb{R}^n$, $n \leq 3$, with C^2 boundary and let $Q = \Omega \times (0,T)$. By $u(x,t)$ we denote the temperature at $x \in \Omega$ and time $t \in [0,T]$. The phase function denoted by $\varphi(x,t)$ describes the phase of the medium. The medium is purely solid if $\varphi(x,t) = -1$ and it is purely liquid if $\varphi(x,t) = +1$. The boundary between the two phases can be described by

$$\Gamma(x,t) = \{(x,t) \in Q \mid \varphi(x,t) = 0\}.$$

The relation between the two quantities u and φ is given through a boundary value problem with two coupled semilinear parabolic differential equations.

$$\begin{aligned} u_t + \tfrac{l}{2}\varphi_t &= \kappa\Delta u + f \\ \tau\varphi_t &= \xi^2\Delta\varphi + g(\varphi) + 2u \end{aligned} \qquad \text{on } Q = \Omega \times (0,T). \tag{2.1}$$

Here g is a nonlinear function of the type

$$g(z) = az + bz^2 - cz^3, \quad a,b,c \in C(\bar{\Omega}),\ a,b,c \geq 0.$$

Typically values like $a = c = 0.5$, $b = 0$, are used in [4], [8]. The constants κ, ℓ, τ, and ξ denote the heat conductivity, the latent heat, the relaxation time, and the length scale of the interface, respectively. The function f represents the control input for the system. The boundary conditions are of Neumann type

$$\frac{\partial}{\partial n}u = 0, \quad \frac{\partial}{\partial n}\varphi = 0 \qquad \text{on } \partial\Omega \times (0,T) \tag{2.2}$$

and the initial conditions are

$$u(x,0) = u_0(x), \quad \varphi(x,0) = \varphi_0(x) \qquad x \in \Omega. \tag{2.3}$$

We assume that the initial conditions satisfy

$$u_0, \varphi_0 \in W^2_\infty(\Omega), \quad \frac{\partial}{\partial n}u_0 = \frac{\partial}{\partial n}\varphi_0 = 0. \tag{2.4}$$

Next we formulate the optimal control problem. Suppose that a certain desired phase function φ_d and a temperature u_d are given. The goal is to drive the melting process using f in such a way that the actual phase and temperature is as close as possible to the desired ones. We also include a cost term for the control f in the objective function $J(f)$

$$J(f) = \frac{\alpha}{2}\|u - u_d\|^2_{L^2(Q)} + \frac{\beta}{2}\|\varphi - \varphi_d\|^2_{L^2(Q)} + \frac{\gamma}{2}\|f\|^2_{L^2(Q)}. \tag{2.5}$$

In this paper we also impose pointwise constraints on the control. Given $f_{low}, f_{upp} \in L^\infty(Q)$ the set of admissible controls is defined by

$$F_{ad} = \{f \in L^\infty(Q) \mid f_{low} \leq f \leq f_{upp} \text{ on } Q\}.$$

This leads to the formulation of the optimal control problem using (2.1) – (2.5):

Optimal Control Problem (OCP)

Find $f^* \in F_{ad}$ such that $J(f^*) \leq J(f)$ for all $f \in F_{ad}$.

It has been shown in [11] that the state equation (2.1) – (2.3) has a unique solution and that the solution operator $T(f) = (u, \varphi)$ corresponding to (2.1) – (2.3) satisfies

$$T : L^\infty(Q) \to W_p^{2,1}(Q) \times W_p^{2,1}(Q), \qquad p < \infty.$$

The existence of optimal controls for the unconstrained problem has been shown in [11]. Analogous arguments can be used to prove existence of optimal controls for (OCP).

In the next section we look at the necessary optimality conditions which will lead to the construction of an efficient algorithm.

3. Necessary Optimality Conditions

If the objective function $J(f)$ is Fréchet-differentiable then it is well known that the necessary optimality condition can be expressed as

$$J'(f^*)\,(f - f^*) \geq 0 \quad \text{for all } f \in F_{ad}. \tag{3.1}$$

It is shown in [11] that the solution map T is Fréchet-differentiable with Fréchet-derivative $T'(f^*)(f) = (u, \varphi)$ given by the linearized system of (2.1):

$$\begin{array}{rcl} u_t + \frac{l}{2}\varphi_t & = & \kappa\Delta u + f \\ \tau\varphi_t & = & \xi^2\Delta\varphi + g'(\varphi^*)\varphi + 2u \end{array} \quad \text{on } Q \tag{3.2}$$

with boundary conditions (2.2) and homogeneous initial conditions. In equation (3.2) (u^*, φ^*) denotes the solution of (2.1), (2.2), (2.3) with $f = f^*$, i.e. $T(f^*) = (u^*, \varphi^*)$. In order to write $J'(f^*)f$ in a gradient notation we need to introduce the adjoint equation which is also a system of linear parabolic differential equations. For given f^* and $T(f^*) = (u^*, \varphi^*)$ let (p^*, ψ^*) solve the system of adjoint equations:

$$\begin{array}{rcl} -p_t^* & = & \kappa\Delta p^* + 2\psi^* + \alpha(u^* - u_d) \\ -\tau\psi_t^* - \frac{l}{2}p_t^* & = & \xi^2\Delta\psi^* + g'(\varphi^*)\psi^* + \beta(\varphi^* - \varphi_d) \end{array} \quad \text{on } Q, \tag{3.3}$$

with boundary and final conditions

$$\begin{array}{c} \frac{\partial}{\partial n}p^* = 0, \qquad \frac{\partial}{\partial n}\psi^* = 0 \quad \text{on } \partial\Omega \times (0,T), \\ p^*(x,T) = 0, \quad \psi^*(x,T) = 0 \quad \text{on } \Omega. \end{array} \tag{3.4}$$

Then we can show that the gradient $\nabla J(f^*)$ can be computed by solving (2.1) and (3.3) and is given by

$$\nabla J(f^*) = p^* + \gamma f^*.$$

Note that due to the form of F_{ad} the necessary optimality condition (3.1) can be rewritten as

$$f^* = P(f^* - (p^* + \gamma f^*)), \tag{3.5}$$

where the projection $P : L^\infty(Q) \to L^\infty(Q)$ is defined as

$$P(f)(x,t) = \begin{cases} f_{upp}(x,t) & \text{if} \quad f(x,t) > f_{upp}(x,t), \\ f_{low}(x,t) & \text{if} \quad f(x,t) < f_{low}(x,t), \\ f(x,t) & \text{else.} \end{cases}$$

If we use the variable transformation

$$u(x,t) \to u(x,t) + u_0(x), \quad \varphi(x,t) \to \varphi(x,t) + \varphi_0(x)$$

to obtain a problem with homogeneous initial and boundary conditions for u and φ and if we use (3.5), then we can write (3.1) as a system of nonlinear equations:

Find $z = (u, \varphi, p, \psi, f)^T$ such that

Necessary Optimality Conditions (NOC)

$$\begin{aligned} u_t + \tfrac{l}{2}\varphi_t - \kappa\Delta u &= \kappa\Delta u_0 + f, \\ \tau\varphi_t - \xi^2\Delta\varphi &= \xi^2\Delta\varphi_0 + g(\varphi + \varphi_0) + 2(u + u_0) \\ -p_t - \kappa\Delta p &= 2\psi + \alpha(u + u_0 - u_d), \\ -\tau\psi_t - \tfrac{l}{2}p_t - \xi^2\Delta\psi &= g'(\varphi + \varphi_0)\psi + \beta(\varphi + \varphi_0 - \varphi_d), \\ f &= P(f - (p + \gamma f)) \end{aligned}$$

$$\tfrac{\partial}{\partial n}u = 0, \ \tfrac{\partial}{\partial n}\varphi = 0, \ \tfrac{\partial}{\partial n}p = 0, \ \tfrac{\partial}{\partial n}\psi = 0 \ \text{ on } \partial\Omega \times (0,T),$$
$$u(x,0) = 0, \ \varphi(x,0) = 0, \ p(x,T) = 0, \ \psi(x,T) = 0 \text{ on } \Omega.$$

To rewrite the problem (NOC) into a form more suitable for our approach, we introduce a linear operator $S : L^p(Q)^4 \to W_p^{2,1}(Q)^4, p > 1$, which is the solution operator of a boundary value problem. For $w = (w_1, w_2, w_3, w_4)^T \in L^p(Q)^4, p > 1$, let $Sw = (u, \varphi, p, \psi)^T$ be the solution of

$$\begin{aligned} u_t + \tfrac{l}{2}\varphi_t - \kappa\Delta u &= w_1, \\ \tau\varphi_t - \xi^2\Delta\varphi &= w_2, \\ -p_t - \kappa\Delta p &= w_3, \\ -\tau\psi_t - \tfrac{l}{2}p_t - \xi^2\Delta\psi &= w_4, \end{aligned} \tag{3.6}$$

where the solution has to satisfy the homogeneous boundary, initial and final conditions from (NOC). Note that the system (3.6) is decoupled and $Sw = (u, \varphi, p, \psi)^T$

can be computed by solving four simple parabolic equations. For example, the second equation (3.6) only depends on φ. If φ is computed, then it can be substituted into the first equation which then can be solved for u. Similarly, one can use the third and fourth equation to compute p and ψ. This block triangular structure of S will also be used for the numerical solution, cf. Section 5.

Later we need an extension S_e of S which is defined as

$$S_e(w,f) = (Sw, f), \quad (w,f) \in L^p(Q)^5.$$

If we define a nonlinear operator $G : L^\infty(Q)^5 \to L^\infty(Q)^5$ which includes the nonlinearities of the differential equation and takes care of the inhomogeneous terms by

$$G(z) = \begin{pmatrix} \kappa \Delta u_0 + f \\ \xi^2 \Delta \varphi_0 + g(\varphi + \varphi_0) + 2(u + u_0) \\ 2\psi + \alpha(u + u_0 - u_d) \\ g'(\varphi + \varphi_0)\psi + \beta(\varphi + \varphi_0 - \varphi_d) \\ P(f - (p + \gamma f)) \end{pmatrix}, \tag{3.7}$$

then the equation (NOC) can be rewritten as a fixed point equation to find $z = (u, \varphi, p, \psi, f)^T$ such that

$$F(z) = z - S_e G(z) = 0 \tag{3.8}$$

with a nonlinear operator $F : X = L^\infty(Q)^5 \to L^\infty(Q)^5$.

4. A Projected Newton Method

For nonlinear problems of the type (3.8), Newton's method can be used for the numerical solution in a very efficient way, see e.g. [10]. A complicating factor is the appearance of the projection term in the last component of (3.8) besides the fact that systems of parabolic boundary value problems are included. The projection P destroys the Fréchet-differentiability of the operator F. Therefore we have to modify Newton's method in a proper way. A similar problem with ordinary differential equations has been considered in [12]. Here we extend this method to the case of partial differential equations.

If one considers only the last component of G, then a solution technique by successive approximation would lead to a method of the projected gradient. This is a first order method and converges rather slowly. The improvement of this method by using second order information is not straightforward. It was shown in [2] that a mere replacement of the gradient, in our case $p + \gamma f$, by a Newton or Newton–like direction does no longer guarantee descent and is therefore inappropriate. However, it was also shown in [2] that a fast convergent algorithm can be obtained if one considers for the computation of the Newton direction certain index sets which exclude subsets of the active indices. This was the motivation in [12] to restrict the indices in the projection to the inactive one when using a Newton step for the solution of the nonlinear equation (3.8).

Following [12] we define the sets

$$\begin{aligned} A_{upp}(z) &= \{(x,t) \in Q \mid p(x,t) + \gamma f(x,t) > \|F(z)\|_X^r\}, \\ A_{low}(z) &= \{(x,t) \in Q \mid p(x,t) + \gamma f(x,t) < -\|F(z)\|_X^r\} \end{aligned}$$

depending on the size of the residual $\|F(z)\|_X$ and an exponent $r \in (0,1)$, and

$$A(z) = A_{upp}(z) \cup A_{low}(z), \qquad I(z) = Q \setminus A(z).$$

At the optimal point $z^* = (u^*, \varphi^*, p^*, \psi^*, f^*)^T$ it holds that $F(z^*) = 0$ and, see (3.5),

$$\begin{aligned} p^*(x,t) + \gamma f^*(x,t) > 0 &\implies (x,t) \in A^*_{upp} = \{(x,t) \in Q \mid f^*(x,t) = f_{upp}(x,t)\}, \\ p^*(x,t) + \gamma f^*(x,t) < 0 &\implies (x,t) \in A^*_{low} = \{(x,t) \in Q \mid f^*(x,t) = f_{low}(x,t)\}. \end{aligned}$$

Thus,

$$A_{upp}(z^*) \subset A^*_{upp}, \quad A_{upp}(z^*) \subset A^*_{low}.$$

This motivates to use $A(z)$ as an estimate for the active set at the point z. However, this set is not necessarily equal to the true active set.

Since Newton's method is locally convergent we assume that the obvious choice to set f equal to the maximal and minimal value on $A_{upp}(z)$, and $A_{low}(z)$, respectively, is correct. Therefore, for given $z = (u, \varphi, p, \psi, f)^T$ we reset

$$f = f_{upp} \text{ on } A_{upp}(z), \qquad f = f_{low} \text{ on } A_{low}(z)$$

if these identities did not hold already. The main step is the computation of a correction of f on $I(z)$ with the Newton step. However, $I(z)$ is considered an approximation of the set of inactive indices. On this set the projection usually is the identity, so that we approximate on $I(z)$ the nonlinear maps F and G by $\tilde{F}$ and $\tilde{G}$, respectively.

$$\tilde{F}(z) = z - S_e \tilde{G}(z), \qquad \tilde{G}(z) = \begin{pmatrix} \kappa \Delta u_0 + f \\ \xi^2 \Delta \varphi_0 + g(\varphi + \varphi_0) + 2(u + u_0) \\ 2\psi + \alpha(u + u_0 - u_d) \\ g'(\varphi + \varphi_0)\psi + \beta(\varphi + \varphi_0 - \varphi_d) \\ f - (p + \gamma f) \end{pmatrix}.$$

It can be shown [10] that $\tilde{G}$ is Fréchet-differentiable and admits the representation

$$\tilde{G}'(z)(\delta z) = \begin{pmatrix} \delta f \\ g'(\varphi+\varphi_0)\delta\varphi + 2\delta u \\ 2\delta\psi + \alpha\delta u \\ g''(\varphi+\varphi_0)\delta\varphi\ \psi + g'(\varphi+\varphi_0)\delta\psi + \beta\delta\varphi \\ \delta f - (\delta p + \gamma\delta f) \end{pmatrix}, \quad \delta z = \begin{pmatrix} \delta u \\ \delta\varphi \\ \delta p \\ \delta\psi \\ \delta f \end{pmatrix}.$$

Since S_e is linear we have that

$$\tilde{F}'(z)\delta z = \delta z - S_e\tilde{G}'(z)\delta z.$$

To describe the Newton step more precisely, let

$$Q_I z = (u, \varphi, p, \psi, \chi_I f)^T \text{ for } z = (u, \varphi, p, \psi, f)^T$$

with the characteristic function χ_I on a set $I \subset Q$. Then the Newton step $\delta z \in L^\infty(Q)^5$ is obtained from the solution of

$$Q_{I(z)}\tilde{F}'(z)Q_{I(z)}\delta z = -Q_{I(z)}\tilde{F}(z). \tag{4.1}$$

The equation (4.1) can be written as

$$Q_{I(z)}\delta z - Q_{I(z)}S_e\tilde{G}'(z)Q_{I(z)}\delta z = -Q_{I(z)}z + Q_{I(z)}S_e\tilde{G}(z)$$

or

$$Q_{I(z)}(z+\delta z) = Q_{I(z)}S_e[\tilde{G}'(z)Q_{I(z)}(z+\delta z) - \tilde{G}'(z)Q_{I(z)}z + \tilde{G}(z)]. \tag{4.2}$$

With the definition of the solution operator S in (3.6) we can rewrite (4.2) as a system of boundary value problems. We denote the new iterate by a superscript $+$. The new iterate is given as the projection of $z + Q_{I(z)}\delta z$ onto the feasible set:

$$z^+ = \tilde{P}(z + Q_{I(z)}\delta z),$$

where $\tilde{P}(z) = (p, \psi, u, \varphi, P(f))^T$. Since the projection is only applied to the fifth component, the first four components of the new iterate z^+ are already obtained from (4.2), which can be rewritten as

$$
\begin{aligned}
u_t^+ + \frac{l}{2}\varphi_t^+ - \kappa\Delta u^+ &= \kappa\Delta u_0 + \chi_{I(z)}\delta f + f,\\
\tau\varphi_t^+ - \xi^2\Delta\varphi^+ &= \xi^2\Delta\varphi_0 + g'(\varphi+\varphi_0)(\varphi^+ - \varphi)\\
&\quad + 2(u^+ + u_0) + g(\varphi+\varphi_0),\\
-p_t^+ - \kappa\Delta p^+ &= 2\psi^+ + \alpha(u^+ + u_0 - u_d), \qquad (4.3)\\
-\tau\psi_t^+ - \frac{l}{2}p_t^+ - \xi^2\Delta\psi^+ &= -g''(\varphi+\varphi_0)(\varphi^+ - \varphi)\psi + g'(\varphi+\varphi_0)\psi^+\\
&\quad + \beta(\varphi^+ + \varphi_0 - \varphi_d),\\
\chi_{I(z)}(p^+ + \gamma(f+\delta f)) &= 0,
\end{aligned}
$$

with homogeneous boundary and initial conditions of the type (2.2) and (3.4). The fifth component of z^+ is given by $f^+ = P(f + \chi_{I(z)}\delta f)$, where $\chi_{I(z)}\delta f$ is obtained as the fifth component of the solution of (4.3).

As in the unconstrained case we note that the last equation can be omitted by eliminating $\chi_{I(z)}\delta f$. From the last equation we obtain that

$$\chi_{I(z)}\delta f = -\chi_{I(z)}f - \frac{1}{\gamma}\chi_{I(z)}p^+,$$

which is substituted into the first equation in (4.3).

In summary, the conceptual algorithm looks as follows. Here we have included the inhomogeneous data again into the initial conditions.

Algorithm

Given $r \in (0,1)$ and a current iterate $z^c = (p^c, \psi^c, u^c, \varphi^c, f^c)^T$.

Step 1 *Determine the estimate for the active set*

$$
\begin{aligned}
A_{upp}(z^c) &= \{(x,t) \in Q \mid p^c(x,t) + \gamma f^c(x,t) > \|F(z^c)\|_X^r\},\\
A_{low}(z^c) &= \{(x,t) \in Q \mid p^c(x,t) + \gamma f^c(x,t) < -\|F(z^c)\|_X^r\},\\
A^c &= A(z^c) = A_{low}(z^c) \cup A_{upp}(z^c),\\
I^c &= I(z^c) = Q \setminus A(z^c),
\end{aligned}
$$

and overwrite f^c by

$$f^c(x,t) = \begin{cases} f_{upp}(x,t) & \text{on } A_{upp}(z^c),\\ f_{low}(x,t) & \text{on } A_{low}(z^c).\end{cases}$$

Step 2 *Solve the linear boundary value problem*

$$\begin{aligned}
u_t^+ + \tfrac{l}{2}\varphi_t^+ - \kappa\Delta u^+ &= \chi_{A^c} f^c - \tfrac{1}{\gamma}\chi_{I^c} p^+,\\
\tau\varphi_t^+ - \xi^2\Delta\varphi^+ &= g'(\varphi^c)(\varphi^+ - \varphi^c) + 2u^+ + g(\varphi^c),\\
-p_t^+ - \kappa\Delta p^+ &= 2\psi^+ + \alpha(u^+ - u_d),\\
-\tau\psi_t^+ - \tfrac{l}{2}p_t^+ - \xi^2\Delta\psi^+ &= -g''(\varphi^c)(\varphi^+ - \varphi^c)\psi^c + g'(\varphi^c)\psi^+ + \beta(\varphi^+ - \varphi_d),
\end{aligned}$$

$$\begin{array}{rclrcll}
u^+(x,0) &=& u_0(x), & \varphi^+(x,0) &=& \varphi_0(x), & x \in \Omega,\\
\frac{\partial}{\partial n}u^+(x,t) &=& 0, & \frac{\partial}{\partial n}\varphi^+(x,t) &=& 0, & x \in \partial\Omega,\ t \in (0,T),\\
p^+(x,T) &=& 0, & \psi^+(x,T) &=& 0, & x \in \Omega,\\
\frac{\partial}{\partial n}p^+(x,t) &=& 0, & \frac{\partial}{\partial n}\psi^+(x,t) &=& 0 & x \in \partial\Omega,\ t \in (0,T).
\end{array}$$

Step 3 *Set*

$$f^+ = \chi_{A^c} f^c + P(\chi_{I^c}(f^c + \delta f^c)).$$

The system in Step 2 has been rewritten in a way that is very similar to the Newton system in the unconstrained case [10] and can be solved efficiently using multilevel approaches. A formulation of the system in Step 2 as a linear fixed point problem using the solution operator S yields a decoupling of the system and allows an efficient solution using multilevel approaches. The decoupling and the application of multilevel methods as well as further implementation details are described in more detail in the following section.

5. Implementation and Numerical Results

In each step of the projected Newton method introduced in the previous section we have to solve a system of four linear parabolic equations in the reduced set of variables $\bar{z} = (u, \varphi, p, \psi)^T$. With the solution operator S of (3.6) and

$$B = \begin{pmatrix} 0 & 0 & -\frac{1}{\gamma}\chi_{I^c} & 0 \\ 2 & g'(\varphi^c) & 0 & 0 \\ \alpha & 0 & 0 & 2, \\ 0 & -g''(\varphi^c)\psi^c + \beta & 0 & g'(\varphi^c) \end{pmatrix}, \quad b = \begin{pmatrix} \chi_{A^c} f^c, \\ -g'(\varphi^c)\varphi^c + g(\varphi^c), \\ -\alpha u_d, \\ +g''(\varphi^c)\varphi^c\psi^c - \beta\varphi_d \end{pmatrix},$$

the system arising in Step 2 of the algorithm can be written as

$$(I - K)\bar{z}^+ = b, \tag{5.1}$$

where $K = SB$. Due to the block triangular structure of the solution operator S the application of K to $\bar{z}^+$ requires the sequential solution of a decoupled system of differential equations. For example, given $\bar{z}^+ = (u^+, \varphi^+, p^+, \psi^+)^T$ the product $(u, \varphi, p, \psi)^T = K\bar{z}^+$ can be evaluated by first solving the second equation in (3.6) with $w_2 = 2u^+ + g'(\varphi^c)\varphi^+$ and corresponding boundary conditions. Then the solution φ can be inserted into the first equation and the first equation with

$w_1 = -\frac{1}{\gamma}\chi_{I^c}p^+$ can be solved for u. The third and fourth components p and ψ can be computed similarly. It should also be noted that for the sequential solution we only need to solve heat equations. There exist several efficient methods to perform this task numerically, such as multigrid methods, domain decomposition methods, or multilevel preconditioned conjugate gradient methods. If these methods are used, then the time needed to evaluate $K\bar{z}^+$ for given $\bar{z}^+$ is roughly linear in the number of variables introduced by the discretization.

Moreover, using the block triangular structure of S, the smoothing properties of the heat equation and Sobolev imbedding theorems one can show that K is a compact operator from $L^\infty(Q)^4$ to $L^\infty(Q)^4$, [10]. Therefore (5.1) is a compact fixed point problem and one can use multigrid methods [1], [3], [9] for the solution of (5.1). In our implementations we use the multilevel method from [1], [3] which uses a fine and a coarse grid for the approximation of $(I-K)^{-1}$. Given an approximation $\bar{z}^+$ for the solution of (5.1) the correction $\delta\bar{z}^+$ yielding the exact solution $\bar{z}^+ + \delta\bar{z}^+$ satisfies

$$(I-K)\delta\bar{z}^+ = b - (I-K)\bar{z}^+.$$

Using

$$(I-K)^{-1} = I + (I-K)^{-1}K \approx I + (I_l - K_l)^{-1}K,$$

where the index l denotes a coarse grid and $(I_l - K_l)^{-1}$ is the inverse of a coarse grid operator, one finds that

$$\delta\bar{z}^+ \approx [I + (I_l - K_l)^{-1}K](b - (I-K)\bar{z}^+).$$

Thus the iteration used to solve the system of four linear parabolic equations in Step 2 of the algorithm is given by

$$\bar{z}^+ \longleftarrow \bar{z}^+ + [I + (I_l - K_l)^{-1}K](b - (I-K)\bar{z}^+). \tag{5.2}$$

In our implementation we start this iteration using the current Newton iterate $(u^c, \varphi^c, p^c, \psi^c)^T$. One iteration requires the application of the fine grid operator K, which can be done very efficiently as explained previously and the solution of a coarse grid system with matrix $(I_l - K_l)$. Although the coarse grid system is considerably smaller than the fine grid system, it is still large. In the numerical experiment described below, the coarse grid system has 891 variables. Therefore it also has to be solved iteratively. Since K_l is nonsymmetric, we use GMRES [14]. In our numerical experiments one iteration (5.2) with starting value $(u^c, \varphi^c, p^c, \psi^c)^T$ was sufficient to compute the solution $\bar{z}^+ = (u^+, \varphi^+, p^+, \psi^+)^T$ with required accuracy.

For the discretization of the parabolic partial differential equations we use linear finite elements in space and backward Euler in time. The elliptic differential equations arising in each time step are solved using a hierarchical basis preconditioned conjugate gradient method [16].

The example we use to demonstrate the projected Newton multilevel method is taken from [6]. The following parameters and functions are used.

$$\Omega = (0,1)^2, T = 0.05, a = \frac{1}{2}, b = 0, c = \frac{1}{2}, \kappa = \ell = 1, \tau = 0.05, \xi^2 = 10^{-2},$$

and

$$\begin{aligned}
\varphi_d &= (x-\frac{1}{2})^2 + (y-\frac{1}{2})^2 - \frac{3}{4}\left((x-\frac{1}{2})\cos\theta + (y-\frac{1}{2})\sin\theta\right)^2 \\
&- \left(R_0^2 + \frac{t}{T}(R_1^2 - R_0^2)\right), \\
u_d &= \frac{1}{2}\left(\tau\varphi_t - \xi_x^2\varphi_{xx} - \xi_y^2\varphi_{yy} + (\varphi^3 - \varphi)\right),
\end{aligned}$$

where $\theta = \frac{\pi}{4}(2 - \frac{t}{T})$, $R_0 = 0.05$, $R_1 = 0.15$.

The weighting parameters in the objective function are chosen to be

$$\alpha = 50, \ \beta = 50, \ \gamma = 10^{-2}.$$

We made two tests runs. One for the unconstrained case $f_{low} = -\infty, f_{upp} = \infty$ and one for a constrained problem where we used $f_{low} = -1.3, f_{upp} = 1.3$. In both cases we used a nested approach, starting on a coarse grid and using the interpolation of the coarse grid solution as a starting point on the next finer grid.

As mentioned previously, we use a finite element discretization in space and backward Euler in time. The triangulation of the spatial domain is obtained by dividing x– and y–axis into equidistant subintervals of length h and then dividing each of the resulting subsquares into two triangles. The grid is refined by doubling the number of subintervals in the x– and y–axis and doubling the number of time steps. Thus the number of triangles per time step increases by a factor of four and the total number of variables roughly increases by a factor of eight. We use hierarchical basis in our finite element spaces [16]. The level $\ell = 0$ of the spatial grid conists of two triangles and four vertices. The process of increasing the level by one is described above and increases the number of triangles by a factor of four. For our discretization the level ℓ spatial grid consists of $2^{(2\ell+1)}$ triangles and $(2^\ell + 1)^2$ vertices.

In the tables below ℓ refers to the spatial grid level, *ntime* denotes the number of time steps (without the initial time), and *node* is the number of nodes in the space–time–grid. The coarse grid is given by $\ell = 3$, $ntime = 10$, which means that the number of nodes in the space–time–grid is equal to $(2^3 + 1)^2 \cdot 11 = 891$. Since the variable in our problem is given by $z = (u, \varphi, p, \psi, f)^T$, the number of variables in the coarse grid is given by $5 \cdot 891 = 4455$.

The column *itg* shows the number of GMRES iterations needed to solve the coarse grid system in the multilevel approach. The times given show the accumulated time needed to solve the problem on the specified grid.

Nested Multilevel Newton Method
(Unconstrained Problem, $f_{low} = -\infty, f_{upp} = \infty$)

ℓ	*ntime*	*nodes*	*iter*	$\|\|F(z)\|\|$	$\|\|s\|\|$	*itg*	*time*
3	10	891	0	$0.608E-01$	$0.467E+00$	7	
			1	$0.316E-01$	$0.351E+00$	9	
			2	$0.101E-01$	$0.593E-01$	8	
			3	$0.112E-02$	$0.108E-01$	12	
			4	$0.151E-03$			75
4	20	6069	0	$0.145E+00$	$0.139E+00$	17	
			1	$0.984E-02$			73
5	40	44649	0	$0.448E-01$	$0.851E-01$	15	
			1	$0.759E-02$			324

TABLE 1

Nested Multilevel Projected Newton Method
(Constrained Problem, $f_{low} = -1.3, f_{upp} = 1.3, r = 0.9$)

ℓ	*ntime*	*nodes*	*iter*	$\|\|F(z)\|\|$	$\|\|s\|\|$	$\|\bar{A}(z)\|$	*itg*	*time*
3	10	891	0	$0.608E-01$	$0.467E+00$	10	7	
			1	$0.316E-01$	$0.359E+00$	291	9	
			2	$0.102E-01$	$0.103E-01$	436	11	
			3	$0.152E-02$	$0.285E-01$	463	11	
			4	$0.110E-03$				77
4	20	6069	0	$0.150E+00$	$0.140E+00$	18	17	
			1	$0.987E-02$		2439		72
5	40	44649	0	$0.475E-01$	$0.523E-01$	6990	17	
			1	$0.624E-02$		18502		354

TABLE 2

Figures 1 and 2 show the computed controls at time $t = 0\ 05$

Computed Control, $f_{low} = -\infty, f_{upp} = \infty$**,** $t = 0\ 005$

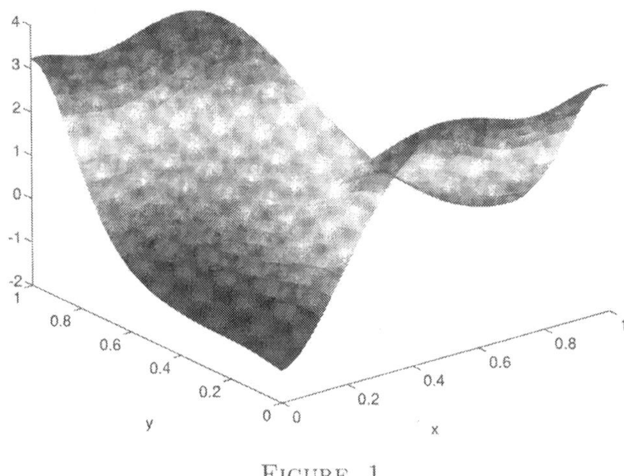

FIGURE 1

Computed Control, $f_{low} = -1\ 3\ \ f_{upp} = 1\ 3$**,** $t = 0\ 005$

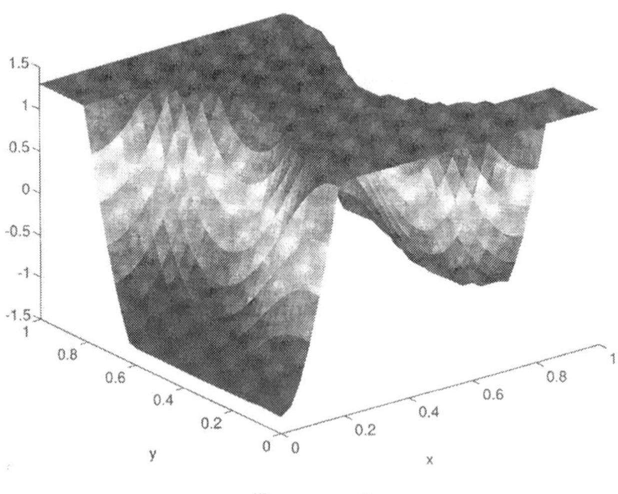

FIGURE 2

Figures 3 and 4 show the contours $\{\varphi = 0\}$ of the phase function for $t = 0, 0.005, 0.01, \ldots, 0.05$ computed in the unconstrained and constrained case. As noted in the introduction, these contours indicate the phase boundary.

Contours of the Phase Function φ
$(f_{low} = -\infty, f_{upp} = \infty)$

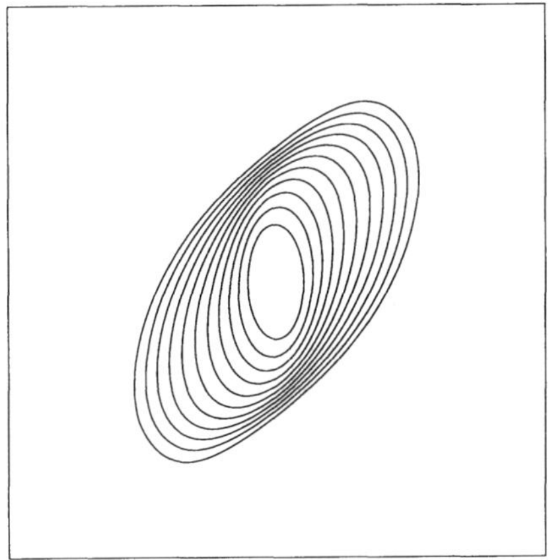

FIGURE 3

Contours of the Phase Function φ
$(f_{low} = -1.3, f_{upp} = 1.3)$

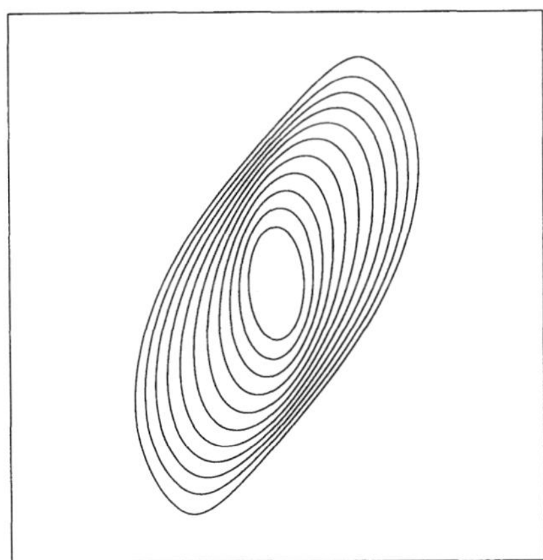

FIGURE 4

Acknowledgement. The authors appreciate the help of Robert Scholz in the section on the numerical results

References

1 K Atkinson The numerical evaluation of fixed points for completely continuous operators *SIAM J Numer Anal*, 10 799–807, 1973

2 D B Bertsekas Projected Newton methods for optimization problems with simple constraints *SIAM J Control Optim*, 20 221–246, 1982

3 H Brakhage Uber die numerische Behandlung von Integralgleichungen nach der Quadraturformelmethode *Numer Math*, 2 183–196, 1960

4 G Caginalp An analysis of a phase field model of a free boundary *Archive for Rational Mechanics and Analysis*, 92 205–245, 1986

5 Z Chen *Das Phasenfeldmodell fur das Problem der Phasenubergange Analysis, Numerik und Steuerung* PhD thesis, Institut fur Mathematik, Universitat Augsburg, Germany, 1991

6 Z Chen and K -H Hoffmann Numerical solutions of the optimal control problem governed by a phase field model In W Desch, F Kappel, and K Kunisch, editors, *Optimal Control of Partial Differential Equations*, Basel, Boston, Berlin, 1991 Birkhauser Verlag

7 J B Collins and H Levine Diffusive interface model of diffusion limited crystal growth *Physica Review B*, 31 6119–6122, 1985

8 G Fix Phase field models for free boundary problems In A Fasano and M Primicerio, editors, *Free Boundary Problems Theory and Applications, Vol II*, pages 580–589 Pitman, Boston, New York, Melbourne, 1983

9 W Hackbusch *Multi–Grid Methods and Applications* Springer–Verlag, Berlin, Heidelberg, New York, Tokyo, 1985

10 M Heinkenschloss Numerical solution of a semilinear parabolic control problem Technical report, Forschungsbericht, Universitat Trier, FB IV – Mathematik, Postfach 3825, D–54286 Trier, 1993

11 K -H Hoffmann and L Jiang Optimal control problem of a phase field model for solidification *Numer Funct Anal*, 13 11–27, 1992

12 C T Kelley and E W Sachs Solution of optimal control problems by a pointwise projected Newton method *Department of Mathematics, North Carolina State University, Raleigh, NC*, 1993

13 O Penrose and P C Fife Thermodynamically consistent models of phase–field type for kinetics of phase transitions *Physica D*, 43 44–62, 1990

14 Y Saad and M H Schultz GMRES a generalized minimal residual algorithm for solving nonsymmetric linear systems *SIAM J Sci Stat Comp*, 7 856–869, 1986

15 A A Wheeler, B T Murray, and R J Schaefer Computation of dentrites using a phase field model *Physica D*, 66 243–262, 1993
16 H Yserentant On the multi–level splitting of finite element spaces *Numer Math* , 49 379–412, 1986

M Heinkenschloss
Interdisciplinary Center for Applied Mathematics
and Department of Mathematics
Virginia Polytechnic Institute and State University
Blacksburg, VA 24061–0123, USA
e–mail na heinken@na–net ornl gov
E W Sachs
FB IV – Mathematik
Universitat Trier
D–54286 Trier, Federal Republic of Germany
e–mail sachs@uni–trier de

UNIFORM STABILIZABILITY OF NONLINEARLY COUPLED KIRCHHOFF PLATE EQUATIONS

MARY ANN HORN* AND IRENA LASIECKA**

School of Mathematics
University of Minnesota

Department of Applied Mathematics
University of Virginia

ABSTRACT A system of two Kirchhoff plate equations with nonlinear coupling through both the boundary and the interior is considered For this problem, it is proven that by appropriately choosing feedback controls, the energy of the system decays at a uniform rate This result extends previous results in a number of directions *(i)* it does not require any geometric hypotheses to be imposed on the domain, *(ii)* it allows for the presence of *nonlinear* coupling terms, *(iii)* it does not require the control functions to satisfy any growth conditions at the origin

1991 *Mathematics Subject Classification* 35, 93

Key words and phrases Uniform stabilization, boundary feedback, coupled plates, Kirchhoff plate

1. Introduction

As more results become available for single plates, attention has naturally shifted to more complex problems, such as those involving connected plates. Because one of the motivating factors in considering plate equations is to model more complicated vibrating flexible structures, our goal is to consider a coupled system, where the actions of one plate are allowed to affect those of the other.

The problem we consider is the following. Let Ω be an open bounded domain in R^2 with sufficiently smooth boundary, Γ. In Ω, we consider the following system of nonlinearly coupled Kirchhoff's equations.

$$\begin{array}{ll} v_{tt} - \gamma^2 \Delta v_{tt} + \Delta^2 v = k_1(u-v) & \\ u_{tt} - \gamma^2 \Delta u_{tt} + \Delta^2 u = k_2(v-u) & \text{in } Q_\infty = (0,\infty)\times\Omega \end{array} \tag{1.1.a}$$

*This material is based upon work partially supported under a National Science Foundation Mathematical Sciences Postdoctoral Research Fellowship

**Partially supported by National Science Foundation Grant NSF DMS-9204338

with initial conditions

$$\text{(1.1.b)} \qquad \left. \begin{array}{l} v(0,\cdot) = v_0 \in H^2(\Omega),\ v_t(0,\cdot) = v_1 \in H^1(\Omega) \\ u(0,\cdot) = u_0 \in H^2(\Omega),\ u_t(0,\cdot) = u_1 \in H^1(\Omega) \end{array} \right\} \quad \text{in } \Omega$$

and boundary conditions on $\Sigma_\infty \equiv \Gamma \times (0,\infty)$,

$$\text{(1.1.c)} \qquad \begin{array}{l} \Delta v + (1-\mu)\mathcal{B}_1 v = g_1 \\ \frac{\partial}{\partial\nu}\Delta v + (1-\mu)\mathcal{B}_2 v - \gamma^2 \frac{\partial}{\partial\nu} v_{tt} - v + \frac{\partial}{\partial\tau} h_1(\frac{\partial}{\partial\tau}(v-u)) - l_1(v-u) \\ \quad = g_2 - \frac{\partial}{\partial\tau} g_3 \end{array}$$

$$\text{(1.1.d)} \qquad \begin{array}{l} \Delta u + (1-\mu)\mathcal{B}_1 u = g_4 \\ \frac{\partial}{\partial\nu}\Delta u + (1-\mu)\mathcal{B}_2 u - \gamma^2 \frac{\partial}{\partial\nu} u_{tt} - u + \frac{\partial}{\partial\tau} h_2(\frac{\partial}{\partial\tau}(u-v)) - l_2(u-v) \\ \quad = g_5 - \frac{\partial}{\partial\tau} g_6. \end{array}$$

Here, $\gamma > 0$ is a constant proportional to the thickness of the plate, $0 < \mu < \frac{1}{2}$ is Poisson's ratio, the operators $\mathcal{B}_1$ and $\mathcal{B}_2$ are given by

$$\text{(1.2)} \qquad \begin{array}{l} \mathcal{B}_1 w = 2n_1 n_2 w_{xy} - n_1^2 w_{yy} - n_2^2 w_{xx} \\ \mathcal{B}_2 w = \frac{\partial}{\partial\tau}[(n_1^2 - n_2^2) w_{xy} + n_1 n_2 (w_{yy} - w_{xx})], \end{array}$$

The functions $k_i(s)$, $h_i(s)$, $l_i(s)$ are assumed to be differentiable with the property that for $s \in R$,

$$\text{(H-1)} \qquad \left\{ \begin{array}{l} s k_i(s) \geq 0 \\ s h_i(s) \geq 0 \\ s l_i(s) \geq 0 \\ |h(s)| \leq C|s|^K \quad \text{for some } K > 0 \text{ and } |s| > 1. \end{array} \right.$$

Control functions are denoted by $g_i \in L_2(\Sigma)$ and physically represent moments and shears applied to the edge of the plate (see [5]). One could also consider an addition of nonlinear coupling in the first boundary condition in (1.1.c)-(1.1.d). Since this will not add extra difficulties, for simplicity we do not consider the additional coupling.

The control model for a single Kirchhoff plate (without coupling) was introduced and analyzed in [5]. Systems of equations of the type as in (1.1) are motivated by problems arising in the modelling of parallelly coupled plates (see [11], where a system of two one-dimensional wave equations with linear boundary conditions was considered). Indeed, parallel connection of plates introduces a force acting upon the system which is proportional to the difference of the displacement. If the plates are connected by, for example, springs, then linear theory gives the *constants* k and l. In the above system, a more general situation is assumed. Coupling of the plates is allowed to produce nonlinear forces. Also, we allow for the fact that the coupling may occur only on the boundary, in which case $k_i = 0$ and $l_i > 0$.

The goal of this paper is to show that for appropriately chosen feedback controls, the energy of system (1.1) decays to zero at a uniform rate. Following

[5], we choose the nonlinear feedback controls of the form

$$
\begin{aligned}
g_1 &= -f_1(\tfrac{\partial}{\partial \nu} v_t) \\
g_2 &= g(v_t) \\
g_3 &= f_2(\tfrac{\partial}{\partial \tau} v_t) \\
g_4 &= -f_1(\tfrac{\partial}{\partial \nu} u_t) \\
g_5 &= g(u_t) \\
g_6 &= f_2(\tfrac{\partial}{\partial \tau} u_t).
\end{aligned} \tag{1.3}
$$

The following hypotheses are assumed on the nonlinear functions g and f_i: The functions g and f_i are continuous, monotone, zero at the origin and subject to the following growth conditions

$$
\begin{cases} m|s| \le |f_i(s)| \le M|s| & \text{for } |s| > 1,\ i = 1,2 \\ |g(s)| \le M|s|^r & \text{for } |s| > 1 \end{cases} \tag{H-2}
$$

for some positive constants m and M.

To define the energy of the system, we introduce the bilinear form

$$
\begin{aligned}
a(w,v) = &\int_\Omega (\Delta w \Delta v + (1-\mu)(2w_{xy}v_{xy} - w_{xx}v_{yy} - w_{yy}v_{xx}))d\Omega \\
&+ \int_\Gamma wv d\Gamma.
\end{aligned} \tag{1.4}
$$

The energy functional is defined by $E(t) \equiv E_v(t) + E_u(t)$, where

$$
E_w(t) \equiv \frac{1}{2}\int_\Omega \{|w_t|^2 + \gamma^2|\nabla w_t|^2\}d\Omega + \frac{1}{2}a(w,w). \tag{1.5}
$$

The main result of our paper is the following.

Theorem 1.1. *Assume (H-1) and (H-2) and consider equation* (1.1) *with* (1.3). *I. For any initial data* $v_0, u_0 \in H^2(\Omega)$, $v_1, u_1 \in H^1(\Omega)$, *there exists a solution* $(v,u) \in C(0,\infty, H^2(\Omega)) \cap C^1(0,\infty; H^1(\Omega))$ *such that*

$$
\nabla v_t|_\Gamma \in L_2(\Sigma_\infty),\ \nabla u_t|_\Gamma \in L_2(\Sigma_\infty). \tag{1.6}
$$

II. Assuming, in addition, that

$$
\begin{cases} m \le f_i'(s) \le M \\ g'(s) \le M(|s|^{r-1} + 1), \end{cases} \tag{H-3}
$$

then there exists constants C, $\omega > 0$ *such that*

$$
E(t) \le Ce^{-\omega t},\ t \ge 0, \tag{1.7}
$$

where the constants C, ω *may depend on* $E(0)$, *but they are independent of* $t > 0$ *and* $\gamma > 0$.

Exponential decay rates for a system of one dimensional wave equations with distributed linear coupling and linear boundary dissipation was proven recently in [10] and [11]. The methods of [10] and [11] rely on spectral analysis which is typically restricted to one dimensional problems. Moreover, [10] and [11] state explicitly that other methods such as multipliers (see [3]) or direct methods (see [4]), which are tailored to multi-dimensional stability analysis, fail even in the case of linear coupling of a one dimensional wave equation with linear dissipation. We shall see that a different and more general approach based on intrinsic comparison of asymptotic behavior of the energy functional with the solution to a certain nonlinear ordinary differential equation problem (introduced in [7]) prove successful in this more complicated situation.

A critical role in the proof of Theorem 1.1 is played by a result on stabilizability of a semilinear Kirchhoff plate with nonlinear boundary conditions. Since this result is of independent interest, we shall state it below.

Let a_i (resp., b) be given elements in $L_2(\Sigma_\infty)$ (resp. $b \in C(0,\infty; L_{2r}(\Gamma))$). Consider the following equation of a nonautonomous semilinear Kirchhoff plate.

$$(1.8)\quad \begin{cases} w_{tt} - \gamma^2 \Delta w_{tt} + \Delta^2 w = -k(w) & \text{in } Q_\infty \\ \quad w(0) = w_0 \in H^2(\Omega);\ w_t(0) = w_1 \in H^1(\Omega) & \text{in } \Omega \\ \Delta w + (1-\mu)\mathcal{B}_1 w = -[f_1(\frac{\partial}{\partial \nu} w_t + a_1) - f_1(a_1)] & \text{on } \Sigma_\infty \\ \frac{\partial}{\partial \nu}\Delta w + (1-\mu)\mathcal{B}_2 w - \gamma^2 \frac{\partial}{\partial \nu} w_{tt} - w + \frac{\partial}{\partial \tau} h(\frac{\partial}{\partial \tau} w) - l(w) & \\ \quad = g(w_t + b) - g(b) - \frac{\partial}{\partial \tau}[f_2(\frac{\partial}{\partial \tau} w_t + a_2) - f_2(a_2)] & \text{on } \Sigma_\infty. \end{cases}$$

Under the assumption that the functions k, h, l satisfy (H-1) and the functions g, f_i satisfy (H-2), one can show (by techniques identical to those in [7]), that for all initial data $w_0 \in H^2(\Omega)$, $w_1 \in H^1(\Omega)$, there exists a solution w (not necessarily unique) to problem (1.8) such that

$$(1.9)\qquad E_w(t) < C(E_w(0)),$$

$$(1.10)\qquad \nabla w_t|_\Gamma \in L_2(\Sigma_\infty).$$

Theorem 1.2. *Assume that (H-1) and (H-3) hold true. Then there exist constants C, $\omega > 0$ such that every solution to (1.8) of finite energy satisfies*

$$(1.11)\qquad E_w(t) \le Ce^{-\omega t}.$$

The constants C and ω may depend on the initial energy $E_w(0)$ and $\|b\|_{C(0,\infty;L_{2r}(\Gamma))}$.

In the special case when $a_i = b = 0$ (or they are constant), a more general version of Theorem 1.2 holds true.

Theorem 1.3. *Let $a_i = b = 0$. Assume (H-1) and (H-2). Then every solution of finite energy which satisfies (1.10) obeys the following estimate. For some $T_0 > 0$,*

$$(1.12)\qquad E_w(t) \le \mathcal{S}(\frac{t}{T_0} - 1) \ for\ t > T_0,$$

where $\mathcal{S}(t) \to 0$ *as* $t \to \infty$ *and is the solution (contraction semigroup) of the differential equation*

$$\begin{cases} \frac{d}{dt}\mathcal{S}(t) + q(\mathcal{S}(t)) = 0 \\ \mathcal{S}(0) = E_w(0), \end{cases} \tag{1.13}$$

and $q(x)$ *is a strictly monotone function constructed in terms of* g *and* f_i *(see Remark below).*

Remark 1.1: To construct the function $q(x)$, we proceed as follows: Let the function $h(x)$ be defined by:

$$h(x) \equiv h_1(x) + h_2(x), \tag{1.14}$$

where $h_i(x)$ are concave, strictly increasing functions with $h_i(0) = 0$ such that

$$h_i(sf_i(s)) \geq s^2 + f_i^2(s) \; |s| \leq 1 \; i = 1, 2. \tag{1.15}$$

(Such functions can be easily constructed. See [7].) Then $h(x)$ enjoys the same properties, i.e., it is concave, strictly increasing, and $h(0) = 0$. Define

$$\tilde{h}(x) \equiv h(\frac{x}{mes\ \Sigma_T}). \tag{1.16}$$

Since $\tilde{h}$ is monotone increasing, for every $c \geq 0$, $cI + \tilde{h}$ is invertible. Setting

$$p(x) \equiv (cI + \tilde{h})^{-1}(Kx), \tag{1.17}$$

where $c = \frac{1}{mes \Sigma_T}(m^{-1} + M)$, the constant K will generally depend on $E_w(0)$ unless $r = 1$ and h, k, l are linearly bounded. We then define $q(x)$ by

$$q(x) \equiv x - (I + p)^{-1}(x), \; x > 0. \tag{1.18}$$

Remark 1.2: In the special case when the growth at the origin of the nonlinear boundary feedbacks is specified, one can compute explicitly the governing decay rates for the energy function, $E_w(t)$. Indeed, this can be easily accomplished by constructing the appropriate function q and solving the nonlinear ODE problem, (1.13). For instance, if the nonlinear functions f_i and g have a linear growth at the origin, then (1.12) specializes to

$$E_w(t) \leq C(E_w(0))e^{-\omega t} \text{ for some } \omega > 0. \tag{1.19}$$

If, instead, these nonlinearities are of polynomial growth at the origin (e.g. $\sim x^p$, $p > 1$), then $E_w(t) \leq C(E_w(0))t^{\frac{2}{1-p}}$.

Remark 1.3: The result of Theorem 1.3 should be compared to a recent result on uniform decay rates obtained in [12] (and references therein) for a linear Euler-Bernoulli model with nonlinear boundary dissipation. Indeed, in [12], the uniform (polynomial) decay rates were obtained under the following hypotheses: *(i)* Ω is star-shaped, *(ii)* nonlinear functions f_i and g are subject to a linear growth at infinity and *polynomial* growth at the origin (hypotheses (1.25), (1.26) in [12]), *(iii)* $k = 0$, $h = 0$, *(iv)* l is linear. Under the above hypotheses, it is shown in [12] that *smooth*[1] solutions decay polynomially to zero.

In view of this, Theorem 1.3 extends the above result in several directions: *(i)* it does not require any geometric hypotheses to be imposed on Ω, *(ii)* the result of Theorem 1.3 remains valid for all weak solutions, *(iii)* it allows the presence of nonlinear terms h, l on the boundary together with the nonlinear interior term k, and, finally, *(iv)* it does not require any growth conditions at the origin.

Needless to say, the techniques leading to the proof of Theorem 1.3 are very different (unfortunately more complicated) from those in [12]. Indeed, Lyaponov's function method used in [12] runs into well recognized difficulties when dealing with the level of generality presented in Theorem 1.3.

The outline of our paper is as follows. In section 2, we prove the result of Theorem 1.2 which is critical to the proof of Theorem 1.1 in section 4. The proof of Theorem 1.3 is relegated to section 3.

2. Proof of Theorem 1.2

2.1 Preliminaries

Our goal is to prove energy decay rates for problem (1.8). In order to do this, one needs to perform certain partial differential equation calculations on the problem. These calculations require regularity of the solutions higher than is available. Since our nonlinear problem may not have a sufficiently regular solution (even if the initial data are smooth), we resort to an approximation argument (this argument was used in the context of wave equations in [7]). In fact, the idea here is to approximate solutions to the nonlinear problem (1.8) by solutions to different (linear) problems. Since this linear problem admits regular solutions for smooth initial data, the partial differential equation calculations can be performed on this problem. Final passage to the limit on the approximation problem allows us to obtain needed energy identities for the original nonlinear problem.

To follow our program, we start by defining appropriate approximations for the quantities in equation (1.8). To do this, we introduce the following notation:

$$\hat{f}_i(s) \equiv f_i(s + a_i) - f_i(a_i); \; \hat{g}(s) \equiv g(s + b) - g(b). \tag{2.1}$$

Note that by virtue of our hypothesis (H-3), $\hat{f}_i$ and $\hat{g}$ are strictly monotone functions and

$$m \leq \hat{f}_i'(s) \leq M. \tag{2.2}$$

[1]A smoothness assumption was not stated in [12], but implicitly assumed (to validate arguments leading to decay estimates)

Corollary 2.1. *Let w be a solution to (1.8) such that (1.9) and (1.10) hold true. Let $T > 0$. Then*

$$k(w) \in L_2(0,T; L_2(\Omega)), \tag{2.3}$$

$$\frac{\partial}{\partial \tau} h(\frac{\partial}{\partial \tau} w) \in L_2(0,T; H^{-1}(\Gamma)); \; l \in L_2(0,T; L_2(\Gamma)), \tag{2.4}$$

$$\hat{f}_1(\frac{\partial}{\partial \nu} w_t) \in L_2(0,T; L_2(\Gamma)), \tag{2.5}$$

$$\hat{g}(w_t) - \frac{\partial}{\partial \tau} \hat{f}_2(\frac{\partial}{\partial \tau} w_t) \in L_2(0,T; H^{-1}(\Gamma)). \tag{2.6}$$

Proof of Corollary 2.1: Regularity in (2.3), (2.4) follows using Sobolev's Imbeddings from (1.9), (H-1).

Hypothesis (H-3) together with (1.10) implies

$$\begin{aligned} f_1(\tfrac{\partial}{\partial \nu} w_t) &\in L_2(\Sigma_T) \\ f_2(\tfrac{\partial}{\partial \tau} w_t) &\in L_2(\Sigma_T). \end{aligned} \tag{2.7}$$

Hence,

$$\frac{\partial}{\partial \tau} f_2(\frac{\partial}{\partial \tau} w_t) \in L_2(0,T; H^{-1}(\Gamma)). \tag{2.8}$$

On the other hand, with $\phi \in L_2(0,T; H^1(\Gamma))$,

$$\begin{aligned} \textstyle\int_0^T \int_\Gamma |g(w_t(t,x))\phi(t,x)| dx dt &\leq C \textstyle\int_0^T \|\phi(t)\|_{H^1(\Gamma)} \int_\Gamma |w_t(t,x)|^r dx dt \\ &\leq C \textstyle\int_0^T \|\phi(t)\|_{H^1(\Gamma)} \|w_t(t)\|^r_{L_p(\Gamma)} dt \\ &\leq C \textstyle\int_0^T \|\phi(t)\|_{H^1(\Gamma)} E_w(t)^{r/2} dt \\ &\leq C E_w(0)^{r/2} \textstyle\int_0^T \|\phi(t)\|_{H^1(\Gamma)} dt, \end{aligned} \tag{2.9}$$

where the first inequality follows from hypothesis (H-2), the second and third follow from Sobolev Imbeddings and the boundedness of Γ, and the fourth from trace theory. Hence,

$$g(w_t) \in L_2(0,T; H^{-1}(\Gamma)), \tag{2.10}$$

which, together with (2.8) proves (2.6). □

Let w be the solution of the original problem (1.8). By using the regularity properties in (1.10), (2.3)-(2.6), along with density of approximate (see below)

Sobolev spaces, we are in a position to define

$$(2.11)\quad f_n \in H^{1,1}(Q_T); \qquad \|f_n + k(w)\|_{L_2(0,T;L_2(\Omega))} \longrightarrow 0$$

$$(2.12)\quad f_{1n} \in H^{1,1}(\Sigma_T); \qquad \|f_{1n} - \hat{f}_1(\frac{\partial}{\partial\nu}w_t)\|_{L_2(\Sigma_T)} \longrightarrow 0$$

$$(2.13)\quad f_{2n} \in H^{1,1}(\Sigma_T); \qquad \|f_{2n} - [\hat{g}(w_t) - \frac{\partial}{\partial\tau}\hat{f}_2(\frac{\partial}{\partial\tau}w_t) - \frac{\partial}{\partial\tau}h(\frac{\partial}{\partial\tau}w) + l(w)]\|_{L_2(0,T;H^{-1}(\Gamma))} \longrightarrow 0$$

$$(2.14)\quad \alpha_n \in H^{1,1}(\Sigma_T); \qquad \|\alpha_n - \frac{\partial}{\partial\nu}w_t\|_{L_2(\Sigma_T)} \longrightarrow 0$$

$$(2.15)\quad \beta_n \in H^{1,1}(\Sigma_T); \qquad \|\beta_n - (w_t - \frac{\partial^2}{\partial\tau^2}w_t)\|_{L_2(0,T;H^{-1}(\Gamma))} \longrightarrow 0,$$

where $Q_T \equiv \Omega \times (0,T)$ and $\Sigma_T \equiv \Gamma \times (0,T)$. We consider the following approximating problem:

$$(2.16)\quad \begin{cases} w_{n,tt} - \gamma^2\Delta w_{n,tt} + \Delta^2 w_n = f_n \\ w_n(0) = w_{n,0};\ w_{n,t}(0) = w_{n,1} \\ \Delta w_n + (1-\mu)\mathcal{B}_1 w_n + \frac{\partial}{\partial\nu}w_{n,t}|_\Gamma = -f_{1n} + \alpha_n \\ \frac{\partial}{\partial\nu}\Delta w_n + (1-\mu)\mathcal{B}_2 w_n - \gamma^2\frac{\partial}{\partial\nu}w_{n,tt} - w_n - w_{n,t} + \frac{\partial^2}{\partial\tau^2}w_{n,t}|_\Gamma \\ \qquad = f_{2n} - \beta_n, \end{cases}$$

where

$$(2.17)\quad \|w_{n,0} - w_0\|_{H^2(\Omega)} \to 0;\ \|w_{n,1} - w_1\|_{H^1(\Omega)} \to 0,$$

and $(w_{n,0}, w_{n,1}) \in \mathcal{D}$, where $\mathcal{D}$, as dense set of $\mathcal{H}$, consists of $w_{n,0} \in H^4(\Omega)$, $w_{n,1} \in H^3(\Omega)$, where $w_{n,0}, w_{n,1}$ satisfy the appropriate compatibility conditions on the boundary. By standard linear semigroup methods, one easily shows that the linear problem, (2.16), admits a classical solution,

$$(2.18)\quad w_n \in C(0,T;H^4(\Omega)) \cap C^1(0,T;H^3(\Omega)).$$

The following proposition plays a critical role in our development.

Proposition 2.1. *Let w_n (respectively, w) be a solution of (2.16) (respectively, (1.8)). Then as $n \to \infty$, the following convergence holds.*

$$(2.19)\quad w_n \to w \text{ in } C(0,T;H^2(\Omega)) \cap C^1(0,T;H^1(\Omega))$$

$$(2.20)\quad \nabla w_{n,t}|_\Gamma \to \nabla w_t \text{ in } L_2(\Sigma_T).$$

Proof: Consider the equation satisfied by the difference $w_n - w_m$. Taking the inner product of this equation with $w_{n,t} - w_{m,t}$ and integrating the result from 0 to T yields

$$\begin{aligned} & E_{w_n - w_m}(T) + \int_0^T \int_\Gamma [\hat{w}_t^2 + |\nabla \hat{w}_t|^2] d\Gamma dt \\ & = \int_0^T \int_\Omega (f_n - f_m)\hat{w}_t d\Omega dt + \int_0^T \int_\Gamma (f_{1n} + \alpha_n - f_{1m} - \alpha_m) \frac{\partial}{\partial \nu}\hat{w}_t d\Gamma dt \\ & \quad + \int_0^T \int_\Gamma (f_{2n} + \beta_n - f_{2m} - \beta_m)\hat{w}_t d\Gamma dt + E_{w_n - w_m}(0), \end{aligned} \tag{2.21}$$

where $\hat{w} \equiv w_n - w_m$. Hence,

$$\begin{aligned} & C_0 \|\hat{w}(T)\|^2_{H^2(\Omega)} + \|\hat{w}_t(T)\|^2_{H^1(\Omega)} + \|\nabla \hat{w}_t\|^2_{L_2(\Sigma_T)} + \|\hat{w}_t\|^2_{L_2(\Sigma_T)} \\ & \leq \frac{1}{2}\|f_n - f_m\|^2_{L_2(0,T,H^{-1}(\Omega))} + \frac{1}{2}\int_0^T \|\hat{w}_t(t)\|^2_{H^1(\Omega)} dt \\ & \quad + \frac{1}{2}\|f_{1n} - \alpha_n - f_{1m} + \alpha_m\|^2_{L_2(\Sigma_T)} + \frac{1}{2}\|\frac{\partial}{\partial \nu}\hat{w}_t\|^2_{L_2(\Sigma_T)} \\ & \quad + \frac{1}{2}\|f_{2n} - \beta_n - f_{2m} + \beta_m\|^2_{L_2(0,T,H^{-1}(\Omega))} \\ & \quad + \frac{1}{2}\|\hat{w}_t\|^2_{L_2(0,T\ H^1(\Omega))} + E_{w_n - w_m, 1}(0), \end{aligned} \tag{2.22}$$

and

$$\begin{aligned} & \|\hat{w}(T)\|^2_{H^2(\Omega)} + \|\hat{w}_t(T)\|^2_{H^1(\Omega)} + \|\nabla \hat{w}_t\|^2_{L_2(\Sigma_T)} \\ & \leq C[\|f_n - f_m\|^2_{L_2(0,T,H^{-1}(\Omega))} + \|f_{1n} - f_{1m}\|^2_{L_2(\Sigma_T)} \\ & \quad + \|\alpha_n - \alpha_m\|^2_{L_2(\Sigma_T)} + \|f_{2n} - f_{2m}\|^2_{L_2(0,T,H^{-1}(\Omega))} \\ & \quad + \|\beta_n - \beta_m\|^2_{L_2(0,T,H^{-1}(\Omega))} + E_{w_n - w_m, 1}(0)] \\ & \longrightarrow 0 \ as\ n \to \infty, \end{aligned} \tag{2.23}$$

where the limit follows by using (2.11)-(2.15). Thus, by (2.8) and Corollary 2.1,

$$\begin{aligned} w_n \to w^* \ in\ C(0,T;H^2(\Omega)) \cap C^1(0,T;H^1(\Omega)) \\ \nabla w_{n,t}|_\Gamma \to \nabla w_t^*|_\Gamma \ in\ L_2(\Sigma_T). \end{aligned} \tag{2.24}$$

This allows us to pass with the limit on the linear equation, (2.16). We obtain

$$\left\{ \begin{aligned} & w_{tt}^* - \gamma^2 \Delta w_{tt}^* + \Delta^2 w^* + k(w) = 0 \\ & w^*(0) = w_0; \ w_t^*(0) = w_1 \in H^2(\Omega) \\ & \Delta w^* + (1-\mu)B_1 w^* + \frac{\partial}{\partial \nu} w_t^*|_\Gamma = -\hat{f}_1(\frac{\partial}{\partial \nu} w_t) + \frac{\partial}{\partial \nu} w_t \\ & \frac{\partial}{\partial \nu}\Delta w^* + (1-\mu)B_2 w^* - \gamma^2 \frac{\partial}{\partial \nu} w_{tt}^* - w^* - w_t^* + \frac{\partial^2}{\partial \tau^2} w_t^*|_\Gamma \\ & \quad = \hat{g}(w_t) - \frac{\partial}{\partial \tau}\hat{f}_2(\frac{\partial}{\partial \tau} w_t) - w_t + \frac{\partial^2}{\partial \tau^2} w_t + l(w) - \frac{\partial}{\partial \tau} h(\frac{\partial}{\partial \tau} w). \end{aligned} \right. \tag{2.25}$$

Since w satisfies (2.25) and the solution to (2.25) is unique, we infer that $w \equiv w^*$ and

$$\begin{aligned} w_n \to w \ in\ C(0,T;H^2(\Omega)) \cap C^1(0,T;H^1(\Omega)) \\ \nabla w_{n,t}|_\Gamma \to \nabla w_t|_\Gamma \ in\ L_2(\Sigma_T), \end{aligned} \tag{2.26}$$

as desired. □

Now we are in a position to prove the fundamental energy relation for problem (1.8).

Lemma 2.1 (Energy Identity). *Let w be the solution to (1.8). Then the following energy identity holds*

$$\hat{E}_w(T) - \hat{E}_w(0) + \int_{\Sigma_T} [\hat{g}(w_t)w_t + \hat{f}_1(\frac{\partial}{\partial\nu}w_t)\frac{\partial}{\partial\nu}w_t + \hat{f}_2(\frac{\partial}{\partial\tau}w_t)\frac{\partial}{\partial\tau}w_t]d\Gamma dt = 0, \tag{2.27}$$

where

$$\hat{E}_w(t) \equiv E_w(t) + \int_\Omega \mathcal{K}(w(t))d\Omega + \int_\Gamma [\mathcal{H}(\frac{\partial}{\partial\tau}w(t)) + \mathcal{L}(w(t))]d\Gamma,$$

and $\mathcal{K}$, $\mathcal{H}$ and $\mathcal{L}$ are antiderivatives of k, h and l.

Proof: We first prove this energy identity for the solution, w_n, of the approximation problem, (2.16). Indeed, by applying a standard energy argument (valid due to the smoothness of solutions w_n) to (2.16), we obtain

$$\begin{aligned} &E_{w_n}(T) - E_{w_n}(0) + \int_{\Sigma_T} |\tfrac{\partial}{\partial\nu}w_{n,t}|^2 d\Gamma dt \\ &\quad + \int_{\Sigma_T}(w_{n,t})^2 d\Gamma dt + \int_{\Sigma_T} |\tfrac{\partial}{\partial\tau}w_{n,t}|^2 d\Gamma dt \\ &\quad = \int_0^T \int_\Omega f_n w_{n,t} d\Omega dt - \int_{\Sigma_T}(f_{1n} - \alpha_n)\tfrac{\partial}{\partial\nu}w_{n,t} d\Gamma dt \\ &\quad - \int_{\Sigma_T}(f_{2n} - \beta_n)w_{n,t} d\Gamma dt. \end{aligned} \tag{2.28}$$

Using convergence properties (2.11)-(2.15) and the result of Proposition 2.1, we obtain

$$\begin{aligned} &E_w(T) - E_w(0) + \int_{\Sigma_T} |\tfrac{\partial}{\partial\nu}w_t|^2 d\Gamma dt \\ &\quad + \int_{\Sigma_T}(w_t)^2 d\Gamma dt + \int_{\Sigma_T} |\tfrac{\partial}{\partial\tau}w_t|^2 d\Gamma dt \\ &\quad = -\int_0^T \int_\Omega k(w)w_t d\Omega dt + \int_{\Sigma_T}[-\hat{f}_1(\tfrac{\partial}{\partial\nu}w_t) + \tfrac{\partial}{\partial\nu}w_t]\tfrac{\partial}{\partial\nu}w_t d\Gamma dt \\ &\quad - \int_{\Sigma_T}[\hat{g}(w_t) - \tfrac{\partial}{\partial\tau}\hat{f}_2(\tfrac{\partial}{\partial\tau}w_t)]w_t d\Gamma dt + \int_{\Sigma_T}(w_t^2 + |\tfrac{\partial}{\partial\tau}w_t|^2)d\Gamma dt \\ &\quad - \int_{\Sigma_T}[l(w)w_t + h(\tfrac{\partial}{\partial\tau}w)\tfrac{\partial}{\partial\tau}w_t]d\Gamma dt. \end{aligned} \tag{2.29}$$

After canceling boundary terms and taking into account the definition of $\hat{E}_w(t)$, we obtain (2.27).

2.2 A Priori Estimates

To proof Theorem 1.2, we first show the following inequality holds.

Theorem 2.1. *Let w be the solution to (1.8) and T be sufficiently large. Then there exist constants, $C > 0$ and $C_T(E_w(0)) > 0$, such that the following inequality holds:*

$$\begin{aligned} &\int_0^T \hat{E}_w(t)dt - C\hat{E}_w(T) \\ &\quad \le C_T(E_w(0), \|b\|_{C(0\ \infty\ L_{2r}(\Gamma))}) \int_{\Sigma_T}(\hat{g}(w_t)w_t + \hat{f}_1(\tfrac{\partial}{\partial\nu}w_t)\tfrac{\partial}{\partial\nu}w_t \\ &\quad + \hat{f}_2(\tfrac{\partial}{\partial\tau}w_t)\tfrac{\partial}{\partial\tau}w_t)d\Gamma dt. \end{aligned} \tag{2.30}$$

2.2.1 Estimates for the Approximated Problem

To prove Theorem 2.1, we begin by using a multiplier method on the approximation problem (2.16) to prove the following preliminary estimate.

Lemma 2.2. *Let* $(w_0, w_1) \in \mathcal{D}$ *and* $0 < \alpha < T/2$. *Then the energy of system* (2.16) *satisfies the following estimate:*

$$
\begin{aligned}
&\int_\alpha^{T-\alpha} E_{w_n}(t)dt - C_1(1+\gamma^2)E_{w_n,1}(T-\alpha) - C_2(1+\gamma^2)E_{w_n,1}(\alpha) \\
&\le C_{T,\alpha,\epsilon}\{\|f_{1n}\|^2_{L_2(\Sigma_T)} + \|f_{2n}\|^2_{L_2(0,T,H^{-1}(\Gamma))} + \|\alpha_n - \tfrac{\partial}{\partial\nu}w_{n,t}\|^2_{L_2(\Sigma_T)} \\
&\quad + \|\beta_n - w_{n,t} + \tfrac{\partial^2}{\partial\tau^2}w_{n,t}\|^2_{L_2(0,T,H^{-1}(\Gamma))} + \|w_{n,t}\|^2_{L_2(\Sigma_T)} \\
&\quad + (1+\gamma^2)\|\nabla w_{n,t}\|^2_{L_2(\Sigma_T)} + \|f_n\|^2_{L_2(0,T,L_2(\Omega))} + l.o.(w_n)\},
\end{aligned} \tag{2.31}
$$

where

$$
l.o.(w_n) \equiv \|w_n\|^2_{L_2(0,T,H^{2-\epsilon}(\Omega))}, \tag{2.32}
$$

$0 < \epsilon < 1/2$, *and* $\vec{h} \equiv x - x_0$ *for some* $x_0 \in R^2$.

Proof of Lemma 2.2: *Step 1: Identities.* From [5] (p. 84, (4.5.17)), with adjustments to take both the nonhomogeneous right-hand side of (2.16) and the boundary conditions into account, we have

$$
\begin{aligned}
&\int_0^T E_{w_n,1}(t)dt \\
&\le -\tfrac{1}{2}[(w_{n,t}, \vec{h}\cdot\nabla w_n)_{L_2(\Omega)} + \gamma^2(\nabla w_{n,t}, \nabla(\vec{h}\cdot\nabla w_n))_{L_2(\Omega)}]_0^T \\
&+\tfrac{1}{2}[(w_{n,t}, w_n)_{L_2(\Omega)} + \gamma^2(\nabla w_{n,t}, \nabla w_n)_{L_2(\Omega)}]_0^T \\
&+\tfrac{1}{2}(\tfrac{\partial}{\partial\nu}w_n, f_{1n})_{L_2(\Sigma_T)} + \tfrac{1}{2}(w_n, f_{2n})_{L_2(\Sigma_T)} \\
&+\tfrac{1}{2}(\tfrac{\partial}{\partial\nu}w_n, \alpha_n - \tfrac{\partial}{\partial\nu}w_{n,t})_{L_2(\Sigma_T)} + \tfrac{1}{2}(w_n, \beta_n - w_{n,t} + \tfrac{\partial^2}{\partial\tau^2}w_{n,t})_{L_2(\Sigma_T)} \\
&-\int_{\Sigma_T}[\tfrac{\partial}{\partial\nu}(\vec{h}\cdot\nabla w_n)f_{1n} + (\vec{h}\cdot\nabla w_n)f_{2n}]d\Gamma dt \\
&+\int_{\Sigma_T}[\tfrac{\partial}{\partial\nu}(\vec{h}\cdot\nabla w_n)(\alpha_n - \tfrac{\partial}{\partial\nu}w_{n,t}) \\
&-(\vec{h}\cdot\nabla w_n)(\beta_n - w_{n,t} + \tfrac{\partial^2}{\partial\tau^2}w_{n,t})]d\Gamma dt \\
&-\int_{\Sigma_T}(\vec{h}\cdot\nabla w_n)w_n d\Gamma dt + \tfrac{1}{2}\int_{\Sigma_T}\vec{h}\cdot\nu(w_{n,t}^2 + \gamma^2|\nabla w_{n,t}|^2)d\Gamma dt \\
&-\tfrac{1}{2}\int_{\Sigma_T}\vec{h}\cdot\nu[w_{n,xx}^2 + w_{n,yy}^2 + 2\mu w_{n,xx}w_{n,yy} + 2(1-\mu)w_{n,xy}^2]d\Gamma dt \\
&+\int_{Q_T} f_n\vec{h}\cdot\nabla w_n d\Omega dt - \tfrac{1}{2}\int_{Q_T} f_n w_n d\Omega dt + \tfrac{1}{2}(b(x)w_n, w_n)_{L_2(\Omega)}|_0^T.
\end{aligned} \tag{2.33}
$$

Notice that the regularity of the solution given by (2.18) allows us to justify the calculations in [5].

Step 2: Bounding Linear Terms. All terms which need to be evaluated at 0 and T, including the first and second lines and the last term on the right-hand side of (2.33), can be bounded by

$$
C_1(1+\gamma^2)E_{w_n,1}(T) + C_2(1+\gamma^2)E_{w_n,1}(0). \tag{2.34}
$$

Finally, by using duality to split the terms involving α_n and β_n, noting that all boundary terms involving second derivatives of the solution can be bounded by second-order traces of the solution and $l.o.(w_n)$, and taking into account the above estimates, we obtain:

$$
\begin{aligned}
\int_0^T E_{w_n}(t)dt &\le C\{(1+\gamma^2)E_{w_n}(T) + (1+\gamma^2)E_{w_n}(0) \\
&+\|f_{1n}\|^2_{L_2(\Sigma_T)} + \|f_{2n}\|^2_{L_2(0,T,H^{-1}(\Gamma))} \\
&+\|\alpha_n - \tfrac{\partial}{\partial\nu}w_{n,t}\|^2_{L_2(\Sigma_T)} + \|\beta_n - w_{n,t} + \tfrac{\partial^2}{\partial\tau^2}w_{n,t}\|^2_{L_2(0,T,H^{-1}(\Gamma))} \\
&+\|w_{n,t}\|^2_{L_2(\Sigma_T)} + (1+\gamma^2)\|\nabla w_{n,t}\|^2_{L_2(\Sigma_T)} \\
&+\int_{\Sigma_T}\{|\tfrac{\partial^2}{\partial\nu^2}w_n|^2 + |\tfrac{\partial^2}{\partial\tau^2}w_n|^2 + |\tfrac{\partial^2}{\partial\nu\partial\tau}w_n|^2\}d\Gamma dt \\
&+\int_{Q_T} f_n\vec{h}\cdot\nabla w_n d\Omega dt - \tfrac{1}{2}\int_{Q_T} f_n w_n d\Omega dt.
\end{aligned}
\tag{2.35}
$$

Step 3: Bounding Second-Order Traces. Our next step is to estimate the second-order traces of the function w_n on the boundary. To accomplish this, it is critical to use the following "regularity" result obtained by microlocal analysis methods.

Proposition 2.2 ([8], Theorem 2.1). *Let $p(t,x)$ be a solution to the following linear problem (in the sense of distributions)*

$$
\begin{cases}
p_{tt} - \gamma^2\Delta p_{tt} + \Delta^2 p = F & \text{in } Q_T \\
p(0,\cdot) = p_0;\ p_t(0,\cdot) = p_1 & \text{in } \Omega \\
\Delta p + (1-\mu)B_1 p = g_1 & \text{on } \Sigma_T \\
\frac{\partial}{\partial\nu}\Delta p + (1-\mu)B_2 p - \gamma^2\frac{\partial}{\partial\nu}p_{tt} - p = g_2 & \text{on } \Sigma_T.
\end{cases}
\tag{2.36}
$$

For every $T > \alpha > 0$ and $\frac{1}{2} > \epsilon > 0$, the following estimate holds:

$$
\begin{aligned}
&\int_\alpha^{T-\alpha}\int_\Gamma(|\tfrac{\partial^2 p}{\partial\tau^2}|^2 + |\tfrac{\partial^2 p}{\partial\nu^2}|^2 + |\tfrac{\partial^2 p}{\partial\nu\partial\tau}|^2)d\Gamma dt \\
&\le C_{T,\alpha}\{\|F\|^2_{L_2([0,T],H^{-\epsilon}(\Omega))} + \|g_1\|^2_{L_2(\Sigma_T)} + \|g_2\|^2_{L_2([0,T],H^{-1}(\Gamma))} \\
&+\|p_t\|^2_{L_2(\Sigma_T)} + \|\tfrac{\partial}{\partial\nu}p_t\|^2_{L_2(0,T,H^{-1}(\Gamma)} + \|p\|^2_{L_2([0,T],H^{3/2+\epsilon}(\Omega))}\}.
\end{aligned}
\tag{2.37}
$$

Using the result of Proposition 2.2, we shall prove

Proposition 2.3. *Let w_n be the solution to (2.16). Then for any $T/2 > \alpha > 0$ and $\frac{1}{2} > \epsilon > 0$, w_n satisfies the following inequality:*

$$
\begin{aligned}
&\int_\alpha^{T-\alpha}\int_\Gamma(|\tfrac{\partial^2 w_n}{\partial\tau^2}|^2 + |\tfrac{\partial^2 w_n}{\partial\nu^2}|^2 + |\tfrac{\partial^2 w_n}{\partial\nu\partial\tau}|^2)d\Gamma dt \\
&\le C_{T,\alpha,\epsilon}\{\|f_n\|_{L_2(0,T,H^{-\epsilon}(\Omega))} + \|\alpha_n - \tfrac{\partial}{\partial\nu}w_{n,t}\|^2_{L_2(\Sigma_T)} + \|f_{1n}\|^2_{L_2(\Sigma_T)} \\
&+\|\beta_n - w_{n,t} + \tfrac{\partial^2}{\partial\tau^2}w_{n,t}\|^2_{L_2(0,T,H^{-1}(\Gamma))} + \|f_{2n}\|^2_{L_2(0,T,H^{-1}(\Gamma))} \\
&+\|w_{n,t}\|^2_{L_2(\Sigma_T)} + \|\nabla w_{n,t}\|^2_{L_2(\Sigma_T)} + l.o.(w_n)\}.
\end{aligned}
\tag{2.38}
$$

Proof: We apply the result of Proposition 2.2 to system (2.16) with

$$
\begin{aligned}
F &\equiv f_n \\
g_1 &\equiv \alpha_n - \tfrac{\partial}{\partial\nu}w_{n,t} - f_{1n} \\
g_2 &\equiv -\beta_n + w_{n,t} - \tfrac{\partial^2}{\partial\tau^2}w_{n,t} + f_{2n}.
\end{aligned}
\tag{2.39}
$$

□

Completion of Proof of Lemma 2.2: Now the result of Lemma 2.2 follows by combining (2.35) applied on $[\alpha, T-\alpha]$ with Proposition 2.3 and using duality to bound the terms involving f_n. □

Recalling Proposition 2.1, we take the limit of (2.31) as $n \to \infty$ to obtain a similar inequality for the solution to (1.8), which we state in the following lemma.

2.2.2 Estimates for the Original Problem

Lemma 2.3. *Let $(w_0, w_1) \in \mathcal{H}$. Then the energy of system* (1.8) *satisfies the following estimate:*

$$\begin{aligned} &\textstyle\int_0^T \hat{E}_w(t)dt - C_1(1+\gamma^2)\hat{E}_w(0) \\ &\quad \leq C_{T,\epsilon}(E_w(0), \|b\|_{C(0,T,L_{2r}(\Gamma))})\{\|w_t\|^2_{L_2(\Sigma_T)} \\ &\qquad +(1+\gamma^2)\|\nabla w_t\|^2_{L_2(\Sigma_T)} + l.o.(w)\}. \end{aligned} \tag{2.40}$$

Proof: *Step 1: Approximation Results.* Taking the limit as $n \to \infty$ in (2.31), by virtue of (2.11)-(2.15) and Proposition 2.1, we find

$$\begin{aligned} &\textstyle\int_\alpha^{T-\alpha} E_w(t)dt - C_1(1+\gamma^2)E_{w,1}(T-\alpha) - C_2(1+\gamma^2)E_{w,1}(\alpha) \\ &\leq C_{T,\alpha,\epsilon}\{\|\hat{f}_1(\tfrac{\partial}{\partial\nu}w_t)\|^2_{L_2(\Sigma_T)} + \|\hat{g}(w_t) - \tfrac{\partial}{\partial\tau}\hat{f}_2(\tfrac{\partial}{\partial\tau}w_t)\|^2_{L_2(0,T,H^{-1}(\Gamma))} \\ &\quad +\|w_t\|^2_{L_2(\Sigma_T)} + (1+\gamma^2)\|\nabla w_t\|^2_{L_2(\Sigma_T)} \\ &\quad +\|k(w)\|^2_{L_2(0,T,L_2(\Omega))} + \|l(w)\|^2_{L_2(0,T,H^{-1}(\Gamma))} \\ &\quad +\|h(\tfrac{\partial}{\partial\tau}w)\|^2_{L_2(\Sigma_T)} + l.o.(w)\}. \end{aligned} \tag{2.41}$$

Step 2: The following bounds can be obtained in a straightforward manner by using Sobolev's Imbeddings and (H-1).

$$\|k(w)\|_{L_2(\Omega)} \leq C(\|w\|_{H^{3/2+\epsilon}(\Omega)})\|w\|_{L_2(\Omega)} \tag{2.42}$$

$$\|l(w)\|_{L_2(\Gamma)} \leq C(\|w\|_{H^{3/2+\epsilon}(\Omega)})\|w\|_{L_2(\Gamma)} \tag{2.43}$$

$$\|h(\frac{\partial}{\partial\tau}w)\|\|_{L_2(\Gamma)} \leq C(\|w\|_{H^2(\Omega)})\|\frac{\partial}{\partial\tau}w\|_{L_2(\Gamma)} \tag{2.44}$$

Hence by Lemma 2.2,

$$\begin{aligned} \|k(w)\|^2_{L_2(0,T,L_2(\Omega))} + \|l(w)\|^2_{L_2(0,T,H^{-1}(\Gamma))} + \|h(\frac{\partial}{\partial\tau}w)\|^2_{L_2(\Sigma_T)} \\ \leq C(E_w(0))l.o.(w). \end{aligned} \tag{2.45}$$

Step 3: To estimate the first two boundary terms in (2.41), we notice that by virtue of (2.2),

$$\|\hat{f}_1(\frac{\partial}{\partial\nu}w_t)\|_{L_2(\Sigma_T)} \leq M\|\frac{\partial}{\partial\nu}w_t\|^2_{L_2(\Sigma_T)}, \tag{2.46}$$

$$\|\frac{\partial}{\partial\tau}\hat{f}_2(\frac{\partial}{\partial\tau}w_t)\|_{L_2(0,T,H^{-1}(\Gamma))} \leq M\|\frac{\partial}{\partial\tau}w_t\|^2_{L_2(\Sigma_T)}. \tag{2.47}$$

As for the term with $\hat{g}(w_t$, we use hypothesis (H-3). Indeed, from the imbedding $H^{-1}(\Gamma) \subset L_1(\Gamma)$,

$$
\begin{aligned}
\|\hat{g}(w_t)\|^2_{L_2(0,T;H^{-1}(\Gamma))} &\leq C\|\hat{g}(w_t)\|^2_{L_2(0,T;L_1(\Gamma))} \\
&\leq C\int_0^T \{\int_\Gamma |w_t|(|w_t|^r + |b|^r + 1)d\Gamma\}^2 dt \\
&\leq C\int_0^T \int_\Gamma w_t^2 d\Gamma\{\int_\Gamma (|w_t|^r + |b|^r + 1)d\Gamma\}dt \\
&\leq C_T\{E_w^r(0) + \|b\|^{2r}_{C(0,T,L_{2r}(\Gamma))} + 1\}\int_{\Sigma_T} w_t^2 d\Gamma dt,
\end{aligned}
\tag{2.48}
$$

where we have used the imbedding $H^{1/2-\epsilon}(\Gamma) \subset L_{2r}(\Gamma)$ together with the Trace Theorem and the result of Lemma 2.2.

Step 4: Noticing that

$$
\begin{aligned}
\int_\alpha^{T-\alpha} \hat{E}_w(t)dt &= \int_\alpha^{T-\alpha} E_w(t)dt \\
&+ \int_\alpha^{T-\alpha}\int_\Gamma \{\mathcal{L}(w) + \mathcal{H}(\frac{\partial}{\partial\tau}w)\}d\Gamma dt + \int_\alpha^{T-\alpha}\int_\Omega \mathcal{K}(w)d\Omega dt,
\end{aligned}
\tag{2.49}
$$

where $\mathcal{K}$, $\mathcal{L}$, $\mathcal{H}$ satisfy (2.42)-(2.44), and applying estimates (2.45)-(2.48) to (2.41) yields

$$
\begin{aligned}
&\int_\alpha^{T-\alpha} \hat{E}_w(t)dt - C_1\hat{E}_w(T-\alpha) - C_2\hat{E}_w(\alpha) \\
&\leq C_{T,\alpha,\epsilon}(E_w(0), \|b\|_{C(0,T;L_{2r}(\Gamma))})\{\int_{\Sigma_T} (w_t^2 + |\nabla w_t|^2)d\Gamma dt + l.o.(w)\}.
\end{aligned}
\tag{2.50}
$$

From (2.27) and (H-3),

$$
\int_0^\alpha \hat{E}_w(t)dt + \int_{T-\alpha}^T \hat{E}_w(t)dt \leq 2\alpha\hat{E}_w(t)(0). \tag{2.51}
$$

Applying now the result of Lemma 2.2 to the term $\hat{E}_w(T-\alpha)$ together with (2.51) yields the conclusion of Lemma 2.3. □

2.2.3 Nonlinear Compactness-Uniqueness Argument

Lemma 2.4. *Let $T > 0$ be sufficiently large. Then*

$$
l.o.(w) \leq C_T(E_w(0), \|b\|_{C(0,T;L_{2r}(\Gamma))})\int_{\Sigma_T} \{w_t^2 + |\nabla w_t|^2 + \hat{g}(w_t)w_t\}d\Gamma dt. \tag{2.52}
$$

Here, the function $C_T(E_w(0), \|b\|_{C(0,T,L_{2r}(\Gamma))})$ does not depend on $\gamma > 0$.

Proof: Identical to that in [2]. $\square$

2.2.4 Completion of the Proof of Theorem 2.1

Combine the result of Lemma 2.3 with that of Lemma 2.4 and use (2.2) together with $\|w\|_{L_2(\Gamma)} \leq C\|\frac{\partial}{\partial\tau}w\|_{L_2(\Gamma)}$. $\square$

2.3 Final Estimates: Proof of Theorem 1.2

Denoting $\mathcal{F} \equiv \int_{\Sigma_T}\{\hat{g}(w_t)w_t + \hat{f}_1(\frac{\partial}{\partial\nu}w_t)\frac{\partial}{\partial\nu}w_t + \hat{f}_2(\frac{\partial}{\partial\tau}w_t)\frac{\partial}{\partial\tau}w_t\}d\Gamma dt$, w e obtain from Theorem 2.1,

$$\int_0^T \hat{E}_w(t)dt - C_1\hat{E}_w(T) \leq C_{T,\alpha,\epsilon}(E_w(0), \|b\|_{C(0\ \infty\ L_{2r}(\Gamma))})\mathcal{F}, \tag{2.53}$$

and by Lemma 2.1,

$$\begin{aligned} &\textstyle\int_0^T \hat{E}_w(t)dt \leq C_{T,\epsilon}(E_w(0), \|b\|_{C(0\ \infty\ L_{2r}(\Gamma))})\mathcal{F} + C_1\hat{E}_w(T) \\ &\Longrightarrow (T - C_1)\hat{E}_w(T) \leq C_{T,\epsilon}(E_w(0), \|b\|_{C(0\ \infty\ L_{2r}(\Gamma))})\mathcal{F} \\ &\Longrightarrow \hat{E}_w(T) \leq C_T(E_w(0), \|b\|_{C(0\ \infty\ L_{2r}(\Gamma))})\mathcal{F}. \end{aligned} \tag{2.54}$$

Hence, recalling (2.27),

$$\frac{\hat{E}_w(T)}{C_T(E_w(0), \|b\|_{C(0\ \infty\ L_{2r}(\Gamma))})} \leq \mathcal{F} = \hat{E}_w(0) - \hat{E}_w(T). \tag{2.55}$$

Setting

$$p(s) \equiv \frac{s}{C_T(E_w(0), \|b\|_{C(0\ \infty\ L_{2r}(\Gamma))})}, \tag{2.56}$$

we have proven the following proposition.

Proposition 2.4. *Let w be the solution to (1.1) and $E_w(t)$ be the corresponding energy at time t. If T is sufficiently large, then there exists a monotone increasing function, p, such that*

$$p(\hat{E}_w(T)) + \hat{E}_w(T) \leq \hat{E}_w(0). \tag{2.57}$$

To arrive at the conclusion of Theorem 1.2, we apply a (much more general than needed here) result of Lemma 3.3 in [7].

Lemma 2.5 ([7], Lemma 3.3). *Let p be a positive, increasing function such that $p(0) = 0$. Since p is increasing, we can define a function q such that $q(x) = x - (I+p)^{-1}(x)$. Notice that q is also an increasing function. Consider a sequence s_n of positive numbers which satisfy:*

$$s_{m+1} + p(s_{m+1}) \leq s_m. \tag{2.58}$$

Then $s_m \le \mathcal{S}(m)$, *where* $\mathcal{S}(t)$ *is a solution of a differential equation*

$$\left\{ \begin{array}{l} \frac{d}{dt}\mathcal{S}(t) + q(\mathcal{S}(t)) = 0 \\ \mathcal{S}(0) = s_0. \end{array} \right. \tag{2.59}$$

Moreover, if $p(x) > 0$ *for* $x > 0$, *then* $\lim_{t\to\infty} \mathcal{S}(t) = 0$.

Applying the result of Proposition 2.4 and noticing that the energy $\hat{E}_w(t)$ is decreasing, we obtain

$$\hat{E}_w(m(T+1)) + p(\hat{E}_w(m(T+1))) \le \hat{E}_w(mT), \tag{2.60}$$

for $m = 0, 1, \ldots$ Thus, applying Lemma 2.5 with

$$s_m \equiv \hat{E}_w(mT), \tag{2.61}$$

yields

$$\hat{E}_w(mT) \le \mathcal{S}(m), \ \ m = 0, 1, \ldots \tag{2.62}$$

Setting $t = mT + \tau$, $0 \le \tau < T$,

$$\hat{E}_w(t) \le \hat{E}_w(mT) \le \mathcal{S}(m) \le \mathcal{S}(\frac{t-\tau}{T}) \le \mathcal{S}(\frac{t}{T} - 1) \ \textit{for}\ t > T, \tag{2.63}$$

and noting that in our case since $q(x)$ is linear, $\mathcal{S}(t) \sim e^{-\omega t}$, completes the proof of Theorem 1.2. □

3. Proof of Theorem 1.3

By arguments similar to those used for the proof of Theorem 2.1 in [2], we obtain the following counterpart of Theorem 2.1.

Theorem 3.1. *Let* w *be a solution to* (1.8) *with regularity properties* (1.9), (1.10) *and let* $T > 0$ *be sufficiently large. Then*

$$\int_0^T \hat{E}_w(t)dt - C\hat{E}_w(0) \le C_T(E_w(0))\{\int_{\Sigma_T} [|w_t|^2 + |\nabla w_t|^2 + g(w_t)w_t] d\Gamma dt\}. \tag{3.1}$$

To proceed with the proof of Theorem 1.3, let the functions $h(x)$, $h_i(x)$, $i = 1, 2$, and $\tilde{h}(x)$ be defined as in (1.14), (1.15). By the hypotheses imposed on functions $h_i(x)$, we obtain

$$\int_{\Sigma_T} |f_1(\frac{\partial}{\partial \nu} w_t)|^2 d\Gamma dt = \int_{\Sigma_{A_1}} |f_1(\frac{\partial}{\partial \nu} w_t)|^2 d\Gamma dt + \int_{\Sigma_{B_1}} |f_1(\frac{\partial}{\partial \nu} w_t)|^2 d\Gamma dt, \tag{3.2}$$

where $\Sigma_{A_1} \equiv \{(t,x) \in \Sigma_T : |\frac{\partial}{\partial\nu}w_t| \leq 1\}$ and $\Sigma_{B_1} \equiv \Sigma_T \backslash \Sigma_{A_1}$. Hence, using hypothesis (H-2) on Σ_{B_1}, we find

$$\begin{aligned}
&\int_{\Sigma_T} |\tfrac{\partial}{\partial\nu}w_t|^2 d\Gamma dt + \int_{\Sigma_T} |f_1(\tfrac{\partial}{\partial\nu}w_t)|^2 d\Gamma dt \\
&\quad \leq \int_{\Sigma_{A_1}} [|\tfrac{\partial}{\partial\nu}w_t|^2 + |f_1(\tfrac{\partial}{\partial\nu}w_t)|^2] d\Gamma dt \\
&\qquad + (M + \tfrac{1}{m}) \int_{\Sigma_{B_1}} f_1(\tfrac{\partial}{\partial\nu}w_t)\tfrac{\partial}{\partial\nu}w_t d\Gamma dt \\
&\quad \leq \int_{\Sigma_{A_1}} h_1(\tfrac{\partial}{\partial\nu}w_t f_1(\tfrac{\partial}{\partial\nu}w_t)) d\Gamma dt \\
&\qquad + (M + \tfrac{1}{m}) \int_{\Sigma_T} f_1(\tfrac{\partial}{\partial\nu}w_t)\tfrac{\partial}{\partial\nu}w_t d\Gamma dt.
\end{aligned} \tag{3.3}$$

Similarly, the same argument applied to f_2 yields

$$\begin{aligned}
&\int_{\Sigma_T} |\frac{\partial}{\partial\tau}w_t|^2 d\Gamma dt + \int_{\Sigma_T} |f_2(\frac{\partial}{\partial\tau}w_t)|^2 d\Gamma dt \\
&\quad \leq \int_{\Sigma_{A_2}} h_2(\frac{\partial}{\partial\tau}w_t f_2(\frac{\partial}{\partial\tau}w_t)) d\Gamma dt + (M + \frac{1}{m}) \int_{\Sigma_T} f_2(\frac{\partial}{\partial\tau}w_t)\frac{\partial}{\partial\tau}w_t d\Gamma dt.
\end{aligned} \tag{3.4}$$

Recall

$$\tilde{h}_\imath(x) \equiv h_\imath(\frac{x}{mes\ \Sigma_T}). \tag{3.5}$$

Then, by Jensen's inequality,

$$\begin{aligned}
&\int_{\Sigma_T} \{|w_t|^2 + |\nabla w_t|^2 + g(w_t)w_t\} d\Gamma dt \\
&\leq C_1 \int_{\Sigma_T} \{g(w_t)w_t + f_1(\tfrac{\partial}{\partial\nu}w_t)\tfrac{\partial}{\partial\nu}w_t + f_2(\tfrac{\partial}{\partial\tau}w_t)\tfrac{\partial}{\partial\tau}w_t\} d\Gamma dt \\
&\quad + C_2 \left[\tilde{h}_1(\int_{\Sigma_T} f_1(\tfrac{\partial}{\partial\nu}w_t)\tfrac{\partial}{\partial\nu}w_t d\Gamma dt) + \tilde{h}_2(\int_{\Sigma_T} f_2(\tfrac{\partial}{\partial\tau}w_t)\tfrac{\partial}{\partial\tau}w_t d\Gamma dt)\right] \\
&\leq C_1 \int_{\Sigma_T} \{g(w_t)w_t + f_1(\tfrac{\partial}{\partial\nu}w_t)\tfrac{\partial}{\partial\nu}w_t + f_2(\tfrac{\partial}{\partial\tau}w_t)\tfrac{\partial}{\partial\tau}w_t\} d\Gamma dt \\
&\quad + C_2 \sum_{\imath=1}^2 \tilde{h}_\imath(\int_{\Sigma_T} \{g(w_t)w_t + f_1(\tfrac{\partial}{\partial\nu}w_t)\tfrac{\partial}{\partial\nu}w_t + f_2(\tfrac{\partial}{\partial\tau}w_t)\tfrac{\partial}{\partial\tau}w_t\} d\Gamma dt,
\end{aligned} \tag{3.6}$$

where the last inequality follows from the monotonicity of the functions $\tilde{h}_\imath$.

Denoting $\mathcal{F} \equiv \int_{\Sigma_T} \{g(w_t)w_t + f_1(\frac{\partial}{\partial\nu}w_t)\frac{\partial}{\partial\nu}w_t + f_2(\frac{\partial}{\partial\tau}w_t)\frac{\partial}{\partial\tau}w_t\} d\Gamma dt$, we obtain from Theorem 3.1 and (3.6),

$$\int_\alpha^{T-\alpha} E_w(t)dt - C_1 E_w(0) \leq C_{T,\alpha,\epsilon}(E_w(0))[\mathcal{F} + \tilde{h}(\mathcal{F})]. \tag{3.7}$$

Since

$$\int_0^\alpha E_w(t)dt + \int_{T-\alpha}^T E_w(t)dt \leq 2\alpha E_w(0), \tag{3.8}$$

we find

$$\int_0^T E_w(t)dt - C_{1,\alpha}E_w(0) \leq C_{T,\alpha,\epsilon}(E_w(0))[\mathcal{F} + \tilde{h}\mathcal{F}], \tag{3.9}$$

and by Lemma 2.1,

$$
\begin{aligned}
&\int_0^T E_w(t)dt \le C_{T,\alpha,\epsilon}(E_w(0))[\mathcal{F} + \tilde{h}\mathcal{F}] + C_{1,\alpha}E_w(0) \\
&\Longrightarrow (T - C_{1,\alpha})E_w(T) \le C_{T,\alpha,\epsilon}(E_w(0))[\mathcal{F} + \tilde{h}(\mathcal{F})] \\
&\Longrightarrow E_w(T) \le C_T(E_w(0))[\mathcal{F} + \tilde{h}(\mathcal{F})].
\end{aligned} \tag{3.10}
$$

Hence, recalling (2.27),

$$(I + \tilde{h})^{-1}\left(\frac{E_w(T)}{C_T(E_w(0))}\right) \le \mathcal{F} = E_w(0) - E_w(T). \tag{3.11}$$

Setting

$$p(s) \equiv (I + \tilde{h})^{-1}\left(\frac{s}{C_T(E_w(0))}\right), \tag{3.12}$$

we have the result of Proposition 2.4 valid with the above function $p(s)$.

As in section 2.3, applying the result of Proposition 2.4, we obtain

$$E_w(m(T+1)) + p(E_w(m(T+1))) \le E_w(mT), \tag{3.13}$$

for $m = 0, 1, ...$ Thus, applying Lemma 2.5 with

$$s_m \equiv E_w(mT), \tag{3.14}$$

yields

$$E_w(mT) \le \mathcal{S}(m), \ m = 0, 1, ... \tag{3.15}$$

Setting $t = mT + \tau$, $0 \le \tau < T$,

$$E_w(t) \le E_w(mT) \le \mathcal{S}(m) \le \mathcal{S}\left(\frac{t-\tau}{T}\right) \le \mathcal{S}\left(\frac{t}{T} - 1\right) \ for \ t > T, \tag{3.16}$$

which completes the proof of Theorem 1.3. □

4. Proof of Theorem 1.1

The first part of Theorem 1.1 (existence) follows by applying the result of the main theorem in [6] within the framework described in the first section of [1] where a system of coupled plate equations is considered. The details are omitted for lack of space. We shall concentrate on the second part of Theorem 1.1, i.e., estimate (1.7).

4.1 Change of Variables

We introduce a change of variables, $w \equiv v - u$. Then w satisfies

$$(4.1)\quad \begin{cases} w_{tt} - \gamma^2 \Delta w_{tt} + \Delta^2 w = k_1(-w) - k_2(w) \\ \quad w(0,\cdot) = v_0 - u_0;\ w_t(0,\cdot) = v_1 - u_1 \\ \Delta w + (1-\mu)\mathcal{B}_1 w = -[f_1(\frac{\partial}{\partial \nu} w_t + \frac{\partial}{\partial \nu} u_t) - f_1(\frac{\partial}{\partial \nu} u_t)] \\ \frac{\partial}{\partial \nu}\Delta w + (1-\mu)\mathcal{B}_2 w - \gamma^2 \frac{\partial}{\partial \nu} w_{tt} - w + \frac{\partial}{\partial \tau} h_1(\frac{\partial}{\partial \tau} w) - \frac{\partial}{\partial \tau} h_1(-\frac{\partial}{\partial \tau} w) \\ \qquad -l_1(w) + l_2(-w) \\ \qquad = g(w_t + u_t) - g(u_t) - \frac{\partial}{\partial \tau} f_2(\frac{\partial}{\partial \tau} w_t + \frac{\partial}{\partial \tau} u_t) + \frac{\partial}{\partial \tau} f_2(\frac{\partial}{\partial \tau} u_t). \end{cases}$$

By virtue of (1.6),

$$(4.2)\qquad \frac{\partial}{\partial \nu} u_t,\ \frac{\partial}{\partial \tau} u_t \in L_2(\Sigma_\infty).$$

Moreover,

$$(4.3)\qquad \|u_t\|_{C(0,\infty,H^1(\Omega))} \le C(E_u(0), E_v(0)).$$

Hence, by Sobolev's Imbeddings, for any $r \ge 1$,

$$(4.4)\qquad \|u_t\|_{C(0,\infty,L_{2r}(\Gamma))} \le C(E_u(0), E_v(0)).$$

Setting $a_1 \equiv \frac{\partial}{\partial \nu} u_t$, $a_2 \equiv \frac{\partial}{\partial \tau} u_t$, $b \equiv u_t$, We conclude that $a_i \in L_2(\Sigma_\infty)$ and $b \in C(0,\infty; L_{2r}(\Gamma))$. On the other hand, functions defined by

$$\begin{aligned} k(s) &\equiv k_2(s) - k_1(-s) \\ l(s) &\equiv l_1(s) - l_2(-s) \\ h(s) &\equiv h_1(s) - h_2(-s) \end{aligned}$$

comply with hypothesis (H-1).

Thus we are in a position to apply the result of Theorem 1.2 to equation (4.1). This yields

Theorem 4.1. *Let $w(t)$ be any solution of finite energy corresponding to system (4.1). Then there exist constants C, $\omega > 0$ such that*

$$(4.5)\qquad \|(u-v)(t)\|_{H^2(\Omega)} + \|(u_t - v_t)(t)\|_{H^1(\Omega)} \le Ce^{-\omega t},$$

where C, ω may depend on $\|u_0\|_{H^2(\Omega)}$, $\|v_0\|_{H^2(\Omega)}$, $\|u_1\|_{H^1(\Omega)}$, $\|v_1\|_{H^1(\Omega)}$.

4.2 Analysis of "u" Equation

Using the variable w, we see that the equation for u (see (1.1)) is equivalent to

$$
(4.6)\quad \begin{cases} u_{tt} - \gamma^2 \Delta u_{tt} + \Delta^2 u = k_2(w) & \text{in } Q_\infty \\ \quad u(0,\cdot) = u_0;\ u_t(0,\cdot) = u_1 & \text{in } \Omega \\ \Delta u + (1-\mu)\mathcal{B}_1 u = -f_1(\frac{\partial}{\partial \nu} u_t) & \text{on } \Sigma_\infty \\ \frac{\partial}{\partial \nu}\Delta u + (1-\mu)\mathcal{B}_2 u - \gamma^2 \frac{\partial}{\partial \nu} u_{tt} - u & \\ \qquad = g(u_t) - \frac{\partial}{\partial \tau} f_2(\frac{\partial}{\partial \tau} u_t) + l_2(-w) - \frac{\partial}{\partial \tau} h_2(\frac{\partial}{\partial \tau}(-w)) & \text{on } \Sigma_\infty, \end{cases}
$$

where from Theorem 4.1, $w(t)$ satisfies

$$
(4.7)\qquad \|w(t)\|_{H^2(\Omega)} \le Ce^{-\omega t},
$$

with the constants C and ω depending on $E_v(0)$, $E_u(0)$, $\|u_0\|_{H^2(\Omega)}$, $\|v_0\|_{H^2(\Omega)}$, $\|u_1\|_{H^1(\Omega)}$, $\|v_1\|_{H^1(\Omega)}$.

Proposition 4.1. *Let $E_v(0) \le R$, $E_u(0) \le R$. Then there exist constants C, $\omega > 0$ such that for all $t \ge 0$,*

$$
\begin{aligned}
&(4.8) & \|k_2(w)(t)\|_{L_2(\Omega)} &\le Ce^{-\omega t} \\
&(4.9) & \|l_2(w)(t)\|_{L_2(\Gamma)} &\le Ce^{-\omega t} \\
&(4.10) & \|\frac{\partial}{\partial \tau} h_2(\frac{\partial}{\partial \tau}(-w(t))\|_{H^{-1}(\Gamma)} &\le Ce^{-\omega t}.
\end{aligned}
$$

Proof: Inequalities (4.8) and (4.9) follow direct from (4.7), hypothesis (H-1) and the imbedding $H^2(\Omega) \subset C(\Omega)$. As for (4.10), we have

$$
(4.11)\quad \begin{aligned} \|\tfrac{\partial}{\partial \tau} h_2(\tfrac{\partial}{\partial \tau}(-w(t))\|^2_{H^{-1}(\Gamma)} &\le C\|h_2(\tfrac{\partial}{\partial \tau}(-w(t))\|^2_{L_2(\Gamma)} \\ &\le C\int_\Gamma \{|\tfrac{\partial}{\partial \tau} w(t)|^{2k} + |\tfrac{\partial}{\partial \tau} w(t)|^2\} d\Gamma \\ &\le C\{\|w(t)\|^2_{H^2(\Omega)} + \|w(t)\|^{2k}_{H^2(\Omega)}\}, \end{aligned}
$$

where the second inequality follows from the imbedding $H^{1/2}(\Gamma) \subset L_{2k}(\Gamma)$ followed by the Trace Theorem. (4.7) together with (4.11) implies (4.10). □

We next consider the following nonautonomous linear problem.

$$
(4.12)\quad \begin{cases} u_{tt} - \gamma^2 \Delta w_{tt} + \Delta^2 u = k(t,x) & \text{in } Q_\infty \\ \quad u(0,\cdot) = u_0;\ u_t(0,\cdot) = u_1 & \text{in } \Omega \\ \Delta u + (1-\mu)\mathcal{B}_1 u = -f_1(t,x)\frac{\partial}{\partial \nu} u_t & \text{on } \Sigma_\infty \\ \frac{\partial}{\partial \nu}\Delta u + (1-\mu)\mathcal{B}_2 u - \gamma^2 \frac{\partial}{\partial \nu} u_{tt} - u & \\ \qquad = g(t,x)u_t - \frac{\partial}{\partial \tau} f_2(t,x)\frac{\partial}{\partial \tau} u_t + l(t,x) & \text{on } \Sigma_\infty, \end{cases}
$$

where

$$\|k(t)\|_{L_2(\Omega)} \leq C_0 e^{-\omega_0 t} \tag{4.13}$$

$$\|l(t)\|_{H^{-1}(\Gamma)} \leq C_0 e^{-\omega_0 t} \tag{4.14}$$

$$m \leq f_\imath(t,x) \leq M \quad (t,x) \in \Sigma_\infty \tag{4.15}$$

$$g(t,x) \geq 0 \quad (t,x) \in \Sigma_\infty \tag{4.16}$$

$$\|g(t,\cdot)\|_{L_2(\Gamma)} \leq M \tag{4.17}$$

Notice that a homogeneous system (4.12) with $k \equiv l \equiv 0$ is exponentially stable. Hence, one can show by using *linear* evolution methods, that this stability is preserved for a system with nonhomogeneous, but exponentially decaying terms. This is stated below.

Lemma 4.1. *Let u be a solution to* (4.12) *subject to assumptions* (4.13)-(4.17). *Then there exist constants C, $\omega > 0$ depending on C_0, ω_0, M, m such that for all $t \geq 0$,*

$$\|u(t)\|_{H^2(\Omega)} + \|u_t(t)\|_{H^1(\Omega)} \leq Ce^{-\omega t}(\|u_0\|_{H^2(\Omega)} + \|u_1\|_{H^1(\Omega)} + 1). \tag{4.18}$$

We note that the solution u to (4.7) satisfies (4.12) with

$$k(t,x) \equiv k_2(w(t,x)),\ l(t,x) \equiv l_2(-w(t,x)) - \frac{\partial}{\partial\tau}h_2(-\frac{\partial}{\partial\tau}w(t,x)) \tag{4.19}$$

$$f_1(t,x) \equiv \frac{f_1(\frac{\partial}{\partial\nu}u_t(t,x))}{\frac{\partial}{\partial\nu}u_t(t,x)} \tag{4.20}$$

$$f_2(t,x) \equiv \frac{f_2(\frac{\partial}{\partial\tau}u_t(t,x))}{\frac{\partial}{\partial\tau}u_t(t,x)} \tag{4.21}$$

$$g(t,x) \equiv \frac{g(u_t(t,x))}{u_t(t,x)}. \tag{4.22}$$

Hypotheses assumed in (H-1) together with the result of Proposition 4.1 imply that (4.13)-(4.17) hold true.

Thus we are in a position to apply the result of Lemma 4.1 to the solution u of equation (4.1). This yields

Theorem 4.2. *Let u be a solution to* (4.5). *Then there exist constants C, $\omega > 0$ depending on $E_v(0)$, $E_u(0)$ such that*

$$\|u(t)\|_{H^2(\Omega)} + \|u_t(t)\|_{H^1(\Omega)} \leq Ce^{-\omega t}. \tag{4.23}$$

Combining together the results of Theorem 4.1 and Theorem 4.2 and recalling $v \equiv w + u$ yields the final conclusion, (1.7) in Theorem 1.1. □

References

1 A Favini and I Lasiecka Well-posedness and regularity of second-order abstract equations arising in hyperbolic-like problems with nonlinear boundary conditions *Osaka Journal of Mathematics* To appear

2 M A Horn and I Lasiecka Global stabilization of a dynamic von Kármán plate with nonlinear boundary feedback *Applied Mathematics and Optimization* To appear

3 V Komornik Rapid boundary stabilization of the wave equation *SIAM Journal of Control and Optimization*, 29 (1) 197-208 January 1991

4 V Komornik and E Zuazua A direct method for the boundary stabilization of the wave equation *J Math Pures et Appl* , 69 33-54, 1990

5 J E Lagnese *Boundary Stabilization of Thin Plates* Society for Industrial and Applied Mathematics, Philadelphia, 1989

6 I Lasiecka Existence and uniqueness of the solutions to second order abstract equations with nonlinear and nonmonotone boundary conditions *Journal of Nonlinear Analysis, Methods and Applications* To appear

7 I Lasiecka and D Tataru Uniform boundary stabilization of semilinear wave equations with nonlinear boundary damping *Differential and Integral Equations*, 6(3) 507-533, 1993

8 I Lasiecka and R Triggiani Sharp trace estimates of solutions to Kirchoff and Euler-Bernoulli equations *Applied Mathematics and Optimization*, 28 277-306, 1993

9 W Littman Boundary control theory for beams and plates In *Proceedings of 24th Conference on Decision and Control*, pages 2007-2009, 1985

10 M Najafi, G Sarhangi, and H Wang The study of the stabilizability of the coupled wave equations under various end conditions In *Proceedings of the 31nd IEEE Conference on Decision and Control*, Tucson, Arizona, 1992

11 M Najafi and H Wang Exponential stabilty of wave equations coupled in parallel by viscous damping, In *Proceedings of the 32nd IEEE Conference on Decision and Control*, San Antonio, Texas, 1993

12 B Rao Stabilization of a Kirchhoff plate equation in star-shaped domain by nonlinear boundary feedback *Nonlinear Analysis*, 20(6) 605-626, 1993

Mary Ann Horn
School of Mathematics
University of Minnesota
Minneapolis, Minnesota 55455, USA

Irena Lasiecka
Department of Applied Mathematics
University of Virginia
Charlottesville, Virginia 22903, USA

BOUNDARY TEMPERATURE CONTROL FOR THERMALLY COUPLED NAVIER-STOKES EQUATIONS

KAZUFUMI ITO

Center for Research in Scientific Computation
Department of Mathematics
North Carolina State University

ABSTRACT In this paper the optimal control problem for the thermally coupled incompressible Navier-Stokes equations by the Dirichelet boundary temperature control is discussed Well-posedness and existence of the optimal control for the finite-time horizon problem and optimal control problem for the stationary equations are established Necessary optimality conditions are also obtained

1991 *Mathematics Subject Classification* 76D05, 93C20, 49B22

Key words and phrases Boussinesq equation, boundary temperature control, necessary optimality condition

1. Introduction

In this paper we discuss the optimal control problem of the thermally coupled incompressible Navier-Stokes equations. Consider the following optimal control problem

$$\text{(1.1)} \qquad \begin{aligned} &\text{minimize} \quad J(g) = \int_0^T \left(\varphi(u(t), T(t) - T_0) + \frac{\beta}{2}\, |g(t) - T_0|^2_{L^2(\Gamma)} \right) dt \\ &\qquad\qquad \text{over } g(t) \in \mathcal{C} \end{aligned}$$

subject to

$$\text{(1.2)} \qquad \begin{aligned} &\frac{\partial}{\partial t} u + u \cdot \nabla u + \nabla p = \nu\, \Delta u + \gamma\, (T - T_0)\, e_d + f \\ &\nabla \cdot u = 0, \quad u|_\Gamma = 0, \\ &\frac{\partial}{\partial t} T + u \cdot \nabla T = \nabla \cdot (\kappa\, \nabla T), \quad T = g \text{ on } \Gamma \end{aligned}$$

where $f \in L^2(\Omega)^d$ is a source field, u, p, T stand for the nondimensionalized velocity vector in R^d with $d = 2$, 3, pressure, and temperature, respectively and

$\mathcal{C}$ is the closed convex set in $L^2(\Gamma)$ such that

$$\bar{T}_1 \leq g(x) \leq \bar{T}_2 \quad \text{a.e. } x \in \Gamma \tag{1.3}$$

Here, $\bar{T}_1 \leq T_0 \leq \bar{T}_2$ and $\gamma = \dfrac{\bar{g}}{T_0}$ where $T_0 > 0$ is a constant reference temperature and $\bar{g}$ is the gravitational constant. the vector e_d denotes the $d - th$ unit vector of R^d. Throughout this paper we assume that Ω is sufficiently smooth, ν, κ are positive constants and $\theta(t,\cdot) = T(t,\cdot) - T_0$. This control problem and the corresponding problem for the stationary flow (see, section 2) are motivated from control of the transport process in the high pressure vapor transport (HPVT) reactor [IST1]. For example we consider the Scholz geometry depicted in Figure 1. The source material and the growing crystal are sealed in a fused silica ampoule that is heated by a furnace liner at its outer cylindrical surface. The substrate Γ_0 (the single crystal) is located on a fused silica window (W) which is cooled by a jet of helium gas from the outer surface. HPVT processes are based on physical vapor transport and can be described very roughly as proceeding via evaporation at the polycrystalline source and condensation at the surface of the cooler substrate. The system of equations (1.2) is called the Boussinesq equation where we assume that the flow is incompressible and the transport phenomena of a single (carrier) gas is modeled. The objective of our control problem includes the uniformity of the temperature in a neighborhood of the substrate (W). For example, the performance index φ appearing (1.1) is given by

$$\varphi(u(t), T(t) - T_0) = \frac{1}{2}\int_\Omega \left(|u(t,x) - u_d(x)|^2 + |T(t,x) - T_d(x)|^2\right) dx,$$

where the pair $(u_d(x), T_d(x))$ is the desired state. We refer [IST1],[IST2] for the specific choice of (u_d, T_d).

Also we consider a family of control problems parameterized by $\epsilon > 0$:

$$\text{minimize} \quad J(g) \quad \text{over } g(t) \in \mathcal{C} \tag{1.4}$$

subject to

$$\begin{aligned} &\frac{\partial}{\partial t}u + u\cdot\nabla u + \nabla p = \nu\,\Delta u + \gamma\,(T - T_0)\,e_d + f \\ &\nabla\cdot u = 0, \quad u|_\Gamma = 0, \\ &\frac{\partial}{\partial t}T + u\cdot\nabla T = \nabla\cdot(\kappa\,\nabla T), \\ &\kappa\, n\cdot\nabla T(t,\cdot) = \frac{1}{\epsilon}\,(g(t,\cdot) - T(t,\cdot)) \text{ on } \Gamma. \end{aligned} \tag{1.5}$$

where n is the outward normal vector at the boundary Γ. The boundary condition for T is given by the Newton's law of cooling. Problem (1.4) – (1.5) is discussed in [IST2] and is much nicer to be dealt with both in the theoretical and numerical point of view since it is naturally formulated under the Gelfand-triple framework in [LM],[IST2]. In this paper we develop a weak formulation of the Dirichlet boundary control problem in the thermal equation of (1.2) (see, Definitions 2.1 and 3.1) which is based on the transposition of the Gelfand triple. A solution to equation (1.2) is defined by an asymptotic limit of solutions to equation (1.5) when $\epsilon \to 0^+$. This limiting procedure is necessary since our weak formulation does not allow us to use the (weak) maximum principle to obtain *a priori* bound of solutions. Thus we analyze problem (1.1)–(1.2) as the limiting case of problem (1.1) and (1.5).

The outline of the paper is as follows. In section 2 we discuss the corresponding control problems (2.1)-(2.2) for the stationary flow. Basic functional space framework and the definition of weak solutions to the steady-state equation are given. The existence of weak solutions then is shown by the limiting procedure described above where *a priori* L^∞-bound of the thermal component θ is used in an essential way. Necessary optimality condition is obtained by applying the Lagrange multiplier theory [MZ]. In section 3 we discuss the finite time horizon problem (1.1)-(1.2). The existence of weak solutions to (1.2) and necessary optimality condition for $d = 2$ are established by the limiting procedure.

2. Stationary Case

In this section we consider control problems for the stationary flow:

$$\text{(2.1)} \qquad \text{minimize} \quad J(g) = \varphi(u, T - T_0) + \frac{\beta}{2}\,|g - T_0|^2_{L^2(\Gamma)} \quad \text{over } g \in \mathcal{C}$$

subject to

$$\text{(2.2)} \qquad \begin{aligned} &-\nu\,\Delta u + u\cdot\nabla u + \nabla p = \gamma\,(T - T_0)\,e_d + f \\ &\nabla\cdot u = 0, \quad u|_\Gamma = 0, \\ &-\kappa\,\Delta T + u\cdot\nabla T = 0, \quad T = g \ \text{ on } \Gamma. \end{aligned}$$

2.1. Wellposedness. In this section we discuss existence and regularity of solutions to (2.2). Let V_0 be the divergence free subspace of $(H^1_0(\Omega))^d$ [GR] and H_0 is defined by

$$H_0 = \{\phi \in L^2(\Omega)^d :\ \nabla\cdot\phi = 0 \text{ and } n\cdot\phi = 0 \text{ on } \Gamma\}.$$

H_0 is equipped with the natural L^2-norm and V_0 is equipped with $|u|^2_{V_0} = |\nabla u|^2_{L^2(\Omega)}$. Let $H_1 = L^2(\Omega)$ equipped with the natural L^2-norm and $V_1 = H^1_0(\Omega)$ equipped with $|\phi|^2_{V_1} = |\nabla\phi|^2_{L^2(\Omega)}$. If we identify H_1 with H_1^* then $V_1 \subset H_1 = H_1^* \subset V_1^*$. Let Δ denote the Laplacian and $\Delta_0 = \Delta$ with $\text{dom}(\Delta_0) = H^2(\Omega)\cap H^1_0(\Omega)$. Then $-\Delta_0$ is positive selfadjoint operator on H_1, and $-\Delta_0 \in \mathcal{L}(V_1, V_1^*)$ with

$$\langle -\Delta_0\phi,\ \psi\rangle = (\nabla\phi,\ \nabla\psi)_{L^2} \quad \text{for } \phi,\ \psi \in V_1$$

where $V_1^* = H^{-1}(\Omega)$ and $\langle\cdot,\cdot\rangle$ denotes the dual product of $V_1^* \times V_1$. Moreover, V_1^* can be equipped with

$$|\phi|^2_{V_1^*} = \langle\phi,\, (-\Delta_0)^{-1}\phi\rangle, \tag{2.3}$$

i.e., $(-\Delta_0)^{-1} \in \mathcal{L}(V_1^*, V_1)$ is the Riesz map of V_1^*. Consider the second equation of (2.2) when $u \in V_0$ is fixed. By Green's formula we have from the second equation of (2.2)

$$(u\cdot\nabla\theta - \kappa\,\Delta\theta,\, \psi) = \kappa\,(\theta,\phi) + \kappa\,(g - T_0, n\cdot\nabla\psi)_\Gamma - (\theta,\; u\cdot\nabla\psi) = 0$$

for $\psi = (-\Delta_0)^{-1}\phi$, $\phi \in H_1$, where we assumed that $\theta = T - T_0$ is sufficiently smooth and $(f,g)_\Gamma = \displaystyle\int_\Gamma f(s)g(s)\,ds$. Thus, the second equation of (2.2) can be written as

$$\kappa\,(\theta,\,\phi)_{L^2} - (\theta,\, u\cdot\nabla\psi)_{L^2} + \kappa\,(g - T_0,\, n\cdot\nabla\psi)_\Gamma = 0, \quad \text{where } \psi = (-\Delta_0)^{-1}\phi \tag{2.4}$$

for $\phi \in H_1 = L^2(\Omega)$. Here, we assume that the domain Ω is sufficiently smooth so that

$$|(-\Delta_0)^{-1}\phi|_{H^2(\Omega)} \le \alpha\,|\phi|_{L^2(\Omega)}, \quad \phi \in L^2(\Omega)$$

for some $\alpha > 0$. Hence, we have $\nabla\psi \in H^1(\Omega)$ and $n\cdot\nabla\psi \in H^{\frac{1}{2}}(\Gamma)$ for $\psi = (-\Delta_0)^{-1}\phi$, $\phi \in L^2(\Omega)$. Let W be the completion of $L^2(\Omega)$ with respect to norm $|(-\Delta_0)^{-1}\phi|_{L^2}$. Then, $W = (H_1)^*$ where V_1^* is taken to be the pivoting space equipped with inner product (2.3). The dual product on $W \times H_1$ is defined as

$$\langle\theta,\phi\rangle_{W\times H_1} = ((-\Delta_0)^{-1}\theta,\phi), \quad \theta \in W \text{ and } \phi \in H_1.$$

Thus, $H_1 \subset V_1^* = (V_1^*)^* \subset W$ defines the Gelfand triple and results in the transposition of the standard triple: $H_0^1(\Omega) \subset L^2(\Omega) = L^2(\Omega)^* \subset H^{-1}(\Omega)$ by one Sobolev index. Moreover, (2.4) is equivalently written as

$$-\kappa\,\Delta\theta + \nabla\cdot(u\theta) + r = 0 \quad \text{in } W$$

where $r \in W$ is defined by

$$(r,\phi)_{W\times H_1} = \kappa\,(g - T_0,\, n\cdot\nabla\psi)_\Gamma, \quad \text{with } \psi = (-\Delta_0)^{-1}\phi,\ \phi \in H_1.$$

Now, we define the weak solution to (2.2) by

Definition 2.1. The pair $(u,\theta) \in V_0 \times L^2(\Omega)$ is a weak solution to (2.2) if

$$\nu\,(\nabla u, \nabla v) + b(u,u,v) = (\gamma\,\theta e_d + f, v) \tag{2.5a}$$

for all $v \in V_0$ and

$$\kappa\,(\theta,\,\phi)_{L^2} - (\theta,\, u\cdot\nabla\psi)_{L^2} + \kappa\,(g - T_0,\, n\cdot\nabla\psi)_\Gamma = 0, \quad \text{where } \psi = (-\Delta_0)^{-1}\phi \tag{2.5b}$$

for all $\phi \in L^2(\Omega)$, where the trilinear form b on $(V_0)^3$ is defined by

$$b(u,v,w) = \int_\Omega (u\cdot\nabla v)\cdot w\,dx \quad \text{for } u,v,w \in V_0.$$

It follows from [Te1] that

Lemma 2.2. *The trilinear form b satisfies*

$$(a)\quad |b(u,v,w) \le |u|_{L^4}|\nabla v|_{V_0}|w|_{L^4} \le M_1\,|u|_{V_0}|v|_{V_0}|w|_{V_0}$$

$$(b)\quad b(u,v,w)+b(u,w,v)=0 \quad \textit{in particular}\; b(u,v,v)=0$$

$$(c)\quad |b(u,v,w)| \le M_2|u|_{H_0}^{1/2}|u|_{V_0}^{1/2}|v|_{H_0}^{1/2}|v|_{V_0}^{1/2}|w|_{V_0}\; \textit{for}\; d=2$$

$$(d)\quad |b(u,v,w)| \le M_2|u|_{H_0}^{1/4}|u|_{V_0}^{3/4}|v|_{H_0}^{1/4}|v|_{V_0}^{3/4}|w|_{V_0}\; \textit{for}\; d=3$$

for $u,v,w\in V_0$.

The sesquilinear form on $H_1\times H_1$ defined by

$$\kappa\,(\phi_1,\,\phi_2)_{L^2}-(\phi_1,\,u\cdot\nabla\psi)_{L^2}\quad \text{where}\ \psi=(-\Delta_0)^{-1}\phi_2$$

for $\phi_i,\ i=1,2\in H_1$ and $u\in V_0$ given, is bounded but is not necessarily H_1-coercive unless $|u|_{V_0}$ is sufficiently small. Thus, one cannot show that (2.5) has a solution $(u,\theta)\in V_0\times H_1$ in general. We overcome this difficulty by an approach in [It] in which a solution to (2.5) is defined by a weak limit of $(u_\epsilon,\theta_\epsilon)$ where $(u_\epsilon,\theta_\epsilon)\in V_0\times H^1(\Omega)$ is a weak solution to

$$\begin{aligned}
&-\nu\,\Delta u_\epsilon+u\cdot\nabla u_\epsilon+\nabla p=\gamma\,\theta_\epsilon\,e_d+f\\
&-\kappa\,\Delta\theta_\epsilon+u_\epsilon\cdot\nabla\theta_\epsilon=0,\\
&\kappa\,n\cdot\theta_\epsilon=\frac{1}{\epsilon}\,(g-T_0-\theta_\epsilon)\ \text{on}\ \Gamma.
\end{aligned}\tag{2.6}$$

A variational form of (2.6) is given by (2.5a) and

$$\kappa\,(\nabla\theta,\,\nabla\psi)_{L^2}+(u\cdot\nabla\theta,\psi)+\frac{1}{\epsilon}\,(\theta,\psi)_\Gamma=\frac{1}{\epsilon}\,(g-T_0,\psi)_\Gamma\tag{2.7}$$

for $\psi\in H^1(\Omega)$. Then we have

Lemma 2.3. *For* $\epsilon>0$ *and* $g\in L^2(\Gamma)$ *equation (2.5a) and (2.7) has a solution* $(u_\epsilon,\theta_\epsilon)$ *in* $V_0\times H^1(\Omega)$. *Moreover if* $\bar T_1\le g\le \bar T_2$ *a.e. in* Γ *then* $\bar T_1-T_0\le\theta_\epsilon\le\bar T_2-T_0$ *a.e in* Ω.

Proof: Let $\bar u\in V_0$ be given. Note that $(\bar u\cdot\nabla\psi,\,\psi)=-\frac{1}{2}\,(\nabla\cdot\bar u,\,|\psi|^2)=0$. Thus, it follows from Lemma 2.2 that

$$\begin{aligned}
\sigma((u_1,\psi_1),(u_2,\psi_2))&=\nu\,(\nabla u_1,\nabla u_2)+b(\bar u,u_1,u_2)\\
&+\kappa\,(\nabla\psi_1,\,\nabla\psi_2)_{L^2}+\frac{1}{\epsilon}\,(\psi_1,\,\psi_2)_\Gamma+(\bar u\cdot\nabla\psi_1,\,\psi_2)
\end{aligned}$$

for (u_i, ψ_i), $i = 1, 2$, defines a bounded, elliptic sesquilinear form on $(V_0 \times H^1(\Omega))^2$ with

$$\sigma((u,\psi),(u,\psi)) = \nu\,|\nabla u|^2 + \kappa\,|\nabla\psi|^2 + \frac{1}{\epsilon}\,|\psi|_\Gamma^2,$$

where we used the fact that for $\epsilon > 0$

$$|\psi|_{H^1}^2 = \kappa\,|\nabla\psi|^2 + \frac{1}{2\epsilon}\,|\psi|_\Gamma^2$$

defines the equivalent norm on $H^1(\Omega)$. Thus, it follows from Lax-Milgram theorem that given $(\bar u, \bar\theta) \in V_0 \times H^1(\Omega)$

$$\sigma((u,\theta),(v,\psi)) = (\gamma\,\bar\theta e_d + f, v) + \frac{1}{\epsilon}\,(g - T_0, \psi)_\Gamma, \quad \bar\theta \in H^1(\Omega)$$

has a unique solution (u, θ) in $V_0 \times H^1(\Omega)$. Moreover,

$$|\theta|_{H^1}^2 = |\nabla\theta|_{L^2}^2 + \frac{1}{2\epsilon}\,|\theta|_{L^2(\Gamma)}^2 \le \frac{1}{2\epsilon}\,|g - T_0|_{L^2(\Gamma)}^2.$$

and

$$|u|_{V_0} \le \frac{M_3}{\nu}\,(\gamma|\bar\theta|_{L^2} + |f|_{L^2}),$$

where $|\phi|_{H_0} \le M_3\,|\phi|_{V_0}$, $\phi \in V_0$. Define a solution map Φ on $V_0 \times H^1(\Omega)$ by $\Phi(\bar u, \bar\theta) = (u, \theta)$. Let C be the closed convex subspace of $V_0 \times H^1(\Omega)$, defined by

$$C = \{(\phi,\psi) : |\phi|_{V_0} \le \frac{M_3}{\nu}(\frac{\gamma M_4}{\sqrt{2\epsilon}}\,|g - T_0|_{L^2(\Gamma)} + |f|_{L^2}) \text{ and } |\psi|_{H^1} \le \frac{1}{\sqrt{2\epsilon}}\,|g - T_0|_{L^2(\Gamma)}\},$$

where $|\theta|_{L^2} \le M_4\,|\theta|_{H^1}$, $\theta \in H^1(\Omega)$. It follows fron the above estimates that Φ maps from C into C Moreover, it is shown in [IST2] that Φ is continuous and compact. Hence by Schauder fixed point theorem Φ has a fixed point in $V_0 \times H^1(\Omega)$ that defines a solution to (2.5a) and (2.7).

Moreover, we assume that $\bar T_1 \le g \le \bar T_2$ a.e. in Γ. Let $\psi = \sup(\theta_\epsilon, \bar T_2 - T_0)$. Then $\psi \in H^1(\Omega)$ [Tr] and $(u_\epsilon \cdot \nabla\theta_\epsilon, \psi) = -\frac{1}{2}(\nabla \cdot u_\epsilon, |\psi|^2) = 0$. From (2.7)

$$\kappa\,|\nabla\psi|^2 + \frac{1}{\epsilon}\,(\theta_\epsilon - (g - T_0), \psi)_\Gamma,$$

where $(\theta_\epsilon - (g - T_0))\,\psi \ge 0$ a.e. on Γ. Hence $|\nabla\psi|_{L^2}^2 = 0$ and thus $\theta_\epsilon \le \bar T_2 - T_0$. Similarly, we can prove that $\theta_\epsilon \ge \bar T_1 - T_0$ by choosing $\psi = \inf(\bar T_1 - T_0, \theta_\epsilon)$. □

Lemma 2.4. *Suppose $\bar T_1 \le g \le \bar T_2$ a.e. in Γ. Let $(u_\epsilon, \theta_\epsilon) \in V_0 \times H^1(\Omega)$ be the solution to (2.5a) and (2.7). Then there exists a sequence $\{\epsilon\}$ of positive numbers such that u_ϵ converges weakly to u in V_0 and θ_ϵ converges weakly * to θ in $L^\infty(\Omega)$, where $(u, \theta) \in V_0 \times L^\infty(\Omega)$ satisfies (2.5) with $\bar T_1 - T_0 \le \theta \le \bar T_2 - T_0$ a.e. in Ω.*

Proof: Setting $\psi = \theta_\epsilon$ in (2.7), we obtain

$$\epsilon\kappa\,|\nabla\theta_\epsilon|_{L^2}^2 + \frac{1}{2}\,|\theta_\epsilon|_\Gamma^2 \le \frac{1}{2}\,|g - T_0|_\Gamma^2 \tag{2.8}$$

Hence $|\theta_\epsilon|^2_\Gamma$ is uniformly bounded in $\epsilon > 0$. Note that from (2.5a)

$$|u_\epsilon|_{V_0} \leq \frac{M_3}{\nu}\,(\gamma\,|\theta_\epsilon|_{L^2} + |f|_{L^2}).$$

It follows from Lemma 2.3 that $\bar{T}_1 - T_0 \leq \theta_\epsilon \leq \bar{T}_2 - T_0$. Hence $|\theta_\epsilon|_{L^2} \leq \sqrt{\text{meas}(\Omega)} \times \max(|\bar{T}_1 - T_0|, |\bar{T}_2 - T_0|)$ and $|u_\epsilon|_{V_0}$ is uniformly bounded in $\epsilon > 0$. Thus there exist $(u, \theta, \hat{g}) \in V_0 \times H^1(\Omega) \times L^2(\Gamma)$ and a sequence $\{\epsilon\}$ of positive numbers such that the trace of $\theta_\epsilon + T_0$ on Γ converges weakly to $\hat{g}$ in $L^2(\Gamma)$, θ_ϵ converges to θ weakly * in $L^\infty(\Omega)$ and u_ϵ converges to u weakly in V_0, as $\epsilon \to 0^+$. From (2.7), we have for $\psi \in H^1(\Omega)$

$$(\theta_\epsilon - (g - T_0), \psi)_\Gamma + \epsilon\,(\kappa\,(\nabla\theta_\epsilon,\,\nabla\psi) + (u_\epsilon\cdot\nabla\theta_\epsilon,\,\psi)) = 0. \tag{2.9}$$

where it follows from (2.8) that

$$\epsilon\,|\nabla\theta_\epsilon|^2_{L^2} \quad \text{is uniformly bounded in } \epsilon > 0$$

and

$$|(u_\epsilon\cdot\nabla\theta_\epsilon,\,\psi)| \leq M\,|u_\epsilon|_{V_0}|\nabla\theta_\epsilon|_{L^2}|\psi|_{H^1}$$

for some $M > 0$. Thus, from (2.9)

$$|((\theta_\epsilon + T_0) - g), \psi)_\Gamma| \leq \tilde{M}\,\sqrt{\epsilon}|\psi|_{H^1}$$

for some $\tilde{M} > 0$ and all $\psi \in H^1(\Omega)$, which implies that $\hat{g} = g$ and $\theta_\epsilon \to g - T_0$ as $\epsilon \to 0^+$ in $H^{-1/2}(\Gamma)$. Setting $\psi = (-\Delta_0)^{-1}\phi$, $\phi \in H_1 = L^2(\Omega)$ in (2.7) and by Green's formula we have

$$\kappa\,(\theta_\epsilon,\,\phi)_{L^2(\Omega)} - (\theta_\epsilon,\,u_\epsilon\cdot\nabla\psi)_{L^2(\Omega)} + \kappa\,(\theta_\epsilon,\,n\cdot\nabla\psi)_\Gamma = 0. \tag{2.10}$$

Note that $\nabla\psi \in H^1(\Omega)$, $n\cdot\nabla\psi \in H^{1/2}(\Gamma)$. and

$$|(\theta,\,u\cdot\nabla\psi)_{L^2}| \leq |\theta|_{L^2}|u|_{L^4}|\nabla\psi|_{L^4}$$

Since $H^1(\Omega)$ is compactly embedded into $L^4(\Omega)$ it follows from (2.10) that $(u, \theta) \in V_0 \times L^\infty(\Omega)$ satisfies (2.5b). Since $(u_\epsilon, \theta_\epsilon)$ converges weakly to (u, θ) in $V_0 \times L^2(\Omega)$ and $H^1(\Omega)$ is compactly embedded into $L^4(\Omega)$, it follows from Lemma 2.2 that $(u, \theta) \in V_0 \times L^\infty(\Omega)$ satisfies (2.5a). □

2.2. Necessary Optimality Condition. Let us denote by $S(g)$, the solution set of (2.2) for $g \in L^2(\Gamma)$. Then, we have the existence of minimizer for (2.1)–(2.2).

Theorem 2.5. *Consider the minimization problem* (2.1) – (2.2) *which is equivalently written as*

$$\textit{minimize} \quad J(u, T - T_0, g) = \varphi(u, T - T_0) + \tfrac{\beta}{2}\,|g - T_0|^2_{L^2(\Gamma)}$$

$$\textit{over} \ (u, T - T_0) \in S(g) \ \textit{and} \ g \in \mathcal{C},$$

where $\mathcal{C}$ is a closed convex set in $L^2(\Gamma)$. Assume that $\varphi(u,\theta)$ is coercive in $\theta \in H_1 = L^2(\Omega)$ and satisfies

$$\varphi(z) : z = (u, T - T_0) \in V_0 \times L^2(\Omega) \to R^+ \text{ is convex and lower semicontinous}$$
$$\text{and } 0 \leq \varphi(z) \leq b_1 \, |z|^2_{V_0 \times L^2(\Omega)} + b_2 \text{ for } b_1,\ b_2 \in R^+.$$

Then Problem (2.1)–(2.2) has a solution.

Proof: It follows from Lemma 2.4 that $S(g)$ is nonempty for $g \in L^\infty(\Gamma)$. Let $(u_k, \theta_k) \in S(g_k)$, $g_k \in \mathcal{C}$ be a minimizing sequence. Since $\beta > 0$ $|g_k - T_0|_{L^2(\Gamma)}$ is uniformly bounded in k. From the assumption on φ we have $|\theta_k|_{L^2}$ is uniformly bounded. Setting $v = u_k$ in (2.5a), we obtain from Lemma 2.2

$$\nu \, |u_k|^2_{V_0} \leq (\gamma \, |\theta_k|_{L^2} + |f|_{L^2}) \, |u_k|_{H_0}.$$

Hence, $|u_k|_{V_0}$ is uniformly bounded and thus there exists a subsequence of $\{k\}$, which will be denoted by the same index, such that (u_k, θ_k, g_k) converges weakly to $(u, \theta, g) \in V_0 \times H_1 \times \mathcal{C}$ since $V_0 \times H_1 \times L^2(\Gamma)$ is a Hilbert space and $\mathcal{C}$ is closed and convex. Since $H^1(\Omega)$ is compactly embedded into $L^4(\Omega)$, it follows from Lemma 2.2 that $b(u_k, u_k, v) \to b(u, u, v)$ for $v \in V_0$. Note that

$$|(\theta,\, u \cdot \nabla\psi)| \leq |\theta|_{L^2} |u|_{L^4} |\nabla\psi|_{L^4} \leq M \, |\theta|_{L^2} |u|_{L^4} |\phi|_{L^2}$$

for $\psi = (-\Delta_0)^{-1}\phi$. Thus, $(\theta_k,\, u_k \cdot \nabla\psi) \to (\theta,\, u \cdot \nabla\psi)$ for $\psi = (-\Delta_0)^{-1}\phi$ and $(u, \theta) \in S(g)$. Now, since φ is convex and lower semicontinous it follows from [ET] that (u, θ, g) minimizes (2.1). □

Recall that W is the completion of $H_1 = L^2(\Omega)$ with respect to norm $|(-\Delta_0)^{-1}\phi|_{H_1}$. Problem (2.1)–(2.2) is equivalently written as a constrained minimization on $x = (u, T - T_0, g) \in X = V_0 \times H_1 \times L^2(\Gamma)$ with

$$\text{minimize} \quad J(x) = \varphi(u, T - T_0) + \tfrac{\beta}{2} \, |g - T_0|^2_{L^2(\Gamma)} \quad \textit{over } x \in X$$

$$\text{subject to } e(x) = 0 \text{ and } g \in \mathcal{C}$$

where the equality constraint $e : X \to Y = V_0^* \times W$ is defined by

$$\tag{2.11} \begin{aligned} \langle e(x),\, (v, \phi) \rangle &= \nu \, (\nabla u,\, \nabla v) + b(u, u, v) - \gamma \, (\theta e_d,\, v) \\ &\quad + \kappa \, (\theta,\, \phi) - (\theta,\, u \cdot \nabla\psi) + \kappa \, (g - T_0, n \cdot \nabla\psi)_\Gamma. \end{aligned}$$

for $(v, \phi) \in V_0 \times H_1$ with $\psi = (-\Delta_0)^{-1}\phi$. Assume that $x^* = (u^*, \theta^*, g^*)$ denotes the optimal solution of (2.1)–(2.2). Then we have

Theorem 2.6. *Assume that x^* is a regular point in the sense [MZ] that*

$$\tag{2.12} 0 \in \textit{int}\, \{e'(x^*)(v, \phi, h) : (v, \phi) \in V_0 \times H_1 \textit{ and } h \in \mathcal{C} - g^*\} \quad \textit{in } (V_0)^* \times W.$$

Then there exists a Lagrange multiplier $(\lambda^*, \mu^*) \in V_0 \times \big(H^2(\Omega) \cap H_0^1(\Omega)\big)$ *such that*

$$\begin{aligned} &\nu\,(\nabla\lambda^*,\,\nabla v) + b(v,u^*,\lambda^*) + b(u^*,v,\lambda^*) \\ &\qquad -(\theta^*\nabla\mu^*,v) + \langle\varphi_u(u^*,\theta^*),\,v\rangle = 0 \qquad (2.13)\\ &\kappa\,(\nabla\mu^*,\,\nabla\psi) - \langle u^*\cdot\nabla\mu^*,\,\psi\rangle - \gamma\,(\lambda^*,\,\psi e_d) + (\varphi_\theta(u^*,\theta^*),\,\psi) = 0 \end{aligned}$$

for $v \in V_0$ *and* $\psi \in H_0^1(\Omega)$ *and*

$$(\beta\,(g^* - T_0) + \kappa\, n\cdot\nabla\mu^*,\; h - g^*)_\Gamma \geq 0 \quad \textit{for all}\; h \in \mathcal{C}. \qquad (2.14)$$

Proof: It follows from Lemma 2.2 and (2.11) that e is Fréchet differentiable and the F–derivative $e'(x^*)(v,h)$ is given by

$$\begin{aligned} \langle e'(x^*)(v,\phi,h),\psi\rangle &= \nu\,(\nabla v,\,\nabla\psi_1) + b(v,u^*,\psi_1) + b(u^*,v,\psi_1) - \gamma\,(\phi e_d,\,\psi_1) \\ &\quad + \kappa\,(\phi,\,\psi_2) - (\theta^*,\,v\cdot\nabla\psi) - (\phi,\,u^*\cdot\nabla\psi) + \kappa\,(h,\,n\cdot\psi)_\Gamma \end{aligned}$$

for $(v,\phi,h) \in X$ and $\psi = (\psi_1,\psi_2) \in V_0 \times H_1$, where $\psi = (-\Delta_0)^{-1}\psi_2$. Since x^* is regular, it then follows from [MZ] that there exists a Lagrange multiplier $\lambda = (\lambda^*,\eta^*) \in Y^* = V_0 \times H_1$ such that

$$\langle\varphi'(u^*,\theta^*),\,(v,\phi)\rangle + \beta\,(g^* - T_0,\,h - g^*)_\Gamma + \langle e'(x^*)(v,\phi,h-g^*),\,\lambda\rangle \geq 0 \qquad (2.15)$$

for all $(v,\phi) \in V_0 \times H_1$ and $h \in \mathcal{C}$. Setting $(v,\phi) = 0$, we obtain (2.14) if we define

$$\mu^* - (-\Delta_0)^{-1}\eta^*.$$

Next, setting $h = g^*$ in (2.15), we have

$$\langle\varphi'(u^*,\theta^*),\,(v,\phi)\rangle + \langle e'(x^*)(v,\phi,0),\,\lambda\rangle = 0$$

for all $(v,\phi) \in V_0 \times H_1$. Thus we obtain

$$\nu\,(\nabla\lambda^*,\,\nabla v) + b(v,u^*,\lambda^*) + b(u^*,v,\lambda^*) - (\theta^*\,\nabla\mu^*,\,v) + \langle\varphi_u(u^*,\theta^*),\,v\rangle = 0$$

for all $v \in V_0$ and

$$\kappa\,(-\Delta_0\mu^*,\,\phi)_{L^2} - (u^*\cdot\nabla\mu^*,\phi) - \gamma\,(\lambda^*,\,\phi\, e_d) + (\varphi_\theta(u^*,\theta^*),\,\phi) = 0,$$

for all $\phi \in H_1 = L^2(\Omega)$ which implies (2.13). □

Note that if the linear operator $E \in \mathcal{L}(V_0 \times H_1, V_0^* \times W)$, defined by $E(v,\phi) = e'(x^*)(v,\phi,0)$, is surjective, then the regular point condition (2.12) is satisfied. Moreover, we have the following lemma.

Lemma 2.7. *If $g^* \in int\ (\mathcal{C})$ then the regular point condition (2.12) is equivalent to that equation for $(\lambda, \mu) \in V_0 \times \big(H^2(\Omega) \cap H_0^1(\Omega)\big)$*

$$\nu\,(\nabla\lambda,\, \nabla v) + b(v, u^*, \lambda) + b(u^*, v, \lambda) - (\theta^*\,\nabla\mu,\, v) = 0 \quad \textit{for}\; v \in V_0$$

$$\kappa\,(\nabla\mu,\, \nabla\psi) - (u^* \cdot \nabla\mu,\, \psi) - \gamma\,(\lambda,\, \psi\, e_d) = 0. \quad \textit{for}\; \psi \in H_0^1(\Omega) \tag{2.16}$$

$$\textit{and} \quad n \cdot \nabla\mu = 0 \quad \textit{on}\; \Gamma,$$

implies $(\lambda, \mu) = 0$.

Proof: If $g^* \in \mathrm{int}(\mathcal{C})$ then (2.12) is equivalent to that $G = e'(x^*)$ is surjective. Define a linear map $C \in \mathcal{L}(X, V_0 \times H_1)$ by $C(v, \phi, h) = (U, \Theta)$ where $(U, \Theta) \in V_0 \times H_1$ is the unique solution to

$$\nu\,(\nabla U,\, \nabla\psi_1) + b(v, u^*, \psi_1) + b(u^*, v, \psi_1) - \gamma\,(\phi\, e_d,\, \psi_1) = 0$$

$$\kappa\,(\Theta,\, \psi_2) - (\phi, u^* \cdot \nabla\psi) - (\theta^*, v \cdot \nabla\psi) + \kappa\,(h,\, n \cdot \nabla\psi)_\Gamma = 0$$

for $(\psi_1, \psi_2) \in V_0 \times H_1$ and $\psi = (-\Delta_0)^{-1}\psi_2$. Since $H^1(\Omega)$ is compactly embedded int $L^4(\Omega)$, it follows from Lemma 2.2 that C is compact. It thus follows from Banach closed range and Riesz-Schauder theorems that $e'(x^*)$ is surjective if and only if $ker(G^*) = \{0\}$ [DI], which is equivalent to (2.16). □

3. Finite-time horizon problem

In this section we discuss the finite-time horizon problem (1.1)-(1.2). First, we formulate the weak form of equation (1.1)–(1.2). By Green's formula we have from (1.2)

$$\begin{aligned} (\frac{d}{dt}\theta(t),\, \psi) &= (-u(t) \cdot \nabla\theta(t) + \kappa\,\Delta\theta(t),\, \psi) \\ &= -\kappa\,(\theta(t), \phi) - \kappa\,(g(t) - T_0, n \cdot \nabla\psi)_\Gamma + (\theta(t),\; u(t) \cdot \nabla\psi) \end{aligned}$$

for $\psi = (-\Delta_0)^{-1}\phi$, $\phi \in H_1 = L^2(\Omega)$, where we assumed that $\theta(t)$ is sufficiently smooth. We define the weak solution of (1.1)–(1.2) by

Definition 3.1. The pair $(u, \theta) \in L^2(0, T; V_0 \times H_1) \cap W^{1,s}(0, T; V_0^* \times W)$ with $s > 1$ is a weak solution to (1.1)–(1.2) if

$$\langle \frac{d}{dt}u(t),\, v\rangle_{V_0^* \times V_0} + \nu\,(\nabla u(t), \nabla v) + b(u(t), u(t), v) - \gamma\,(\theta(t) e_d,\, v) = 0 \tag{3.1a}$$

for all $v \in V_0$,

$$\langle \frac{d}{dt}\theta(t),\, \phi\rangle_{W \times H_1} - \kappa\,(\theta(t), \phi) - (\theta(t),\; u \cdot \nabla\psi) + \kappa\,(g(t) - T_0, n \cdot \nabla\psi)_\Gamma = 0 \tag{3.1b}$$

for all $\phi \in H_1$, where $\psi = (-\Delta_0)^{-1}\phi$.

Recall that $H_1 = L^2(\Omega)$, $V_1^* = H^{-1}(\Omega)$ and $V_0 \times H_1 \subset H_0 \times V_1^* = (H_0 \times V_1^*)^* \subset V_0^* \times W$ defines the Gelfand triple where $H_0 \times V_1^*$ is the pivoting space and V_1^* is equipped with inner product (2.3).

Again, we construct a solution to (3.1) by the limiting procedure for solutions to (1.5) as $\epsilon \to 0^+$. To this end we consider equation (1.5). Let $V = V_0 \times H^1(\Omega)$ and $H = H_0 \times L^2(\Omega)$ For $\epsilon > 0$ define a sesquilinear form $\tilde{a}$ on $V \times V$ by

$$\tilde{a}(\phi, \psi) = \nu (\nabla\phi_1, \nabla\psi_1) - \gamma (\phi_2 e_d, \psi_1) + \kappa (\nabla\phi_2, \nabla\psi_2) + \frac{1}{\epsilon} (\phi_2, \psi_2)_\Gamma \tag{3.2}$$

for $\phi = (\phi_1, \phi_2)$, $\psi = (\psi_1, \psi_2) \in V$. Then, $\tilde{a}$ satisfies

$$\begin{aligned} &|\tilde{a}(\phi, \psi)| \le M\, |\phi|_V |\psi|_V \text{ for } \phi,\ \psi \in V \quad \text{and} \\ &\tilde{a}(\phi, \phi) \ge \omega\, |\phi|_V^2 - \gamma\, |\phi|_H^2 \text{ for } \phi \in V, \end{aligned} \tag{3.3}$$

for $\omega > 0$ and $M \ge 1$, where we used the fact that

$$\kappa\, |\nabla\theta|_{L^2(\Omega)}^2 + \frac{1}{2\epsilon} |\theta|_{L^2(\Gamma)}^2 \ge \omega\, |\theta|_{H^1(\Omega)}^2 \tag{3.4}$$

for $\epsilon \le 1$. Define the tri-linear form $\tilde{b}$ on V^3 by

$$\tilde{b}((u, \theta_1), (v, \theta_2), (w, \theta_3)) = b(u, v, w) + (u \cdot \nabla\theta_2, \theta_3)$$

Since

$$(u \cdot \nabla\theta_1,\ \nabla\theta_2) + (u \cdot \nabla\theta_1,\ \nabla\theta_2) = -(\nabla \cdot u,\ \theta_1\theta_2) = 0 \tag{3.5}$$

for $u \in V_0$ and $\theta_1,\ \theta_2 \in H^1(\Omega)$, it follows that $\tilde{b}$ satisfies Lemma 3.2 in which $V_0,\ H_0$ are replaced by $V,\ H$, respectively. The weak form of equation (1.5) is then given by

$$\langle \frac{d}{dt} u(t),\ \phi\rangle_{V_0^* \times V_0} + \nu (\nabla u(t), \nabla\phi) + b(u(t), u(t), \phi) = (\gamma\theta(t) e_d + f, \phi), \tag{3.6a}$$

for $\phi \in V_0$ and

$$\langle \frac{d}{dt} \theta(t),\ \psi\rangle_{(H^1)^* \times H^1} + \kappa (\nabla\theta(t), \nabla\psi) + \frac{1}{\epsilon} (\theta(t) - (g(t) - T_0), \psi)_\Gamma = 0 \tag{3.6b}$$

for $\psi \in H^1(\Omega)$, where $\theta(t) = T(t, \cdot) - T_0$, or equivalently

$$\langle \frac{d}{dt} z(t),\ \psi\rangle_{V^* \times V} + \tilde{a}(z(t), \psi) + \tilde{b}(z(t), z(t), \psi) = (f, \psi_1) + \frac{1}{\epsilon} (g(t) - T_0,\ \psi_2)_\Gamma,$$

for $\psi = (\psi_1, \psi_2) \in V$, where $z(t) = (u(t, \cdot), \theta(t))$. Note that

$$|(g,\ \psi)_\Gamma| \le M\, |g|_{L^2(\Gamma)} |\psi|_{H^1(\Omega)}$$

for $g \in L^2(\Gamma)$ and $\psi \in H^1(\Omega)$. Hence, by using the standard arguments in [Te1],[CF] we have

Theorem 3.2. *If $d = 2$ then for any $z(0) \in H$ and $g \in U = L^2(0, T; L^2(\Gamma))$, equation (1.3)–(1.4) has a unique solution $z(t) = (u(t, \cdot), \theta(t) = T(t, \cdot) - T_0) \in$*

$C(0,T;H) \cap L^2(0,T;V) \cap H^1(0,T;V^*)$ satisfying (3.6) a.e. in $(0,T)$ for all T. If $d = 3$ then equation (1.3)–(1.4) has a weak solution $z(t) \in C_w(0,T;H) \cap L^2(0,T;V) \cap W^{1,4/3}(0,T;V^*)$ for any $z(0) \in H$ and $g \in L^2(0,T;L^2(\Gamma))$ with bound

$$
\begin{aligned}
&\frac{1}{2}|z(t)|_H^2 + \int_0^t \nu\, |u(s)|_{V_0}^2 + \kappa\, |\nabla\theta(s)|^2 + \frac{1}{\epsilon}\, |\theta(s)|_\Gamma^2\, ds \\
&\le \frac{1}{2}|z(0)|_H^2 + \int_0^t (\gamma\theta(s)\, e_d + f, u(s)) + \frac{1}{\epsilon}\, (\theta(s),\ (g(s) - T_0) - \theta(s))_\Gamma\, ds
\end{aligned}
\tag{3.7}
$$

for $t \ge 0$. In (3.7) equality holds instead of the inequality when $d = 2$.

Let K be the closed convex set in $L^2(0,T : L^2(\Gamma))$ defined by

$$K = \{g(t) \in L^2(0,T;L^2(\Gamma)) : g(t) \in \mathcal{C} \text{ a.e. in } (0,T)\}. \tag{3.8}$$

Then, the following corollary is proved in [IST2].

Corollary 3.3. *Assume that $g \in K$, i.e., $\bar{T}_1 - T_0 \le g(t,\cdot) \le \bar{T}_2 - T_0$ a.e in $(0,T) \times \Omega$ and $\bar{T}_1 \le T(0,\cdot) \le \bar{T}_2$ a.e in Ω. Then, there exists a solution to (3.6) such that $\bar{T}_1 - T_0 \le \theta(t,\cdot) \le \bar{T}_2 - T_0$ a.e in $(0,T) \times \Omega$. For such a solution $\theta(t) = T(t) - T_0 \in C(0,T;L^2(\Omega)) \cap L^2(0,T;H^1(\Omega)) \cap H^1(0,T;H^1(\Omega)^*)$.*

For $0 < \epsilon \le 1$ let us denote by $(u_\epsilon, \theta_\epsilon)$ the solution to (3.6) in the sense of Theorem 3.2 and Corollary 3.3. Next, we show that $(u_\epsilon, \theta_\epsilon)$ converges to a solution (3.1) in the sense of Definition 3.1.

Theorem 3.4. *Suppose $g \in K$ and $\bar{T}_1 \le T(0,\cdot) \le \bar{T}_2$ a.e. in Ω. Then there exists a sequence $\{\epsilon\}$ of positive numbers such that u_ϵ converges to u weakly in $L^2(0,T;V_0)$ and $W^{1,s}(0,T;V_0^*)$ ($s = 2$ if $d = 2$ and $s = 4/3$ if $d = 3$) and strongly in $L^2(0,T;H_0)$, and θ_ϵ converges to θ in weakly * in $L^\infty((0,T) \times \Omega)$ and strongly in $L^2(0,T;H^{-1}(\Omega))$ as $\epsilon \to 0^+$, where the pair (u,θ) is a solution to (3.1) with $\bar{T}_1 - T_0 \le \theta(t,\cdot) \le \bar{T}_2 - T_0$ a.e in $(0,T) \times \Omega$.*

Proof: The proof is given in several steps.
Step 1. We establish the uniform bound for $(u_\epsilon, \theta_\epsilon)$ and the trace of $\theta_\epsilon(t)$ on Γ. Since for $\theta(t) \in L^2(0,T;H^1(\Omega)) \cap H^1(0,T;H^1(\Omega)^*)$

$$\frac{1}{2}\frac{d}{dt}|\theta(t)|_{L^2}^2 = \langle \frac{d}{dt}\theta(t),\ \theta(t)\rangle_{(H^1)^* \times H^1} \quad \text{a.e. } t \in (0,T), \tag{3.9}$$

it follows from Corollary 3.3 and (3.6b) that

$$\frac{1}{2}\frac{d}{dt}|\theta_\epsilon(t)|_{L^2}^2 + \kappa\, |\nabla\theta_\epsilon(t)|^2 + \frac{1}{2\epsilon}\, |\theta_\epsilon(t)|_\Gamma^2 \le \frac{1}{2\epsilon}\, |g(t) - T_0|_\Gamma^2$$

Integrating this with respect to t, we obtain

$$|\theta_\epsilon(t)|_{L^2}^2 + \int_0^t 2\kappa\, |\nabla\theta_\epsilon(s)|^2 + \frac{1}{\epsilon}\, |\theta_\epsilon(s)|_\Gamma^2\, ds \le |\theta(0)|_{L^2} + \int_0^T \frac{1}{\epsilon}\, |g(s) - T_0|_\Gamma^2\, ds \tag{3.10}$$

for $t \in [0,T]$. Hence $\int_0^T |\theta_\epsilon(s)|_\Gamma^2\, ds$ is uniformly bounded in $\epsilon > 0$.

From (3.7) and (3.9) we have

$$|u_\epsilon(t)|_{H_0}^2 + \int_0^t \nu\,|u_\epsilon(s)|_{V_0}^2\, ds \le \frac{M_3}{\nu}\int_0^t \gamma|\theta_\epsilon(s)|_{L^2}^2 + |f|_{L^2})^2\, ds.$$

where $|\phi|_{H_0} \le M_3|\phi|_{V_0}$. It follows from Corollary 3.3 that

$$|\theta_\epsilon(t)|_{L^\infty(\Omega)} \le \max(\bar{T}_1 - T_0, \bar{T}_2 - T_0). \tag{3.11}$$

Hence $\int_0^T |\theta_\epsilon(s)|_{L^2(\Omega)}^2\, ds$ is uniformly bounded and so is

$$\sup_{t\in[0,T]} |u_\epsilon(t)|_{H_0}^2 + \int_0^T \nu\,|u_\epsilon(t)|_{V_0}^2\, dt$$

in $\epsilon > 0$. It follows from Lemma 2.2 that

$$|b(u(t),u(t),\phi)| \le M_2|u(t)|_{V_0}|u(t)|_{H_0}|\phi|_{V_0} \quad \text{for } d = 2$$

and

$$|b(u(t),u(t),\phi)| \le M_2|u(t)|_{V_0}^{3/2}|u(t)|_{H_0}^{1/2}|\phi|_{V_0} \quad \text{for } d = 3.$$

Hence, from (3.6a)

$$|\frac{d}{dt}u_\epsilon(t)|_{V_0^*} \le \nu\,|u_\epsilon(t)|_{V_0} + M_2\,|u_\epsilon(t)|_{V_0}^{3/2}|u_\epsilon(t)|_{H_0}^{1/2} + M_3(\gamma|\theta_\epsilon(t)|_{L^2} + |f|_{L^2}).$$

and $\int_0^T |\frac{d}{dt}u_\epsilon(t)|_{V_0^*}^{4/3}\, dt$ is uniformly bounded for $d = 3$. Similarly, for $d = 2$ we have $\int_0^T |\frac{d}{dt}u_\epsilon(t)|_{V_0^*}^2\, dt$.

Setting $\psi = (-\Delta_0)^{-1}\phi$, $\phi \in H_1 = L^2(\Omega)$ in (3.6b) and by Green's formula we obtain

$$\langle\frac{d}{dt}\theta_\epsilon(t),\phi\rangle_{W\times H_1} + \kappa\,(\theta_\epsilon(t),\phi)_{L^2} - (\theta_\epsilon(t), u_\epsilon(t)\cdot\nabla\psi)_{L^2} + \kappa\,(\theta_\epsilon(t), n\cdot\nabla\psi)_\Gamma = 0. \tag{3.12}$$

Since

$$|(\theta_\epsilon(t), u_\epsilon(t)\cdot\nabla\psi)_{L^2}| \le M\,|\theta_\epsilon(t)|_{L^2}|u_\epsilon(t)|_{V_0}|\phi|_{L^2}$$

for some $M > 0$ it follows from (3.12) that $\int_0^T |\frac{d}{dt}\theta_\epsilon(t)|_W^2\, dt$ is uniformly bounded.

Step 2. We show that $(u_\epsilon(t), \theta_\epsilon(t))$ and the trace of $\theta_\epsilon(t) + T_0$ on Γ have the appropriate limit. It follows from Aubin's lemma [CF] that there exist $g(t) \in$

$L^2(0,T;L^2(\Gamma))$, $u(t) \in L^2(0,T;V_0) \times W^{1,s}(0,T;V_0^*)$, $\theta(t) \in L^\infty((0,T)\times\Omega) \times H^1(0,T;W)$ and a subsequence $\{\epsilon\}$ of positive numbers such that

$$(\theta_\epsilon(t)+T_0)|_\Gamma \text{ converges weakly to } \hat{g}(t) \text{ in } L^2(0,T;L^2(\Gamma)),$$

$$u_\epsilon(t) \to u(t) \quad \text{weakly in } L^2(0,T;V_0) \text{ and strongly in } L^2(0,T;H_0)$$

$$\theta_\epsilon(t) \to \theta(t) \quad \text{weakly * in } L^\infty((0,T)\times\Omega) \text{ and strongly in } L^2(0,T;H^{-1}(\Omega))$$

From (3.6b), for $\psi(\cdot) \in L^2(0,T;H^1(\Omega))$

$$\epsilon\langle \frac{d}{dt}\theta_\epsilon(t),\,\psi(t)\rangle + (\theta_\epsilon(t)-(g(t)-T_0),\psi(t))_\Gamma$$
$$+\epsilon\kappa\,(\nabla\theta_\epsilon(t),\nabla\psi(t)) + \epsilon\,(u\cdot\nabla\theta_\epsilon(t),\,\psi(t)) = 0$$

where from (3.10)

$$\epsilon\int_0^T |\frac{d}{dt}\theta_\epsilon(t)|^2_{(H^1)^*} + \kappa\,|\nabla\theta_\epsilon(t)|^2_{L^2}\,dt \quad \text{is uniformly bounded in } \epsilon>0$$

and from (3.5) and (3.11)

$$|\int_0^T (u_\epsilon(t)\cdot\nabla\theta_\epsilon(t),\,\psi(t))\,dt| \le \beta\,|u_\epsilon|_{L^2(0,T;H_0)}|\psi|_{L^2(0,T;H^1)}.$$

with $\beta = \max(|\bar{T}_1 - T_0|, |\bar{T}_2 - T_0|)$. Thus,

$$\left|\int_0^T (\theta_\epsilon(t)+T_0-g(t),\psi(t))_\Gamma\,dt\right| \le \tilde{M}\,\sqrt{\epsilon}|\psi(t)|_{L^2(0,T;H^1)} \tag{3.13}$$

for some $\tilde{M}>0$ and all $\psi \in L^2(0,T;H^1(\Omega))$, which implies that $\hat{g}(t) = g(t)$ and $\theta_\epsilon(t)+T_0 \to g(t)$ as $\epsilon \to 0^+$ in $L^2(0,T;H^{-1/2}(\Gamma))$.

Step 3. We show that $(u(t),\theta(t))$ is a weak solution of (1.2) in the sense of Definition 3.1. From (3.12)

$$\begin{aligned}&((-\Delta_0)^{-1}(\theta_\epsilon(t)-\theta(0)),\,\phi)_{L^2}\\ &+\int_0^t \kappa\,(\theta_\epsilon(t),\,\phi)_{L^2} - (\theta_\epsilon(t),u_\epsilon(t)\cdot\nabla\psi)_{L^2} + \kappa\,(\theta_\epsilon(t),\,n\cdot\nabla\psi)_\Gamma\,dt = 0\end{aligned} \tag{3.14}$$

for $\psi = (-\Delta_0)^{-1}\phi$, $\phi \in L^2(\Omega)$. Since every L^2-convergent sequence has an almost everywhere pointwise convergent subsequence, we can assume that

$$(u_\epsilon(t),(-\Delta_0)^{-1}\theta_\epsilon(t)) \to (u(t),(-\Delta_0)^{-1}\theta(t)) \quad \text{in } L^2(\Omega)^d \times L^2(\Omega)$$

for $t \in (0,T] \setminus E$, where E has zero Lebesgue measure. From (3.6a)

$$\begin{aligned}&(u_\epsilon(t) - u(0),\, v)\\ (3.15)\qquad &+ \int_0^t \nu\,(\nabla u_\epsilon(s), \nabla v) + b(u_\epsilon(s), u_\epsilon(s), v) - (\gamma\theta_\epsilon(s)e_d + f,\, v)\, ds = 0\end{aligned}$$

for $v \in V_0$. Note that

$$\begin{aligned}&\int_0^t |b(u_\epsilon(s) - u(s), u_\epsilon(s), v)|\, ds\\ &\quad\le M \int_0^t |u_\epsilon(s) - u(s)|_{H_0}^{1/2} |u_\epsilon(s) - u(s)|_{V_0}^{1/2} |u_\epsilon(s)|_{V_0} |v|_{V_0}\, ds\\ &\quad\le M\, |u_\epsilon(s) - u(s)|_{L^2(0,T,H_0)}^{1/2} |u_\epsilon(s) - u(s)|_{L^2(0,T,V_0)}^{1/2} |u_\epsilon(s)|_{L^2(0,T\, V_0)} |v|_{V_0}\end{aligned}$$

where we used the fact that $|\phi|_{L^6} \le c\,|\phi|_{H^1}$ and $|\phi|_{L^3} \le c\,|\phi|_{L^2}^{1/2} |\phi|_{H^1}^{1/2}$ for $\phi \in H^1(\Omega)$ and the Hölder inequality. Similarly, we have

$$\begin{aligned}&\int_0^t |(\theta_\epsilon(s), (u_\epsilon(s) - u(s)) \cdot \nabla\psi)|\, ds\\ &\quad\le M\, |u_\epsilon(s) - u(s)|_{L^2(0,T,H_0)}^{1/2} |u_\epsilon(s) - u(s)|_{L^2(0,T,V_0)}^{1/2} |\theta_\epsilon(s)|_{L^2(0,T,L^2)} |\psi|_{H^2}.\end{aligned}$$

Hence, we can pass the limit of $\epsilon \to 0^+$ in (3.14) and (3.15) to obtain

$$\begin{aligned}&(u(t) - u(0),\, v)\\ &\quad+ \int_0^t \nu\,(\nabla u(s), \nabla v) + b(u(s), u(s), v) - (\gamma\theta(s)e_d + f,\, v)\, ds = 0\\ (3.16)\qquad &(\theta(t) - \theta(0),\, \psi)\\ &\quad+ \int_0^t \kappa\,(\theta(t),\, \phi)_{L^2} - (\theta(t), u(t) \cdot \nabla\psi) + \kappa\,(\theta(t),\, n \cdot \nabla\psi)_\Gamma\, dt = 0\end{aligned}$$

for $v \in V_0$, $\psi = (-\Delta_0)^{-1}\phi$, $\phi \in L^2(\Omega)$ and $t \in (0,T] \setminus E$. Since the integrands appearing in (3.16) are integrable, (3.16) holds for all $t \in (0,T]$. Therefore (u,θ) is a weak solution to (1.2). □

Corollary 3.5. *Suppose $u(0) \in V_0$ and $d = 2$. Then the solution to (1.1)–(1.2) in the sense of Definition 3.1 is unique and $(u_\epsilon, \theta_\epsilon) \to (u, \theta)$ in $L^2(0,T; V_0 \times L^2(\Omega)) \cap H^1(0,T; V_0^* \times W)$.*

Proof: Since $\theta(t) \in L^2(0,T;L^2(\Omega))$ it follows from [Te1],[WvW] that we have $u(t) \in C(0,T;V_0) \cap L^2(0,T;H^2(\Omega) \cap V_0)$. Note that

$$
\begin{aligned}
&|(\theta_1, u_1 \cdot \nabla(-\Delta_0)^{-1}\phi) - (\theta_2, u_2 \cdot \nabla(-\Delta_0)^{-1}\phi)| \\
&\qquad \le |\theta_1 - \theta_2|_{L^2} |u_2|_{H^2} |\phi|_{H^{-1}} + |\theta_1|_{L^2} |u_2 \\
&\qquad\qquad - u_1|_{V_0}^{1/2} |u_2 - u_1|_{H_0}^{1/2} |\phi|_{H^{-1}}^{1/2} |\phi|_{L^2}^{1/2}
\end{aligned} \tag{3.17}
$$

for $\theta_i \in L^2(\Omega)$, $u_i \in H^2(\Omega) \cap V_0$, $i = 1, 2$ and $\phi \in L^2(\Omega)$, where we used the fact that $|\psi|_{L^4} \le c\,|\psi|_{L^2}^{1/2} |\psi|_{H^1}^{1/2}$ for $\psi \in H^1(\Omega)$. Suppose (u_i, θ_i), $i = 1, 2$ is two solutions to (3.1). Since

$$
|(\theta(t), u(t) \cdot \nabla(-\Delta_0)^{-1}\phi)| \le M\, |\theta(t)|_{L^2} |u(t)|_{V_0} |\phi|_{L^2},
$$

we have $\theta_i(t) \in H^1(0,T;W)$. Then, it follows from Lemma 2.2 and (3.17) that

$$
\begin{aligned}
&|u_1(t) - u_2(t)|_{H_0}^2 + |\theta_1(t) - \theta_2(t)|_{H^{-1}}^2 \\
&\le M \int_0^t (1 + |u_1(s)|_{V_0}^2 + |\theta_1(s)|_{L^2}^2) |u_1(s) - u_2(s)|_{H_0}^2 + |u(s)|_{H^2}^2 |\theta_1 - \theta_2(s)|_{H^{-1}}^2\, ds.
\end{aligned}
$$

Hence, by Gronwall's inequality we have $(u_1(t), \theta_1(t)) = (u_2(t), \theta_2(t))$ for every $t \in [0, T]$.

Let $v(t) = u_\epsilon(t) - u(t)$ and $w(t) = \theta_\epsilon - \theta(t)$. Since $(v(t), w(t)) \in L^2(0,T;V_0 \times L^2(\Omega)) \cap H^1(0,T;V_0^* \times W)$, it follows from (3.1) and (3.12) that

$$
\frac{1}{2}\frac{d}{dt}|v(t)|_{H_0}^2 + \nu\, |v(t)|_{V_0}^2 + b(v(t), u(t), v(t)) = \gamma\, (w(t) e_d, v(t))
$$

$$
\begin{aligned}
&\frac{1}{2}\frac{d}{dt}|w(t)|_{H^{-1}}^2 + \kappa\, |w(t)|_{L^2}^2 + (\theta_\epsilon(t), v(t) \cdot \nabla(-\Delta_0)^{-1} w(t)) \\
&\quad + (w(t), u_\epsilon(t) \cdot \nabla(-\Delta_0)^{-1} w(t)) + \kappa\, (\theta_\epsilon + T_0 - g(t),\, n \cdot \nabla(-\Delta_0)^{-1} w(t))_\Gamma = 0.
\end{aligned}
$$

Note that

$$
|(\theta_\epsilon(t) + T_0 - g(t),\, n \cdot \nabla(-\Delta_0)^{-1} w(t))_\Gamma| \le M\, |(\theta_\epsilon(t) + T_0) - g(t)|_{H^{-1/2}} |w(t)|_{L^2}
$$

and that from (3.13) $\theta_\epsilon + T_0 \to g(t)$ in $L^2(0,T;H^{-1/2}(\Gamma))$. Hence, from (3.17)

$$
\begin{aligned}
&|v(t)|_{H_0}^2 + |w(t)|_{H^{-1}}^2 + \int_0^t \nu\, |v(s)|_{V_0}^2 + \kappa\, |w(s)|_{L^2}^2\, ds \\
&\quad \le M \int_0^t (1 + |u_\epsilon(s)|_{V_0}^2 + |\theta_\epsilon(s)|_{L^2}^2)\, |v(s)|_{H_0}^2 + |u(s)|_{H^2}^2\, |w(s)|_{H^{-1}}^2 \\
&\qquad\qquad + \kappa\, |(\theta_\epsilon(s) + T_0) - g(s)|_{H^{-1/2}(\Gamma)}^2\, ds
\end{aligned}
$$

The desired convergence property follows from Gronwall's inequality. $\square$

Let us denote by $S(g)$, the solution set of (1.2) for $g \in L^2(\Gamma)$ in the sense of Definition 3.1. Then, we have

Theorem 3.6. *Consider the finite-time horizon problem (1.1)–(1.2) which is equivalently written as*

$$\text{minimize} \quad J(u, T-T_0, g) = \int_0^T \varphi(u(t), T(t)-T_0) + \frac{\beta}{2}\,|g(t)-T_0|^2_{L^2(\Gamma)}\,dt$$

$$\text{over } (u(t), T(t)-T_0) \in S(g) \text{ and } g(t) \in K.$$

where K is a closed convex set in $U = L^2(0,T;L^2(\Gamma))$. Assume that $\varphi(u,\theta)$ is coercive in $\theta \in H_1 = L^2(\Omega)$ and satisfies

$$\varphi(z) : z = (u, T-T_0) \in V_0 \times L^2(\Omega) \to R^+ \text{ is convex and lower semicontinous and } 0 \le \varphi(z) \le b_1\,|z|^2_{V_0 \times L^2} + b_2 \text{ for } b_1,\ b_2 \in R^+.$$

Then Problem (1.1)–(1.2) has a solution.

Proof: It follows from Lemma 3.4 that $S(g)$ is nonempty for $g \in L^\infty((0,T)\times\Gamma)$. Let $(u_k, g_k) \in S(g_k)\times K$ be a minimizing sequence. Since $\beta > 0$ $|g_k(t)-T_0|_U$ is uniformly bounded in k. It follows from the assumption on φ that $|\theta_k(t)|_{L^2(0,T,L^2(\Omega))}$ is uniformly bounded. From (3.1a) we have

$$|u_k(t)|^2_{H_0} + \int_0^t \nu\,|u_k(s)|^2_{V_0}\,ds \le \frac{M_3^2}{\nu}\int_0^t \gamma^2\,|\theta_k(s)|^2_{L^2} + |f|^2_{L^2}\,ds$$

for $t \in [0,T]$. Hence $\sup_{t\in[0,T]} |u_k(t)|^2_{H_0} + \int_0^T |u_k(t)|^2_{V_0}\,dt$ is uniformly bounded. This implies that from Lemma 2.2 $(u_k, \theta_k) \in W^{1,4/3}(0,T;V_0^* \times W)$. It thus follows from Aubin's lemma that there exists a subsequence of $\{k\}$, which will be denoted by the same index, such that (u_k, θ_k, g_k) converges to (u, θ, g) weakly in $L^2(0,T;V_0)\times L^2(0,T;L^2(\Omega)) \times K$ and strongly in $L^2(0,T;H_0 \times H^{-1})$ since K is closed and convex. Hence, using the same arguments as in the proof of Theorem 3.4, it can be shown that $(u(t), \theta(t), g(t))$ satisfies (3.1). Now, since φ is convex and lower semicontinous it follows from [ET] that (u, θ, g) minimizes (1.1). □

We have the following necessary optimality condition.

Theorem 3.7. *Let $d = 2$, K be given by (3 8) and $u(0) \in V_0$. If $(u^*, \theta^*) \in S(g^*)$, $g^* \in K$ minimizes (1.1), then there exists a Lagrange multiplier* $(\lambda(t), \mu(t)) \in L^2(0,T;V_0)\cap H^1(0,T;V_0^*) \times L^2(0,T;H^2(\Omega)\cap H_0^1(\Omega))\cap H^1(0,T;L^2(\Omega))$, *such that*

$$(3.18)\qquad \begin{aligned} &\langle -\frac{d}{dt}\lambda(t),\, v\rangle + \nu\,(\nabla\lambda(t),\, \nabla v) + b(v, u^*(t), \lambda(t)) + b(u^*(t), v, \lambda(t)) \\ &\quad -(\theta(t)\nabla\mu(t),\, v) + \langle\varphi_u(u^*(t), \theta^*(t)),\, v\rangle = 0 \quad \textit{with } \lambda(T) = 0 \\ &\langle -\frac{d}{dt}\mu(t),\, \psi\rangle + \kappa\,(\nabla\mu(t),\, \nabla\psi) - \langle u^*(t)\cdot\nabla\mu(t),\, \psi\rangle \\ &\quad -\gamma\,(\lambda(t), \psi\, e_d) + (\varphi_\theta(u^*(t), \theta^*(t)),\, \psi) = 0 \quad \textit{with } \mu(T) = 0 \end{aligned}$$

for $v \in V$ *and* $\psi \in H_0^1(\Omega)$ *and*

$$(3.19)\qquad (\beta\,(g^*(t) - T_0) + \kappa\, n\cdot\nabla\mu(t),\, h(t) - g^*(t))_U \geq 0 \quad \textit{for all } h \in K.$$

Proof: Consider a family of control problems

$$(3.20)\qquad \text{minimize} \quad \tilde{J}(u, \theta, g) = J(u, \theta, g) + |g - g^*|_U^2 \quad \text{subject to (1.5)}$$

for $\epsilon > 0$. It is shown in [IST2] that (3.20) has a solution $(u_\epsilon^*, \theta_\epsilon^*, g_\epsilon^*)$ where $u_\epsilon^* \in L^2(0,T;V_0)\cap H^1(0,T;V_0^*)$, $\theta_\epsilon^* \in L^\infty((0,T)\times\Omega)$ and $g_\epsilon^* \in K$ and they satisfy the necessary optimality condition

$$(3.21)\qquad \begin{aligned} &\langle -\frac{d}{dt}\lambda_\epsilon(t),\, v\rangle + \nu\,(\nabla\lambda_\epsilon(t),\, \nabla v) + b(v, u_\epsilon^*(t), \lambda_\epsilon(t)) + b(u_\epsilon^*(t), v, \lambda_\epsilon(t)) \\ &\quad -(\theta_\epsilon^*(t)\nabla\mu_\epsilon(t), v) + \langle\varphi_u(u_\epsilon^*(t), \theta_\epsilon^*,\, v\rangle = 0 \quad \text{with } \lambda_\epsilon(T) = 0 \\ &\langle -\frac{d}{dt}\mu_\epsilon(t),\, \psi\rangle + \kappa\,(\nabla\mu_\epsilon,\, \nabla\psi) + \frac{1}{\epsilon}\,(\mu_\epsilon(t), \psi)_\Gamma - \langle u_\epsilon^*(t)\cdot\nabla\mu_\epsilon(t),\, \psi\rangle \\ &\quad -\gamma\,(\lambda_\epsilon(t),\, \psi e_d) + (\varphi_\theta(u_\epsilon^*(t), \theta_\epsilon^*(t)),\, \psi) = 0 \quad \text{with } \mu_\epsilon(T) = 0 \end{aligned}$$

for $v \in V$ and $\psi \in H^1(\Omega)$ and

$$(3.22)\quad (\beta\,(g_\epsilon^*(t) - T_0) + (g_\epsilon^*(t) - g^*(t)) + \kappa\, n\cdot\nabla\mu_\epsilon,\, h(t) - g_\epsilon^*(t))_U \geq 0 \quad \text{for all } h \in K.$$

Here we used the fact that $\mu_\epsilon(t) \in L^2(0,T;H^2(\Omega))$ and $\kappa\, n\cdot\nabla\mu_\epsilon(t) + \frac{1}{\epsilon}\mu_\epsilon(t) = 0$ on Γ. It follows from the proof of Theorem 3.4 that there exists $u(t) \in L^2(0,T;V_0)\times H^1(0,T;V_0^*)$, $\theta \in L^\infty((0,T)\times\Omega)$, $g(t) \in K$ and a sequence ϵ of positive numbers such that $\theta_\epsilon^*(t)$ converges to $\theta(t)$ weakly * in $L^\infty((0,T)\times\Omega)$, $u_\epsilon(t)$ converges to $u(t)$ weakly in $L^2(0,T;V_0)$ and strongly in $L^2(0,T;H_0)$, and $g_\epsilon^*(t)$ converges to $g(t)$ weakly in U where $(u(t), \theta(t)) \in S(g)$. Note that

$$J(u_\epsilon^*, \theta_\epsilon^*, g_\epsilon^*) \leq J(u_\epsilon, \theta_\epsilon, g^*),$$

where $(u_\epsilon, \theta_\epsilon)$ is the solution to (1.5) for $g(t) = g^*(t)$. Then, it follows from Corollary 3.5 that $(u_\epsilon, \theta_\epsilon) \to (u^*, \theta^*)$, strongly in $L^2(0,T; V_0 \times L^2(\Omega))$. Since J is lower semicontinous, we obtain

$$J(u, \theta, g) + |g - g^*|_U^2 \le J(u^*, \theta^*, g^*)$$

which implies $g(t) = g^*(t)$ and $g_\epsilon^*(t) \to g^*(t)$, strongly in $L^2(0,T; L^2(\Gamma))$.

Now we show that $(\lambda_\epsilon(t), \mu_\epsilon(t))$ converges to $(\lambda(t), \mu(t))$, weakly in $L^2(0,T; V_0 \times H^2(\Omega))$ and strongly in $L^2(0,T; H_0 \times H^1(\Omega))$. Let $V_2 = \{\psi \in H^2(\Omega) : n \cdot \nabla\psi + \frac{1}{\epsilon}\psi = 0 \text{ on } \Gamma\}$ and consider the sesquilinear form $\tilde\sigma$ on $V_2 \times V_2$ defined by

$$\tilde\sigma(t, \psi_1, \psi_2) = \kappa\,(\Delta\psi_1, \Delta\psi_2) + (u_\epsilon^*(t) \cdot \nabla\psi_1,\, \Delta\psi_2)$$

Then, $\tilde\sigma$ satisfies

$$\sup_{\psi_2 \in V_2} |\tilde\sigma(t, \psi_1, \psi_2)| \le \kappa|\Delta\psi_1| + c\,|u_\epsilon^*(t)|_{H^2}|\nabla\psi_1|_{L^2}$$

and

$$\tilde\sigma(t, \psi, \psi) \ge \frac{\kappa}{2}\,|\Delta\psi|_{L^2}^2 - \rho_2(t)\,|\nabla\psi|_{L^2}^2 \quad \text{for } \psi \in V_2, \tag{3.23}$$

where $\rho_2(t) = c\,|u_\epsilon^*(t)|_{H^2}^2 \in L^1(0,T)$. Let $H^1(\Omega)$ be equipped with

$$|\psi|_{H^1}^2 = |\nabla\psi|_{L^2(\Omega)}^2 + \frac{1}{\epsilon}\,|\psi|_{L^2(\Gamma)}^2.$$

Then, $L^2(\Omega)$ is the dual space of V_2 when $H^1(\Omega)$ is identified with its dual. Since for $\lambda \in V_2$, $\psi \in H^1(\Omega)$

$$\kappa\,(\nabla\lambda,\, \nabla\psi) + \frac{1}{\epsilon}\,(\lambda,\, \psi)_\Gamma = -\kappa\,(\Delta\lambda,\, \psi),$$

The second equation of (3.21) can be written as

$$\begin{aligned} &\langle -\tfrac{d}{dt}\mu_\epsilon(t),\, \psi\rangle_{L^2 \times V_2} + \tilde\sigma(t, \mu_\epsilon(t), \psi) - \gamma\,(\lambda_\epsilon(t), -\Delta\psi\, e_d) \\ &\qquad + (\varphi_\theta(u_\epsilon^*(t), \theta_\epsilon^*(t)),\, -\Delta\psi) = 0 \end{aligned} \tag{3.24}$$

for $\psi \in V_2$, where the dual product on $L^2(\Omega) \times V_2$ is defined as

$$\langle \phi,\, \psi\rangle_{L^2 \times V_2} = (\phi, -\Delta\psi) \quad \text{for } \phi \in L^2(\Omega),\ \psi \in V_2.$$

It can be shown, using the Galerkin method in [LM] (see, [IST2] for details), that equation the first equation of (3.21) coupled with (3.24) has a unique solution $(\lambda_\epsilon(t), \mu_\epsilon(t)) \in L^2(0,T; V_0 \times V_2) \times H^1(0,T; V_0^* \times L^2(\Omega))$. Moreover, we have

$$\begin{aligned} &\frac{d}{dt}|\mu_\epsilon(t)|_{H^1}^2 + \kappa\,|\Delta\mu_\epsilon(t)|_{L^2}^2 \\ &\qquad \le 2\rho_2(t)|\mu_\epsilon(t)|_{H^1}^2 + \frac{1}{\kappa}\,(\gamma|\lambda_\epsilon(t)|_{L^2} + |\varphi_\theta(u_\epsilon^*(t), \theta_\epsilon^*(t))|_{L^2})^2. \end{aligned}$$

Hence, by the Gronwall's inequality $\sup_{t\in[0\ T]}\ |\mu_\epsilon(t)|^2_{H^1} + \int_0^T \kappa\,|\Delta\mu_\epsilon(t)|^2_{L^2}\,dt$ is uniformly bounded in $0 < \epsilon \leq 1$, which implies $\mu_\epsilon \to 0$ a.e. in Γ uniformly in $t \in [0,T]$. Thus, it is not difficult to show that $(\lambda_\epsilon(t), \mu_\epsilon(t))$ converges to $(\lambda(t), \mu(t))$, weakly in $L^2(0,T;V_0\times H^2(\Omega))$ and strongly in $L^2(0,T,H_0\times H^1(\Omega))$ and $(\lambda(t), \mu(t))$ satisfies (3 18) □

References

[AT] F Abergel and R Temam, On some control problems in fluid mechanics, Theoretical and Computational Fluid Mechanics, 1 (1990), 303-325

[CF] P Constantin and C Foias, *Navier-Stokes Equations*, The University of Chicago Press, Chicago, 1988

[DI] M C Desai and K Ito, Optimal Control of Navier-Stokes Equations, SIAM J Control & Optim , to appear (1994)

[ET] I Ekeland and R Temam, *Convex Analysis and Variational Problems*, North Holland, Amsterdam (1976)

[Gl] R Glowinski, *Numerical Methods for Nonlinear Variational Problems*, Springer-Verlag, Berlin, 1984

[GR] V Girault and P A Raviart, *Finite Element Methods for Navier-Stokes Equations*, Springer-Verlag, Berlin, 1984

[IST1] K Ito, J S Scroggs and H T Tran, Mathematical issues in optimal design of a vapor transport reactor, Proceeding of IMA workshop on Flow Control, ed by M Gunzburger, (1993)

[IST2] K Ito, J S Scroggs and H T Tran, Optimal Control of Thermally Coupled Navier Stokes Equations, submitted to SIAM J Control & Optim, (1993)

[LM] J L Lions and E Magenes,*Non-homogeneous Boundary Value Problems and Applications*, Vol I,II, Springer-Verlag, New York, 1972

[MZ] H Maurer and J Zowe, First and second-order necessary and sufficient optimality conditions for infinite-dimensional programming problems, Math Programming, 16 (1979), 98-110

[Ta] H Tanabe, *Equations of Evolution*, Pitman, San Francisco, 1979

[Te1] R Temam, *Navier-Stokes Equations and Nonlinear Functional Analysis*, SIAM, Philadelphia, 1983

[Te2] R Temam, *Infinite Dimensional Dynamical Systems in Mechanics and Pysics*, Appl Math Sci 68, Springer-Verlag, New York, 1988

[Tr] G M Troianiello, *Elliptic Differential Equations and Obstacle Problems*, Plenum Press, New York, 1987

[WvW] Wolf von Wahl, *The Equations of Navier-Stokes and Abstract Parabolic Equations*, Vieweg, Braunschweig 1985

[Yo] K Yosida, *Functional Analysis*, Springer-Verlag, New York

Kazufumi Ito
Center for Research in Scientific Computation
Department of Mathematics
North Carolina State University
Raleigh, NC 27695, USA

International Series of Numerical Mathematics Vol 118 © 1994 Birkhauser Verlag Basel

ADAPTIVE ESTIMATION OF NONLINEAR DISTRIBUTED PARAMETER SYSTEMS

JOSEPH KAZIMIR AND I G ROSEN

Center for Applied Mathematical Sciences
Department of Mathematics
University of Southern California

ABSTRACT The adaptive (on-line) estimation of parameters for a class of nonlinear distributed parameter systems is considered A combined state and parameter estimator is constructed as an initial value problem for an infinite dimensional evolution equation State convergence is established via a Lyapunov-like estimate The finite dimensional notion of persistence of excitation is extended to the infinite dimensional case and used to establish parameter convergence A finite dimensional approximation theory is presented and a convergence result is proven An example involving the identification of a nonlinear heat equation is discussed and results of a numerical study are presented

1991 *Mathematics Subject Classification* 93B30, 93C25, 93C20, 65J10

Key words and phrases On-line estimation, adaptive identification, parameter convergence, persistence of excitation, distributed parameter systems, infinite dimensional systems, finite dimensional approximation

1. Introduction

In this brief note we consider the adaptive, or on-line, identification of unknown parameters in nonlinear infinite dimensional, or distributed parameter, systems. Our estimator takes the form of an, in general, infinite dimensional nonautonomous *linear* dynamical system. This system is nonautonomous as a result of its dependence upon the state of the plant, which is assumed to be available for measurement in its entirety at all times $t > 0$. The estimator is integrated in real time producing estimates for both the plant state and the unknown parameters. The estimator equations are constructed so as to force the derivative of an associated *energy* or Lyapunov functional to be non-positive. It is not, however, negative definite. As a result of this, it is possible to argue, via a modification of a result known as Barbalat's Lemma (a uniformly continuous summable function must tend to zero asymptotically, see [14]), that the state error converges to zero as time tends to infinity. To establish parameter convergence, however, requires an additional *richness* condition on the measured plant data. In the adaptive control literature, this condition is referred to as *persistence of excitation*. We define a

This research was supported in part by the Air Force Office of Scientific Research under grant AFOSR 91-0076

particular notion of richness or persistence of excitation, and then use it to argue that the parameter error converges to zero as time tends to infinity. Finally, since the estimator is infnite dimensional, its integration requires some form of finite dimensional approximation. Consequently, we also develop an associated finite dimensional approximation and convergence theory.

The approach we take here is motivated by the finite dimensional treatments by Morgan and Narendra in [11] and Narendra and Kudva in [12]. We are not the first researchers to suggest that finite dimensional adaptive techniques be applied to infinite dimensional systems. In [1] Alt, Hoffmann, and Sprekels, and in [8] Hoffmann and Sprekels considered an asymptotic embedding method for the identification of linear elliptic partial differential equations. The elliptic equation is embedded in an evolution equation so that the elliptic equation's solution is an asymptotic steady state of the evolution equation. In [9] (see also [7]) Hoffmann and Sprekels provide an abstract functional analytic framework for their earlier results and extend them to certain classes of stationary elliptic and evolutionary parabolic nonlinear variational inequalities. Our treatment here is most closely related to the work of Baumeister and Scondo in [2] (see also [3]) and the results in Scondo's Ph.D. thesis [16]. The primary difference between these efforts and ours is that theirs is a *two-space* theory, while ours is set in only a single Hilbert space. More precisely, their state estimator is defined via a Gelfand triple of spaces (i.e. $V \hookrightarrow H \hookrightarrow V^*$) and a strongly V-coercive operator. Our state estimator, on the other hand, is set in the same Hilbert space, H, as the plant and is governed by an H-accretive operator. To establish convergence in our case, we had to appropriately modify the requisite regularity and richness conditions on the plant. In our definitions of admissibility and persistence of excitation, the interaction between the plant and the estimator dynamics is more explicit than it is in the two-space theory. Interaction between the plant and estimator dynamics in the two-space theory is certainly present. However, it is somewhat subtler. The one-space theory is in some ways more versatile than the corresponding two-space theory. This is particularly true in the case of hereditary or delay systems where in general there is no natural and simple choice for an energy space.

This paper represents our first reporting of our findings and is by no means complete. For example, a careful examination and characterization of the persistence of excitation condition given in Section 3 in the spirit of the one given for the two-space theory in [6] must be carried out. The same is true for the admissiblity condition given in Section 2. The significance of such studies stems from the fact that in actual practice, with the possible exception of only the simplest plants and estimators, these conditions are extremely difficult, if not impossible, to check. Our efforts in this direction and others are currently underway. When these studies have beeen completed, our findings will be reported on elsewhere.

An outline of the remainder of the paper is as follows. In Section 2 we define the plant and the estimator. The notion of an *admissible* plant is defined and the estimator equations are shown to be well posed. In Section 3 the convergence of the state error to zero is established. We define the notion of a plant being persistently exciting, and use this to establish parameter convergence. In Section 4 we develop our abstract approximation theory, and provide sufficient conditions for an approximation scheme to be convergent. Finally, in a fifth section, to illustrate the application of our general approach, we consider the identification of a gradient dependent thermal conductivity in a one dimensional nonlinear heat equation with Dirichlet boundary conditions. An estimator in the form of a simple constant coefficient heat operator with Neumann boundary conditions is defined. A linear spline based approximation scheme is developed. Numerical results are presented.

2. An Adaptive or On-Line Estimator

Let H be a real Hilbert space with inner product $\langle\cdot,\cdot\rangle$ and corresponding induced norm $|\cdot|$, and let Q be a real Hilbert space with inner product $\langle\cdot,\cdot\rangle_Q$ and corresponding induced norm $|\cdot|_Q$. For each $q \in Q$ let $A(q) : Dom(A(q)) \subset H \to H$ be a possibly nonlinear operator on H, where $Dom(A(q)) = \{\varphi \in H : A(q)\varphi \in H\}$. We assume that the map $q \to A(q)\varphi$, $\varphi \in Dom(A(q))$ is linear from Q into H. Let $\overline{f} : [0,\infty) \to H$, and let $\overline{x}_0 \in H$. Let $\overline{q} \in Q$ and consider the dynamical system in H given by

$$\dot{\overline{x}}(t) + A(\overline{q})\overline{x}(t) = \overline{f}(t), \qquad t > 0, \tag{2.1}$$

$$\overline{x}(0) = \overline{x}_0. \tag{2.2}$$

We assume that $\overline{q}$ is unknown and that the state of the plant, (2.1), (2.2), $\overline{x}(t)$ can be measured and is available at all times $t > 0$. We want to define a linear dynamical system on H which uses the measurement of $\overline{x}$ to asymptotically (i.e. as $t \to \infty$) estimate $\overline{q}$.

Let $A_0 : Dom(A_0) \subset H \to H$ be a linear operator on H satisfying the following two assumptions.

($A1$) (Accretivity) There exists $\alpha_0 \geq 0$ for which

$$\langle A_0\varphi, \varphi\rangle \geq \alpha_0|\varphi|^2, \qquad \varphi \in Dom(A_0).$$

($A2$) (Surjectivity) The operator $\lambda + A_0 : Dom(A_0) \subset H \to H$ is surjective for some $\lambda > 0$.

It follows (see, for example, [10], [13] or [17]) that $\overline{Dom(A_0)} = H$, and that $-A_0 \in G(H, 1, -\alpha_0)$. That is, $-A_0$ is the infinitesimal generator of a C_0-semigroup of bounded linear operators on H, $\{T_0(t) : t \geq 0\}$, satisfying

$$|T_0(t)| \leq e^{-\alpha_0 t}, \qquad t \geq 0. \tag{2.3}$$

Our on-line estimator is then defined to be the dynamical system in H given by

$$\dot{x}(t) + A_0 x(t) + A(q(t))\overline{x}(t) = \overline{f}(t) + A_0\overline{x}(t), \qquad t > 0, \tag{2.4}$$

$$\langle \dot{q}(t), p\rangle_Q - \langle A(p)\overline{x}(t), x(t)\rangle = -\langle A(p)\overline{x}(t), \overline{x}(t)\rangle, \qquad p \in Q,\ t > 0, \tag{2.5}$$

$$x(0) = x_0 \in H, \qquad q(0) = q_0 \in Q. \tag{2.6}$$

The definition of our estimator given in (2.4)-(2.6) above is primarily motivated by the analogous definition in finite dimensions (see [12]). In the finite dimensional setting, the construction of the estimator in this form is based upon the desire for linearity, homogeneous error equations, and the ability to obtain a particular Lyapunov estimate. This estimator can also be derived via a gradient argument (see [16]).

We assume that $\overline{x}$ is a strong solution to the initial value problem (2.1), (2.2) in the sense that $\overline{x} \in C^1([0,\infty); H)$ and $\overline{x}(t) \in Dom(A(\overline{q}))$ for all $t \geq 0$. We make the additional regularity assumption on the data, $\overline{x}$, that for all $t \geq 0$, $\overline{x}(t) \in Dom(A(q))$, for all $q \in Q$. We note that if the operator A and the parameter space Q are sufficiently regular in the sense that $Dom(A(q)) = Dom(A)$ is independent of $q \in Q$, then this additional regularity assumption on the data $\overline{x}$ can be eliminated.

For $t \geq 0$, define the linear map $B(t) : Q \to H$ by

$$B(t)q = A(q)\overline{x}(t), \qquad q \in Q.$$

To establish that the estimator (2.4)-(2.6) is well posed, and to establish our convergence results in the next section, we require the following definition.

Definition 2.1. *The triple $\{\overline{q}, \overline{x}_0, \overline{f}\}$ will be called A_0-admissible if*

(i) *for each $t \geq 0$, the operator $B(t)$ is bounded (i.e. $B(t) \in \mathcal{L}(Q,H)$) and there exists $\alpha > 0$ such that*

$$|B(t)q| \leq \alpha |q|_Q, \qquad q \in Q,$$

(ii) *the operators $B(t)^* \in \mathcal{L}(H,Q)$, $t \geq 0$, are strongly continuously differentiable on $[0,\infty)$,*

(iii) *$\overline{x}(t) \in Dom(A_0)$, for $t \geq 0$, and $A_0\overline{x} \in C^1([0,\infty); H)$,*

(iv) *$\overline{f} \in C^1([0,\infty); H)$.*

We note that when $B(t) \in \mathcal{L}(Q,H)$, its Hilbert space adjoint, $B(t)^* \in \mathcal{L}(H,Q)$ is given by

$$\langle B(t)^*\varphi, q\rangle_Q = \langle A(q)\overline{x}(t), \varphi\rangle_H, \qquad \varphi \in H,\ q \in Q.$$

Let $Z = H \times Q$ with inner product $\langle \cdot,\cdot \rangle_Z$ given by

$$\langle (\varphi,q),(\psi,p)\rangle_Z = \langle \varphi,\psi\rangle + \langle q,p\rangle_Q, \qquad (\varphi,q),(\psi,p) \in Z, \tag{2.7}$$

with the corresponding induced norm being denoted by $|\cdot|_Z$. For $t \geq 0$, define the linear operator $\mathcal{A}(t) : Dom(\mathcal{A}) \subset Z \to Z$ by

$$\mathcal{A}(t) = \begin{bmatrix} A_0 & B(t) \\ -B(t)^* & 0 \end{bmatrix}$$

with $Dom(\mathcal{A}) = Dom(A_0) \times Q$, define $F(t) \in Z$ by

$$F(t) = \begin{bmatrix} \overline{f}(t) + A_0\overline{x}(t) \\ -B(t)^*\overline{x}(t) \end{bmatrix},$$

and set $z_0 = (x_0, q_0) \in Z$. Note that $Dom(\mathcal{A})$ is dense in Z. The estimator (2.4)-(2.6) can then be written as the abstract initial value problem in Z given by

$$\dot{z}(t) + \mathcal{A}(t)z(t) = F(t), \qquad t > 0, \tag{2.8}$$

$$z(0) = z_0. \tag{2.9}$$

Assume that the triple $\{\overline{q}, \overline{x}_0, \overline{f}\}$ is A_0-admissible, and define the operator $\mathcal{A}_0 : Dom(\mathcal{A}) \subset Z \to Z$ by

$$\mathcal{A}_0 = \begin{bmatrix} A_0 & 0 \\ 0 & 0 \end{bmatrix},$$

and for $t \geq 0$ define the operator $\mathcal{B}(t) \in \mathcal{L}(Z, Z)$ by

$$\mathcal{B}(t) = \begin{bmatrix} 0 & B(t) \\ -B(t)^* & 0 \end{bmatrix}.$$

Then $\mathcal{A}(t) = \mathcal{A}_0 + \mathcal{B}(t)$, $t \geq 0$, and it is easily argued that for $t \geq 0$

$$|\langle \mathcal{B}(t)z_1, z_2\rangle_Z| \leq \alpha |z_1|_Z |z_2|_Z, \qquad z_1, z_2 \in Z,$$

and thus that $|\mathcal{B}(t)|_{\mathcal{L}(Z,Z)} \leq \alpha$, $t \geq 0$. The fact that $-A_0 \in G(H, 1, -\alpha_0)$ implies that $-\mathcal{A}_0$ is the infinitesimal generator of a C_0-semigroup of contractions, $\{\mathcal{T}_0(t) : t \geq 0\}$, on Z. That is, $-\mathcal{A}_0 \in G(Z, 1, 0)$. It follows that (see, for example, [10], Theorem IX.2.1) for each $t \geq 0$, $-\mathcal{A}(t) \in G(Z, 1, \alpha)$. Consequently, the family of operators $\{-\mathcal{A}(t) : 0 \leq t \leq T\}$ is *stable* (see [18], Definition 4.3) for all $T > 0$.

The fact that it was assumed that $\overline{x}$ is a strong solution to the plant equations, (2.1), (2.2), and that the triple $\{\overline{q}, \overline{x}_0, \overline{f}\}$ is A_0-admissible, implies that the map $t \to -\mathcal{A}(t)z$ is strongly continuously differentiable in Z for each $z \in Dom(\mathcal{A})$. It follows from the corollary to Theorem 4.4.2 in [18] that there exists a unique evolution system, $\{U(t, s) : 0 \leq s \leq t\}$, with $U(t, s) \in \mathcal{L}(Z, Z)$, $0 \leq s \leq t$ satisfying

(a) $U(t, s)$ is strongly continuous in s and t, $U(s, s) = I$, and $|U(t, s)| \leq e^{\alpha(t-s)}$,
(b) $U(t, s) = U(t, r)U(r, s)$, for $0 \leq s \leq r \leq t$,
(c) $U(t, s)Dom(\mathcal{A}) \subset Dom(\mathcal{A})$, for $0 \leq s \leq t$,
(d) for $z \in Dom(\mathcal{A})$, $U(t, s)z$ is strongly continuously differentiable on Z in both s and t, and

$$\frac{\partial}{\partial t}U(t, s)z = -\mathcal{A}(t)U(t, s)z, \qquad 0 < s < t,$$

$$\frac{\partial}{\partial s}U(t, s)z = U(t, s)\mathcal{A}(s)z, \qquad 0 < s < t,$$

with both sides of these equations being strongly continuous in Z.

Continuing to assume that the triple $\{\overline{q}, \overline{x}_0, \overline{f}\}$ is A_0-admissible, define the function $z \in C([0,T]; Z)$, for all $T > 0$ by

$$z(t) = U(t,0)z_0 + \int_0^t U(t,s)F(s)ds, \qquad t \geq 0. \tag{2.10}$$

The following theorem and its corollary are then immediate consequences of Theorem 4.5.3 in [18].

Theorem 2.2. *Suppose that the triple $\{\overline{q}, \overline{x}_0, \overline{f}\}$ is A_0-admissible and that $z_0 = (x_0, q_0) \in Dom(\mathcal{A})$. Then the Z-valued function z given by (2.10) is an element in $C^1([0,T]; Z)$, for all $T > 0$, and satisfies (2.8) and (2.9).*

Corollary 2.3. *Suppose that the triple $\{\overline{q}, \overline{x}_0, \overline{f}\}$ is A_0-admissible, that $q_0 \in Q$ and $x_0 \in Dom(A_0)$, and for $t \geq 0$, we write $z(t)$ given by (2.10) as $z(t) = (x(t), q(t))$. Then $x \in C^1([0,T]; H)$ and $q \in C^1([0,T]; Q)$, for all $T > 0$, and the pair (x, q) satisfies (2.4)-(2.6).*

For $t \geq 0$ set $e(t) = x(t) - \overline{x}(t)$ and $r(t) = q(t) - \overline{q}$. It then follows that e and r satisfy

$$\dot{e}(t) + A_0 e(t) + A(r(t))\overline{x}(t) = 0, \qquad t > 0, \tag{2.11}$$

$$\langle \dot{r}(t), p \rangle_Q - \langle A(p)\overline{x}(t), e(t) \rangle = 0, \qquad p \in Q, \ t > 0, \tag{2.12}$$

$$e(0) = e_0 \in H, \qquad r(0) = r_0 \in Q, \tag{2.13}$$

where $e_0 = x_0 - \overline{x}_0$ and $r_0 = q_0 - \overline{q}$. The system (2.11)-(2.13) may equivalently be written

$$y(t) + \mathcal{A}(t)y(t) = 0, \qquad t > 0,$$

$$y(0) = y_0$$

where for $t \geq 0$, $y(t) = (e(t), r(t)) \in Z$ and $y_0 = (e_0, r_0) \in Z$. It follows that

$$y(t) = U(t,s)y(s), \qquad 0 \leq s \leq t.$$

If the triple $\{\overline{q}, \overline{x}_0, \overline{f}\}$ is A_0-admissible and $e_0 \in Dom(A_0)$ and $r_0 \in Q$, then $y \in C^1([0,T]; Z)$, $e \in C^1([0,T]; H)$, and $r \in C^1([0,T]; Q)$, for all $T > 0$. In the next section we argue that $\lim_{t\to\infty} |e(t)| = 0$ and, under an appropriate additional richness assumption on the plant data, $\overline{x}(t)$, $t \geq 0$, (i.e. *A_0-persistence of excitation*) that $\lim_{t\to\infty} |r(t)|_Q = 0$.

3. State and Parameter Error Convergence

Let e and r be the solutions to the initial value problem (2.11)-(2.13) and define

$$E(t) = \frac{1}{2}\{|e(t)|^2 + |r(t)|_Q^2\}, \qquad t \geq 0.$$

Lemma 3.1. *Suppose that the triple $\{\overline{q}, \overline{x}_0, \overline{f}\}$ is A_0-admissible and that $x_0 \in Dom(A_0)$. Then*

$$E(t) + \alpha_0 \int_0^t |e(s)|^2 ds \leq \zeta, \qquad t \geq 0, \tag{3.1}$$

where $\zeta = E(0) = \frac{1}{2}\{|e_0|^2 + |r_0|_Q^2\}$.

Proof. Using (2.11) and (2.12) we obtain

$$\begin{aligned} \frac{d}{dt}E(t) &= \frac{d}{dt}\frac{1}{2}\{\langle e(t), e(t)\rangle + \langle r(t), r(t)\rangle_Q\} \\ &= \langle \dot{e}(t), e(t)\rangle + \langle \dot{r(t)}, r(t)\rangle_Q \\ &= -\langle A_0 e(t) + B(t)r(t), e(t)\rangle + \langle B(t)^* e(t), r(t)\rangle_Q \\ &= -\langle A_0 e(t), e(t)\rangle \\ &\leq -\alpha_0 |e(t)|^2. \end{aligned} \tag{3.2}$$

Integrating both sides of the above expression from 0 to t, we obtain (3.1). □

The convergence of the state estimator is established in the next theorem. The proof we give is a variant of the argument used to establish Barbalat's Lemma in [14].

Theorem 3.2. *If the triple $\{\overline{q}, \overline{x}_0, \overline{f}\}$ is A_0-admissible and $x_0 \in Dom(A_0)$, then E is nonincreasing and $\lim_{t\to\infty} |e(t)| = 0$.*

Proof. That E is nonincreasing follows immediately from (3.2). Let $t_2 \geq t_1 \geq 0$ and note that Assumption ($A1$), Definition 2.1, and (3.1) imply that

$$\begin{aligned} |e(t_2)|^2 - |e(t_1)|^2 &= \int_{t_1}^{t_2} \frac{d}{dt}|e(t)|^2 dt \\ &= 2\int_{t_1}^{t_2} \langle \dot{e}(t), e(t)\rangle dt \\ &= -2\int_{t_1}^{t_2} \langle A_0 e(t), e(t)\rangle dt - 2\int_{t_1}^{t_2} \langle B(t)r(t), e(t)\rangle dt \\ &\leq -2\alpha_0 \int_{t_1}^{t_2} |e(t)|^2 dt + 2\alpha \int_{t_1}^{t_2} |r(t)|_Q |e(t)| dt \\ &\leq 2\alpha \int_{t_1}^{t_2} \frac{1}{2}\{|e(t)|^2 + |r(t)|_Q^2\} dt \\ &\leq 2\alpha\zeta(t_2 - t_1). \end{aligned} \tag{3.3}$$

Now suppose that $\lim_{t\to\infty}|e(t)| \neq 0$. Then there exist $\varepsilon > 0$ and a sequence $\{t_i\}_{i=1}^{\infty}$ with $\lim_{i\to\infty} t_i = \infty$ for which

$$|e(t_i)|^2 > \varepsilon, \qquad i = 1, 2, \ldots . \tag{3.4}$$

It follows from (3.3) and (3.4) that for $\delta > 0$ and $i = 1, 2, \ldots$

$$\begin{aligned} \int_{t_i-\delta}^{t_i} |e(t)|^2 dt &= \int_{t_i-\delta}^{t_i} |e(t_i)|^2 dt - \int_{t_i-\delta}^{t_i} \{|e(t_i)|^2 - |e(t)|^2\} dt \\ &> \varepsilon\delta - 2\alpha\zeta \int_{t_i-\delta}^{t_i} (t_i - t) dt \\ &= \varepsilon\delta - \alpha\zeta\delta^2. \end{aligned}$$

Choosing $\delta = \varepsilon/(2\alpha\zeta)$, we obtain

$$\int_{t_i-\delta}^{t_i} |e(t)|^2 dt \geq \frac{\varepsilon\delta}{2}, \qquad i = 1, 2, \ldots .$$

This contradicts the fact that Lemma 3.1 implies that

$$\alpha_0 \int_0^{\infty} |e(t)|^2 dt \leq \zeta < \infty,$$

and consequently $\lim_{t\to\infty}|e(t)|^2 = 0$, and the theorem is proved. □

In order to establish parameter convergence, we require the following definition.

Definition 3.3. *The triple $\{\overline{q}, \overline{x}_0, \overline{f}\}$ is said to be either A_0-persistently exciting or A_0-rich if there exist positive constants τ_0, δ_0, and ε_0 such that for each $q \in Q$ with $|q|_Q = 1$ and $t \geq 0$ sufficiently large, there exists $\tilde{t} \in [t, t+\tau_0]$ such that*

$$\left|\int_{\tilde{t}}^{\tilde{t}+\delta_0} T_0(\tilde{t} + \delta_0 - s) B(s) q ds\right| \geq \varepsilon_0.$$

Parameter convergence is demonstrated in the next theorem. The proof is a variant on an argument used in the finite dimensional case by Baumeister and Scondo in [2].

Theorem 3.4. *Suppose that the triple $\{\overline{q}, \overline{x}_0, \overline{f}\}$ is A_0-admissible and A_0-persistently exciting and that $x_0 \in Dom(A_0)$. Then $\lim_{t\to\infty}|r(t)|_Q = 0$.*

Proof. We begin by noting that for $t_2 \geq t_1$, (2.11) implies that

$$e(t_2) = T_0(t_2 - t_1)e(t_1) - \int_{t_1}^{t_2} T_0(t_2 - s) B(s) r(s) ds.$$

It then follows from (2.3) that

$$
\begin{aligned}
\left|\int_{t_1}^{t_2} T_0(t_2-s)B(s)r(s)ds\right| &\leq |e(t_2)| + |T_0(t_2-t_1)e(t_1)| \\
&\leq |e(t_2)| + e^{-\alpha_0(t_2-t_1)}|e(t_1)|. \qquad (3.5)
\end{aligned}
$$

Also, from (2.12) and Definition 2.1 we obtain

$$
\begin{aligned}
|r(t_2)-r(t_1)|_Q &= \sup_{|q|_Q\leq 1} |\langle r(t_2)-r(t_1), q\rangle_Q| = \sup_{|q|_Q\leq 1} |\langle \int_{t_1}^{t_2} \dot{r}(t)dt, q\rangle_Q| \\
&= \sup_{|q|_Q\leq 1} \left|\int_{t_1}^{t_2} \langle \dot{r}(t), q\rangle_Q dt\right| = \sup_{|q|_Q\leq 1} \left|\int_{t_1}^{t_2} \langle B(t)^* e(t), q\rangle_Q dt\right| \qquad (3.6) \\
&= \sup_{|q|_Q\leq 1} \left|\int_{t_1}^{t_2} \langle A(q)\overline{x}(t), e(t)\rangle dt\right| \leq \sup_{|q|_Q\leq 1} \int_{t_1}^{t_2} \alpha |q|_Q |e(t)| dt \\
&\leq \alpha \int_{t_1}^{t_2} |e(t)|dt \leq \alpha(t_2-t_1)^{\frac{1}{2}} \left\{\int_{t_1}^{t_2} |e(t)|^2 dt\right\}^{\frac{1}{2}}.
\end{aligned}
$$

Now, from Theorem 3.2 we know that E is nonincreasing. Since $E(t) \geq 0$, $t \geq 0$, it follows that $\lim_{t\to\infty} E(t)$ must exist. Also from Theorem 3.2 we know that $\lim_{t\to\infty} |e(t)| = 0$. Consequently, we must have that $\lim_{t\to\infty} |r(t)|_Q = \rho$, for some $\rho \geq 0$. Suppose that $\rho \neq 0$. Then there exist a sequence of positive numbers $\{t_k\}_{k=1}^{\infty}$ with $\lim_{k\to\infty} t_k = \infty$ and $\delta > 0$ for which $|r(t_k)|_Q > \delta$, $k = 1, 2, \ldots$. For each positive integer k sufficiently large, let $\tilde{t}_k$ be as in Definition 3.3 corresponding to $q = r(t_k)/|r(t_k)|_Q$. Then, using (3.5) and (3.6) we obtain

$$
\begin{aligned}
&\left|\int_{\tilde{t}_k}^{\tilde{t}_k+\delta_0} T_0(\tilde{t}_k+\delta_0-t)B(t)r(t_k)dt\right| \\
&\leq \left|\int_{\tilde{t}_k}^{\tilde{t}_k+\delta_0} T_0(\tilde{t}_k+\delta_0-t)B(t)r(t)dt\right| \\
&+ \left|\int_{\tilde{t}_k}^{\tilde{t}_k+\delta_0} T_0(\tilde{t}_k+\delta_0-t)B(t)(r(t_k)-r(t))dt\right| \\
&\leq |e(\tilde{t}_k+\delta_0)| + e^{-\alpha_0\delta_0}|e(\tilde{t}_k)| \\
&+ \left|\int_{\tilde{t}_k}^{\tilde{t}_k+\delta_0} e^{-\alpha_0(\tilde{t}_k+\delta_0-t)}\alpha^2(t-t_k)^{\frac{1}{2}}\left\{\int_{t_k}^{t} |e(s)|^2 ds\right\}^{\frac{1}{2}} dt\right| \\
&\leq |e(\tilde{t}_k+\delta_0)| + e^{-\alpha_0\delta_0}|e(\tilde{t}_k)| + \alpha^2(\tilde{t}_k+\delta_0-t_k)^{\frac{1}{2}}\left\{\int_{t_k}^{\tilde{t}_k+\delta_0} |e(s)|^2 ds\right\}^{\frac{1}{2}}\delta_0 \\
&\leq |e(\tilde{t}_k+\delta_0)| + e^{-\alpha_0\delta_0}|e(\tilde{t}_k)| + \alpha^2\delta_0\sqrt{\tau_0+\delta_0}\left\{\int_{t_k}^{t_k+\tau_0+\delta_0} |e(s)|^2 ds\right\}^{\frac{1}{2}}.
\end{aligned}
$$

It then follows from Definition 3.3 and the above estimate that for all positive integers k sufficiently large, we have

$$0<\varepsilon_0\delta \leq |r(t_k)|_Q|\int_{\tilde{t}_k}^{\tilde{t}_k+\delta_0} T_0(\tilde{t}_k+\delta_0-t)B(t)\frac{r(t_k)}{|r(t_k)|_Q}dt|$$

$$\leq |e(\tilde{t}_k+\delta_0)|+e^{-\alpha_0\delta_0}|e(\tilde{t}_k)|+\alpha^2\delta_0\sqrt{\tau_0+\delta_0}\{\int_{t_k}^{t_k+\tau_0+\delta_0}|e(t)|^2dt\}^{\frac{1}{2}}. \tag{3.7}$$

Letting $k\to\infty$ in (3.7), and recalling Lemma 3.1 and Theorem 3.2, we obtain a contradiction. Thus we must have that $\lim_{t\to\infty}|r(t)|_Q=0$, and the theorem is proved. □

4. Approximation Theory

The estimator, (2.4)-(2.6) is, in general, infinite dimensional. Consequently its implementation requires some form of finite dimensional approximation and a corresponding convergence theory.

For each $n=1,2,\ldots$, let H^n and Q^n be finite dimensional subspaces of H and Q, respectively. Let P_H^n and P_Q^n denote the corresponding orthogonal projections of H onto H^n and Q onto Q^n, respectively. Let $A_0^n\in\mathcal{L}(H^n,H^n)$. In order to establish a convergence result, we will require the following three assumptions.

($A3$) (Subspace Approximation) For each $\varphi\in H$ and $q\in Q$ we have

$$\lim_{n\to\infty}|P_H^n\varphi-\varphi|=0 \qquad \text{and} \qquad \lim_{n\to\infty}|P_Q^nq-q|_Q=0.$$

($A4$) (Uniform Accretivity) For $\alpha_0\geq 0$ as in Assumption ($A1$), we have

$$\langle A_0^n\varphi^n,\varphi^n\rangle\geq\alpha_0|\varphi^n|^2, \qquad \varphi^n\in H^n.$$

($A5$) (Strong Convergence in the Generalized Sense) For some λ with $Re\,\lambda>-\alpha_0$, we have

$$\lim_{n\to\infty}|(\lambda+A_0^n)^{-1}P_H^n\varphi-(\lambda+A_0)^{-1}\varphi|=0, \qquad \varphi\in H.$$

Let $\{T_0^n(t):t\geq 0\}$ denote the C_0-semigroup of bounded linear operators on H^n with infinitesimal generator $-A_0^n$. That is,

$$T_0^n(t)=\exp(-A_0^nt), \qquad t\geq 0.$$

It follows from Assumption ($A4$) above that

$$|T_0^n(t)|\leq e^{-\alpha_0t}, \qquad t\geq 0,$$

and from Assumptions ($A3$) - ($A5$) and the Trotter-Kato theorem (see, for example, [10] or [13]) that

$$\lim_{n\to\infty}|T_0^n(t)P_H^n\varphi-T_0(t)\varphi|=0, \qquad \varphi\in H, \tag{4.1}$$

with the convergence being uniform in t on compact t-intervals.

For each $t \geq 0$ define $B^n(t) \in \mathcal{L}(Q^n, H^n)$ by

$$B^n(t)q^n = P_H^n B(t)q^n = P_H^n A(q^n)\overline{x}(t), \qquad q^n \in Q^n.$$

If the triple $\{\overline{q}, \overline{x}_0, \overline{f}\}$ is A_0-admissible, then $|B^n(t)| \leq \alpha$, $t \geq 0$.

We consider the following approximating estimator,

$$\dot{x}^n(t) + A_0^n x^n(t) + B^n(t)q^n(t) = P_H^n \overline{f}(t) + P_H^n A_0 \overline{x}(t), \qquad t > 0, \tag{4.2}$$

$$\dot{q}^n(t) - B^n(t)^* x^n(t) = -P_Q^n B(t)^* \overline{x}(t), \qquad t > 0, \tag{4.3}$$

$$x^n(0) = x_0^n = P_H^n x_0, \qquad q^n(0) = q_0^n = P_Q^n q_0, \tag{4.4}$$

and argue that

$$\lim_{n\to\infty} |x^n(t) - x(t)| = 0 \qquad \text{and} \qquad \lim_{n\to\infty} |q^n(t) - q(t)|_Q = 0, \tag{4.5}$$

uniformly in t on compact t-intervals, where x^n and q^n are the solutions to the system (4.2)-(4.4) and x and q are the solutions to the system (2.4)-(2.6).

To establish the convergence in (4.5), we rewrite the initial value problem (4.2)-(4.4) as an equivalent system in the product space $Z^n = H^n \times Q^n$. Note that Z^n is a finite dimensional subspace of Z. For $t \geq 0$, define the linear operator $\mathcal{A}^n(t) \in \mathcal{L}(Z^n, Z^n)$ by

$$\mathcal{A}^n(t) = \begin{bmatrix} A_0^n & B^n(t) \\ -B^n(t)^* & 0 \end{bmatrix}.$$

For each $n = 1, 2, \ldots$, let $\mathcal{P}^n \in \mathcal{L}(Z, Z^n)$ denote the orthogonal projection of Z onto Z^n with respect to the inner product given in (2.7). It follows that

$$\mathcal{P}^n = \begin{bmatrix} P_H^n & 0 \\ 0 & P_Q^n \end{bmatrix}.$$

For each $t \geq 0$ set $F^n(t) = \mathcal{P}^n F(t) \in Z^n$; that is

$$F^n(t) = \begin{bmatrix} P_H^n \overline{f}(t) + P_H^n A_0 \overline{x}(t) \\ -P_Q^n B(t)^* \overline{x}(t) \end{bmatrix}.$$

We set $z_0^n = (x_0^n, q_0^n) \in Z^n$. The estimator (4.2)-(4.4) can then be written as the initial value problem in Z^n given by

$$\dot{z}^n(t) + \mathcal{A}^n(t) z^n(t) = F^n(t), \qquad t > 0, \tag{4.6}$$

$$z^n(0) = z_0^n, \tag{4.7}$$

where $z^n(t) = (x^n(t), q^n(t))$, $t \geq 0$.

Assume that the triple $\{\overline{q}, \overline{x}_0, \overline{f}\}$ is A_0-admissible, and let $\{U^n(t,s) : 0 \leq s \leq t\}$ be the fundamental operator solution to the initial value problem (4.6), (4.7). It then follows that

$$|U^n(t,s)| \leq e^{\alpha(t-s)}, \qquad 0 \leq s \leq t,$$

and that

$$z^n(t) = U^n(t,0)z_0^n + \int_0^t U^n(t,s)F^n(s)ds, \qquad t \geq 0, \tag{4.8}$$

and we want to show that

$$\lim_{n\to\infty} |z^n(t) - z(t)|_Z = 0,$$

uniformly in t on compact t-intervals, where z^n is given by (4.8) and z is given by (2.10).

Define the operator $\mathcal{A}_0^n \in \mathcal{L}(Z^n, Z^n)$ by

$$\mathcal{A}_0^n = \begin{bmatrix} A_0^n & 0 \\ 0 & 0 \end{bmatrix},$$

and for $t \geq 0$ define the operator $\mathcal{B}^n(t) \in \mathcal{L}(Z^n, Z^n)$ by

$$\mathcal{B}^n(t) = \begin{bmatrix} 0 & B^n(t) \\ -B^n(t)^* & 0 \end{bmatrix}.$$

Then $\mathcal{A}^n(t) = \mathcal{A}_0^n + \mathcal{B}^n(t)$, $t \geq 0$, $\mathcal{B}^n(t) = \mathcal{P}^n\mathcal{B}(t)$, $t \geq 0$, and it is easily argued that for $t \geq 0$, $|\mathcal{B}^n(t)| \leq \alpha$. Let $\{\mathcal{T}_0^n(t) : t \geq 0\}$ be the C_0-semigroup of contractions on Z^n with infinitesimal generator $-\mathcal{A}_0^n$. That is $\mathcal{T}_0^n(t) = \exp(-\mathcal{A}_0^n t)$, $t \geq 0$, with $|\mathcal{T}_0^n(t)| \leq 1$, $t \geq 0$. It also follows that

$$\mathcal{T}_0^n(t) = \begin{bmatrix} T_0^n(t) & 0 \\ 0 & I \end{bmatrix}, \qquad t \geq 0,$$

and from (4.1), that

$$\lim_{n\to\infty} |\mathcal{T}_0^n(t)\mathcal{P}^n\varphi - \mathcal{T}_0(t)\varphi|_Z = 0, \qquad \varphi \in Z, \tag{4.9}$$

uniformly in t on compact t-intervals, where $\{\mathcal{T}_0(t) : t \geq 0\}$ is the C_0-semigroup of contractions on Z with generator $-\mathcal{A}_0$.

Theorem 4.1. *Suppose that the Assumptions (A1)-(A5) hold and that the triple $\{\overline{q}, \overline{x}_0, \overline{f}\}$ is A_0-admissible. Then*

$$\lim_{n\to\infty} |z^n(t) - z(t)|_Z = 0,$$

uniformly in t on compact t-intervals, where z^n is given by (4.8) and z is given by (2.10).

Proof. From (2.10) and (4.8), it is clear that we need only argue that

$$\lim_{n\to\infty} |U^n(t,s)\mathcal{P}^n\varphi - U(t,s)\varphi|_Z = 0, \qquad \varphi \in Z, \tag{4.10}$$

uniformly on the triangle $\Delta_T = \{(s,t) : 0 \le s \le t \le T\}$, where $T > 0$ is fixed but arbitrary. Now, for $(s,t) \in \Delta_T$ we have (see [4], Theorem 5.2)

$$U^n(t,s)\mathcal{P}^n\varphi = \mathcal{T}_0^n(t-s)\mathcal{P}^n\varphi + \int_s^t \mathcal{T}_0^n(t-r)\mathcal{B}^n(r)U^n(r,s)\mathcal{P}^n\varphi dr, \qquad \varphi \in Z,$$

and

$$U(t,s)\varphi = \mathcal{T}_0(t-s)\varphi + \int_s^t \mathcal{T}_0(t-r)\mathcal{B}(r)U(r,s)\varphi dr, \qquad \varphi \in Dom(\mathcal{A}).$$

Consequently, for $\varphi \in Dom(\mathcal{A})$ and $(s,t) \in \Delta_T$, we have

$$\begin{aligned}
&|U^n(t,s)\mathcal{P}^n\varphi - U(t,s)\varphi|_Z \\
&\le |\{\mathcal{T}_0^n(t-s)\mathcal{P}^n\varphi - \mathcal{T}_0(t-s)\varphi\} \\
&\qquad + \int_s^t \{\mathcal{T}_0^n(t-r)\mathcal{B}^n(r)U^n(r,s)\mathcal{P}^n\varphi - \mathcal{T}_0(t-r)\mathcal{B}(r)U(r,s)\varphi\}dr|_Z \\
&\le |\mathcal{T}_0^n(t-s)\mathcal{P}^n\varphi - \mathcal{T}_0(t-s)\varphi|_Z \\
&\qquad + \int_s^t |\mathcal{T}_0^n(t-r)\mathcal{P}^n\mathcal{B}(r)\{U^n(r,s)\mathcal{P}^n\varphi - U(r,s)\varphi\}|_Z dr \\
&\qquad + \int_s^t |\{\mathcal{T}_0^n(t-r)\mathcal{P}^n - \mathcal{T}_0(t-r)\}\mathcal{B}(r)U(r,s)\varphi|_Z dr \\
&\le |\mathcal{T}_0^n(t-s)\mathcal{P}^n\varphi - \mathcal{T}_0(t-s)\varphi|_Z \\
&\qquad + \int_s^t |\{\mathcal{T}_0^n(t-r)\mathcal{P}^n - \mathcal{T}_0(t-r)\}\mathcal{B}(r)U(r,s)\varphi|_Z dr \\
&\qquad + \alpha\int_s^t |U^n(r,s)\mathcal{P}^n\varphi - U(r,s)\varphi|_Z dr.
\end{aligned}$$

The convergence result given in (4.9), the strong continuity of the mappings $\tau \mapsto \mathcal{B}(\tau)$ and $(\sigma,\tau) \mapsto U(\tau,\sigma)$, Δ_T compact, and an application of the Gronwall inequality then imply the convergence in (4.10) for $\varphi \in Dom(\mathcal{A})$. The uniform exponential bound on U and U^n for all $n = 1, 2, \ldots$, together with the fact that $Dom(\mathcal{A})$ is dense in Z then yield the desired result for all $\varphi \in Z$. □

5. An Example and Numerical Results

In this section we illustrate the application of our scheme to the estimation of the thermal conductivity in a one dimensional nonlinear (strictly speaking, quasi-linear) heat equation. We note that we do not attempt to verify A_0-admissibility or A_0-persistence of excitation for the example presented below. The A_0-admissibility is determined by the regularity of the plant. Assumptions, albeit rather restrictive ones, can be made which would guarantee that the state of the plant is sufficiently regular to yield A_0-admissibility. As for A_0-persistence of excitation, the condition specified in Definition 3.3 is rather difficult, if not impossible to check. It is possible, however, to study this condition and its characterization in some very simple cases. Such a study provides valuable insight into tuning our adaptive estimator when it is applied in the context of more complex examples. In the case of the two-space theory detailed in [5] and [16], an investigation of the persistence of excitation condition was reported on in [6]. An analogous study for the one-space theory treated here is currently under way and when completed, will be reported on elsewhere.

We consider the identification of the thermal conductivity, $\overline{q}$, in the one dimensional quasi-linear heat equation

$$(5.1)\quad \frac{\partial \overline{x}}{\partial t}(t,\eta) - \frac{\partial}{\partial \eta}\{\overline{q}(|\frac{\partial \overline{x}}{\partial \eta}(t,\eta)|)\frac{\partial \overline{x}}{\partial \eta}(t,\eta)\} = \overline{f}(t,\eta), \qquad 0<\eta<1,\ t>0,$$

together with the Dirichlet boundary conditions

$$(5.2)\qquad \overline{x}(t,0)=0 \quad \text{and} \quad \overline{x}(t,1)=0, \qquad t>0,$$

and the initial conditions

$$(5.3)\qquad \overline{x}(0,\eta)=\overline{x}_0(\eta), \qquad 0\le\eta\le 1.$$

We assume that $\overline{x}_0 \in L_2(0,1)$ and $\overline{f}(t,\cdot) \in L_2(0,1)$.

Let $H = L_2(0,1)$ be endowed with the standard inner product, and define the Hilbert space Q as follows. Let

$$\hat{Q} = \{\varphi : \varphi \in H^1_{Loc}(\mathbb{R}^+) \text{ and } \varphi, D\varphi \in L_\infty(\mathbb{R}^+)\}.$$

Define the inner product, $\langle\cdot,\cdot\rangle_Q$, on $\hat{Q}$ by

$$(5.4)\quad \langle\varphi,\psi\rangle_Q = \int_0^\infty \omega_0(\theta)\varphi(\theta)\psi(\theta)d\theta + \int_0^\infty \omega_1(\theta)D\varphi(\theta)D\psi(\theta)d\theta, \qquad \varphi,\psi\in\hat{Q},$$

where $\omega_0,\omega_1 \in L_1(\mathbb{R}^+)$ are positive weighting functions. Let $|\cdot|_Q$ denote the norm induced by the inner product given in (5.4), and define the Hilbert space Q to be the completion of the inner product space $\{\hat{Q},\langle\cdot,\cdot\rangle_Q,|\cdot|_Q\}$. Let $Dom(A) = H^2(0,1)\cap H^1_0(0,1)$ and for $q\in Q$ and $\varphi \in Dom(A)$, set $A(q)\varphi = -D\{q(D\varphi)D\varphi\}$.

For our estimator dynamics, we use the sum of a linear constant coefficient heat conduction operator and a decay term with Neumann (insulated) boundary conditions. That is, we define $A_0 : Dom(A_0) \subset H \to H$ by

$$A_0\varphi = \{-\alpha_1 D^2 + \alpha_0\}\varphi, \qquad \varphi\in Dom(A_0),$$

where $\alpha_0,\alpha_1>0$, with

$$Dom(A_0) = \{\varphi : \varphi\in H^2(0,1), D\varphi(0)=D\varphi(1)=0\}.$$

It follows that for $\varphi \in Dom(A_0)$ we have

$$\langle A_0\varphi, \varphi\rangle \geq \alpha_0|\varphi|^2.$$

It is also easily argued that $\lambda + A_0 : Dom(A_0) \subset H \to H$ is surjective for any $\lambda > -\alpha_0$.

We approximate using linear B-splines. For $n = 1, 2, \ldots$, let $\{\varphi_j^n\}_{j=0}^n$ be the standard linear B-splines on the interval $[0,1]$ defined with respect to the uniform mesh $\{0, \frac{1}{n}, \frac{2}{n}, \ldots, 1\}$. That is, for $i = 0, 1, 2, \ldots, n$

$$\varphi_i^n(x) = \begin{cases} 1 - |nx - i|, & x \in [\frac{i-1}{n}, \frac{i+1}{n}], \\ 0, & \text{elsewhere on } [0,1]. \end{cases}$$

Set $H^n = \text{span}\{\varphi_j^n\}_{j=0}^n$. For each $n = 1, 2, \ldots$, let P_H^n denote the orthogonal projection of H onto H^n. Standard approximation results for spline functions (see [15]) can be used to establish that the strong convergence to the identity of P_H^n required by Assumption $(A3)$ is satisfied.

Let $x_n(t) \in \mathbb{R}^{n+1}$ be the coordinate vector for $x^n(t)$ with respect to the basis $\{\varphi_j^n\}_{j=0}^n$. That is,

$$x^n(t) = \sum_{j=0}^n x_n(t)_j \varphi_j^n.$$

Let M_n denote the Gram matrix corresponding to the basis $\{\varphi_j^n\}_{j=0}^n$. We have

$$M_n = [M_n]_{i,j} = \left[\int_0^1 \varphi_i^n(x)\varphi_j^n(x)dx\right] = \frac{1}{6n}\begin{bmatrix} 2 & 1 & 0 & & & & & 0 \\ 1 & 4 & 1 & 0 & & & & \\ 0 & 1 & 4 & 1 & & & & \\ & 0 & \cdot & \cdot & \cdot & & & \\ & & & \cdot & \cdot & & 0 & \\ & & & & 1 & 4 & 1 & 0 \\ & & & & 0 & 1 & 4 & 1 \\ 0 & & & & & 0 & 1 & 2 \end{bmatrix}.$$

Also, let K_n be the $(n+1) \times (n+1)$ matrix given by

$$K_n = [K_n]_{i,j} = \left[\int_0^1 D\varphi_i^n(x)D\varphi_j^n(x)dx\right]$$

$$= n\begin{bmatrix} 1 & -1 & 0 & & & & & 0 \\ -1 & 2 & -1 & 0 & & & & \\ 0 & -1 & 2 & -1 & & & & \\ & 0 & \cdot & \cdot & \cdot & & & \\ & & & \cdot & \cdot & \cdot & 0 & \\ & & & & -1 & 2 & -1 & 0 \\ & & & & 0 & -1 & 2 & -1 \\ 0 & & & & & 0 & -1 & 1 \end{bmatrix}.$$

We approximate the operator A_0, by the operator A_0^n constructed using a standard Galerkin approach. For this type of approximation, it is easily argued that Assumptions (A4) and (A5) are satisfied. It is also easily established that the matrix representation for the operator A_0^n with respect to the basis $\{\varphi_j^n\}_{j=0}^n$, A_{0n}, is given by

$$A_{0n} = M_n^{-1}(\alpha_1 K_n + \alpha_0 M_n) = \alpha_1 M_n^{-1} K_n + \alpha_0 I_n,$$

where I_n denotes the $(n+1)\times(n+1)$ identity matrix.

We also use linear B-splines to discretize Q. For each $m = 1, 2, \ldots$, and each $r > 0$, let $\{\hat{\psi}_j^{m,r}\}_{j=0}^m$ be the standard linear B-splines on the interval $[0, r]$ defined with respect to the uniform mesh $\{0, \frac{r}{m}, \frac{2r}{m}, \ldots, r\}$. Let $Q^{m,r} = \mathrm{span}\{\psi_j^{m,r}\}_{j=0}^m$, where

$$\psi_j^{m,r} = \begin{cases} \hat{\psi}_j^{m,r}, & j = 0, 1, 2, \ldots, m-1, \\ \hat{\psi}_m^{m,r} + \chi_{[r,\infty)}, & j = m, \end{cases}$$

with χ_J denoting the characteristic function for the interval J. If we let $P_Q^{m,r}$ denote the orthogonal projection of Q onto $Q^{m,r}$, the requisite strong convergence to the identity can be demonstrated. Consequently, Assumption (A3) holds. Let $\Omega_{m,r}$ denote the $(m+1)\times(m+1)$ Gram matrix corresponding to the basis $\{\psi_j^{m,r}\}_{j=0}^m$. That is

$$\begin{aligned} \Omega_{m,r} = [\Omega_{m,r}]_{i,j} &= \langle \psi_i^{m,r}, \psi_j^{m,r} \rangle_Q \\ &= \int_0^\infty \omega_0(\theta)\psi_i^{m,r}(\theta)\psi_j^{m,r}(\theta)d\theta + \int_0^\infty \omega_1(\theta) D\psi_i^{m,r}(\theta) D\psi_j^{m,r}(\theta)d\theta. \end{aligned}$$

Let $q_{m,r}(t) = [q_{m,r}(t)_0, \ldots, q_{m,r}(t)_m]^T \in \mathbb{R}^{m+1}$ denote the coordinate vector for the approximating estimate $q^{m,r}(t)$ with respect to the basis $\{\psi_j^{m,r}\}_{j=0}^m$. That is

$$q^{m,r}(t) = \sum_{j=0}^m q_{m,r}(t)_j \psi_j^{m,r}, \qquad t \geq 0.$$

There is some practical advantage to be gained by using a finite dimensional approximation to the plant data, $\overline{x}(t)$, $t \geq 0$, in the estimator equations. The convergence theory given in Section 4 can be easily modified to permit this. (Our results on this will be reported on elsewhere.) To do this we use the orthogonal projection, P^n, of $H_0^1(0,1)$ onto the subspace of H^n, H_0^n, defined by $H_0^n = \mathrm{span}\{\varphi_j^n\}_{j=1}^{n-1}$. If

$$\overline{x}^n(t) = P^n\overline{x}(t) = \sum_{j=1}^{n-1} \overline{x}_n(t)_j \varphi_j^n,$$

(i.e. let $\overline{x}_n(t) \in \mathbb{R}^{n-1}$ be the coordinate vector for $\overline{x}^n(t)$ with respect to the basis $\{\varphi_j^n\}_{j=1}^{n-1}$) then

$$\overline{x}_n(t) = L_n^{-1} h_n(\overline{x}(t)),$$

where for $\varphi \in H_0^1(0,1)$, $h_n(\varphi) \in \mathbb{R}^{n-1}$ is given by

$$h_n(\varphi)_j = \int_0^1 D\varphi(\eta) D\varphi_j^n(\eta) d\eta, \qquad j = 1, 2, \ldots, n-1,$$

and $L_n \in \mathbb{R}^{(n-1)\times(n-1)}$ is given by

$$L_n = [L_n]_{i,j} = \left[\int_0^1 D\varphi_i^n(x) D\varphi_j^n(x) dx\right] = n \begin{bmatrix} 2 & -1 & 0 & & & & 0 \\ -1 & 2 & \cdot & & & & \\ 0 & \cdot & \cdot & \cdot & & & \\ & & & \cdot & \cdot & \cdot & 0 \\ & & & & \cdot & \cdot & -1 \\ 0 & & & & 0 & -1 & 2 \end{bmatrix}.$$

It is easily verified that for $\varphi \in H_0^1(0,1)$

$$h_n(\varphi)_j = -n\Delta^2_{\frac{1}{n}} \varphi(\frac{j-1}{n}) \quad = \quad -n\left\{\varphi\left(\frac{j+1}{n}\right) - 2\varphi\left(\frac{j}{n}\right) + \varphi\left(\frac{j-1}{n}\right)\right\},$$
$$j = 1, 2, \ldots, n-1.$$

Thus the approximating estimator (i.e. (4.3) - (4.5) with $\overline{x}$ replaced by $\overline{x}^n$) does not require spatially distributed data. For a given value of n, $\overline{x}$ need only be spatially sampled at the $n-1$ nodal points $\frac{1}{n}, \frac{2}{n}, \ldots, \frac{n-1}{n}$.

For each $t \geq 0$, define the $(n+1) \times (m+1)$ matrix $B_{n,m,r}(t)$ by

$$[B_{n,m,r}(t)]_{i,j} \quad = \quad \int_0^1 \psi_j^{m,r}(D_\eta \overline{x}^n(t,\eta)) D_\eta \overline{x}^n(t,\eta) D\varphi_i^n(\eta) d\eta,$$
$$j = 0, 1, 2, \ldots, m, \quad i = 0, 1, 2, \ldots, n.$$

Using the fact that $\overline{x}^n(t) \in \mathrm{span}\{\varphi_j^n\}_{j=1}^{n-1} \subset \mathrm{span}\{\varphi_j^n\}_{j=0}^n$, and the fact that $D\varphi_j^n$ is piecewise constant, and adopting the convention that $\overline{x}_n(t)_0 = \overline{x}_n(t)_n = 0$, we obtain that

$$\begin{aligned} [B_{n,m,r}(t)]_{i,j} \quad &= \quad n\psi_j^{m,r}(n\{\overline{x}_n(t)_i - \overline{x}_n(t)_{i-1}\})\{\overline{x}_n(t)_i - \overline{x}_n(t)_{i-1}\} \\ &\quad -n\psi_j^{m,r}(n\{\overline{x}_n(t)_{i+1} - \overline{x}_n(t)_i\})\{\overline{x}_n(t)_{i+1} - \overline{x}_n(t)_i\}, \\ &\quad i = 1, 2, \ldots, n-1, \quad j = 0, 1, 2, \ldots, m, \end{aligned}$$

and

$$[B_{n,m,r}(t)]_{i,j} = 0, \qquad i = 0, n, \quad j = 0, 1, 2, \ldots, m.$$

The matrix form of the approximating estimator, (4.3)-(4.5), is then given by

$$\text{(5.5)} \quad \dot{x}_n(t) + A_{0n} x_n(t) + M_n^{-1} B_{n,m,r}(t) q_{m,r}(t) = M_n^{-1} \overline{f}_n(t) + A_{0n} \overline{x}_n(t), \quad t > 0,$$

$$\text{(5.6)} \quad \dot{q}_{m,r}(t) - \Omega_{m,r}^{-1} B_{n,m,r}(t)^T x_n(t) = -\Omega_{m,r}^{-1} B_{n,m,r}(t)^T \overline{x}_n(t), \qquad t > 0,$$

$$\text{(5.7)} \qquad x_n(0) = M_n^{-1} x_{0n}, \qquad q_{m,r}(0) = \Omega_{m,r}^{-1} q_{0m,r},$$

where the $(n+1)$-vectors $\overline{f}_n(t)$ and x_{0n} and the $(m+1)$-vector $q_{0m,r}$ are given by

$$\overline{f}_n(t)_i = \int_0^1 \overline{f}(t,\eta)\varphi_i^n(\eta)d\eta, \qquad i = 0,1,2,\dots,n,$$

$$[x_{0n}]_i = \int_0^1 x_0(\eta)\varphi_i^n(\eta)d\eta, \qquad i = 0,1,2,\dots,n,$$

and

$$[q_{0m,r}]_i = \langle q_0, \psi_i^{m,r}\rangle_Q =$$
$$\int_0^\infty \omega_0(\theta)q_0(\theta)\psi_i^{m,r}(\theta)d\theta + \int_0^\infty \omega_1(\theta)Dq_0(\theta)D\psi_i^{m,r}(\theta)d\theta, \quad i = 0,1,\dots,m,$$

respectively.

We set

$$\overline{q}(\theta) = .9(1 - \frac{1}{2}e^{-\frac{1}{2}\theta^2}), \qquad \theta \geq 0,$$

$$\overline{f}(t,\eta) = \{sin(4\pi t) + 10^{-3}t^2\}\chi_{[\,215,\ 315]}(\eta), \qquad 0 < \eta < 1,\ \ t > 0,$$

and

$$\overline{x}_0(\eta) = 0, \qquad 0 < \eta < 1,$$

and simulated the plant (equations (5.1)-(5.3)) using the IMSL routine DMOLCH, a double precision Hermite polynomial based method of lines PDE solver. In our estimator, we set $\alpha_0 = 10^{-2}$, $\alpha_1 = 5 \times 10^{-3}$, $r = 3.5$,

$$\omega_0(\theta) = \omega_1(\theta) = \begin{cases} 1 & 0 \leq \theta < r \\ \frac{1}{2}e^{-20\theta} & r < \theta < \infty, \end{cases}$$

$$x_0(\eta) = 0, \qquad 0 < \eta < 1,$$

and

$$q_0(\theta) = 1, \qquad 0 < \theta < \infty.$$

We integrated the approximating estimator (i.e. equations (5.5)-(5.7) using the NAG Library stiff system solver, D02NBF. The requisite integrals were computed using a composite two point Gauss-Legendre quadrature rule. All codes were written in FORTRAN and run on a Sun SPARCstation 10 under UNIX. In Figure 5.1 we have plotted our final (i.e. at time $t = 100$) estimates for $\overline{q}$ for various values of n and m. In Figure 5.2 we have plotted the estimates for $\overline{q}$ at various times. These estimates were generated with $n = 32$ and $m = 16$.

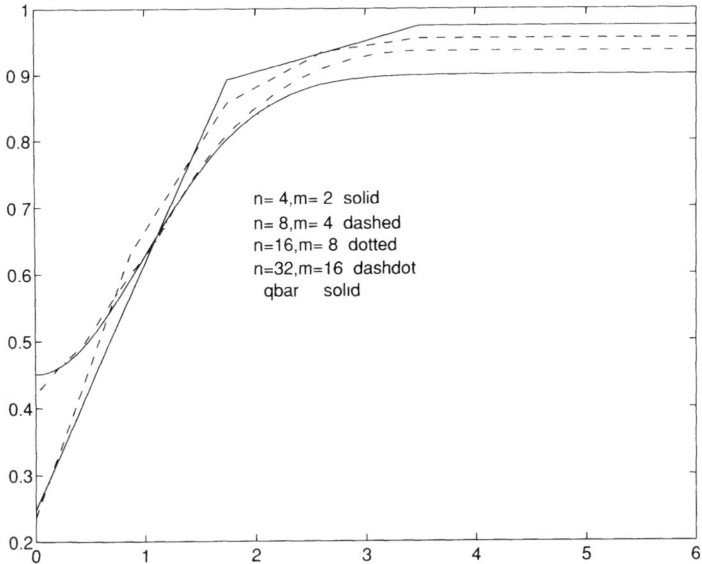

FIGURE 5.1. Final ($t = 100$) estimates for $\overline{q}$ for $n = 2^{j+1}$, $m = 2^j$ for $j = 1, 2, 3, 4$.

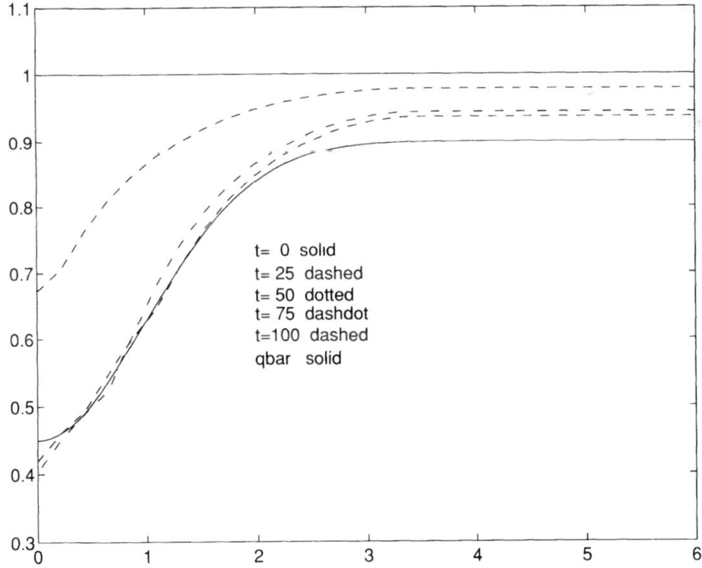

FIGURE 5.2. Estimates for $\overline{q}$ at various times generated with $n = 32$ and $m = 16$.

References

1. H. W. Alt and K.-H. Hoffmann and J. Sprekels, *A numerical procedure to solve certain identification problems*, Intern. Series Numer. Math., 68 (1984), pp. 11–43.
2. J. Baumeister and W. Scondo, *Adaptive methods for parameter identification*, Methoden und Verfahren der Mathematischen Physik, (1987), pp. 87–116.
3. ———, *Asymptotic embedding methods for parameter estimation*, Proceedings of the 26th IEEE Conference on Decision and Control, (1987), pp. 170–174.
4. A. Belleni-Morante, *Applied Semigroups and Evolution Equations*, Clarendon Press, Oxford, 1979.
5. M. A. Demetriou and I. G. Rosen, *On-line parameter estimation for infinite dimensional dynamical systems*, CAMS Report 93-2, Center for Applied Mathematical Sciences, Department of Mathematics, University of Southern California, Los Angeles, CA 90089-1113, February 1993.
6. ———, *On the persistence of excitation in the adaptive identification of distributed parameter systems*, CAMS Report 93-4, Center for Applied Mathematical Sciences, Department of Mathematics, University of Southern California, Los Angeles, CA 90089-1113, March 1993 and *IEEE Transactions on Automatic Control*, to appear.
7. K. -H. Hoffmann and J. Sprekels, *The method of asymptotic regularization and restricted parameter identification problems in variational inequalities*, in Free Boundary Problems: Application and Theory, IV, Maubuisson, (1984), pp. 508–513.
8. ———, *On the identification of coefficients of elliptic problems by asymptotic regularization*, Numer. Funct. Anal. and Optimiz. 7 (1984-85), pp. 157–177.
9. ———, *On the identification of parameters in general variational inequalities by asymptotic regularization*, SIAM J. Math. Anal. 17 (1986), pp. 1198–1217.
10. T. Kato, *Perturbation Theory for Linear Operators*, Second Edition, Springer-Verlag, New York, 1984.
11. A. P. Morgan and K. S. Narendra, *On the stability of nonautonomous differential equations* $\dot{x} = [A + B(t)]x$, *with skew symmetric matrix* $B(t)$, SIAM J. Control and Optimization 15 (1977), pp. 163–176.
12. K. S. Narendra and P. Kudva, *Stable adaptive schemes for system identification and control, Parts I and II*, IEEE Trans. Systems, Man and Cybernetics, SMC-4 (1974), pp. 542–560.
13. A. Pazy, *Semigroups of Linear Operators and Applications to Partial Differential Equations*, Springer-Verlag, New York, 1983.
14. V. M. Popov, *Hyperstability of Control Systems*, Springer-Verlag, Berlin, Heidelberg, and New York, 1973.
15. M. H. Schultz, *Spline Analysis*, Prentice Hall, Englewood Clffis, N.J., 1973.
16. W. Scondo, *Ein Modellabgleichsverfahren zur adaptiven Parameteridentifikation in Evolutionsgleichungen*, PhD thesis, Johann Wolfgang Goethe-Universitat zu Frankfurt am Main, Frankfurt am Main, Germany, 1987.

17 R E Showalter, *Hılbert Space Methods for Partıal Dıfferentıal Equatıons*, Pıtman, London, 1977
18 H Tanabe, *Equatıons of Evolutıon*, Pıtman, London, 1979

Joseph Kazımır
Center for Applıed Mathematıcal Scıences
Department of Mathematıcs
Unıversıty of Southern Calıfornıa
Los Angeles, Calıfornıa 90089-1113, USA
I G Rosen
Center for Applıed Mathematıcal Scıences
Department of Mathematıcs
Unıversıty of Southern Calıfornıa
Los Angeles, Calıfornıa 90089-1113, USA

DECAY ESTIMATES FOR THE WAVE EQUATION WITH INTERNAL DAMPING

VILMOS KOMORNIK

Institut de Recherche Mathématique Avancée
Université Louis Pasteur et C.N.R.S.

ABSTRACT. We obtain optimal estimates for the solution of an integral inequality related to many stabilization problems. Then it is applied to improve some recent results concerning the energy decay of the wave equation with internal nonlinear feedback. Unlike the earlier works, our method also applies in the case of bounded feedback functions.

1991 *Mathematics Subject Classification.* 35L05,93D15

Key words and phrases. Wave equation, feedback stabilization, integral inequality.

1. Introduction

Let Ω be a bounded domain in $\mathbb{R}^n$ $(n \geq 1)$ having a boundary Γ of class C^∞ and let $g : \mathbb{R} \to \mathbb{R}$ be a non-decreasing continuous function satisfying $g(0) = 0$. Consider the problem

$$u'' - \Delta u + g(u') = 0 \quad \text{in} \quad \Omega \times \mathbb{R}_+ \tag{1.1}$$

$$u = 0 \quad \text{on} \quad \Gamma \times \mathbb{R}_+ \tag{1.2}$$

$$u(0) = u_0 \quad \text{and} \quad u'(0) = u_1 \quad \text{on} \quad \Omega \tag{1.3}$$

where we use the notation $\mathbb{R}_+ := [0, +\infty)$. We define the energy of the solutions of this system by

$$E = E(t) := \frac{1}{2} \int_\Omega u'(t)^2 + |\nabla u(t)|^2 \, \mathrm{dx}, \ t \in \mathbb{R}_+.$$

This problem is well-posed and dissipative. Indeed, we have the following theorem due to Haraux; it improved some earlier results of Lions–Strauss [11] and Brézis [1].

Theorem 1.1 (Haraux [4], [5]). — *a) Given* $(u_0, u_1) \in H_0^1(\Omega) \times L^2(\Omega)$ *arbitrarily, the problem* (1.1)–(1.3) *has a unique solution (defined in a suitable weak* *sense) satisfying*

$$u \in C(\mathbb{R}_+; H_0^1(\Omega)) \cap C^1(\mathbb{R}_+; L^2(\Omega)) \tag{1.4}$$

and

$$u' g(u') \in L^1_{loc}(\mathbb{R}_+; L^1(\Omega)). \tag{1.5}$$

Moreover, its energy is non-increasing.

b) If $(u_0, u_1) \in \left(H^2(\Omega) \cap H_0^1(\Omega)\right) \times H_0^1(\Omega)$ *and* $g(u_1) \in L^2(\Omega)$*, then the solution of* (1.1)–(1.3) *has the following regularity properties:*

$$u \in L^\infty(\mathbb{R}_+; H^2(\Omega) \cap H_0^1(\Omega)), \tag{1.6}$$
$$u' \in L^\infty(\mathbb{R}_+; H_0^1(\Omega)), \tag{1.7}$$
$$u'' \in L^\infty(\mathbb{R}_+; L^2(\Omega)), \tag{1.8}$$
$$g(u') \in L^\infty(\mathbb{R}_+; L^2(\Omega)). \tag{1.9}$$

Moreover, the function

$$\mathbb{R}_+ \ni t \mapsto \int_\Omega |\Delta u(t,x)|^2 + |\nabla u'(t,x)|^2 \, dx$$

is non-increasing. Finally, the energy function is locally absolutely continuous and we have

$$E' = -\int_\Omega u' g(u') \; dx \leq 0 \tag{1.10}$$

almost everywhere in $\mathbb{R}_+$.

The purpose of this paper is to study the energy decay rate under suitable growth assumptions on g. In the rest of the paper we assume that there exist three numbers $p > 1$, $q \geq 1 \geq r \geq 0$ and four positive constants c_i, $1 \leq i \leq 4$ such that

$$c_1|x|^p \leq |g(x)| \leq c_2|x|^{1/p} \quad \textit{if} \quad |x| \leq 1 \tag{1.11}$$

and

$$c_3|x|^r \leq |g(x)| \leq c_4|x|^q \quad \textit{if} \quad |x| > 1. \tag{1.12}$$

Let us observe that if the inequalities of (1.11) (resp. of (1.12)) are satisfied only in some smaller neighbourhood of 0 (resp. of $\pm\infty$), then they are also satisfied (maybe with other constants c_i) in the neighbourhood $|x| \leq 1$ (resp. $|x| > 1$) by the non-decreasing property of g.

Theorem 1.2. — *Assume* (1.11) *and* (1.12) *with*

$$(1.13) \qquad p>1, \quad r=1 \quad \text{and} \quad q\geq 1 \quad \text{such that} \quad (n-2)q\leq n+2.$$

Then for every $(u_0,u_1)\in H_0^1(\Omega)\cap L^2(\Omega)$ *the solution of the problem* (1.1) - (1.3) *satisfies the estimate*

$$(1.14) \qquad E(t)\leq Ct^{-2/(p-1)} \quad \forall t>0,$$

where C *is a constant depending only on the initial energy* $E(0)$ *in acontinuous way.*

Remarks. — There is an analogous result if we assume (1.11) with $p=1$: then an easy adaptation of our proof given below leads to the estimate

$$(1.15) \qquad E(t)\leq e^{1-\omega t}E(0) \quad \forall t>0,$$

where ω is a positive constant, *independent* of the initial data.

Theorem 1.2 improves and simplifies several earlier theorems of Nakao [13], Haraux and Zuazua [6], Zuazua [14], Conrad, Leblond and Marmorat [2] by weakening their growth assumptions on g near zero.

It is easy to give examples of functions g satisfying (1.11)–(1.13). For example, choose two numbers $\alpha>0$, $\beta\geq 0$ and consider the function

$$(1.16) \qquad g(x):=\begin{cases}|x|^\alpha \operatorname{sgn} x, & \text{if } |x|\leq 1,\\ |x|^\beta \operatorname{sgn} x, & \text{if } |x|>1.\end{cases}$$

If $n>4$ and $\beta>(n+4)/(n-4)$, then we cannot apply theorem 1.1 because there is no q satisfying (1.12) and (1.13). We cannot apply this theorem either if $\beta<1$: then the first inequality in (1.12) is not satisfied with $r=1$. In the remaining cases, i.e. when $\beta\geq 1$ and $(n-4)\beta\leq n+4$, we may apply theorem 1.1 with $p=\max\{\alpha,1/\alpha\}$, $q=\beta$ and $r=1$; for $p=\alpha=1$ we rather apply the complement of the theorem mentioned above (cf.(1.15)).

As F. Conrad kindly pointed out to the author, theorem 1.2 and the previous results all have a drawback from the point of view of applications: they never apply if the function g is bounded. Indeed, no bounded function can verify the first inequality in (1.12) if $r>0$. On the other hand, in many applications the function g is bounded, as for example the function introduced in (1.16) if $\beta=0$. The following result applies in such situations as well, at least for smooth initial data.

Theorem 1.3. — *Assume* (1.11), (1.12) *with*

$$(1.17) \qquad p>1, \quad 0\leq r\leq 1 \quad \text{and} \quad q\geq 1 \quad \text{such that} \quad (n-4)q\leq n+4.$$

Furthermore, assume that

$$(1.18) \qquad 2p \geq n + (2-n)r \quad \textit{and} \quad 2p \geq n - 2 - nq^{-1}.$$

Then for every $(u_0, u_1) \in \big(H^2(\Omega) \cap H_0^1(\Omega)\big) \times H_0^1(\Omega)$ *satisfying* $g(u_1) \in L^2(\Omega)$ *there exists a constant* C *such that the solution of the problem* (1.1) – (1.3) *satisfies the estimate*

$$(1.19) \qquad E(t) \leq Ct^{-2/(p-1)} \quad \forall t > 0.$$

Remarks. — The technical condition (1.18) is easy to satisfy. Indeed, if (1.11), (1.12) and (1.17) are satisfied, then replacing p by

$$(1.20) \qquad \max\{p, (n + (2-n)r)/2, (n - 2 - nq^{-1})/2\}$$

(1.11), (1.12), (1.17) remain valid (with the same constants c_i) and (1.18) will be satisfied, too. Note that condition (1.18) is automatically satisfied if $n \leq 2$.

Consider for example the function g defined by the formula (1.16) with $\alpha > 0$, $\beta \geq 0$. If $n > 4$ and $\beta > (n+4)/(n-4)$, then there is no q satisfying (1.12) and (1.17). On the other hand, if $(n-4)\beta \leq n+4$, then we may apply theorem 1.3 with $r = \min\{\beta, 1\}$, $q = \max\{\beta, 1\}$ and with any $p > 1$ satisfying $p \geq \max\{\alpha, 1/\alpha\}$ and (1.18). For example, if $\alpha \neq 1$ and $\beta = 0$, then we may choose $r = 0$, $q = 1$ and $p = \max\{\alpha, 1/\alpha, n/2\}$.

Another example is given by

$$g(x) := \begin{cases} \sqrt{x}, & \text{if } x \geq 0, \\ -x^2, & \text{if } x \leq 0. \end{cases}$$

In this case theorem 1.2 cannot be applied because (1.12) is not satisfied with $r = 1$. If $n > 12$, then theorem 1.3 cannot be applied either because there is no q satisfying (1.12) and (1.17). On the other hand, in case $n \leq 12$ theorem 1.3 applies with $r = 1/2$, $q = 2$ and with $p = 2$ if $n \leq 6$, $p = (n+2)/4$ if $6 \leq n \leq 12$. A more general example is given by the formula

$$g(x) := \sum_{i=1}^{M} c_i (x_+)^{p_i} - \sum_{i=1}^{N} d_i (x_-)^{q_i}$$

where $x_+ = \max(x, 0)$, $x_- = \max(-x, 0)$, c_i, d_i, p_i, q_i are given positive numbers.

There is a result analogous to theorem 1.3 if (1.11), (1.18) are satisfied with $p = 1$ and q, r satisfy the conditions in (1.12), (1.17); if $n \geq 2$, then we also need the assumption $r = 1$. If these conditions are satisfied, then an easy adaptation of the proof of theorem 1.3 leads to the estimate (1.15) where ω is a positive constant *depending* on the initial data.

Let us note that theorem 1.3 implies an earlier theorem of Nakao [13]; his original, stronger growth assumptions on g excluded the bounded functions.

Our method of proof differs from those of the earlier works. The idea is to prove that the energy function satisfies the integral inequalities

$$\int_t^{\infty} E^{\alpha+1}(s)\, ds \le TE(0)^{\alpha} E(t), \quad \forall t \in \mathbb{R}_+ \tag{1.21}$$

with $\alpha = (p-1)/2$, for a suitable positive constant A. Then we can conclude by applying the following result which improves some earlier theorems of Haraux [3] and Lagnese [10].

Theorem 1.4. — *Let $E : \mathbb{R}_+ \to \mathbb{R}_+$ be a non-increasing function and assume that there are two constants $\alpha > 0$ and $T > 0$ such that (1.21) is satisfied. Then we have*

$$E(t) \le E(0) \frac{T+\alpha t}{T+\alpha T}^{-1/\alpha}, \quad \forall t \ge T. \tag{1.22}$$

Remarks. — Note that the inequality (1.22) is obviously satisfied for $0 \le t < T$ too: then it is weaker then the trivial inequality $E(t) \le E(0)$. There is an analogous (and simpler) result if (1.21) is satisfied with $\alpha = 0$: then we have

$$E(t) \le E(0) e^{1-t/T}, \quad \forall t \ge T, \tag{1.23}$$

instead of (1.22).

Our approach seems to be simpler than that of Nakao [13] because we apply simpler and more natural integral inequalities and also simpler than the Liapunov type method of Haraux and Zuazua [6] because we do not have to introduce artificial Liapunov functions and a corresponding artificial parameter.

Let us also mention that strong stability results (without decay estimates) were obtained earlier (by applying LaSalle's invariance principle) by Marcati [12].

The author is grateful to F. Conrad and A. Haraux for fruitful discussions and to the organizers of the 1993 Vorau Conference on Control Theory for their invitation.

2. Proof of Theorem 1.4

If $E(0) = 0$, then $E \equiv 0$ and there is nothing to prove. Otherwise, replacing the function E by the function $E/E(0)$ we may assume that $E(0) = 1$ and we have to prove the estimate

$$E(t) \le \frac{T+\alpha t}{T+\alpha T}^{-1/\alpha}, \quad \forall t \ge T. \tag{2.1}$$

Introduce the function

$$F : \mathbb{R}_+ \to \mathbb{R}_+, \qquad F(t) = \int_t^\infty E^{\alpha+1}\, ds.$$

It is non-increasing and locally absolutely continuous. Differentiating and using (1.21) we find that

$$-F' \geq T^{-\alpha-1} F^{\alpha+1} \quad \text{a.e. in} \quad (0, +\infty)$$

whence

$$(F^{-\alpha})' \geq \alpha T^{-\alpha-1} \quad \text{a.e. in} \quad (0, B), \quad B := \sup\{t : \ E(t) > 0\}.$$

(Observe that $F^{-\alpha}(t)$ is defined for $t < B$.) Integrating in $[0, s]$ we obtain

$$(F(s)^{-\alpha} - F(0)^{-\alpha}) \geq \alpha T^{-\alpha-1} s \quad \text{for every} \quad s \in [0, B)$$

whence

$$F(s) \leq (F(0)^{-\alpha} + \alpha T^{-\alpha-1} s)^{-1/\alpha} \quad \text{for every} \quad s \in [0, B). \tag{2.2}$$

Since $F(s) = 0$ if $s \geq B$, this inequality is valid in fact for every $s \in \mathbb{R}_+$. Since $F(0) \leq TE(0)^{\alpha+1} = T$ by (1.21), the right-hand side of (2.2) is less than equal to

$$(T^{-\alpha} + \alpha T^{-\alpha-1} s)^{-1/\alpha} = T^{(\alpha+1)/\alpha}(T + \alpha s)^{-1/\alpha}.$$

On the other hand, E being nonnegative and non-increasing, the left-hand side of (2.2) may be estimated as follows:

$$F(s) = \int_s^{+\infty} E^{\alpha+1}\, \mathrm{dt} \geq \int_s^{T+(\alpha+1)s} E^{\alpha+1}\, \mathrm{dt} \geq (T + \alpha s) E(T + (\alpha+1)s)^{\alpha+1}.$$

Therefore we deduce from (2.2) the estimate

$$(T + \alpha s) E(T + (\alpha+1)s)^{\alpha+1} \leq T^{(\alpha+1)/\alpha}(T + \alpha s)^{-1/\alpha}$$

whence

$$E(T + (\alpha+1)s) \leq \left(1 + \frac{\alpha s}{T}\right)^{-1/\alpha}, \quad \forall s \geq 0.$$

Choosing $t := T + (\alpha + 1)s$ hence the inequality(2.1) follows. □

Remark 2.1. — The theorem is optimal in the following sense: given $\alpha > 0$, $T > 0$, $C > 0$ and $t' \geq T$ arbitrarily, there exists a non-increasing function $E : \mathbb{R}_+ \to \mathbb{R}_+$ satisfying (1.21) and such that

$$E(0) = C \quad \text{and} \quad E(t') = E(0) \frac{T + \alpha t'}{T + \alpha T}^{-1/\alpha}.$$

We leave to the reader to verify that the following example has these properties:

$$E(t):=\begin{cases} C\big(1+\alpha C^{-\alpha}t/T\big)^{-1/\alpha}, & \text{if } 0\le t\le (t'-TC^{\alpha})/(\alpha+1),\\ C(1+\alpha)^{1/\alpha}\big(1+\alpha C^{-\alpha}t'/T\big)^{-1/\alpha}, & \text{if } (t'-TC^{\alpha})/(\alpha+1)\le t\le t',\\ 0, & \text{if } t>t'. \end{cases} \tag{2.3}$$

Let us also note that for $t < T$ we cannot say more than the trivial estimate $E(t) \le E(0)$. Indeed, for any given $\alpha > 0$, $T > 0$, $C > 0$ and $t' < T$ the function

$$E(t) := \begin{cases} C, & \text{if } 0 \le t \le T,\\ 0, & \text{if } t > T \end{cases} \tag{2.4}$$

satisfies (1.21) and $E(t') = E(0) = C$.

Remark 2.2. — Assume that E is also continuous. Then the inequalities (1.22) are strict; in particular, $E(T) < E(0)$. We refer to [9] for this result, for a detailed study of integral inequalities of type (1.21) (also for $\alpha \le 0$) and for the study of closely related differential inequalities.

3. Proof of Theorem 1.2

Applying a density and approximation argument as in [4], [5] it is sufficient to consider the case where

$$(u_0, u_1) \in \big(H^2(\Omega) \cap H_0^1(\Omega)\big) \times H_0^1(\Omega) \quad \text{and} \quad g(u_1) \in L^2(\Omega).$$

In this case the solution of (1.1) – (1.3) has the regularity properties (1.6) – (1.9) and this regularity is sufficient to justify all computations that follow. Let us begin by recalling the proof of formula (1.10) in theorem 1.1. Multiplying (1.1) with u', integrating by parts in $\Omega \times (S, T)$ and using (1.2) we obtain easily that

$$E(S) - E(T) = \int_S^T \int_\Omega u'g(u') \,\mathrm{dx}\,\mathrm{dt}, \quad 0 \le S < T < +\infty. \tag{3.1}$$

Since $xg(x) \ge 0$ for all $x \in \mathbb{R}$, it follows that the energy is non-increasing, locally absolutely continuous and

$$E' = -\int_\Omega u'g(u') \,\mathrm{dx} \quad \text{a.e. in } \mathbb{R}_+. \tag{3.2}$$

Next we multiply (1.1) with $E^{(p-1)/2}u$. We have

$$\begin{aligned} 0 &= \int_S^T E^{(p-1)/2} \int_\Omega u(u'' - \Delta u + g(u'))\ \mathrm{dx}\ \mathrm{dt} \\ &= \left[E^{(p-1)/2} \int_\Omega uu'\ \mathrm{dx}\right]_S^T - \frac{p-1}{2} \int_S^T E^{(p-3)/2} E' \int_\Omega uu'\ \mathrm{dx}\ \mathrm{dt} \\ &\quad + \int_S^T E^{(p-1)/2} \int_\Omega |\nabla u|^2 - (u')^2 + ug(u')\ \mathrm{dx}\ \mathrm{dt} \end{aligned}$$

whence

$$\begin{aligned} \int_S^T E^{(p+1)/2}\ \mathrm{dt} &= \left[E^{(p-1)/2} \int_\Omega uu'\ \mathrm{dx}\right]_T^S + \frac{p-1}{2} \int_S^T E^{(p-3)/2} E' \int_\Omega uu'\ \mathrm{dx}\ \mathrm{dt} \\ &\quad + \int_S^T E^{(p-1)/2} \int_\Omega 2(u')^2 - ug(u')\ \mathrm{dx}\ \mathrm{dt} \end{aligned}$$

Using the non-increasing property of E, the Sobolev imbedding $H^1(\Omega) \subset L^2(\Omega)$, the definition of E and the Cauchy-Schwarz inequality we have

$$\left|E^{(p-1)/2} \int_\Omega uu'\ \mathrm{dx}\right| \le cE^{(p+1)/2} \le cE$$

and

$$\left|\int_S^T E^{(p-3)/2} E' \int_\Omega uu'\ \mathrm{dx}\ \mathrm{dt}\right| \le -c \int_S^T E^{(p-1)/2} E'\ \mathrm{dt} \le -c \int_S^T E'\ \mathrm{dt} \le cE(S).$$

(Here and in the sequel we denote by c positive constants which may be different at different occurencies.) Using these estimates we conclude from the above identity that

$$2 \int_S^T E^{(p+1)/2}\ \mathrm{dt} \le cE(S) + \int_S^T E^{(p-1)/2} \int_\Omega 2(u')^2 - ug(u')\ \mathrm{dx}\ \mathrm{dt}. \tag{3.3}$$

In order to estimate the last term of (3.3), we set

$$\Omega_1 := \{x \in \Omega : |u'| \le 1\}, \quad \Omega_2 := \{x \in \Omega : |u'| > 1\}, \tag{3.4}$$

and we remark that the assumption (1.11) implies the following two inequalities (we also use (3.2) and the Hölder inequality):

$$\int_{\Omega_1} (u')^2\ \mathrm{dx} \le c \int_{\Omega_1} (u'g(u'))^{2/(p+1)}\ \mathrm{dx} \le c(-E')^{2/(p+1)}, \tag{3.5}$$

and

$$\int_{\Omega_1} g(u')^2\ \mathrm{dx} \le c \int_{\Omega_1} (u'g(u'))^{2/(p+1)}\ \mathrm{dx} \le c(-E')^{2/(p+1)}. \tag{3.6}$$

Now fix an arbitrarily small $\varepsilon > 0$ (to be chosen later). Using (3.5), (3.6), the boundedness of E^α for any $\alpha \geq 0$, the obvious estimate

$$\|u\|_{L^2(\Omega)} \leq cE^{1/2}$$

and applying the Young inequality we obtain the following estimate:

$$\begin{aligned} E^{(p-1)/2} \int_{\Omega_1} 2(u')^2 - ug(u') \text{ dx dt} &\leq cE^{(p-1)/2}(-E')^{2/(p+1)} + cE^{p/2}(-E')^{1/(p+1)} \\ &\leq \varepsilon E^{(p+1)/2} - c(\varepsilon)E'. \end{aligned}$$

Hence

$$\int_S^T E^{(p-1)/2} \int_{\Omega_1} 2(u')^2 - ug(u') \text{ dx dt} \leq \varepsilon \int_S^T E^{(p+1)/2} \text{ dt} + c(\varepsilon)E(S) \tag{3.7}$$

and substituting this into (3.3) we obtain that

$$(2-\varepsilon)\int_S^T E^{(p+1)/2} \text{ dt} \leq c(\varepsilon)E(S) + \int_S^T E^{(p-1)/2} \int_{\Omega_2} 2(u')^2 - ug(u') \text{ dx dt}. \tag{3.8}$$

Assume for a moment the estimates

$$E^{(p-1)/2} \int_{\Omega_2} (u')^2 \text{ dx} \leq \varepsilon E^{(p+1)/2} - c(\varepsilon)E' \tag{3.9}$$

and

$$E^{(p-1)/2} \int_{\Omega_2} |ug(u')| \text{ dx} \leq \varepsilon E^{(p+1)/2} - c(\varepsilon)E'. \tag{3.10}$$

Then we have (similarly to (3.7))

$$\int_S^T E^{(p-1)/2} \int_{\Omega_2} 2(u')^2 - ug(u') \text{ dx dt} \leq 2\varepsilon \int_S^T E^{(p+1)/2} \text{ dt} + c(\varepsilon)E(S)$$

and we conclude from (3.8) that

$$(2 - 3\varepsilon) \int_S^T E^{(p+1)/2} \text{ dt} \leq c(\varepsilon)E(S). \tag{3.11}$$

Choosing $\varepsilon = 1/3$ and $A = c(\varepsilon)$ we deduce from (3.11) that

$$\int_S^T E^{(p+1)/2} \text{ dt} \leq AE(S), \quad 0 \leq S < T < +\infty,$$

and letting $T \to +\infty$ this yields the following crucial estimate

$$\int_S^\infty E^{(p+1)/2} \text{ dt} \leq AE(S), \quad \forall S \geq 0. \tag{3.12}$$

We may thus complete the proof by applying theorem 1.4.

It remains to prove the estimates (3.9) and (3.10). Since now $r = 1$ by hypothesis (3.9) follows easily from (1.12) and (3.2):

$$E^{(p-1)/2} \int_{\Omega_2} (u')^2 \, \mathrm{dx} \le c_3^{-1} \int_{\Omega_2} u' g(u') \, \mathrm{dx} \le -c_3^{-1} E'.$$

Turning to the proof of (3.10) first we apply the Hölder inequality, the Sobolev imbedding $H^1(\Omega) \subset L^{q+1}(\Omega)$ (this follows from hypothesis (1.13) on q) and the definition of the energy:

$$E^{(p-1)/2} \int_{\Omega_2} |ug(u')| \mathrm{dx} \le E^{(p-1)/2} \|u\|_{q+1} \|g(u')\|_{L^{1+1/q}(\Omega_2)} \le c E^{p/2} \|g(u')\|_{L^{1+1/q}(\Omega_2)}.$$

(Here and in the sequel we denote the norm of $L^\alpha(\Omega)$ by $\|\cdot\|_\alpha$, $1 \le \alpha \le \infty$.) Using (1.12) and (3.2) to estimate the last norm we obtain that

$$E^{(p-1)/2} \int_{\Omega_2} |ug(u')| \, \mathrm{dx} \le c E^{p/2} \Big(\int_{\Omega_2} u' g(u') \, \mathrm{dx} \Big)^{q/(q+1)} \le c E^{p/2} |E'|^{q/(1+q)}.$$

Applying the Young inequality hence we conclude that

$$E^{(p-1)/2} \int_{\Omega_2} |ug(u')| \, \mathrm{dx} \le \varepsilon E^{p(q+1)/2} - c(\varepsilon) E'.$$

Since $p(q+1) \ge p+1$, hence (3.10) follows (with another ε).

4. Proof of Theorem 1.3

Adapting the proof of theorem 1.2 we only have to modify the proof of the estimates (3.9) and (3.10). Let us begin with the proof of (3.9). If $r = 1$, then the previous proof remains valid. If $n = 1$, then $H^1(\Omega) \subset L^\infty(\Omega)$ and we conclude from (1.7) that $u' \in L^\infty(\mathbb{R}_+; L^\infty(\Omega))$. Using the obvious relation

$$|g(x)| \ge \min\{g(1), -g(-1)\} > 0 \quad \text{if} \quad |x| > 1$$

it follows that

$$\begin{aligned} E^{(p-1)/2} \int_{\Omega_2} (u')^2 \, \mathrm{dx} &\le c E^{(p-1)/2} \int_{\Omega_2} |u'| |u' g(u')| \, \mathrm{dx} \le c E^{(p-1)/2} \|u'\|_\infty \|u' g(u')\|_1 \\ &\le c E^{(p-1)/2} (-E') \le -c E'. \end{aligned}$$

We may therefore assume that $0 \le r < 1$ and $n \ge 2$. First we show for every fixed $\varepsilon' > 0$ and $s \in (0,1)$ the inequality

$$E^{(p-1)/2} \int_{\Omega_2} (u')^2 \, \mathrm{dx} \le \varepsilon' E^{(p-1)/(2(1-s))} \|u'\|_{(2-(r+1)s)/(1-s)}^{(2-(r+1)s)/(1-s)} - c(\varepsilon') E'. \tag{4.1}$$

Indeed, we have, putting $\alpha := (2-(r+1)s)/(1-s)$ for brevity,

$$\begin{aligned}E^{(p-1)/2}\int_{\Omega_2}(u')^2\ \mathrm{dx} &\le cE^{(p-1)/2}\int_{\Omega_2}|u'|^{\alpha(1-s)}|u'g(u')|^s\ \mathrm{dx}\\ &\le cE^{(p-1)/2}\||u'|^{\alpha(1-s)}\|_{1/(1-s)}\|(u'g(u'))^s\|_{1/s}\\ &= cE^{(p-1)/2}\|u'\|_\alpha^{\alpha(1-s)}\|u'g(u')\|_1^s\\ &= cE^{(p-1)/2}\|u'\|_\alpha^{\alpha(1-s)}(-E')^s\\ &\le \varepsilon' E^{(p-1)/(2(1-s))}\|u'\|_\alpha^\alpha - c(\varepsilon')E'.\end{aligned}$$

Choosing $s = 2/(p+1)$ the right hand side of (4.1) becomes

$$\varepsilon' E^{(p+1)/2}\|u'\|_{2(p-r)/(p-1)}^{2(p-r)/(p-1)} - c(\varepsilon')E'.$$

Assumption (1.18) on r implies the Sobolev imbedding $H^1(\Omega)\subset L^{2(p-r)/(p-1)}(\Omega)$. (This assumption is needed only if $n\ge 3$: the case $n=2$ is trivial.) Using also (1.7) we thus deduce from (4.1) that

$$E^{(p-1)/2}\int_{\Omega_2}(u')^2\ \mathrm{dx} \le c\varepsilon' E^{(+1)/2} - c(\varepsilon')E'$$

with some constant c which does not depend on ε'. Choosing $\varepsilon' := \varepsilon/c$ hence (3.9) follows.

Turning to the proof of (3.10) we may assume that $q>1$. Indeed, if $q=1$, then we may replace q by $1+\delta$ with a sufficiently small $\delta>0$: then (1.12), (1.17) and (1.18) remain valid. The case $n<4$ is obvious: then we have $H^2(\Omega)\subset L^\infty(\Omega)$ whence $u\in L^\infty(\mathbb{R}_+;L^\infty(\Omega))$ (cf. (1.6)). Therefore

$$\begin{aligned}E^{(p-1)/2}\int_{\Omega_2}|ug(u')|\ \mathrm{dx} &\le E^{(p-1)/2}\|u\|_\infty\|g(u')\|_{L^1(\Omega_2)}\\ &\le cE^{(p-1)/2}\|u\|_\infty\|u'g(u')\|_{L^1(\Omega_2)}\\ &\le cE^{(p-1)/2}\|u\|_\infty(-E') \le -cE'.\end{aligned}$$

Now assume that $n\ge 4$. First we show for any fixed $\varepsilon'>0$ the inequality

$$E^{(p-1)/2}\int_{\Omega_2}|ug(u')|\ \mathrm{dx} \le \varepsilon' E^{(p-1)(1+q)/2}\|u\|_{q+1}^{q+1} - c(\varepsilon')E'. \tag{4.2}$$

Indeed, we have

$$\begin{aligned}E^{(p-1)/2}\int_{\Omega_2}|ug(u')|\ \mathrm{dx} &\le E^{(p-1)/2}\|u\|_{q+1}\|g(u')\|_{L^{1+1/q}(\Omega_2)}\\ &\le cE^{(p-1)/2}\|u\|_{q+1}\|(u'g(u'))^{q/(1+q)}\|_{L^{1+1/q}(\Omega_2)}\\ &= cE^{(p-1)/2}\|u\|_{q+1}\|u'g(u')\|^{q/(1+q)}_{L^1(\Omega_2)}\\ &\le cE^{(p-1)/2}\|u\|_{q+1}(-E')^{q/(1+q)}\\ &\le \varepsilon' E^{(p-1)(1+q)/2}\|u\|^{q+1}_{q+1}-c(\varepsilon')E'.\end{aligned}$$

Next we recall that the interpolational inequality

$$\|u\|_{q+1}\le c\|u\|^t_{2n/(n-2)}\|u\|^{1-t}_s \tag{4.3}$$

is valid for every $s\in[2n/(n-2),+\infty)$ and $t\in[0,1]$ such that

$$\frac{1}{q+1}\ge t\frac{n-2}{2n}+\frac{1-t}{s}. \tag{4.4}$$

Choosing

$$t=\max\left\{\frac{p+1}{q+1}+1-p,0\right\},$$

one can readily verify that $t\in[0,1]$. Furthermore, it is easy to show that in case $n=4$ (4.4) is satisfied if we choose a sufficiently large $s\ge 2n/(n-2)$, while in case $n>4$ it follows from (1.17) and (1.18) that (4.4) is satisfied with $s=2n/(n-4)$. Choosing s in this way, using (4.3), the Sobolev imbeddings $H^1(\Omega)\subset L^{2n/(n-2)}(\Omega)$, $H^2(\Omega)\subset L^s(\Omega)$ and applying (1.6) we obtain that

$$\begin{aligned}E^{(p-1)(1+q)/2}\|u\|^{q+1}_{q+1} &\le cE^{(p-1)(1+q)/2}\|u\|^{t(q+1)}_{2n/(n-2)}\|u\|^{(1-t)(q+1)}_{2n/(n-4)}\\ &\le cE^{(p-1)(1+q)/2}\|u\|^{t(q+1)}_{2n/(n-2)}\le cE^{(p-1+t)(1+q)/2}\\ &\le cE^{(p+1)/2}\end{aligned}$$

(in the last step we used the definition of t). We thus deduce from (4.2) the estimate

$$E^{(p-1)/2}\int_{\Omega_2}|ug(u')|\ \mathrm{dx}\le c\varepsilon' E^{(p+1)/2}-c(\varepsilon')E'$$

with some constant c which does not depend on ε'. Choosing $\varepsilon':=\varepsilon/c$ hence (3.10) follows.

Now the proof may be completed in the same way as that of theorem 1.2 in the preceeding section. □

Remarks. — Unlike the situation of the preceeding section, in the present case the constants in the estimates (3.9), (3.10) do not depend only on the initial energy of the solution. Therefore we cannot apply here a density argument to obtain the estimate (1.19) for all solutions with initial data in $H^1_0(\Omega)\times L^2(\Omega)$.

We mentioned in section 1 that there is a result, analogous to theorem 1.2 if $p = 1$. In this case we can repeat the above proof, excepting that we cannot assume in the case $n \geq 2$ that $r < 1$. Under the additional assumption $r = 1$ we may then prove (3.9) as follows: in the inequality following (4.1) we have $\alpha = 2$ for *any* choice of $s \in (0,1)$. Hence

$$E^{(p-1)/2} \int_{\Omega_2} (u')^2 \, \mathrm{dx} \leq \varepsilon' E^{(p-1)/(2(1-s))} \|u'\|_2^2 - c(\varepsilon')E'$$
$$\leq c\varepsilon' E^{(p-1)/(2(1-s))} E - c(\varepsilon')E' \leq cE - c(\varepsilon')E'$$

and (3.9) follows by choosing $\varepsilon' := \varepsilon/c$. Then we obtain again the integral estimates (3.12) (with $p = 1$) and we conclude by applying the corresponding variant (with $\alpha = 0$) of theorem 1.4.

References

1 H Brézis, *Problèmes unilatéraux*, J Math Pures Appl **51** (1972), 1–168

2 F Conrad, J Leblond and J P Marmorat, *Stabilization of second order evolution equations by unbounded nonlinear feedback*, to appear

3 A Haraux, *Oscillations forcées pour certains systèmes dissipatifs non linéaires*, preprint n° 78010, Laboratoire d'Analyse Numérique, Université Pierre et Marie Curie, Paris, 1978

4 A Haraux, *Comportement à l'infini pour une équation d'ondes non linéaire dissipative*, C R Acad Sci Paris Sér I Math **287** (1978), 507–509

5 A Haraux, "Semilinear hyperbolic problems in bounded domains, Mathematical reports", J Dieudonné editor, Harwood Academic Publishers, Gordon and Breach, 1987

6 A Haraux and E Zuazua, *Decay estimates for some semilinear damped hyperbolic problems*, Arch Rat Mech Anal (1988), 191–206

7 V Komornik, *Estimation d'énergie pour quelques systèmes dissipatifs semilinéaires*, C R Acad Sci Paris Sér I Math **314** (1992), 747–751

8 V Komornik, *Stabilisation non linéaire de l'équation des ondes et d'un modelede plaques*, C R Acad Sci Paris Sér I Math **315** (1992), 55–60

9 V Komornik, *On some integral and differential inequalities*, preprint n^o(1992) 490/P-283, IRMA, Université Louis Pasteur, Strasbourg, France

10 J Lagnese, *Boundary stabilization of thin plates*, SIAM Studies in Applied Mathematics **10**, SIAM, Philadelphia, PA, 1989

11 J-L Lions and W A Strauss, *Some non-linear evolution equations*, Bull Soc Math Fance **93** (1965), 43–96

12 P Marcati, *Decay and stability for nonlinear hyperbolic equations*, J Diff Equations **55** (1984), 30–58

13 M Nakao, *On the decay of solutions of some nonlinear dissipative wave equations in higher dimensions*, Math Z **193** (1986), 227–234

14 E Zuazua, *Stability and decay for a class of nonlinear hyperbolic problems*, Asymptotic Anal **1**, 2 (1988), 161–185

Vilmos Komornik
Institut de Recherche Mathématique Avancée
Université Louis Pasteur et C.N.R.S.
7, rue René Descartes
F-67084 Strasbourg Cedex, France

ON THE CONTROLLABILITY OF THE ROTATION OF A FLEXIBLE ARM

W KRABS

Fachbereich Mathematik
Technische Hochschule Darmstadt

ABSTRACT Considered is the rotation of a flexible arm in a horizontal plane about an axis through the arm's fixed end and driven by a motor whose torque is controlled The model was derived and investigated computationally by Sakawa for the case that the arm is described as a homogeneous Euler beam The resulting equation of motion is a partial differential equation of the type of a wave equation which is linear with respect to the state, if the control is fixed, and non-linear with respect to the control

Considered is the problem of steering the beam, within a given time interval, from the position of rest for the angle zero into the position of rest under a certain given angle

At first we show that, for every L^2-control which is suitably bounded, there is exactly one (weak) solution of the initial boundary value problem which describes the system without the end condition

Then we present an iterative method for solving the problem of controllability and discuss its convergence

1991 *Mathematics Subject Classification* 93B05, 93C10, 93C15

Key words and phrases Controllability, rotation of a flexible arm, homogeneous Euler beam

Introduction

The problem of controllability of a rotating homogeneous Euler beam has been dealt with in [4] and [5] where also references of former work are given. The model being investigated in [5] was slightly modified in [3] and also treated by a method of decomposition, however, of a more direct and simpler way than in [5].

In this paper we consider the model which was derived in [6] and treated numerically based on the method of Galerkin. We are concerned with the homogeneous case which leads to a differential equation of the form (1.8) with boundary conditions (1.3) and initial conditions (1.4) which correspond to the position of rest of the system at the beginning of the movement.

In Section 1 we formulate the problem of controllability to be solved in this paper. Section 2 is devoted to the question of solvability of the model equations (1.8), (1.3), (1.4). Here we also derive an integral differential equation (see (2.7))

which is equivalent to the model equations and show that this has a unique solution, if the control is suitably bounded.

In Section 3 we present an iterative method for solving the problem of controllability and discuss its convergence under suitable conditions.

1. The Model and the Problem of Controllability

We consider the rotation of a flexible arm in a horizontal plane about an axis through the arm's fixed end. We assume this rotation to be driven by a motor whose torque is controlled. The equations of motion of the driving motor are given by

$$\begin{aligned} \dot{\varphi}(t) &= w(t), \\ \dot{w}(t) &= u(t), \quad t \in [0,T], \end{aligned} \tag{1.1}$$

for some given time $T > 0$, where $\varphi = \varphi(t)$ is the angle of rotation, $w = w(t)$ is the angular velocity, and $u(t) = \tau(t)/J$ where $\tau = \tau(t)$ is the torque generated by the motor and J is the moment of inertia of the motor. If the bending vibrations of the arm are small, we can assume J to be constant.

In addition we assume the arm to be homogeneous and of length 1. Following [6] we model the displacement $y = y(t,x)$ of the arm from the rotating zero line by the differential equation

$$\begin{aligned} y_{tt}(t,x) + y_{xxxx}(t,x) - \frac{w(t)^2}{2}\frac{\partial}{\partial x}[(1-x^2)y_x(t,x)] - w(t)^2 y(t,x) \\ = -xu(t) \quad \text{for all} \quad t \in (0,T), \quad x \in (0,1). \end{aligned} \tag{1.2}$$

The left end of the arm being clamped and the right end being free leads to the boundary conditions

$$\left.\begin{aligned} y(t,0) &= y_x(t,0) &= 0 \\ \text{and} \qquad y_{xx}(t,1) &= y_{xxx}(t,1) &= 0 \end{aligned}\right\} \text{for} \quad t \in [0,T]. \tag{1.3}$$

At the beginning of the motion the arm is assumed to be in rest which leads to the initial conditions

$$y(0,x) = y_t(0,x) = 0 \quad \text{for} \quad x \in (0,1) \tag{1.4}$$

and

$$\varphi(0) = \dot{\varphi}(0) = 0. \tag{1.5}$$

On using (1.1) and (1.5) we get

$$\dot{\varphi}(t) = w(t) = \int_0^t u(s)\,ds \tag{1.6}$$

and

$$\varphi(t) = \int_0^t (t-s)u(s)\,ds \tag{1.7}$$

for $t \in [0,T]$ so that (1.2) can be rewritten in the form

$$\begin{aligned} y_{tt}(t,x) + y_{xxxx}(t,x) - \Big(\int_0^t u(s)ds\Big)^2 \Big\{\frac{1}{2}\frac{\partial}{\partial x}[(1-x^2)y_x(t,x)] + y(t,x)\Big\} \\ = -xu(t) \quad \text{for} \quad t \in (0,T), \quad x \in (0,1). \end{aligned} \tag{1.8}$$

Now we are in the position of formulating the

Problem of Controllability: Let some angle $\varphi_T \in \mathbb{R}$ with $\varphi_T \neq 0$ be prescribed. Determine $u \in L^2[0,T]$ such that

$$\int_0^T u(t)\,dt = 0 \quad (\Longleftrightarrow \dot{\varphi}(T) = 0), \tag{1.9}$$

$$-\int_0^T tu(t)\,dt = \varphi_T \quad (\Longleftrightarrow \varphi(T) = \varphi_T), \tag{1.10}$$

and the corresponding solution $y = y(t,x)$ of (1.8), (1.3), and (1.4) satisfies the end condition

$$y(T,x) = y_t(T,x) = 0 \quad \text{for} \quad x \in (0,1). \tag{1.11}$$

In words: The torque of the driving motor is to be determined such that the arm is steered from a position of rest at $t = 0$ with angle zero to a position of rest at $t = T$ with angle φ_T.

2. On the Solvability of the Model Equations

Before dealing with the problem of controllability we investigate the question under which condition, for a given $u \in L^2(0,T)$, there exists a unique weak solution $y = y(t,x)$ of (1.8), (1.3), (1.4).

For this purpose we at first observe that the differential operator

$$L[z](x) = z^{(4)}(x) \quad \text{for almost all} \quad x \in (0,1)$$

is self adjoint and positive definite on

$$D(L) = \{z \in H^4(0,1) \,|\, z(0) = z'(0) = z''(1) = z'''(1) = 0\}$$

and has an inifinite sequence of simple eigenvalues

$$\lambda_j = [(j - \frac{1}{2})\pi + \epsilon_j]^2, \quad j \in \mathbb{N},$$

which are the positive solutions of the transcendental equation

$$\cosh(\lambda^{\frac{1}{4}}) \cos(\lambda^{\frac{1}{4}}) = -1$$

and for which $\lim_{j\to\infty} \epsilon_j = 0$ holds true. The corresponding sequence of orthonormal eigenfunctions $e_j \in D(L)$, $j \in \mathbb{N}$, is given by $(e_j = \varphi_j / \|\varphi_j\|_{L^2(0,1)})_{j\in\mathbb{N}}$ where

$$\varphi_j(x) = \cosh(\lambda_j^{\frac{1}{4}} x) - \cos(\lambda_j^{\frac{1}{4}} x) - \gamma_j(\sinh(\lambda_j^{\frac{1}{4}} x) - \sin(\lambda_j^{\frac{1}{4}} x))$$

and

$$\gamma_j = \frac{\cosh(\lambda_j^{\frac{1}{4}}) + \cos(\lambda_j^{\frac{1}{4}})}{\sinh(\lambda_j^{\frac{1}{4}}) + \sin(\lambda_j^{\frac{1}{4}})}, \quad j \in \mathbb{N}.$$

Now we consider the differential equation

$$\begin{aligned} &y_{tt}(t,x) + y_{xxxx}(t,x) = f(t,x) \\ &\text{for} \quad t \in (0,T) \quad \text{and} \quad x \in (0,1) \end{aligned} \tag{2.1}$$

where $f \in L^2((0,T), H)$ is chosen arbitrarily together with the boundary conditions (1.3) and the initial conditions (1.4). By results in [1] and [2] there is a unique weak solution $y = y(t,x)$ of (2.1), (1.3), (1.4) with $y \in C([0,T], V) \cap C^1([0,T], H)$ which is given by

$$y(t,x) = \sum_{j=1}^{\infty} \frac{1}{\sqrt{\lambda_j}} \int_0^t \int_0^1 f(s,\tilde{x}) e_j(\tilde{x}) d\tilde{x} \sin\sqrt{\lambda_j}(t-s) ds\, e_j(x) \tag{2.2}$$

for $t \in [0,T]$ and $x \in [0,1]$ where $H = L^2(0,T)$ and

$$V = \{v \in H^2(0,1) |\, v(0) = v'(0) = 0\}.$$

Its time derivative reads

$$y_t(t,x) = \sum_{j=1}^{\infty} \int_0^t \int_0^1 f(s,\tilde{x}) e_j(\tilde{x}) d\tilde{x} \cos\sqrt{\lambda_j}(t-s) ds\, e_j(x) \tag{2.3}$$

for $t \in [0,T]$ and $x \in [0,1]$.

If $y = y(t,x)$ is in $C([0,T], V) \cap C^1([0,T], H)$ and is a solution of (1.8), (1.3), (1.4) for some given $u \in L^2(0,T)$, then we put

$$f(t,x) = (\int_0^t u(s) ds)^2 \{\frac{1}{2}\frac{\partial}{\partial x}[(1-x^2) y_x(t,x)] + y(t,x)\} - x u(t) \tag{2.4}$$

for $t \in (0,T)$ and $x \in (0,1)$ and infer that $y = y(t,x)$ is also a weak solution of (2.1), (1.3), (1.4), since $f \in L^2((0,T),H)$.

In order to show that $f \in L^2((0,T),H)$ we define

$$
\begin{aligned}
g(t,x) &= \frac{1}{2}\frac{\partial}{\partial x}[(1-x^2)y_x(t,x)] + y(t,x) \\
&= \frac{1}{2}(1-x^2)y_{xx}(t,x) - xy_x(t,x) + y(t,x)
\end{aligned}
\tag{2.5}
$$

for $t \in (0,T)$ and $x \in (0,1)$. Then it follows, for every $t \in (0,T)$, that

$$
\begin{aligned}
\|g(t,\cdot)\|_H \le{}& \frac{1}{2}\Big(\int_0^1 (1-x^2)^2\, y_{xx}(t,x)^2 dx\Big)^{\frac{1}{2}} \\
&+ \Big(\int_0^1 x^2 y_x(t,x)^2 dx\Big)^{\frac{1}{2}} + \Big(\int_0^1 y(t,x)^2 dx\Big)^{\frac{1}{2}} \\
\le{}& \frac{1}{2}\Big(\int_0^1 y_{xx}(t,x)^2 dx\Big)^{\frac{1}{2}} + 2\Big(\int_0^1 y_{xx}(t,x)^2 dx\Big)^{\frac{1}{2}} + 4\Big(\int_0^1 y_{xx}(t,x)^2 dx\Big)^{\frac{1}{2}} \\
={}& \frac{13}{2}\|y(t,\cdot)\|_V.
\end{aligned}
\tag{2.6}
$$

From this estimate and the fact that $y \in C([0,T],V)$ one can easily infer that f defined by (2.4) is in $L^2((0,T),H)$. Therefore every weak solution $y = y(t,x)$ of (1.8), (1.3), (1.4) for some given $u \in L^2(0,T)$ which is in $C([0,T],V) \cap C^1([0,T],H)$ satisfies the following integral differential equation

$$
\begin{aligned}
y(t,x) = \sum_{j=1}^{\infty} \frac{1}{\sqrt{\lambda_j}} \int_0^t &\Big[\Big(\int_0^s u(\sigma)d\sigma\Big)^2 \int_0^1 g(s,\tilde{x})e_j(\tilde{x})d\tilde{x} \\
&- \int_0^1 \tilde{x}\, e_j(\tilde{x})d\tilde{x}\, u(s)\Big] \sin\sqrt{\lambda_j}(t-s)ds\, e_j(x)
\end{aligned}
\tag{2.7}
$$

for $t \in [0,T]$ and $x \in [0,T]$ where $g = g(t,x)$ is defined by (2.5).

Conversely, if $y = y(t,x)$ is in $C([0,T],V) \cap C^1([0,T],H)$ and satisfies (2.7) with $g = g(t,x)$ being defined by (2.5), then $y = y(t,x)$ is a weak solution of (1.8), (1.3), (1.4) in $C([0,T],V) \cap C^1([0,T],H)$.

For every $u \in L^2(0,T)$ we define an affine linear mapping

$$
S(u,\cdot) : C([0,T],V) \cap C^1([0,T],H) \to C([0,T],V) \cap C^1([0,T],H)
$$

by

$$S(u,y)(t,x) = \sum_{j=1}^{\infty} \frac{1}{\sqrt{\lambda_j}} \int_0^t [(\int_0^s u(\sigma)d\sigma)^2 \int_0^1 g(s,\tilde{x},y)e_j(\tilde{x})d\tilde{x} \tag{2.8}$$
$$-u(s)\int_0^1 \tilde{x}\, e_j(\tilde{x})d\tilde{x}] \sin\sqrt{\lambda_j}(t-s)ds\, e_j(x)$$

for $(t,x) \in [0,T] \times [0,1]$ where $g = g(s,\tilde{x},y)$ is defined by (2.5).

Obviously $y \in C([0,T],V) \cap C^1([0,T],H)$ is a solution of (2.7), if and only if y satisfies the fixed point equation

$$y = S(u,y). \tag{2.9}$$

Let y_1, $y_2 \in C([0,T],V) \cap C^1([0,T],H)$ be chosen arbitrarily. Then, for every $t \in [0,T]$, we obtain

$$\begin{aligned}
&\|S(u,y_1)(t,\cdot) - S(u,y_2)(t,\cdot)\|_V^2 \\
&= \sum_{j=1}^{\infty} (\int_0^t [\int_0^s u(\sigma)d\sigma)^2 \int_0^1 g(s,\tilde{x},y_1-y_2)e_j(\tilde{x})d\tilde{x} \sin\sqrt{\lambda_j}(t-s)]ds)^2 \\
&\leq \sum_{j=1}^{\infty} (\int_0^t (\int_0^s u(\sigma)d\sigma)^4 (\int_0^1 g(s,\tilde{x},y_1-y_2)e_j(\tilde{x})d\tilde{x})^2\, ds \cdot T \\
&\leq T^4 \|u\|_H^4 (\operatorname*{ess\ sup}_{t\in[0,T]} \|g(t,\cdot,y_1-y_2)\|_H)^2 \\
&\leq \frac{169}{4} T^4 \|u\|_H^4 (\max_{t\in[0,1]} \|(y_1-y_2)(t,\cdot)\|_V)^2 \qquad \text{see (2.6).}
\end{aligned}$$

Therefore

$$\begin{aligned}
&\max_{t\in[0,T]} \|S(u,y_1)(t,\cdot) - S(u,y_2)(t,\cdot)\|_V \\
&\leq \frac{13}{2} T^2 \|u\|_H^2 \max_{t\in[0,T]} \|(y_1-y_2)(t,\cdot)\|_V
\end{aligned} \tag{2.10}$$

which implies that equation (2.9) has exactly one solution $y = y_u \in C([0,T],V) \cap C^1([0,T],H)$, if there is some constant $\gamma \in [0,1)$ such that

$$\frac{13}{2} T^2 \|u\|_H^2 \leq \gamma^2$$

which is equivalent to

$$\|u\|_H \leq \sqrt{\frac{2}{13}\frac{\gamma}{T}}. \tag{2.11}$$

Under this condition y_u is also the unique weak solution of (1.8), (1.3), (1.4).

3. Iterative Solution of the Problem of Controllability

In view of the considerations of Section 2 the problem of controllability can be equivalently rewritten as follows: Find $u \in L^2(0,T)$ and $y \in Y = C([0,T],V) \cap C^1([0,T],H)$ such that

$$\int_0^T u(t)dt = 0, \quad \int_0^T tu(t)dt = -\varphi_T, \tag{3.1}$$

$$y - S(u,y) = 0, \tag{3.2}$$

$$y(T,\cdot) = y_t(T,\cdot) = 0 \tag{3.3}$$

where

$$\begin{aligned} S(u,y)(t,x) &= \sum_{j=1}^{\infty} \frac{1}{\sqrt{\lambda_j}} \int_0^t [(\int_0^s u(\sigma)d\sigma)^2 \int_0^1 g(s,\tilde{x},y)e_j(\tilde{x})d\tilde{x} \\ &\quad -u(s) \int_0^1 \tilde{x}e_j(\tilde{x})d\tilde{x}] \sin\sqrt{\lambda_j}(t-s)ds\, e_j(x) \end{aligned} \tag{3.4}$$

for $(t,x) \in [0,T] \times [0,1]$ and

$$g(s,\tilde{x},y) = \frac{1}{2}\frac{\partial}{\partial x}[(1-\tilde{x}^2)y_x(s,\tilde{x})] + y(s,\tilde{x}) \tag{3.5}$$

for $(s,\tilde{x}) \in [0,T] \times [0,1]$.

We have shown that for each $u \in H = L^2(0,T)$ with

$$\|u\|_H \leq \sqrt{\frac{2}{13}\frac{\gamma}{T}} \tag{3.6}$$

for some $\gamma \in (0,1)$ there is exactly one solution $y = y_u \in Y$ of (3.2).

If $u \in H$ with (3.1) and (3.6) is given, then the end conditions (3.3) for the corresponding solution $y = y_u \in Y$ of (3.2) are equivalent to the following system of equations

$$\int_0^T [(\int_0^t u(s)ds)^2 \int_0^1 g(t,\tilde{x},y_u)e_j(\tilde{x})d\tilde{x} - u(t)\int_0^1 \tilde{x}e_j(\tilde{x})d\tilde{x}] \sin\sqrt{\lambda_j}(T-t)dt = 0,$$

$$\int_0^T [(\int_0^t u(s)ds)^2 \int_0^1 g(t,\tilde{x},y_u)e_j(\tilde{x})d\tilde{x} - u(t)\int_0^1 \tilde{x}e_j(\tilde{x})d\tilde{x}] \cos\sqrt{\lambda_j}(T-t)dt = 0$$

for $j \in \mathbb{N}$.

On using integration by parts in connection with

$$\int_0^T u(t)dt = 0$$

these turn out to be equivalent to

$$(3.7)\qquad \begin{cases} \int_0^T [\int_0^t \int_0^s u(\sigma)d\sigma \int_0^1 g(s,\tilde{x},y_u)e_j(\tilde{x})d\tilde{x} \sin\sqrt{\lambda_j}(T-s)ds \\ \qquad + \int_0^1 \tilde{x}e_j(\tilde{x})d\tilde{x} \sin\sqrt{\lambda_j}(T-t)]u(t)dt = 0\ , \\ \int_0^T [\int_0^t \int_0^s u(\sigma)d\sigma \int_0^1 g(s,\tilde{x},y_u)e_j(\tilde{x})d\tilde{x} \cos\sqrt{\lambda_j}(T-s)ds \\ \qquad + \int_0^1 \tilde{x}e_j(\tilde{x})d\tilde{x} \cos\sqrt{\lambda_j}(T-t)]u(t)dt = 0 \end{cases}$$

for $j \in \mathbb{N}$.

This leads to the following iterative method for solving the problem of controllability: At first we determine $u_1 \in H$ such that

$$(3.8)\qquad \int_0^T u_1(t)dt = 0, \quad \int_0^T tu_1(t)dt = -\varphi_T$$

and

$$(3.9)\qquad \int_0^T \sin\sqrt{\lambda_j}(T-t)u_1(t)dt = \int_0^T \cos\sqrt{\lambda_j}(T-t)u_1(t)dt = 0.$$

This u_1 is the first function in a sequence $(u_k)_{k\in\mathbb{N}}$ in H which is constructed as follows: If $u_k \in H$ has been determined, then $y_k \in Y$ is determined as the unique solution of the equation

$$(3.10)\qquad y_k - S(u_k, y_k) = 0.$$

Then $u_{k+1} \in H$ is determined such that

$$(3.11)\qquad \int_0^T u_{k+1}(t)dt = 0, \int_0^T tu_{k+1}(t)dt = -\varphi_T,$$

$$
(3.12)\quad \begin{cases} \int_0^T \Big[\int_0^t \int_0^s u_k(\sigma)d\sigma \int_0^1 g(s,\tilde{x},y_k)e_j(\tilde{x})d\tilde{x}\, \sin\sqrt{\lambda_j}(T-s)ds \\ \qquad + \int_0^1 \tilde{x}e_j(\tilde{x})d\tilde{x}\, \sin\sqrt{\lambda_j}(T-t)\Big]u_{k+1}(t)\,dt = 0\,, \\ \int_0^T \Big[\int_0^t \int_0^s u_k(\sigma)d\sigma \int_0^1 g(s,\tilde{x},y_k)e_j(\tilde{x})\,d\tilde{x}\, \cos\sqrt{\lambda_j}(T-s)ds \\ \qquad + \int_0^1 \tilde{x}e_j(\tilde{x})d\tilde{x}\, \cos\sqrt{\lambda_j}(T-t)\Big]u_{k+1}(t)\,dt = 0 \end{cases}
$$

for $j \in \mathbb{N}$.

Obviously we obtain (3.8) and (3.9) for $k = 0$, of we choose $u_0 \equiv 0$ and $y_0 \equiv 0$.

Let us define. for every $j \in \mathbb{N}$ and $k \in \mathbb{N}_0$,

$$
(3.13)\begin{cases} z^1_{jk}(t) = \int_0^t \int_0^s u_k(\sigma)d\sigma \int_0^1 g(s,\tilde{x},y_k)e_j(\tilde{x})d\tilde{x}\, \sin\sqrt{\lambda_j}(T-s)\,ds \\ \qquad\qquad + \int_0^1 \tilde{x}e_j(\tilde{x})d\tilde{x}\, \sin\sqrt{\lambda_j}(T-t)\,, \\ z^2_{jk}(t) = \int_0^t \int_0^s u_k(\sigma)d\sigma \int_0^1 g(s,\tilde{x},y_k)e_j(\tilde{x})d\tilde{x}\, \cos\sqrt{\lambda_j}(T-s)\,ds \\ \qquad\qquad + \int_0^1 \tilde{x}e_j(\tilde{x})d\tilde{x}\, \cos\sqrt{\lambda_j}(T-t) \end{cases}
$$

(with $u_0 \equiv 0$ and $y_0 \equiv 0$).

Then (3.12) can be written in the form

$$
(3.14)\qquad \int_0^T z^1_{jk}(t)u_{k+1}(t)\,dt = \int_0^T z^2_{jk}(t)u_{k+1}(t)\,dt = 0
$$

for $j \in \mathbb{N}$ and $k \in \mathbb{N}$.

Let W_k be the L^2-closure of the span of $z_{0k} \equiv 1$, and z^1_{jk}, z^2_{jk} for $j \in \mathbb{N}$.

Assumption 1: For every $k \in \mathbb{N}$ the function $f(t) = t$, $t \in [0,T]$ does not belong to W_k.

For $k = 0$ this follows from results in [2].

Under this Assumption there exists exactly one $w_k \in W_k$ such that

$$\|w_k - f\|_{L^2(0,t)} \leq \|w - f\|_{L^2(0,T)} \quad \text{for all} \quad w \in W_k$$

which is characterized by

$$\int_0^T (f(t) - w_k(t))\, w(t)\, dt = 0 \quad \text{for all} \quad w \in W$$

or equivalently by

$$\int_0^T (f(t) - w_k(t))\, dt = 0, \quad \int_0^T (f(t) - w_k(t)) z^i_{jk}(t)\, dt = 0$$
$$\text{for} \quad i = 1, 2 \quad \text{and} \quad j \in \mathbb{N}.$$

Moreover, it follows that

$$\int_0^T (f(t) - w_k(t)) f(t)\, dt = \|f - w_k\|^2_{L^2(0,T)} > 0.$$

If we put

$$u_{k+1} = \frac{\varphi_T}{\|w_k - f\|^2_{L^2(0,T)}}(w_k - f), \tag{3.15}$$

then $u_{k+1} \in H$ and satisfies (3.11) and (3.12).

Assertion: Among all solutions of (3.11) and (3.12) the one given by (3.15) has the smallest possible norm.

Proof: As a result of a well known duality theorem in linear approximation theory it follows that

$$\|f - w_k\|_{L^2(0,T)} = \max\{|\int_0^T f(t)u(t)dt|\, |u \in L^2(0,T), \quad \int_0^T u(t)w(t)dt = 0$$
$$\text{for all} \quad w \in W_k \quad \text{and} \quad \|u\|_{L^2(0,T)} = 1\}.$$

Let $u \in L^2(0,T)$ be any solution of (3.11), (3.12). Then

$$\|u\|_{L^2(0,T)} > 0 \quad \text{and} \quad \int_0^T u(t)w(t)\, dt \quad \text{for all} \quad w \in W_k.$$

Therefore

$$|\int_0^T f(t)\frac{u(t)}{\|u\|_{L^2(0,T}}dt| = \frac{|\varphi_T|}{\|u\|_{L^2(0,T)}} \le \|f - w_k\|_{L^2(0,T)} = \frac{|\varphi_T|}{\|u_{k+1}\|_{L^2(0,T)}}$$

which implies $\|u_{k+1}\|_{L^2(0,T)} \le \|u\|_{L^2(0,T)}$ and completes the proof.

Assumption 2: For every $k \in \mathbb{N}_0$ we have

$$\|u_{k+1}\|_H \le \sqrt{\frac{2}{13}\frac{\gamma}{T}} \tag{3.16}$$

for some $\gamma \in (0,1)$ where u_{k+1} is given by (3.15).

As an immediate consequence of this assumption we infer the existence of a subsequence $(u_{k_\imath})_{\imath\in\mathbb{N}}$ of $(u_k)_{k\in\mathbb{N}}$ which converges weakly to some $\hat{u} \in H$. Since the function $u \to \|u\|_{L^2(0,T)}$ is weakly sequentially lower semi-continuous, it follows that

$$\|\hat{u}\|_H \le \liminf_{\imath\to\infty} \|u_{k_\imath}\|_H \le \sqrt{\frac{2}{13}\frac{\gamma}{T}}.$$

Further it follows from (3.11) that

$$\int_0^T \hat{u}(t)dt = 0 \quad \text{and} \quad \int_0^T t\hat{u}(t)dt = -\varphi_T.$$

Let $y_{\hat{u}} \in Y$ be the corresponding solution of (3.2). Then, for every $k \in \mathbb{N}$ we get

$$y_{\hat{u}} - y_{u_k} = S(\hat{u}, y_{\hat{u}}) - S(u_k, y_{\hat{u}}) + S(u_k, y_{\hat{u})} - S(u_k, y_{u_k})$$

which implies

$$\begin{aligned} \max_{t\in[0,T]} \|(y_{\hat{u}} - y_{u_k})(t,\cdot)\|_V &\le \max_{t\in[0,T]} \|(S(\hat{u}, y_{\hat{u}}) - S(u_k, y_{\hat{u}})(t,\cdot)\|_V \\ &\quad + \max_{t\in[0,t]} \|(S(u_k, y_{\hat{u}}) - S(u_k, y_{u_k})(t,\cdot)\|_V \\ \le \max_{t\in[0,T]} \|(S(\hat{u}, y_{\hat{u}}) &- S(u_k, y_{\hat{u}})(t,\cdot))\|_V + \gamma^2 \max_{t\in[0,T]} \|(y_{\hat{u}} - y_{u_k})(t,\cdot)\|_V. \end{aligned}$$

Hence we obtain the estimate

$$\max_{t\in[0,T]} \|(y_{\hat{u}} - y_{u_k})(t,\cdot)\|_V \le \frac{1}{1-\gamma^2} \max_{t\in[0,T]} \|(S(\hat{u}, y_{\hat{u}}) - S(u_k, y_{\hat{u}}))(t,\cdot)\|_V. \tag{3.17}$$

Similiar to the estimate (2.10) one can show that

$$\begin{aligned}
&\max_{t\in[0,T]} \|(S(\hat{u},y_{\hat{u}}) - S(u_k,y_{\hat{u}}))(t,\cdot)\|_V \\
&\le (\frac{13}{2}T^2\|\hat{u}+u_k\|_H \max_{t\in[0,T]}\|y_{\hat{u}}(t,\cdot)\|_V + \sqrt{\frac{T}{3}})\|\hat{u}-u_k\|_H \qquad (3.18)\\
&\le (\sqrt{26}T \max_{t\in[0,T]}\|y_{\hat{u}}(t,\cdot)\|_V + \sqrt{\frac{T}{3}})\|\hat{u}-u_k\|_H .
\end{aligned}$$

For every $k \in \mathbb{N}$ we now define

$$\begin{aligned}
\tilde{y}_k(t,x) &= \sum_{j=1}^{\infty} \frac{1}{\sqrt{\lambda_j}} \int_0^t [\int_0^s u_{k+1}(\sigma)d\sigma \int_0^s u_k(\sigma)d\sigma \int_0^1 g(s,\tilde{x},y_k)e_j(\tilde{x})d\tilde{x} \\
&\quad - \int_0^1 \tilde{x}e_j(\tilde{x})d\tilde{x}\, u_{k+1}(s)] \sin\sqrt{\lambda_j}(t-s)ds\, e_j(x)
\end{aligned}$$

for $(t,x) \in [0,T]\times[0,1]$.

Then it follows from (3.12) that

$$\tilde{y}_k(T,\cdot) = (\tilde{y}_k)_t(T,\cdot) = 0 \quad \text{a.e.} \qquad (3.19)$$

for all $k \in \mathbb{N}$.

Further we obtain

$$\|(\tilde{y}_k - y_{u_k})(T,\cdot)\|_V \le \frac{13}{2}T^2\,\|u_k\|_H \max_{t\in[0,T]}\|y_{u_k}(t,\cdot)\|_V\,\|u_{k+1}-u_k\|_H \qquad (3.20)$$

and

$$\|(\tilde{y}_k - y_{u_k})(T,\cdot)\|_H \le \frac{13}{2}T^2\,\|u_k\|_H \max_{t\in[0,T]}\|y_{u_k}(t,\cdot)\|_V\,\|u_{k+1}-u_k\|_H .$$

Assumption 3: $\hat{u} = \lim_{k\to\infty} u_k$. Then (3.17) and (3.18) imply

$$\lim_{k\to\infty}\max_{t\in[0,T]}\|(y_{\hat{u}} - y_{u_k})(t,\cdot)\|_V = 0$$

and on using (3.19) and (3.20) we infer

$$y_{\hat{u}}(T,\cdot) = (y_{\hat{u}})_T(T,\cdot) = 0 \quad \text{a.e.}$$

Hence $\hat{u}$ and $\hat{y}$ solve the problem of controllability.

References

1 W Krabs On Time-Optimal Boundary Control of Vibrating Beams In Control Theory for Distributed Parameter Systems and Applications, edited by F Kappel, K Kunisch, W Schappacher Lecture Notes in Control and Information Sciences No 54, Springer-Verlag Berlin – Heidelberg – New York – Tokyo 1983, pp 127 – 137

2 W Krabs On Moment Theory and Controllability of One-Dimensional Vibrating Systems and Heating Processes Lecture Notes in Control and Information Sciences No 173, Springer-Verlag Berlin 1992

3 W Krabs Controllability of a Rotating Beam In Analysis and Optimization of Systems State and Frequency Domain Approaches for Infinite-Dimensional Systems, edited by R F Curtain Lecture Notes in Control and Information Sciences No 185, Springer-Verlag Berlin 1993, pp 447 – 458

4 G Leugering Control and Stabilization of a Flexible Robotarm Dynamics and Stability of Systems 5 (1990), 37 – 46

5 G Leugering On Control and Stabilization of a Rotating Beam by Applying Moments at the Base Only In Optimal Control of Partial Differential Equations, edited by K -H Hoffmann and W Krabs Lecture Notes in Control and Information Sciences No 149, Springer-Verlag Berlin 1991, pp 182 – 191

6 Y Sakawa, R Ito and N Fujii Optimal Control of Rotation of a Flexible Arm In Control Theory for Distributed Parameter Systems and Applications, edited by F Kappel, K Kunisch, W Schappacher Lecture Notes in Control and Information Sciences No 54, Springer-Verlag Berlin 1983, pp 175 – 187

W Krabs
Fachbereich Mathematik
Technische Hochschule Darmstadt
Schlossgartenstrasse 7
D-64289 Darmstadt, West Germany

MODELING AND CONTROLLABILITY OF INTERCONNECTED ELASTIC MEMBRANES

J E LAGNESE

Department of Mathematics
Georgetown Univesity

ABSTRACT A collection of systems of partial differential equations, each of which models the nonlinear deformation of an elastic membrane, is considered The state variables of the various membranes are required to satisfy certain "geometric" and "dynamic" coupling conditions which arise from continuity and balance law considerations The resulting coupled systems then describe the nonlinear deformations of a system of interconnected membranes The question of exact controllability of the linearization of such a network is discussed, where the controls are in the forms of forces applied on the outer edges and in the junction regions (where membranes are connected to one another)

1991 *Mathematics Subject Classification* 35L20, 93B05, 93C20

Key words and phrases Elastic membranes, controllability, junction conditions

1. Modeling of Dynamic Nonlinear Elastic Membranes

Let Ω be a bounded, open, connected set in $\mathbb{R}^2$ with Lipschitz continuous boundary Γ, and let $\mathbf{a}_1, \mathbf{a}_2, \mathbf{a}_3$ be a fixed, right-handed orthonormal basis for $\mathbb{R}^3$. We denote by (x_1, x_2, x_3) (resp., (x_1, x_2) the coordinates of a vector $\mathbf{x} \in \mathbb{R}^3$ (resp., $\mathbf{x} \in \mathbb{R}^2$) with respect to the natural basis of $\mathbb{R}^3$ (resp., $\mathbb{R}^2$). Consider an infinite elastic cylinder in $\mathbb{R}^3$ whose reference configuration is

$$\mathcal{B} = \{\mathbf{p}_0 + x_i \mathbf{a}_i | (x_1, x_2) \in \Omega, \ -\infty < x_3 < \infty\},$$

where $\mathbf{p}_0$ is a fixed vector in $\mathbb{R}^3$, Roman indices take the values 1,2,3, Greek indices take the values 1,2, and summation convention is assumed. We set

$$\mathbf{r}(x_1, x_2, x_3) = \mathbf{p}_0 + x_i \mathbf{a}_i.$$

It is assumed that the motion of the cylinder obeys the following hypothesis.

Research supported by the Air Force Office of Sponsored Research through grant F49620-92-0031

Basic kinematic hypothesis. The position vector at time t to the displaced location of the particle situated at $\mathbf{r}(x_1, x_2, x_3)$ in the reference configuration is

$$\mathbf{R}(x_1, x_2, x_3, t) = \mathbf{r}(x_1, x_2, x_3) + \mathbf{W}(x_1, x_2, t). \tag{1.1}$$

Hypothesis (1.1) means that the deformation of a cross-section orthogonal to the axis of the cylinder is independent of the variable x_3 along the axis of the cylinder. Thus the deformation of the cylinder is completely determined by the deformation of any of its cross-sections. The latter is specified by the vector function $\mathbf{W}$ which measures the *displacement* of a cross-section. Let $\{W_i\}$ denote the components of $\mathbf{W}$ in the $\{\mathbf{a}_i\}$ basis: $\mathbf{W} = W_i\mathbf{a}_i$. We refer to a cross-section as an *elastic membrane* and to the planar region

$$\mathcal{P} = \{\mathbf{p}_0 + x_\alpha \mathbf{a}_\alpha : (x_1, x_2) \in \Omega\}$$

as the *reference cross-section.*

Set $\mathbf{G}_i = \mathbf{R}_{,i}$, where a subscript after a comma means differentiation with respect to the indicated spatial variable. The components ε_{ij} of the *strain tensor* are given by

$$\varepsilon_{ij} = \frac{1}{2}(\mathbf{G}_i \cdot \mathbf{G}_j - \delta_{ij}), \tag{1.2}$$

It follows from (1.1) and (1.2) that

$$\begin{cases} \varepsilon_{\alpha\beta} = \dfrac{1}{2}(W_{\alpha,\beta} + W_{\beta,\alpha} + \mathbf{W}_{,\alpha} \cdot \mathbf{W}_{,\beta}), \\ \varepsilon_{\alpha 3} = \dfrac{1}{2} W_{3,\alpha}, \\ \varepsilon_{33} = 0. \end{cases} \tag{1.3}$$

1.1. Equations of Motion. Let s^{ij} denote the components of the *second Piola-Kirchhoff stress tensor.* This symmetric tensor relates the *stress vectors* $\mathbf{t}^i$ to the quantities $\mathbf{G}_i$ through

$$\mathbf{t}^i = s^{ij}\mathbf{G}_j \tag{1.4}$$

(see [8, Chapter 3]). If $\mathbf{F}$ denotes the body force per unit of undeformed volume, the three-dimensional conditions for the balance of forces is

$$\mathbf{t}^i_{,i} + \mathbf{F} = \rho \mathbf{R}_{,tt}, \tag{1.5}$$

where ρ is the mass of the body per unit of reference volume. Use of (1.1) and (1.4) in (1.5) allows that equation to be written

$$s^{ij}_{,i}\mathbf{a}_j + s^{i\beta}_{,i}\mathbf{W}_{,\beta} + s^{\alpha\beta}\mathbf{W}_{,\alpha\beta} + \mathbf{F} = \rho \mathbf{W}_{,tt}. \tag{1.6}$$

In order to be consistent with our basic kinematic hypothesis, it is assumed that $\mathbf{F}$ and ρ are independent of x_3. Let us further assume that the material in question is *linearly elastic* (hookean), i.e.,

$$s^{ij} = C^{ijkl}\varepsilon_{kl}, \tag{1.7}$$

where the coefficients of elasticity C^{ijkl} *depend only on* x_1 *and* x_2. The symmetry of s^{ij} and ε_{kl} requires that

$$C^{ijkl} = C^{jikl} = C^{ijlk}, \tag{1.8}$$

and we further *assume* that

$$C^{ijkl} = C^{klij} \tag{1.9}$$

Then, in particular, the stresses s^{ij} do not depend on x_3 and (1.6) reduces to

$$s^{\alpha j}_{,\alpha}\mathbf{a}_j + (s^{\alpha\beta}\mathbf{W}_{,\beta})_{,\alpha} + \mathbf{F} = \rho\mathbf{W}_{,tt}. \tag{1.10}$$

This system comprises three scalar equations for the scalar components W_i of $\mathbf{W}$.

1.2. Edge Conditions. Let Λ denote the two-dimensional boundary of the reference cross-section $\mathcal{P}$ and assume that

$$\Lambda = \overline{\Lambda}^C \cup \overline{\Lambda}^D, \tag{1.11}$$

where Λ^C, Λ^D are mutually disjoint, relatively open subsets of Λ and $\Lambda^D \neq \emptyset$. (The superscripts C and D are used to indicate "clamped" and "dynamic," respectively.) The partition (1.11) induces a similar partition of $\Gamma := \partial\Omega$ into

$$\Gamma = \overline{\Gamma}^C \cup \overline{\Gamma}^D \tag{1.12}$$

through the mapping

$$\Gamma \mapsto \Lambda : (x_1, x_2) \mapsto \mathbf{p}_0 + x_\alpha\mathbf{a}_\alpha.$$

Let (ν_1, ν_2) denote a unit exterior normal vector and $(-\nu_2, \nu_1)$ a unit tangent vector at a point (x_1, x_2) of Γ. Since the mapping $(x_1, x_2, x_3) \mapsto x_i\mathbf{a}_i$ is orthogonal with determinant unity, $\boldsymbol{\nu} := \nu_1\mathbf{a}_1 + \nu_2\mathbf{a}_2$ and $\boldsymbol{\tau} := -\nu_2\mathbf{a}_1 + \nu_1\mathbf{a}_2$ are vectors which are normal and tangent, respectively, to Λ at the point $\mathbf{p}_0 + x_\alpha\mathbf{a}_\alpha$ with $\boldsymbol{\nu}$ pointing out of $\mathcal{P}$ and $\boldsymbol{\tau}$ positively oriented with respect to $\mathbf{a}_3$, i.e., $\boldsymbol{\tau} \times \mathbf{a}_3 = \boldsymbol{\nu}$.

The edge conditions along the part of $\partial\mathcal{B}$ corresponding to Λ^C are described by

$$\mathbf{W} = 0 \quad \text{on } \Gamma^C,\ t > 0. \tag{1.13}$$

Let $\mathbf{f}$ denote a given distribution of forces along $\Lambda^D + x_3\mathbf{a}_3$, $-\infty < x_3 < +\infty$ which is *independent of* x_3. The condition for the balance of forces along this part of $\partial\mathcal{B}$ is

$$\nu_\alpha\mathbf{t}^\alpha = \mathbf{f}$$

which, in view of our previous hypotheses, may be written

$$\nu_\alpha s^{\alpha j}\mathbf{a}_j + \nu_\alpha s^{\alpha\beta}\mathbf{W}_{,\beta} = \mathbf{f} \quad \text{on } \Gamma^D,\ t > 0. \tag{1.14}$$

1.3. Hamilton's Principle. Let $\mathbf{W}$ satisfy (1.10) and (1.14), and let $\hat{\mathbf{W}} = \hat{W}_j\mathbf{a}_j$ be a smooth function defined on $\overline{\Omega} \times [0,T]$ such that $\hat{\mathbf{W}}|_{t=0} = \hat{\mathbf{W}}|_{t=T} = 0$, $\hat{\mathbf{W}}|_{\Gamma^C\times(0,T)} = 0$. We form the scalar produce in $\mathbb{R}^3$ of (1.10) with $\hat{\mathbf{W}}$ and integrate over $\Omega \times (0,T)$. After some integrations by parts we obtain

$$\begin{aligned}\int_0^T\int_\Omega (\rho\mathbf{W}_{,t}\cdot\hat{\mathbf{W}}_{,t} - s^{\alpha j}\hat{W}_{j,\alpha} - s^{\alpha\beta}\mathbf{W}_{,\beta}\cdot\hat{\mathbf{W}}_{,\alpha})\,d\Omega \\ + \int_0^T\int_\Omega \mathbf{F}\cdot\hat{\mathbf{W}}\,d\Omega + \int_0^T\int_{\Gamma^D}\mathbf{f}\cdot\hat{\mathbf{W}}\,d\Gamma = 0.\end{aligned} \tag{1.15}$$

Conversely, if $\mathbf{W}$ satisfies (1.15) for all such test functions $\hat{\mathbf{W}}$, then $\mathbf{W}$ satisfies (in an appropriate sense) (1.10) and (1.14). The variational equation (1.15) is known as the *principle of virtual work* for the membrane problem under consideration.

We now recast (1.15) into a form known as *Hamilton's principle.* Let δ denote the first Fréchet derivative with respect to $\mathbf{W}$. We may write

$$s^{\alpha j}\hat{W}_{j,\alpha} + s^{\alpha\beta}\mathbf{W}_{,\beta}\cdot\hat{\mathbf{W}}_{,\alpha} = s^{ij}\langle\delta\varepsilon_{ij},\hat{\mathbf{W}}\rangle, \tag{1.16}$$

where $\langle\delta\varepsilon_{ij},\hat{\mathbf{W}}\rangle$ denotes the value of $\delta\varepsilon_{ij}$ at $\hat{\mathbf{W}}$. For any material obeying (1.7)–(1.9) one has

$$s^{ij}\delta\varepsilon_{ij} = C^{ijkl}\varepsilon_{kl}\delta\varepsilon_{ij} = \frac{1}{2}\delta(C^{ijkl}\varepsilon_{kl}\varepsilon_{ij}) = \frac{1}{2}\delta(s^{ij}\varepsilon_{ij}). \tag{1.17}$$

Under assumption (1.1),

$$s^{ij}\varepsilon_{ij} = s^{\alpha\beta}\varepsilon_{\alpha\beta} + 2s^{\alpha 3}\varepsilon_{\alpha 3}. \tag{1.18}$$

It then follows from (1.16)–(1.18) that (1.15) may be written

$$\delta\int_0^T [\mathcal{K}(t) - \mathcal{U}(t) + \mathcal{W}(t)]\,dt = 0, \tag{1.19}$$

where

$$\mathcal{K}(t) = \frac{1}{2}\int_\Omega \rho|\mathbf{W}_{,t}|^2 d\Omega \tag{1.20}$$

is the *kinetic energy* of the deformation,

$$\begin{aligned}\mathcal{U}(t) &= \frac{1}{2}\int_\Omega (s^{\alpha\beta}\varepsilon_{\alpha\beta} + 2s^{\alpha 3}\varepsilon_{\alpha 3})\,d\Omega \\ &= \frac{1}{2}\int_\Omega (s^{\alpha j}W_{j,\alpha} + \frac{1}{2}s^{\alpha\beta}\mathbf{W}_{,\alpha}\cdot\mathbf{W}_{,\beta})\,d\Omega\end{aligned} \tag{1.21}$$

is the *strain energy*, and

$$\mathcal{W}(t) = \int_\Omega \mathbf{F} \cdot \mathbf{W} \, d\Omega + \int_{\Gamma^D} \mathbf{f} \cdot \mathbf{W} \, d\Gamma \tag{1.22}$$

is the *work of external forces.* The variational formulation (1.19) with $\mathcal{K}, \mathcal{U}$ and $\mathcal{W}$ so defined, is Hamilton's principle for the membrane problem under consideration.

2. Systems of Interconnected Elastic Membranes

For $i = 1, \ldots, n$, let $\mathbf{a}_1^i, \mathbf{a}_2^i, \mathbf{a}_3^i$ be a right-handed orthonormal system in $\mathbb{R}^3$, let $\mathbf{p}_0^i$ be a fixed vector in $\mathbb{R}^3$, and Ω_i be a bounded, open, connected region in $\mathbb{R}^2$ with Lipschitz boundary consisting of a finite number of smooth simple curves. We set

$$\mathbf{p}_i(x_1, x_2) = \mathbf{p}_0^i + \sum_\alpha x_\alpha \mathbf{a}_\alpha^i, \tag{2.1}$$

$$\mathbf{r}_i(x_1, x_2, x_3) = \mathbf{p}_i(x_1, x_2) + x_3 \mathbf{a}_3^i,$$
$$\mathcal{P}_i = \{\mathbf{p}_i(x_1, x_2) | \, (x_1, x_2) \in \Omega_i\}.$$

The mapping $(x_1, x_2) \mapsto \mathbf{p}_i(x_1, x_2)$ is a homeomorphism of $\overline{\Omega}_i$ onto $\overline{\mathcal{P}}_i$ and of $\partial\Omega_i$ onto $\partial\mathcal{P}_i$, with

$$\mathbf{p}_i^{-1}(\mathbf{p}) = ((\mathbf{p} - \mathbf{p}_0^i) \cdot \mathbf{a}_1^i, (\mathbf{p} - \mathbf{p}_0^i) \cdot \mathbf{a}_2^i).$$

Remark 2.1. As in the last section, Greek indices take values 1,2, but, unlike before, we *do not* assume summation convention and Roman indices are not necessarily restricted to set $\{1, 2, 3\}$. The range of such an index will be explicitly indicated whenever it is not clear fom context. We do, however, make the convention that

$$\sum_j := \sum_{j=1}^3.$$

The set $\mathcal{P}_i$ is considered as a reference cross-section of an infinite cylinder whose reference configuration is

$$\mathcal{B}_i = \{\mathbf{r}_i(x_1, x_2, x_3) \, | \, (x_1, x_2) \in \Omega_i, \ |x_3| < \infty\}.$$

We make the following hypotheses regarding the membranes $\mathcal{P}_i$.

$$\overline{\mathcal{P}}_i \cap \mathcal{P}_k = \emptyset, \ \ \forall i \neq k;$$
$$\cup_{i=1}^n \overline{\mathcal{P}}_i \ \text{ is a connected set in } \mathbb{R}^3;$$
$$\overline{\mathcal{P}}_i \cap \overline{\mathcal{P}}_k \ \text{ is either empty or is a finite union of components, } \forall i \neq k.$$

Each of the bodies $\mathcal{B}_i$ is assumed to satisfy the basic kinematic assumption (1.1): the position vector $\mathbf{R}^i(x_1, x_2, x_3)$ to the displaced particle originally at $\mathbf{r}_i(x_1, x_2, x_3)$ is given by

$$\mathbf{R}^i(x_1, x_2, x_3, t) = \mathbf{r}^i(x_1, x_2, x_3) + \mathbf{W}^i(x_1, x_2, t), \qquad (x_1, x_2) \in \Omega_i,\ |x_3| < \infty,\ 1 = 1, \dots, n. \tag{2.2}$$

2.1. Geometric Junction Conditions. We tentatively define a *joint*, or *junction*, of the multi-membrane system as a component (a maximal line segment; by definition, line segments are connected) which is a common edge of two or more reference cross-sections, that is, a maximal line segment where a $\overline{\mathcal{P}}_i$ and $\overline{\mathcal{P}}_k$ intersect for some $k > i$. However, this definition allows the possibility that the set of all such components does not form a mutually disjoint collection. If C_1 and C_2 are components such that $C_1 \cap C_2 \neq \emptyset$, we replace C_1 and C_2 by $C_1 \cap C_2$, $C_1/(C_1 \cap C_2)$ and $C_2/(C_1 \cap C_2)$, whose union is $C_1 \cup C_2$. Each set $C_\alpha/(C_1 \cap C_2)$ is either a single line segment or the union of disjoint segments. In the latter case we consider $C_\alpha/(C_1 \cap C_2)$ to consist of several separate components. In this manner one may eliminate the overlapping of components. Finally, we define the joints of the system to be the collection of disjoint, open line segments which are the *interiors* of the collection of non-overlapping components.

For any joint J of the multi-membrane system define

$$\mathcal{I}(J) = \{i \in [1, \dots, n] | \, J \subset \partial \mathcal{P}_i\}.$$

Thus $\mathcal{I}(J)$ is the index set of those $\mathcal{P}_i$ which share J. The deformation of the overall system is constrained by the following *geometric junction condition* at each joint J:

$$\mathbf{W}^i(\mathbf{p}_i^{-1}(\mathbf{p}), t) = \mathbf{W}^j(\mathbf{p}_j^{-1}(\mathbf{p}), t), \quad \mathbf{p} \in J,\ i, j \in \mathcal{I}(J). \tag{2.3}$$

Condition (2.3) simply requires that each membrane sharing J have the same displacement there.

2.2. Dynamic Conditions. The geometric joint conditions reflect the continuity of the overall structure. In addition, there are *dynamic junction conditions* which represent the balance of linear momentum at the joints. The simplest way to obtain these is to minimize the Hamiltonian associated with the structure. In carrying out the calculation, the variation must be taken with respect to displacements which satisfy the geometric boundary conditions and the geometric joint conditions. Of course, in addition to the dynamic junction conditions Hamilton's principle also yields the equations of motion and dynamic boundary conditions of the structure, as usual.

Set $\Lambda_i = \partial \mathcal{P}_i$. We assume that

$$\Lambda_i = \overline{\Lambda}_i^C \cup \overline{\Lambda}_i^D \cup \overline{\Lambda}_i^J, \tag{2.4}$$

where Λ_i^C, Λ_i^D, Λ_i^J are disjoint and relatively open in Λ_i. This partition of Λ_i induces a similar partition of $\Gamma_i := \partial\Omega_i$ into

$$\Gamma_i = \overline{\Gamma}_i^C \cup \overline{\Gamma}_i^D \cup \overline{\Gamma}_i^J, \tag{2.5}$$

where

$$\Lambda_i^C = \mathbf{p}_i(\Gamma_i^C),\ \ \Lambda_i^D = \mathbf{p}_i(\Gamma_i^D),\ \ \Lambda_i^J = \mathbf{p}_i(\Gamma_i^J).$$

The set Λ_i^C represents the part of Λ_i that is clamped, i.e.,

$$\mathbf{W}^i = 0 \ \text{ on } \Gamma_i^C, \tag{2.6}$$

Λ_i^D is the part of Λ_i along which "dynamic" boundary conditions will hold, and Λ_i^J is that part of Λ_i such that

$$\Lambda_i^J \cap \Lambda_k \neq \emptyset \ \text{ for at least one } k \neq i.$$

Thus Λ_i^J consists of those joints of the multi-membrane system which belong to Λ_i. We label the entire collection of joints by $J_1, J_2, \dots, J_m$.

The total kinetic energy of the system is defined to be

$$\mathcal{K}(t) = \frac{1}{2}\sum_{i=1}^{n}\int_{\Omega_i} \rho_i |\mathbf{W}^i_{,t}|^2 d\Omega,$$

where $\rho_i(x_1, x_2)$ is the mass density per unit of reference area of $\mathcal{P}_i$. Let s_i^{jk} denote the stress tensor associated with the ith membrane. The total strain energy is defined to be

$$\mathcal{U}(t) = \frac{1}{2}\sum_{i=1}^{n}\int_{\Omega_i} \Big(\sum_{\alpha,j} s_i^{\alpha j} W^i_{j,\alpha} + \frac{1}{2}\sum_{\alpha,\beta} s_i^{\alpha\beta}\mathbf{W}^i_{,\alpha}\cdot\mathbf{W}^i_{,\beta}\Big)\, d\Omega.$$

Let $\mathbf{F}^i$ denote an external body force acting on the ith membrane $\mathcal{P}_i$, $\mathbf{f}^i$ be a force acting along Λ_i^D and $\tilde{\mathbf{f}}^k$ be an applied force along the junction region J_k. The work of these external forces is defined to be

$$\begin{aligned}\mathcal{W}(t) &= \sum_{i=1}^{n}\Big\{\int_{\mathcal{P}_i} \mathbf{F}^i\cdot\mathbf{W}^i\, d\Lambda_i + \int_{\Lambda_i^D} \mathbf{f}^i\cdot\mathbf{W}^i\, d\Lambda\Big\} + \sum_{k=1}^{m}\int_{J_k} \tilde{\mathbf{f}}^k\cdot\mathbf{W}^{i_k}\, dJ_k\\ &= \sum_{i=1}^{n}\Big\{\int_{\Omega_i} \mathbf{F}^i\cdot\mathbf{W}^i\, d\Omega_i + \int_{\Gamma_i^D} \mathbf{f}^i\cdot\mathbf{W}^i\, d\Gamma\Big\} + \sum_{k=1}^{m}\int_{\Gamma_{i_k}(J_k)} \tilde{\mathbf{f}}^k\cdot\mathbf{W}^{i_k}\, d\Gamma,\end{aligned}$$

where

$$\Gamma_i(J_k) = \{(x_1, x_2) \in \Gamma_i^J \,|\, \mathbf{p}_i(x_1, x_2) \in J_k\} = \{\mathbf{p}_i^{-1}(\mathbf{p}) \,|\, \mathbf{p} \in J_k\},$$
$$i = 1, \dots, n;\ k = 1, \dots, m,$$

and where we may take

$$i_k = \min\{i \,|\, i \in \mathcal{I}(J_k)\}$$

in accordance with the geometric junction conditions (2.3). In the integrals over $\mathcal{P}_i$ and Λ_i^D, the integrands are evaluated at $\mathbf{p}_i^{-1}(\mathbf{p})$ with $\mathbf{p} \in \mathcal{P}_i$ and $\mathbf{p} \in \Lambda_i^D$, respectively, while the integrand in the integral over J_k is evaluated at $\mathbf{p}_{i_k}^{-1}(\mathbf{p})$, $\mathbf{p} \in J_k$. The dynamic equations of the multi-membrane system are *postulated* to be those associated with the Hamilton's principle

$$\delta \int_0^T [\mathcal{K}(t) - \mathcal{U}(t) + \mathcal{W}(t)]\, dt = 0, \tag{2.7}$$

in which the variation δ is taken with respect to displacements $(\mathbf{W}^1, \ldots, \mathbf{W}^n)$ which respect the geometric boundary conditions (2.6) and geometric junction conditions (2.3) along each joint $J_1, \ldots, J_m$. It is fairly obvious that we will obtain, in particular, the equations of motion (1.10) for each membrane of the system and the dynamic boundary conditions (1.14) along Γ_i^D. Thus we have for each $i = 1, \ldots, n$ the equations of motion

$$\sum_{\alpha,j} s_{i,\alpha}^{\alpha j} \mathbf{a}_j^i + \sum_{\alpha,\beta} (s_i^{\alpha\beta} \mathbf{W}_{,\beta}^i)_{,\alpha} + \mathbf{F}^i = \rho_i \mathbf{W}_{,tt}^i, \quad (x_1, x_2) \in \Omega_i, \quad t > 0, \tag{2.8}$$

and the boundary conditions

$$\sum_{\alpha,j} \nu_\alpha^i s_i^{\alpha j} \mathbf{a}_j^i + \sum_{\alpha,\beta} \nu_\alpha^i s_i^{\alpha\beta} \mathbf{W}_{,\beta}^i = \mathbf{f}^i \quad \text{on } \Gamma_i^D, \quad t > 0. \tag{2.9}$$

There are, in addition, junction conditions given by the following variational equation:

$$\sum_{i=1}^n \int_{\Gamma_i^J} \Big(\sum_{\alpha,j} \nu_\alpha^i s_i^{\alpha j} \mathbf{a}_j^i + \sum_{\alpha,\beta} \nu_\alpha^i s_i^{\alpha\beta} \mathbf{W}_{,\beta}^i \Big) \cdot \hat{\mathbf{W}}^i \, d\Gamma = \sum_{k=1}^m \int_{J_k} \tilde{\mathbf{f}}^k \cdot \hat{\mathbf{W}}^{i_k} \, dJ_k \tag{2.10}$$

for all sufficiently smooth displacements $\hat{\mathbf{W}}^i$, $i = 1, \ldots, n$, which satisfy the geometric boundary and junction conditions.

We have

$$\sum_{i=1}^n \int_{\Gamma_i^J} = \sum_{k=1}^m \sum_{i \in \mathcal{I}(J_k)} \int_{\Gamma_i(J_k)}.$$

The segment J_k may be parameterized as

$$J_k = \{\mathbf{q}_k^0 + s\boldsymbol{\tau}^{i_k} \,|\, a_k \le s \le b_k\},$$

where $\mathbf{q}_k^0 \in J_k$ is fixed but arbitrary. Since $J_k = \mathbf{p}_i(\Gamma_i(J_k))$, this parameterization of J_k induces a parameterization of the segment $\Gamma_i(J_k)$ given by given by

$$\Gamma_i(J_k) = \{\mathbf{x}_{ik}^0 + s(\tau_1, \tau_2) | \, a_k \le s \le b_k\},$$

where $\boldsymbol{\tau}^{ik} = \tau_1 \mathbf{a}_1^{ik} + \tau_2 \mathbf{a}_2^{ik}$ and $\mathbf{x}_{ik}^0$ is some point of $\Gamma_i(J_k)$ depending on both i and k. Therefore

$$\int_{\Gamma_i(J_k)} \left(\sum_{\alpha,j} \nu_\alpha^i s_i^{\alpha j} \mathbf{a}_j^i + \sum_{\alpha,\beta} \nu_\alpha^i s_i^{\alpha\beta} \mathbf{W}^i_{,\beta} \right) \cdot \hat{\mathbf{W}}^i(\mathbf{x})\, d\Gamma$$

$$= \int_{a_k}^{b_k} \left(\sum_{\alpha,j} \nu_\alpha^i s_i^{\alpha j} \mathbf{a}_j^i + \sum_{\alpha,\beta} \nu_\alpha^i s_i^{\alpha\beta} \mathbf{W}^i_{,\beta} \right) \cdot \hat{\mathbf{W}}^i(\mathbf{x})\, ds$$

$$(\mathbf{x} = \mathbf{x}_{ik}^0 + s(\tau_1, \tau_2))$$

$$= \int_{a_k}^{b_k} \left(\sum_{\alpha,j} \nu_\alpha^i s_i^{\alpha j} \mathbf{a}_j^i + \sum_{\alpha,\beta} \nu_\alpha^i s_i^{\alpha\beta} \mathbf{W}^i_{,\beta} \right) \cdot \hat{\mathbf{W}}^i(\mathbf{p}_i^{-1}(\mathbf{p}))\, ds$$

$$(\mathbf{p} = \mathbf{q}_k^0 + s\boldsymbol{\tau}^{ik})$$

$$= \int_{J_k} \left(\sum_{\alpha,j} \nu_\alpha^i s_i^{\alpha j} \mathbf{a}_j^i + \sum_{\alpha,\beta} \nu_\alpha^i s_i^{\alpha\beta} \mathbf{W}^i_{,\beta} \right) \cdot \hat{\mathbf{W}}^i(\mathbf{p}_i^{-1}(\mathbf{p}))\, dJ_k.$$

Thus (2.10) may be expressed as

$$\sum_{k=1}^{m} \sum_{i \in \mathcal{I}(J_k)} \int_{J_k} \left(\sum_{\alpha,j} \nu_\alpha^i s_i^{\alpha j} \mathbf{a}_j^i + \sum_{\alpha,\beta} \nu_\alpha^i s_i^{\alpha\beta} \mathbf{W}^i_{,\beta} \right) \cdot \hat{\mathbf{W}}^i(\mathbf{p}_i^{-1}(\mathbf{p}))\, dJ_k$$
$$= \sum_{k=1}^{m} \int_{J_k} \tilde{\mathbf{f}}^k \cdot \hat{\mathbf{W}}^{i_k}\, dJ_k.$$

Since the J_k's are mutually disjoint and the geometric junction constraints are local to each J_k individually, the last variational equation can hold if and only if

$$\sum_{i \in \mathcal{I}(J_k)} \int_{J_k} \left(\sum_{\alpha,j} \nu_\alpha^i s_i^{\alpha j} \mathbf{a}_j^i + \sum_{\alpha,\beta} \nu_\alpha^i s_i^{\alpha\beta} \mathbf{W}^i_{,\beta} \right) \cdot \hat{\mathbf{W}}^i(\mathbf{p}_i^{-1}(\mathbf{p}))\, dJ_k$$
$$= \int_{J_k} \tilde{\mathbf{f}}^k \cdot \hat{\mathbf{W}}^{i_k}\, dJ_k, \quad k = 1, \ldots, m. \tag{2.11}$$

From (2.3) we have

$$\hat{\mathbf{W}}^i(\mathbf{p}_i^{-1}(\mathbf{p})) = \hat{\mathbf{W}}^{i_k}(\mathbf{p}_{i_k}^{-1}(\mathbf{p})), \quad \forall i \in \mathcal{I}(J_k).$$

Since $\hat{\mathbf{W}}^{i_k}(\mathbf{p}_{i_k}^{-1}(\mathbf{p}))$ is arbitrary, it follows that

$$(2.12) \quad \sum_{i \in \mathcal{I}(J_k)} \left(\sum_{\alpha,j} \nu_\alpha^i s_i^{\alpha j} \mathbf{a}_j^i + \sum_{\alpha,\beta} \nu_\alpha^i s_i^{\alpha\beta} \mathbf{W}_{,\beta}^i \right) (\mathbf{p}_i^{-1}(\mathbf{p}), t) = \tilde{\mathbf{f}}^k(\mathbf{p}_{i_k}^{-1}(\mathbf{p}), t), \quad \mathbf{p} \in J_k.$$

Equation (2.12) is a balance law for forces along the joint J_k and we refer to it as the *dynamic junction condition* there.

2.3. Linearization. The nonlinear equations of motion (2.8), dynamic boundary conditions (2.9) and dynamic junction conditions (2.12) are geometrically exact under the kinematic hypotheses imposed above (i.e., (1.1) and Hooke's Law). It is apparent that their linearizations about the equilibrium $\mathbf{W}^i = 0$, $i = 1, \dots, n$, are the same as the linearizations of the equations

$$(2.13) \quad \sum_{\alpha,j} s_{i,\alpha}^{\alpha j} \mathbf{a}_j^i + \mathbf{F}^i = \rho_i \mathbf{W}_{,tt}^i, \quad (x_1, x_2) \in \Omega_i, \ t > 0;$$

$$(2.14) \quad \sum_{\alpha,j} \nu_\alpha^i s_i^{\alpha j} \mathbf{a}_j^i = \mathbf{f}^i \ \text{ on } \Gamma_i^D, \ t > 0;$$

$$(2.15) \quad \sum_{i \in \mathcal{I}(J_k)} \left(\sum_{\alpha,j} \nu_\alpha^i s_i^{\alpha j} \mathbf{a}_j^i \right) (\mathbf{p}_i^{-1}(\mathbf{p}), t) = \tilde{\mathbf{f}}^k(\mathbf{p}_{i_k}^{-1}(\mathbf{p}), t), \quad \mathbf{p} \in J_k, \ t > 0, \ k = 1, \dots, m.$$

The linearizations of (2.13)–(2.15) are obtained by replacing the nonlinear strain components $\varepsilon_{\alpha\beta}$ in the stress-strain law (1.7) by their linear approximations

$$(2.16) \quad \varepsilon_{\alpha\beta} = \frac{1}{2}(W_{\alpha,\beta} + W_{\beta,\alpha}).$$

Let us consider the case of an elastically isotropic body $\mathcal{B}_i$. Its stress-strain law is then

$$s_i^{jk} = 2\mu_i \varepsilon_{jk}^i + \lambda_i \sum_l \varepsilon_{ll}^i \delta_{jk},$$

where ε_{jk}^i are the linearized strains in $\mathcal{B}_i$ and λ_i and μ_i are its Lamé parameters. It follows from (2.16) that the linearized stress-displacement relation is

$$(2.17) \quad \begin{cases} s_i^{\alpha\beta} = \mu_i (W_{\alpha,\beta}^i + W_{\beta,\alpha}^i) + \lambda_i \sum_\gamma W_{\gamma,\gamma}^i \delta_{\alpha\beta}, \\ s_i^{\alpha 3} = \mu_i W_{3,\alpha}^i. \end{cases}$$

With the stresses given by (2.17), the system (2.13) may be written

$$\text{(2.18)} \qquad \begin{cases} \sum_\alpha s^{\alpha\beta}_{\imath,\alpha} + F^\imath_\beta = \rho_\imath W^\imath_{\beta,tt}, \\ \mu_\imath \Delta W^\imath_3 + F^\imath_3 = \rho_\imath W^\imath_{3,tt} \ \text{ in } \Omega,\ t > 0, \end{cases}$$

where $F^\imath_\jmath = \mathbf{F}^\imath \cdot \mathbf{a}^\imath_\jmath$ and $\Delta W^\imath_3 = \sum_\alpha W^\imath_{3,\alpha\alpha}$. The boundary conditions (2.14) linearize to

$$\text{(2.19)} \qquad \begin{cases} \sum_\alpha \nu^\imath_\alpha s^{\alpha\beta}_\imath = f^\imath_\beta, \\ \mu_\imath \boldsymbol{\nu}^\imath \cdot \nabla W^\imath_3 = f^\imath_3 \ \text{ on } \Gamma^D_\imath,\ t > 0, \end{cases}$$

where $f^\imath_\jmath = \mathbf{f}^\imath \cdot \mathbf{a}^\imath_\jmath$ and $\nabla W^\imath_3 = \sum_\alpha W^\imath_{3,\alpha} \mathbf{a}^\imath_\alpha$, and the dynamic junction condition (2.15) becomes

$$\text{(2.20)} \qquad \sum_{\imath \in \mathcal{I}(J_k)} \sum_\alpha \nu^\imath_\alpha \left(\sum_\beta s^{\alpha\beta}_\imath \mathbf{a}^\imath_\beta + \mu_\imath W_{3,\alpha} \mathbf{a}^\imath_3 \right) (\mathbf{p}^{-1}_\imath(\mathbf{p}), t) = \tilde{\mathbf{f}}^k(\mathbf{p}^{-1}_{\imath_k}(\mathbf{p}), t), \ \ \mathbf{p} \in J_k,\ t > 0.$$

The description of the linear system is completed by adjoining to (2.18)–(2.20) the geometric conditions

$$\text{(2.21)} \qquad \mathbf{W}^\imath = 0 \ \text{ on } \Gamma^C_\imath,\ t > 0,$$

$$\text{(2.22)} \quad \mathbf{W}^\imath(\mathbf{p}^{-1}_\imath(\mathbf{p}), t) = \mathbf{W}^\jmath(\mathbf{p}^{-1}_\jmath(\mathbf{p}), t),\ \mathbf{p} \in J_k;\ i, \jmath \in \mathcal{I}(J_k);\ k = 1, \ldots, m,$$

and the initial conditions

$$\text{(2.23)} \qquad \mathbf{W}^\imath|_{t=0} = \mathbf{W}^\imath_0,\ \ \mathbf{W}^\imath_{,t}|_{t=0} = \mathbf{W}^\imath_1 \ \text{ in } \Omega,\ , \imath = 1, \ldots, n.$$

Equations (2.18)–(2.22), with $s^{\jmath k}_\imath$ given by (2.17), describe the linearized motion of a system of interconnected isotropic elastic membranes. Let us note that the above system is also directly derivable from the variational principle

$$\delta \int_0^T [\mathcal{K}(t) - \mathcal{U}(t) + \mathcal{W}(t)]\, dt = 0,$$

where $\mathcal{K}$ and $\mathcal{W}$ are as above and

$$\mathcal{U}(t) = \frac{1}{2} \sum_{\imath=1}^n \int_{\Omega_\imath} \left(\sum_{\alpha,\beta} s^{\alpha\beta}_\imath W^\imath_{\beta,\alpha} + \mu_\imath |\nabla W^\imath_3|^2 \right) d\Omega.$$

3. Controllability of Linked Isotropic Membranes

In this section we shall consider the question of exact controllability of solutions of the linearized equations of motion of the system of interconnected, elastically isotropic membranes given by (2.18)–(2.23). Because of space limitations, only an overview of this problem can be presented.

It is assumed that all distributed forces vanish. Given a suitable $T > 0$ and "arbitrary" initial data $(\mathbf{W}_0^i, \mathbf{W}_1^i)$ and final data $(\hat{\mathbf{W}}_0^i, \hat{\mathbf{W}}_1^i)$, the above system is *exactly controllable in time* T if there are *control functions* $\mathbf{f}^i = \sum_j f_j^i \mathbf{a}_j^i$, $\tilde{\mathbf{f}}^k$ such that the solution of (2.18)–(2.23) achieves the state $(\hat{\mathbf{W}}_0^i, \hat{\mathbf{W}}_1^i)$ at time T:

$$\mathbf{W}^i|_{t=T} = \hat{\mathbf{W}}_0^i, \quad \mathbf{W}^i_{,t}|_{t=T} = \hat{\mathbf{W}}_1^i, \quad i = 1, \dots, n.$$

Of course, the data and controls must be chosen from appropriate function spaces for which the initial value problem is (at least) well- posed in some sense. Since the problem (2.18)–(2.22) is time reversible, one may assume that either the final state or the initial state vanishes. In the latter situation the exact controllability problem is called the *reachability problem.* Thus the reachability problem is that of showing that starting from the zero state, an "arbitrary state" $(\hat{\mathbf{W}}_0^i, \hat{\mathbf{W}}_1^i)$, $i = 1, \dots, n$, may be achieved in time T through an appropriate choice of control functions.

It is not to be expected that every system of interconnected membranes, no matter how configured, is exactly controllable. Indeed, our purpose is to determine those geometric properties of such a system which will guarantee its exact controllability. It will be fairly apparent that the sufficient conditions for exact controllability discussed below are rather far from necessary.

Another point is that in order to obtain any exact controllability results at all we usually (but not always) find it necessary to employ controls not only along the exterior boundaries L_i^D but also in the junction regions L_i^J. It is not clear to what extent the latter requirement is due to the methods employed or whether there is something intrinsic to wave propagation in general systems of interconnected membranes which necessitates the use of junction based controls.

3.1. Observability Estimates for the Homogeneous Problem. Whether one proceeds via the control-to-state map or employs the (equivalent) methodology of the Hilbert Uniqueness Method, it is well-known that the reachability problem is the "dual" of the *continuous observability problem* for the homogeneous adjoint system (which in the present case coincides with the original system). When dealing with situations in which the controls act on the boundary, the latter problem involves showing that certain traces of the solution on the boundary may be estimated from below by some norm of the initial data. Each such *observability estimate* leads to a reachability result, and conversely. Here we shall describe three such observability estimates.

Let us therefore consider the homogeneous problem describing the motion of a system of interconnected membranes. Let the displacement vector of the i-th membrane be

$$\boldsymbol{\Psi}^i = \sum_j \Psi^i_j \mathbf{a}^i_j = \boldsymbol{\psi}^i + \Psi^i_3 \mathbf{a}^i_3, \quad \boldsymbol{\psi}^i = \sum_\alpha \Psi^i_\alpha \mathbf{a}^i_\alpha. \tag{3.1}$$

We write

$$\varepsilon_{\alpha\beta}(\boldsymbol{\psi}^i) = \frac{1}{2}\left(\boldsymbol{\Psi}^i_{\alpha,\beta} + \boldsymbol{\Psi}^i_{\beta,\alpha}\right),$$

$$s^{\alpha\beta}_i(\boldsymbol{\psi}^i) = 2\mu_i \varepsilon_{\alpha\beta}(\boldsymbol{\psi}^i) + \lambda_i \sum_\gamma \varepsilon_{\gamma\gamma}(\boldsymbol{\psi}^i)\delta_{\alpha\beta}.$$

The system under consideration is then

$$\begin{cases} \sum_\alpha s^{\alpha\beta}_{i,\alpha}(\boldsymbol{\psi}^i) = \rho_i \Psi^i_{\beta,tt}, \\ \mu_i \Delta \Psi^i_3 = \rho_i \Psi^i_{3,tt}, \quad \mathbf{x} \in \Omega_i, \ t > 0; \end{cases} \tag{3.2}$$

$$\begin{cases} \sum_\alpha \nu^i_\alpha s^{\alpha\beta}_i(\boldsymbol{\psi}^i) = 0, \\ \boldsymbol{\nu}^i \cdot \nabla \Psi^i_3 = 0 \quad \mathbf{x} \in \Gamma^D_i, \ t > 0; \end{cases} \tag{3.3}$$

$$\boldsymbol{\Psi}^i = 0, \quad \mathbf{x} \in \Gamma^C_i, \ t > 0; \tag{3.4}$$

$$\boldsymbol{\Psi}^i(\mathbf{p}_i^{-1}(\mathbf{p}), t) = \boldsymbol{\Psi}^j(\mathbf{p}_j^{-1}(\mathbf{p}), t), \quad \mathbf{p} \in J_k, \ i, j \in \mathcal{I}(J_k), \ k = 1, \dots, m; \tag{3.5}$$

$$\sum_{i \subset \mathcal{I}(J_k)} \sum_\alpha \nu^i_\alpha \Big(\sum_\beta s^{\alpha\beta}_i(\boldsymbol{\psi}^i)\mathbf{a}^i_\beta + \mu_i \Psi_{3,\alpha} \mathbf{a}^i_3\Big)(\mathbf{p}_i^{-1}(\mathbf{p}), t) = 0, \ \mathbf{p} \in J_k, \ t > 0; \tag{3.6}$$

$$\boldsymbol{\Psi}^i|_{t=0} = \boldsymbol{\Psi}^i_0, \quad \boldsymbol{\Psi}^i_{,t}|_{t=0} = \boldsymbol{\Psi}^i_1, \quad \mathbf{x} \in \Omega_i. \tag{3.7}$$

We set $\boldsymbol{\Psi}_0 = (\boldsymbol{\Psi}^1_0, \dots, \boldsymbol{\Psi}^n_0)$ and $\boldsymbol{\Psi}_1 = (\boldsymbol{\Psi}^1_1, \dots, \boldsymbol{\Psi}^n_1)$.

Let $\boldsymbol{\Psi} = (\boldsymbol{\Psi}^1, \dots, \boldsymbol{\Psi}^n)$ with $\boldsymbol{\Psi}^i : \Omega_i \mapsto \mathbb{R}^3$ given by (3.1). Let H (resp., V) be the Hilbert space consisting of those $\boldsymbol{\Psi}$ for which $\Psi^i_j \in L^2(\Omega_i)$ (resp., $\Psi^i_j \in H^1(\Omega_i)$), with the norms of $\boldsymbol{\Psi}$ in H and in V given by

$$\|\boldsymbol{\Psi}\|_H = \left(\sum_{i=1}^n \sum_j \|\Psi^i_j\|^2_{L^2(\Omega_i)}\right)^{1/2}$$

and

$$\|\boldsymbol{\Psi}\|_V = \left(\sum_{i=1}^n \sum_j \|\Psi^i_j\|^2_{H^1(\Omega_i)}\right)^{1/2},$$

respectively. Let $\mathcal{V}$ denote the closed subspace of V consisting of those $\mathbf{\Psi} \in V$ satisfying

$$\mathbf{\Psi}^i = 0 \text{ on } \Gamma_i^C,$$
$$\mathbf{\Psi}^i(\mathbf{p}_i^{-1}(\mathbf{p})) = \mathbf{\Psi}^j(\mathbf{p}_j^{-1}(\mathbf{p})), \quad \mathbf{p} \in J_k, \; i, j \in \mathcal{I}(J_k).$$

The space $\mathcal{V}$ is dense in H with compact embedding.

Let $\mathbf{q}_0$ be a point of $\mathbb{R}^3$ and consider the following restrictions on the geometry of the configuration of membranes and on the material parameters:

$$(\mathbf{p} - \mathbf{q}_0) \cdot \boldsymbol{\nu}^i \leq 0 \text{ on } \Gamma_i^C; \tag{3.8}$$

$$\sum_{i \in \mathcal{I}(J_k)} \rho_i (\mathbf{p} - \mathbf{q}_0) \cdot \boldsymbol{\nu}^i \leq 0 \text{ on } J_k, \; k = 1, \dots, m; \tag{3.9}$$

$$\sum_{i \in \mathcal{I}(J_k)} a_i = 0, \quad \sum_{i \in \mathcal{I}(J_k)} a_i^2 = 1 \Longrightarrow \sum_{i \in \mathcal{I}(J_k)} \frac{a_i^2}{\mu_i} (\mathbf{p} - \mathbf{q}_0) \cdot \boldsymbol{\nu}^i < 0 \text{ on } J_k, \; k = 1, \dots, m; \tag{3.10}$$

and

$$\sum_{i \in \mathcal{I}(J_k)} (b_i \boldsymbol{\nu}^i + c_i \mathbf{a}_3^i) = 0, \quad \sum_{i \in \mathcal{I}(J_k)} (b_i^2 + c_i^2) = 1 \Longrightarrow \sum_{i \in \mathcal{I}(J_k)} \left(\frac{b_i^2}{2\mu_i + \lambda_i} + \frac{c_i^2}{\mu_i} \right) (\mathbf{p} - \mathbf{q}_0) \cdot \boldsymbol{\nu}^i < 0 \text{ on } J_k, \; k = 1, \dots, m. \tag{3.11}$$

Further, let us introduce the sets

$$\Lambda_i^{D^+} = \{\mathbf{p} \in \Lambda_i^D | \, (\mathbf{p} - \mathbf{q}_0) \cdot \boldsymbol{\nu}^i \geq 0\}, \quad \Lambda_i^{D^-} = \Lambda_i^{D^+} / \Lambda_i^D,$$

$$\Gamma_i^{D^+} = \mathbf{p}_i^{-1}(\Lambda_i^{D^+}), \quad \Gamma_i^{D^-} = \mathbf{p}_i^{-1}(\Lambda_i^{D^-}) = \Gamma_i^D / \Gamma_i^{D^+}.$$

We have the following observability estimate.

Theorem 3.1. *Assume that q_0 may be chosen so that (3.8)–(3.11) hold and further suppose that either*

(i) $\Gamma_i^C \neq \emptyset$ *for at least one index i, or*

(ii) $\Gamma_i^{D^+} \neq \emptyset$ *for at least one index i.*

If $\Psi = (\Psi^1, \dots, \Psi^n)$ *is a sufficiently regular solution of* (3.2)–(3.7), *then there is a* $T_0 > 0$ *such that for* $T > T_0$ *there is a constant* C_T *such that*

$$\|(\Psi_0, \Psi_1)\|^2_{\mathcal{V}\times H} \le C_T \left\{ \sum_{i=1}^{n} \left[\int_0^T \int_{\Gamma_i^{D+}} (|\Psi^i_{,t}|^2 + |\Psi^i|^2)\, d\Gamma dt + \int_0^T \int_{\Gamma_i^{D-}} \left| \frac{\partial \Psi^i}{\partial \boldsymbol{\tau}^i} \right|^2 d\Gamma dt \right] + \sum_{k=1}^{m} \int_0^T \int_{\Gamma_{i_k}(J_k)} \left| \frac{\partial \Psi^{i_k}}{\partial \boldsymbol{\tau}^{i_k}} \right|^2 d\Gamma dt \right\}. \tag{3.12}$$

Remark 3.1. The problem (3.2)–(3.7) has a natural variational formulation (the principle of virtual work) from which existence and uniqueness of a solution may be proved. The variational solution has components Ψ^i_j which belong to $H^1(\Omega_i)$ and the Ψ^i satisfy the geometric junction and boundary conditions, provided the initial data do likewise. In order to prove Theorem 3.1, however, we need to require that the solution have additional regularity, namely that $\Psi^i_j \in H^s(\Omega_i)$ for some $s > 3/2$ provided that the initial data are sufficiently regular. It is known that even for a single isotropic membrane this degree of regularity depends on the geometry of Ω and the particular boundary conditions. For example, it is true for any Lipschitz domain if $\overline{\Gamma}^C$ and $\overline{\Gamma}^D$ are disjoint. If $\overline{\Gamma}^C \cap \overline{\Gamma}^D \neq \emptyset$, the desired regularity holds if these sets meet in an angle smaller than π (measured in the interior of Ω). In this matter the reader is referred to Grisvard [1], [2], and Nicaise [4]. For systems of interconnected isotropic membranes, from the regularity results for a single membrane it is clear that there is H^s regularity ($s > 3/2$) in a neighborhood of each point of Γ_i^C, of Γ_i^D and of $\overline{\Gamma_i^C} \cap \overline{\Gamma_i^D}$ provided the angle condition just mentioned is satisfied. (Interior regularity is not at issue.) However, the precise regularity possessed by Ψ^i near Γ_i^J or, more specifically, its regularity near points of $\overline{\Gamma_i^J} \cap \overline{\Gamma_i^C}$ and $\overline{\Gamma_i^J} \cap \overline{\Gamma_i^D}$ is, to the knowledge of this author, unsettled and requires further analysis. Let us mention, however, an important contribution of Nicaise [7] towards resolving this issue, wherin the precise singular behavior of solutions of the Laplace and biharmonic equations on networks at mixed corners is described. See also Nicaise [5], [6], where regularity of solutions (among other things) of "transmission problems" for general elliptic problems with general boundary and interface conditions is investigated.

Remark 3.2. A sufficient condition for the validity of (3.8)–(3.11) is that, at each joint J_k, at most one of the $(\mathbf{p}-\mathbf{q}_0)\cdot\boldsymbol{\nu}^i$, say $(\mathbf{p}-\mathbf{q}_0)\cdot\boldsymbol{\nu}^{q_k}$, is positive, $(\mathbf{p}-\mathbf{q}_0)\cdot\boldsymbol{\nu}^i < 0$ if $i \neq q_k$, $i \in \mathcal{I}(J_k)$, and

$$\rho_{q_k}(\mathbf{p}-\mathbf{q}_0)\cdot\boldsymbol{\nu}^{q_k}, \quad (\mathbf{p}-\mathbf{q}_0)\cdot\boldsymbol{\nu}^{q_k}/\mu_{q_k}, \quad (\mathbf{p}-\mathbf{q}_0)\cdot\boldsymbol{\nu}^{q_k}/\lambda_{q_k}$$

are sufficiently small relative to

$$\rho_i|(\mathbf{p}-\mathbf{q}_0)\cdot\boldsymbol{\nu}^i|, \quad |(\mathbf{p}-\mathbf{q}_0)\cdot\boldsymbol{\nu}^i|/\mu_i, \quad |(\mathbf{p}-\mathbf{q}_0)\cdot\boldsymbol{\nu}^i|/\lambda_i,$$

respectively, for all $i \neq q_k$ and $i \in \mathcal{I}(J_k)$. This means that ρ_{q_k} has to be sufficiently small relative to the other ρ_i's and that μ_{q_k}, λ_{q_k} must be sufficiently large relative to the other μ_i's and λ_i's, respectively.

The observability estimate (3.12) leads to a corresponding reachability result (Theorem 3.3 below). Because of the presence of integrals over J_k in (3.12), this reachability result requires controls in the junction regions as well as on the exterior boundaries Γ_i^D. We shall now discuss one particular configuration of connected membranes (there are others as well) for which the integrals over J_k may be eliminated from (3.12), namely the one in which the membranes are *serially connected.* This means that

$$\begin{cases} \mathbf{a}_3^i = \pm \mathbf{a}_3^j, \ \forall i, j = 1, \dots, n, \\ m = n - 1. \end{cases}$$

By reindexing the membranes and joints, if necessary, we may assume that

$$\begin{cases} \overline{\mathcal{P}}_i \cap \overline{\mathcal{P}}_{i+1} = J_i, \quad i = 1, \dots, n-1, \\ \overline{\mathcal{P}}_i \cap \overline{\mathcal{P}}_j = \emptyset, \quad |i - j| > 1. \end{cases}$$

Thus $\mathcal{I}(J_k) = [k, k+1]$. Further, without loss of generality, we may suppose that $\mathbf{a}_3^i = \mathbf{a}_3^j = (0, 0, 1)$ (hence

$$\boldsymbol{\nu}^i = -\boldsymbol{\nu}^{i+1}, \quad \boldsymbol{\tau}^i = -\boldsymbol{\tau}^{i+1} \text{ on } J_i)$$

and that

$$\mathcal{P}_i = \Omega_i, \ \Lambda_i^C = \Gamma_i^C, \ \Lambda_i^D = \Gamma_i^D,$$

$$\begin{cases} \Gamma_k(J_k) = \Gamma_{k+1}(J_k) = J_k, \quad k = 1, \dots, n-1, \\ \Gamma_i(J_k) = \emptyset, \quad |i - k| > 1. \end{cases}$$

If $\mathcal{P}_i$ is serially connected to $\mathcal{P}_{i+1}$, the geometric and dynamic conditions along J_i reduce to, respectively,

$$\boldsymbol{\psi}^i = \boldsymbol{\psi}^{i+1}, \quad \Psi_3^i = \Psi_3^{i+1} \text{ along } J_i, \tag{3.13}$$

and

$$\begin{cases} \sum_{\alpha,\beta} \nu_\alpha^{i+1} s_{i+1}^{\alpha\beta}(\boldsymbol{\psi}^{i+1})\mathbf{a}_\beta^{i+1} = -\sum_{\alpha,\beta} \nu_\alpha^i s_i^{\alpha\beta}(\boldsymbol{\psi}^i)\mathbf{a}_\beta^i, \\ \mu_i \dfrac{\partial \Psi_3^i}{\partial \boldsymbol{\nu}^i} = -\mu_{i+1} \dfrac{\partial \Psi_3^{i+1}}{\partial \boldsymbol{\nu}^{i+1}} \text{ along } J_i, \end{cases} \tag{3.14}$$

so that the transverse motion is not coupled to the in-plane motion, as is to be expected.

When membranes are serially connected, it is possible to eliminate the integrals over J_k in the estimate (3.12), provided the parameters ρ_i, μ_i, λ_i satisfy certain monotonicity conditions.

Theorem 3.2. *Assume that* $\mathbf{x}_0 \in \mathbb{R}^2$ *may be chosen so that*

$$(3.15) \qquad (\mathbf{x} - \mathbf{x}_0) \cdot \boldsymbol{\nu}^i \leq 0 \quad on\ \Gamma_i^C,\ i = 1, \dots, n.$$

Suppose further that for the given choice of $\mathbf{x}_0$ *and each* $i = 1, \dots, n-1$,

$$(3.16) \qquad \rho_i \leq \rho_{i+1}, \quad \mu_i \geq \mu_{i+1}, \quad \lambda_i \geq \lambda_{i+1}$$

if $(\mathbf{x} - \mathbf{x}_0) \cdot \boldsymbol{\nu}^i > 0$ *on* J_i;

$$(3.17) \qquad \rho_i \geq \rho_{i+1}, \quad \mu_i \leq \mu_{i+1}, \quad \lambda_i \leq \lambda_{i+1}$$

if $(\mathbf{x} - \mathbf{x}_0) \cdot \boldsymbol{\nu}^i < 0$ *on* J_i. *If* $\boldsymbol{\Psi} = (\boldsymbol{\Psi}^1, \dots, \boldsymbol{\Psi}^n)$ *is a sufficiently regular solution of* (3.2)–(3.4), (3.13), (3.14) *with initial data* $(\boldsymbol{\Psi}_0, \boldsymbol{\Psi}_1)$, *there is a* $T_0 > 0$ *such that for each* $T > T_0$ *there is a constant* C_T *such that*

$$(3.18) \quad \|(\boldsymbol{\Psi}_0, \boldsymbol{\Psi}_1)\|^2_{\mathcal{V} \times H} \leq C_T \sum_{i=1}^{n} \left\{ \int_0^T \int_{\Gamma_i^{D+}} (|\boldsymbol{\Psi}^i_{,t}|^2 + |\boldsymbol{\Psi}^i|^2)\, d\Gamma dt + \int_0^T \int_{\Gamma_i^{D-}} \left| \frac{\partial \boldsymbol{\Psi}^i}{\partial \boldsymbol{\tau}^i} \right|^2 d\Gamma dt \right\}.$$

Remark 3.3. Either condition (i) or condition (ii) of the previous theorem will automatically be satisfied in the serial case. The quantity $(\mathbf{x} - \mathbf{x}_0) \cdot \boldsymbol{\nu}^i$ must be of constant sign on J_i. If, for a particular index i, $(\mathbf{x} - \mathbf{x}_0) \cdot \boldsymbol{\nu}^i = 0$ on J_i, no relation between the material parameters of $\mathcal{P}_i$ and $\mathcal{P}_{i+1}$ need be assumed. Also, since the choice of an $\mathbf{x}_0$ satisfying (3.15) is not unique, for certain configurations it may be possible to choose distinct values of $\mathbf{x}_0$ such that for certain indices i, $(\mathbf{x} - \mathbf{x}_0) \cdot \boldsymbol{\nu}^i > 0$ on J_i for the first choice of $\mathbf{x}_0$ while, for the second, $(\mathbf{x} - \mathbf{x}_0) \cdot \boldsymbol{\nu}^i < 0$ on J_i. In such situations either (3.16) or (3.17) is sufficient for the conclusion of the theorem. These observations suggest that the above restrictions on the material parameters are due to the method of proof employed and are not intrinsic to the validity of the theorem.

Remark 3.4. Theorem 3.2 is valid also in the more general situation where the J_i's are any Lipschitz continuous curves. Of course, in this situation $(\mathbf{x} - \mathbf{x}_0) \cdot \boldsymbol{\nu}^i$ need not be of constant sign on J_i so that it is necessary to assume *both* $(\mathbf{x} - \mathbf{x}_0) \cdot \boldsymbol{\nu}^i \geq 0$ on J_i and (3.16), or $(\mathbf{x} - \mathbf{x}_0) \cdot \boldsymbol{\nu}^i \leq 0$ on J_i and (3.17). In partcular, Theorem 3.2 is valid for the following *problem of transmission*. Let Ω, Ω_1 be bounded, open, connected sets in $\mathbb{R}^2$ with Lipschitz continuous boundaries and with $\overline{\Omega}_1 \subset \Omega$, and set $\Omega_2 = \Omega / \overline{\Omega}_1$. We consider the above problem with $J = \partial\Omega_1$ being the junction region. We then have $\partial\Omega = \overline{\Gamma}_2^C \cup \overline{\Gamma}_2^D$. Suppose that there is a point $\mathbf{x}_0 \in \mathbb{R}^2$ such that $(\mathbf{x} - \mathbf{x}_0) \cdot \boldsymbol{\nu}^1 > 0$ on J, $(\mathbf{x} - \mathbf{x}_0) \cdot \boldsymbol{\nu}^2 \leq 0$ on Γ_2^C and assume that (3.16) holds. Then the estimate (3.18) is valid. Since, for serial membranes, there is no coupling between in-plane and transverse motions, (3.18) comprises two separate observability estimates, one for transverse motion alone (by setting $\boldsymbol{\psi}^i = 0$) and one for in-plane motion alone (by setting $\Psi_3^i = 0$). The function Ψ_3^i is a solution of the problem

$$\Psi_{3,tt}^i - \mu_i \Delta \Psi_3^i = 0 \ \text{ in } \Omega_i,\ i = 1, 2, \tag{3.19}$$

$$\Psi_3^1 = \Psi_3^2,\ \ \mu_1 \frac{\partial \Psi_3^1}{\partial \boldsymbol{\nu}^1} = -\mu_2 \frac{\partial \Psi_3^2}{\partial \boldsymbol{\nu}^2} \ \text{ on } J, \tag{3.20}$$

$$\Psi_3^2 = 0 \ \text{ on } \Gamma_2^C,\ \ \partial \Psi_3^2 / \partial \boldsymbol{\nu}^2 = 0 \ \text{ on } \Gamma_2^D. \tag{3.21}$$

An observability estimate analogous to (3.18) for the problem consisting of (3.19), (3.20) and the Dirichlet boundary condition $\Psi_3^2 = 0$ on $\partial\Omega$ in place of (3.21), has been proved by Lions [3, Theorem VI.4.2].

The following corollary is an immediate consequence of Theorem 3.2 and the process of "weakening the norm" (see, e.g., [3, p. 200]). Let $\mathcal{V}'$ denote the dual space of $\mathcal{V}$ with respect to H.

Corollary 3.1. *In addition to the assumptions of Theorem* 3.2, *suppose that*

$$(\mathbf{x} - \mathbf{x}_0) \cdot \boldsymbol{\nu}^i \geq 0 \ \text{ on } \Gamma_i^D$$

(i.e., $\Gamma_i^{D^-} = \emptyset$*). Then there is a* $T_0 > 0$ *such that for each* $T > T_0$ *there is a constant* C_T *such that*

$$\|(\boldsymbol{\Psi}_0, \boldsymbol{\Psi}_1)\|^2_{H \times \mathcal{V}'} \leq C_T \sum_{i=1}^{n} \int_0^T \int_{\Gamma_i^D} |\boldsymbol{\Psi}^i_{,t}|^2 \, d\Gamma dt.$$

3.2. The Reachable States. Each of the above observability estimates leads to a reachability result. Here we indicate two such results, one being the "dual" of Theorem 3.1 and the other the dual of Corollary 3.1. Since the observability estimates are known only for "sufficiently regular" solutions, implicit in the theorems below is the assumption that there is a set of initial data which is dense in $\mathcal{V} \times H$ and for which the corresponding solution of (3.2)–(3.7) has the desired regularity. This assumption imposes a further constraint on the configuration of the system of membranes, but we do not know how to quantify this constraint in terms of simple geometric properties of the configuration. (See Remark 3.1 above).

Theorem 3.3. *Let the hypotheses of Theorem* 3.1 *hold and let* $T > T_0$ *and assume that* $\mathbf{W}_0 \in H$, $\mathbf{W}_1 \in \mathcal{V}'$. *Then there are controls* $\mathbf{f}^i$, $\tilde{\mathbf{f}}^k$ *such that the solution of* (2.18)–(2.22) *with* $\mathbf{W}|_{t=0} = \mathbf{W}_{,t}|_{t=0} = 0$ *satisfies*

$$\mathbf{W}|_{t=T} = \mathbf{W}_0,\ \ \mathbf{W}_{,t}|_{t=T} = \mathbf{W}_1.$$

The controls of Theorem 3.3 belong to the following (very weak) spaces:

$$\mathbf{f}^i \in (H^1(0,T;\mathcal{L}^2(\Gamma_i^{D^+})))' \bigoplus L^2(0,T;(\mathcal{H}^1(\Gamma_i^{D^-}))'),$$

$$\tilde{\mathbf{f}}^k \in L^2(0,T;(\mathcal{H}^1(\Gamma_{i_k}(J_k)))'),$$

the primes denoting dual spaces, where

$$\mathcal{L}^2(\Gamma_i^{D^\pm}) = \{\sum_j f_j \mathbf{a}_j^i | \, f_j \in L^2(\Gamma_i), \ \operatorname{supp} f_j \subset \Gamma_i^{D^\pm}\}$$

and

$$\mathcal{H}^1(\Gamma_i^{D^-}) = \{\hat{\Psi} \in \mathcal{L}^2(\Gamma_i^{D^-}) | \, \frac{\partial \hat{\Psi}}{\partial \tau^i} \in \mathcal{L}^2(\Gamma_i^{D^-})\}.$$

The corresponding solution of (2.18)–(2.22) with $\mathbf{W}|_{t=0} = \mathbf{W}_{,t}|_{t=0} = 0$ exists in a very weak sense which may be made precise through the method of transposition, for example.

In the case of serially connected membranes, a more satisfactory result follows from Corollary 3.1.

Theorem 3.4. *Let the hypotheses of Corollary* 3.1 *hold. Then*

$$\mathcal{V} \times H \subset \{(\mathbf{W}(T), \mathbf{W}_{,t}(T)) | \, \mathbf{f}^i \in L^2(0,T;\mathcal{L}^2(\Gamma_i^D)) \ \textit{and} \ \tilde{\mathbf{f}}^k = 0\}.$$

Thus, for serially connected membranes, all finite energy states are reachable using $L^2(\Gamma_i^D \times (0,T))$ controls only.

References

1. P. Grisvard, *Elliptic Problems in Nonsmooth Domains*, Pitman, London, 1985.
2. ———, *Singularités en élasticité"*, J. Math. Pures Appl. **107** (1989), 157–180.
3. J. L. Lions, *Contrôlabilité Exacte, Perturbations et Stabilisation de Systèmes Distribués, Tome I: Contrôlabilité Exacte"*, Collection RMA, vol 8, Masson, Paris, 1988
4. S. Nicaise, *About the Lamé system in a polygonal or a polyhedral domain and a coupled problem between the Lamé system and the plate equation. I: regularity of the solutions*, Ann. Scuola Normale Sup. Pisa **XIX** (1992), 327–361.
5. ———, *General interface problems I*, Math. Methods Appl. Sci., to appear.
6. ———, *General interface problems II*, Math. Methods Appl. Sci., to appear.
7. ———, *Polygonal Interface Problems*, Peter Lang, Frankfurt am Main, 1993.
8. G. Wempner, *Mechanics of Solids with Applications to Thin Bodies*, Sijthoff and Noordhoff, Rockville, Maryland, 1981.

J. E. Lagnese
Department of Mathematics
Georgetown Univesity
Washington, DC 20057, USA

ON FEEDBACK CONTROLS FOR DYNAMIC NETWORKS OF STRINGS AND BEAMS AND THEIR NUMERICAL SIMULATION

GUNTER LEUGERING

Mathematisches Institut
Universitat Bayreuth

ABSTRACT In these notes we want to present some control strategies for dynamic networks of strings and beams in those situations where the classical concepts of exact or approximate controllability fail This is, in particular, the case for networks containing circuits Generically, a residual motion settles in such circuits, even if all nodes are subject to controls In those situations we resort to controls which direct the flux of energy Using such controls in a network, we are able to steer the entire energy to preassigned parts of the structure In practice these parts are more massive and can absorb energy more easily than the fragile elements We also provide numerical evidence for the control strategies discussed in these notes The material is related to joint work with J E Lagnese and E G P G Schmidt, and is essentially included [14]

1991 *Mathematics Subject Classification* 39C20, 39B52, 65M06, 65M60

Key words and phrases Absorbing, directing and nonlinear controls, disappearing solutions, numerical simulation

1. Introduction

In this note we want to give some examples of feedback control strategies applied to dynamic 1-d-networks of strings and Timoshenko beams We will consider examples of scalar 'out-of-the-plane' displacement models as well as planar systems. Systems of the type considered here have been introduced in Schmidt [19], Leugering and Schmidt [15] and Lagnese, Leugering and Schmidt [13], [14] It turns out that, speaking in general terms, tree-like structures of strings and Timoshenko beams are exactly controllable in finite time (the time being related to the longest path in the tree), if the root (any one of the simple nodes to be introduced below) is clamped and all of the other simple vertices are under control. As for Euler-Bernoulli beams it was shown in that a star-like bundle of beams is exactly controllable in finite time, if all simple vertices are being controlled On the negative side, it is known for quite a while, see [4], that already for two strings (scalar displacement) controllability/stabilizability from multiple nodes is sensitive to the choice of boundary conditions for the simple (uncontrolled) nodes. A similar phenomenon is also observed for tree-like networks. Take for instance the case of three strings coupled at one node: if two simple nodes are clamped and the remaining simple node is controlled, not even approximate controllability can be achieved. The situation, however, is much improved if one of the previously clamped simple nodes is released (satisfies Neumann boundary conditions). Then it can be shown that spectral controllability holds. The picture becomes even more pessimistic if the network contains circuits. In this case, approximate controllability can not be expected, even if all nodes are controlled. As real multi-link flexible structures are, indeed, almost exclusively repetitive, the classical control concepts will usually fail. Even if, from a mathematical point of view, properties like 'rational independence of optical lengths' can yield sharper controllability results, those results are only of academic interest, as it is impossible to produce those networks in reality. In addition, boundary conditions are usually subject to various disturbances, so that a control strategy based on the classical concepts might work in one situation and fail in another. The content of these notes is essentially contained in the above mentioned joint work with J.E. Lagnese and E.J.P.G. Schmidt, [14]. The purpose is to introduce control strategies which do not aim at complete absorption in finite time (controllability) or global decay of the total energy (stability) but, rather, make it possible to channel the energy through the network to end up at a set of specified 'safe' nodes where energy is more easily absorbed (for instance at the base of the structure). The major emphasis is on numerical evidence for these concepts, rather than on the formulation of theorems. We will, however, provide the arguments for exemplaric cases.

2. Notation

We consider a nonempty, finite, simple and connected graph $\mathbf{G}$ in $\mathbf{R}^2$ with n vertices, $V(\mathbf{G}) := \{v_i,\ i = 1, ..., n\}$, and N edges $E(\mathbf{G}) := \{\mathbf{k}_j,\ j = 1, ..., N\}$. The edges $\mathbf{k}_j$ are parametrized by $\pi_j : [0, \ell_j] \to \mathbf{R}^2$, where the running variable $x_j \in [0, \ell_j]$ represents the arclength. The maps π_j are assumed to be C^2-smooth. For $i, j \in V(\mathbf{G})$ we set $\mathbf{k}_{s(i,h)} := v_i v_j = v_j v_i$, where $v_i v_j$ signifies the edge joining the vertices v_i, v_j, if $v_i v_j \in E(\mathbf{G})$. In fact, as we will be concerned here only with networks consisting of straight edges, we assign to each edge $\mathbf{k}_{s(i,h)}$ the unit edge vector $\mathbf{e}_{s(i,h)}$ directed along the edges, and its orthogonal complement $\mathbf{e}^{\perp}_{s(i,h)}$. We also introduce the unit vector $\mathbf{n}$ normal to the plane of the graph. To each triad $\mathbf{e}_{s(i,h)},\ \mathbf{e}^{\perp}_{s(i,h)},\ \mathbf{n}$ we associate the corresponding set of orthogonal projections $\mathbf{P}_{s(i,h)},\ \mathbf{P}^{\perp}_{s(i,h)},\ \mathbf{P}_{\mathbf{n}}$. Furthermore, we introduce the incidence matrix

$$d_{ij} = \begin{cases} 1 & \text{if} \quad \pi_j(\ell_j) = v_i \\ -1 & \text{if} \quad \pi_j(0) = v_i \\ 0 & \text{else} \end{cases} \tag{2.1}$$

and the adjacency matrix

$$e_{ih} = \begin{cases} 1 & \text{if} \quad v_i v_h \in E(\mathbf{G}) \\ 0 & \text{else} \end{cases} . \tag{2.2}$$

For each vertex we define the set $\Gamma(v_i) := \{v_h \in V(\mathbf{G}) | e_{ih} = 1\}$ of all vertices adjacent to the vertex v_i, and $d(v_i) := |\Gamma(v_i)|$ the edge degree of the vertex v_i. We can now characterize the network as follows

$$\begin{aligned} \mathbf{G} &= \cup_{j=1}^{N} \mathbf{k}_j \\ \partial V(\mathbf{G}) &= \{v_i \in V(\mathbf{G}) | d(v_i) = 1\}, \\ \overset{\circ}{V}(\mathbf{G}) &= \{v_i \in V(\mathbf{G}) | d(v_i) > 1\}. \end{aligned}$$

Obviously, $\partial V(\mathbf{G})$ signifies the set of simple nodes, where only one edge starts or ends, while $\overset{\circ}{V}(\mathbf{G})$ signifies the set of multiple nodes, where more than 1 edge meet. A function $\mathbf{r} : \mathbf{G} \to \mathbf{R}^r$ can then be viewed as a collection of functions with values along the individual edges by

$$\mathbf{r}_j := \mathbf{r}(\pi_j(x)), \quad x \in [0, \ell_j].$$

3. Networks of strings

In this section we consider a network of strings which, in its reference configuration, coincides with a graph $\mathbf{G}$ as described in the first subsection. Let $\mathbf{r}$ denote the deviation from the reference configuration. By the definition above, $\mathbf{r}_j$ denotes the displacement of the string (edge) with label j. We define

$$\begin{aligned} u_j \mathbf{e}_j &:= \langle \mathbf{r}_j, \mathbf{e}_j \rangle \mathbf{e}_j = \mathbf{P}_j \mathbf{r}_j, \\ w_j \mathbf{e}_j^{\perp} &:= \langle \mathbf{r}_j, \mathbf{e}_j^{\perp} \rangle \mathbf{e}_j^{\perp} = \mathbf{P}_j^{\perp} \mathbf{r}_j, \end{aligned}$$

so that u_j, w_j represent the longitudinal and vertical displacement of the j-th string, respectively. Let p_j^2, q_j^2 denote the longitudinal and flexural rigidity. Furthermore, let $\mathbf{K}_j := p_j^2 \mathbf{P}_j + q_j^2 \mathbf{P}_j^{\perp}$ be the stiffness operator associated with j-th string. Then the equations governing the motion of the network of strings can be written down as follows:

$$\begin{gathered} \ddot{\mathbf{r}}_j - \mathbf{K}_j \mathbf{r}_j'' = 0, \ j = 1, ..., N, \\ \mathbf{r}_{s(i,h)}(v_i) = 0, \ \forall v_i \in \partial_0 V(\mathbf{G}), \\ \mathbf{K}_{s(i,h)} \mathbf{r}'_{s(i,h)}(v_i) = \mathbf{f}_{v_i}, \ \forall v_i \in \partial_1 V(\mathbf{G}), \\ \mathbf{r}_{s(i,h)}(v_i) = \mathbf{r}_{s(i,j)}(v_i), \ v_h, v_j \in \Gamma(v_i), \ v_i \in \overset{\circ}{V}(\mathbf{G}), \\ \sum_{h \ v_h \in \Gamma(v_i)} d_{is(i,h)} \mathbf{K}_{s(i,h)} \mathbf{r}'_{s(i,h)}(\pi_{s(i,h)}^{-1}(v_i)) = \mathbf{g}_{v_i}, \ \forall i : v_i \in \overset{\circ}{V}(\mathbf{G}), \\ \mathbf{r}_j(\cdot, 0) = \mathbf{r}_{j0}, \ \dot{\mathbf{r}}_j(\cdot, 0) = \mathbf{r}_{j1}. \end{gathered}$$

Here $\partial_0 V(\mathbf{G}) \subset \partial V(\mathbf{G})$ denotes the set of simple nodes with clamped boundary conditions, and $\partial_1 V(\mathbf{G})$ is the set of stress free (or externally loaded as in the case where $\mathbf{f}_v \neq 0$) simple nodes. The third equation expresses the continuity of the network, or its connectedness also in the deformed configuration, while the fourth equation represents the balance of forces (with a possible external loading modelled through the functions $\mathbf{g}_v$).

4. Networks of Timoshenko beams

The situation is notationally very similar but mathematically more involved in the case where each edge is considered as a Timoshenko beam in its reference configuration. In addition to the longitudinal and vertical displacement (u_j, w_j) of the centerline of the beam we also have a rotation $(\psi_j := \langle \mathbf{P_n r}_j, \mathbf{n}\rangle)$ of the cross section out of its position perpendicular to the centerline, i.e. we have to account for shearing. We denote the shear moduli by q_j^2 and the flexural rigidity by m_j^2, as before p_j^2 denotes the longitudinal stiffness. We only consider here an initially straight and untwisted linear Timoshenko-Bresse system as described above.

$$
\begin{aligned}
\ddot{u}_j - p_j^2 u_j'' &= 0, \\
\ddot{w}_j - q_j^2 (w_j' + \psi_j)' &= 0, \\
\ddot{\psi}_j - m_j^2 \psi_j'' + q_j^2 (w_j' + \psi_j) &= 0,
\end{aligned} \tag{4.1}
$$

$$
\mathbf{r}_{s(i,h)}(v_i) = \mathbf{r}_{s(i,j)}(v_i), \quad \forall h, j \in ind(\Gamma(v_i)), v_i \in \overset{\circ}{V}(\mathbf{G}), \tag{4.2}
$$

$$
\begin{aligned}
\sum_h d_{is(i,h)} \{ p_{s(i,h)}^2 u_{s(i,h)}' \mathbf{e}_{s(i,h)} + q_{s(i,h)}^2 (w_{s(i,h)}' + \psi_{s(i,h)}) \mathbf{e}_{s(i,h)}^\perp \\
+ m_{s(i,h)}^2 \psi_{s(i,h)}' \mathbf{n} \} (\pi_{s(i,h)}^{-1}(v_i)) = \mathbf{g}_i, \ \forall v_i \in \overset{\circ}{V}(\mathbf{G}),
\end{aligned} \tag{4.3}
$$

$$
\begin{aligned}
\mathbf{r}_{s(i,h)}(v_i) = 0, \ \forall v_i \in \partial_0 V(\mathbf{G}), \\
\{ p_{s(i,h)}^2 u_{s(i,h)}' \mathbf{e}_{s(i,h)} + q_{s(i,h)}^2 (w_{s(i,h)}' + \psi_{s(i,h)}) \mathbf{e}_{s(i,h)}^\perp \} (\pi_{s(i,h)}^{-1}(v_i)) = \mathbf{f}_i, \\
m_{s(i,h)}^2 \psi_{s(i,h)}'(v_i) = h_i, \forall v_i \in \partial_1 V(\mathbf{G}).
\end{aligned} \tag{4.4}
$$

In analogy to the previous section we define the operators

$$
\begin{aligned}
\mathbf{K}_j &:= p_j^2 \mathbf{P}_j + q_j^2 \mathbf{P}_j^\perp + m_j^2 \mathbf{P_n}, \\
\mathbf{M}_j &:= q_i^2 \langle \mathbf{P_n}(\cdot), \mathbf{n} \rangle \mathbf{e}_j^\perp, \\
\mathbf{N}_j &:= q_j^2 \langle \mathbf{P}_j^\perp(\cdot), \mathbf{e}_j^\perp \rangle, \mathbf{n} \\
\mathbf{S}_j &:= \mathbf{N}_j - \mathbf{M}_j \Rightarrow \mathbf{S}_j^* = -\mathbf{S}_j.
\end{aligned}
$$

Then the state equations (4.1) can be written as

$$\ddot{\mathbf{r}}_j - \mathbf{K}_j\mathbf{r}_j'' + \mathbf{S}_j\mathbf{r}_j' + q_j^2\mathbf{P}_\mathbf{n}\mathbf{r}_j = 0. \tag{4.5}$$

The corresponding continuity condition is exactly of the form as in (3.1), with the only difference that the vectors have now three components. Let us now consider the homogeneous situation. The balance of forces requirement can now be expressed as

$$\sum_h d_{is(i,h)}(\mathbf{K}_{s(i,h)}\mathbf{r}'_{s(i,h)} + \mathbf{M}_{s(i,h)}\mathbf{r}_{s(i,h)})(\pi^{-1}_{s(i,h)})(v_i) = 0, \qquad \forall v_i \in V(\mathbf{G}) \setminus \partial_0 V(\mathbf{G}). \tag{4.6}$$

5. Numerical Simulation of controlled networks

5.1. Absorbing controls. In order to give an exemplaric introduction, we will be dealing with the numerical realization of stabilizing feedback controls. Those controls are of more practical interest than open loop controls. In particular, the feedback controls we discuss are of the absorbing type. They can be viewed as being at the interface between controllability concepts and stabilizability concepts. Loosely speaking, an absorbing control is one which passes on the entire energy flux into one direction. The history of these controls goes back right to the beginning of control theory of distributed parameter systems and is connected with D.L. Russell, see [18] for an excellent review. From the mathematical viewpoint it is most easily explained for the 1-d wave equation in its characteristic description ([18]):

Example 5.1. *The 1-d-wave equation.*

Let us write down the scaled classical 1-d-wave equation as follows

$$\begin{cases} \ddot{w} = w'', & \text{on } (0,1)\times(0,T), \\ w(0,t) = 0,\ w'(1,t) = \phi, & t \in (0,T), \\ w(x,0) = w_0(x),\ \dot{w}(x,0) = w_1(x), & x \in [0,1]. \end{cases}$$

Under the transformations

$$u_1 := \dot{w},\ u_2 := w',$$
$$v_1 := \frac{1}{\sqrt{2}}(-u_1 + u_2),\ v_2 = \frac{1}{\sqrt{2}}(u_1 + u_2),$$

we have with $\Gamma = \begin{pmatrix} -1 & 0 \\ 0 & 1 \end{pmatrix}$ the first order system

$$\dot{\mathbf{v}} = \Gamma \mathbf{v}',$$
$$v_1(0) = v_2(0),\ v_1(1) + v_2(1) = \sqrt{2}\phi.$$

In this simple case the total energy and its time derivative are given by

$$E = \int_0^1 \|\mathbf{v}\|^2 dx, \ \dot{E} = \frac{1}{2}\{-v_1(1)^2 + v_2(1)^2\}.$$

It is obvious that the optimal choice of a control is the one leading to $v_2(1) = 0$ for all $t \geq 0$. In fact, it is readily seen that this choice amounts to cancelling the outgoing signals at the boundary point $x = 1$. This means that the control mimics the response of the half-infinite string. This is Russell's philosophy in [18]: "let nature take its course". In terms of feedback controls the goal of cancelling outgoing signals can be achieved by letting $\phi(t) = \frac{1}{\sqrt{2}} v_1(1,t)$ which is equivalent to the closed loop boundary condition $w'(1,t) = -\dot{w}(1,t)$, which, in turn, is the classical *dead beat control*. Similar arguments can be applied to the Timoshenko system. We note, however, that because of the dispersion of waves, we do not have exact cancellation of outgoing signals at the boundary, to the extend that exact controllability by velocity feedback controls cannot be achieved. Nevertheless, exact controllability can clearly be proved for open-loop controls. It is apparent that besides the necessity of discretizing the state equation and the boundary conditions in appropriate way, the total energy as function of time plays a key role. In particular, *it appears to be necessary to use a numerical scheme which conserves the total energy of the discretized system as long as no controls are applied.* This can, for instance, be achieved by the classical *Lax- Friedrichs-scheme* discussed by Le Roux [17]. See also Halpern [9] for various schemes simulating absorbing boundaries. This scheme is applied to the system above in terms of the variables u_i, and it is also appropriate to quasi-linear models. To avoid unnecessary indexing, we write down $u = w'$ $v = \dot{w}$ where w solves

$$\ddot{w} = (\sigma(w'))'$$

for a suitable function σ, such that

$$\dot{u} = v', \quad \dot{v} = u'.$$

Upon introducing the grid $x_i = i \cdot \Delta x, \ i = 0, ..., n, \ t_j = j \cdot \Delta t, \ j = 0, ...m$, where $\Delta x = 1/n, \ \Delta t = q\Delta x$, q being the appropriate mesh ratio, we have ([17]):

$$\begin{aligned} u_{i+1}^{j+1} &= \frac{1}{2}\{(u_{i+1}^j + u_{i-1}^j) + \frac{q}{2}(v_{i+1}^j - v_{i-1}^j)\}, \\ v_{i+1}^{j+1} &= \frac{1}{2}\{(v_{i+1}^j + v_{i-1}^j) + \frac{q}{2}(\sigma(u_{i+1}^j) - \sigma(u_{i-1}^j))\}, \end{aligned} \tag{5.1}$$

and the boundary conditions are represented by

$$\begin{gathered} v_0^{j+1} = 0, \ u_n^{j+1} = 0, \\ u_0^{j+1} = u_1^j + q \cdot v_1^j, \ v_n^{j+1} = v_{n-1}^{j+1} - q \cdot u_{n-1}^j. \end{gathered} \tag{5.2}$$

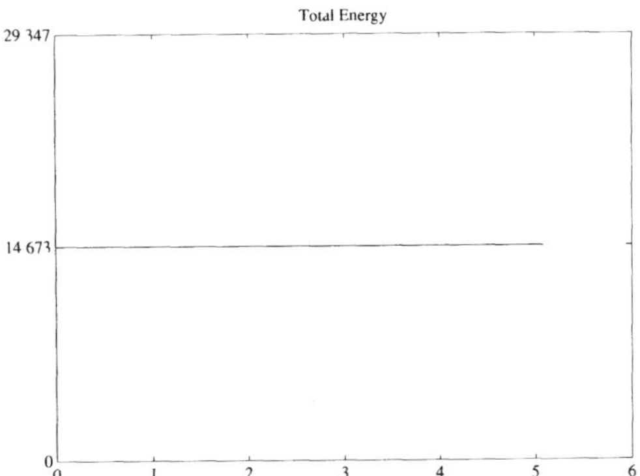

FIGURE 1. Conservation of energy

The Courant-Friedrichs-Lewy condition $q \leq 1$; we always use $q = 1$. This set up is particularily useful if the boundary feedback above is to be implemented, because it prescribes the strain at the boundary $x = 1$. We simply put: $u_n^{j+1} = -v_n^{j+1}$. For further information, in particular on other realizations of absorbing boundary conditions see Halpern [9].

We first look at the uncontrolled string. We take 45 spatial mesh points and 225 time steps, which amounts to time span of $[0, 5]$. As initial conditions we choose $w(x, 0) = \sin(3\pi(x)^2)$, $\dot{w}(x, 0) = 0$. The reason for this choice is that we do not want to look at eigenmodes but rather to a response to rich-enough data. Because of space restrictions, we display only the total energy, which is computed numerically using the trapezoidal rule, see figure 1. We are now going to apply the absorbing boundary control at the boundary $x = 1$. The result is shown in figure 2, where we display the strain, rather than the displacement, simply because it is the canonical variable in the Lax-Friedrichs-scheme. The total energy is displayed in figure 3.

These plots serve as evidence for the remarks above. The numerical procedure can be (and has been) applied also to planar strings and planar Timoshenko beams. For similar simulations, using however the classical direct discretization of the wave equation, see J. Schmidt [20]. □

5.2. Directing controls. We pursue the concept of absorbing controls a bit further and consider now three identical strings and out-of-the-plane-displacement. The fact that we concentrate on this model, rather than on the planar system, is only for the sake of a convenient display of the plots. We will consider planar systems later.

Example 5.2. *Three serial strings.*

We have the following system:

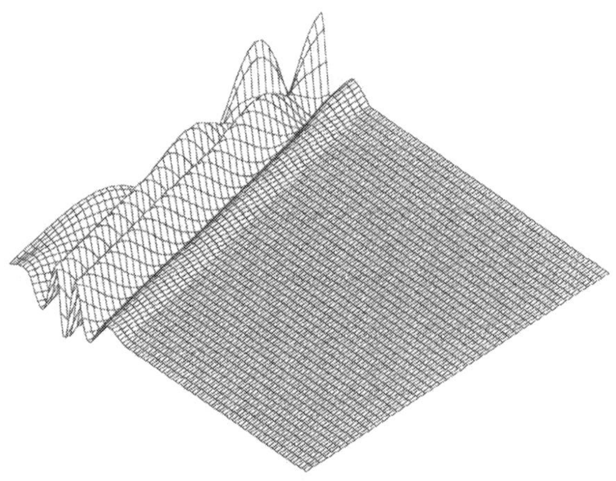

FIGURE 2. Absorbing boundary control

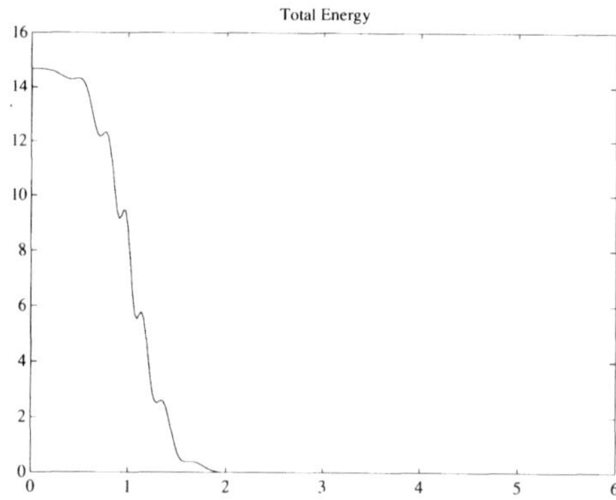

FIGURE 3. Energy plot of disappearing solution

$$
\begin{aligned}
(5.3) \qquad & \ddot{w}^i = (w^i)'' \quad \text{on} \quad (0,1)\times(0,T), \\
& w^1(0)=0,\ w^1(1)=w^2(1),\ (w^1)'(1)+(w^2)'(1)=f^1, \\
& w^3(1)=0,\ w^2(0)=w^3(0),\ (w^2)'(0)+(w^3)'(0)=f^2, \\
& + \text{ initial conditions.}
\end{aligned}
$$

This system describes the motion of three strings with obvious orientation of the running variable x. The orientation is chosen in order to avoid mixed boundary conditions. We perform a succession of transformations as above. Namely,

$$
\begin{aligned}
& u_1^i := \dot{w}^i,\ u_2^i := (w^i)', \\
& v_1^i := \frac{1}{\sqrt{2}}(-u_1^i + u_2^i),\ v_2^i := \frac{1}{\sqrt{2}}(u_1^i + u_2^i), \\
& z_i := v_1^i,\ i = 1,2,3,\quad z_i := (v_2^{i-3})', i = 4,5,6.
\end{aligned}
$$

With these transformations we are able to write the system as

$$
\begin{aligned}
& \dot{z}^i = -(z_i)',\ i = 1,2,3, \\
& \dot{z}^i = (z_i)',\ i = 4,5,6, \\
& z_1(0) = z_4(0),\ z_1(1) - z_4(1) = z_2(1) - z_5(1), \\
& z_3(1) = z_6(1),\ z_2(0) - z_5(0) = z_3(0) - z_6(0), \\
& z_4(1) + z_2(1) = \frac{1}{\sqrt{2}} f^1,\ z_5(0) + z_3(0) = \frac{1}{\sqrt{2}} f^2,
\end{aligned}
$$

together with initial conditions. Our goal now is to steer the energy of the entire system to the second string by applying absorbing controls at the two multiple joints connecting the second string with the first and the third string. We adopt the strategy of P.Hagedorn and J.Schmidt, see [20]. To this end, we define

$$
(5.4) \qquad f^1 := \sqrt{2}z_2(1),\ f^2 := \sqrt{2}z_5(0).
$$

By this choice the outgoing signals $z_3(0)$ at $x = 0$ and $z_4(1)$ at the boundary $x = 1$ are cancelled. For the sake of convenience, we write down the boundary conditions in matrix form also reflecting on incoming and outgoing signals as in Russell [18]:

$$
(5.5) \qquad
\begin{aligned}
\begin{pmatrix} z_1 \\ z_2 \\ z_3 \end{pmatrix}(0) &= \begin{pmatrix} 1 & 0 & 0 \\ 0 & 1 & 1 \\ 0 & 0 & 0 \end{pmatrix} \begin{pmatrix} z_4 \\ z_5 \\ z_6 \end{pmatrix}(0), \\
\begin{pmatrix} z_4 \\ z_5 \\ z_6 \end{pmatrix}(1) &= \begin{pmatrix} 0 & 0 & 0 \\ -1 & 1 & 0 \\ 0 & 0 & 1 \end{pmatrix} \begin{pmatrix} z_1 \\ z_2 \\ z_3 \end{pmatrix}(1).
\end{aligned}
$$

We consider a particular exemplaric path of characteristics, namely, for $t = 0$ we start with the three outgoing signals z_1, z_2, z_3 moving to the boundary $x = 1$ with characteristic speed equal to 1. These characteristics are exactly the incoming signals at that boundary. According to the reflection condition (5.5) at $x = 1$, z_4 is cancelled at $x = 1$. Now, the triple z_4, z_5, z_6 constitutes the set of outgoing signals at $x = 1$, and accordingly, the set of incoming signals at $x = 0$. As z_4 is already equal to zero at $x = 0$ (because of the properties of characteristics), the set of nonvanishing outgoing signals at $x = 0$ reduces to the variable z_2, according to the reflection condition (5.5). Therefore, after reflection and take off to the boundary at $x = 1$, the only nonvanishing incoming signal there is z_2, which, in turn, is reflected to z_5 as the only nonvanishing outgoing signal at $x = 1$. This, however, closes the cycle, because z_5 is converted to z_2 upon reflection at $x = 0$, and so on. We depict this situation in the following diagram:

$$\begin{pmatrix} z_1 \\ z_2 \\ z_3 \end{pmatrix} \xrightarrow{at x=1} \begin{pmatrix} 0 \\ z_5 \\ z_6 \end{pmatrix} \xrightarrow{at x=0} \begin{pmatrix} 0 \\ z_2 \\ 0 \end{pmatrix}$$
$$\xrightarrow{at x=1} \begin{pmatrix} 0 \\ z_5 \\ 0 \end{pmatrix} \xrightarrow{at x=0} \begin{pmatrix} 0 \\ z_2 \\ 0 \end{pmatrix}.$$

This shows that after two reflections at the boundaries there is only a residual motion left in the second string This is what is meant by directing or diode-type controls. We will demonstrate that concept for *planar* frames shortly. Let us first, however, provide some numerical evidence for the situation described above. In figure 4 we take an initial condition distributed along the three strings, which, incidentally, are chosen to have individual length $\frac{1}{3}$, so that the length of the entire system is equal to 1. That the energy remains constant after two reflections (that is at $t = \frac{2}{3}$) is seen in figure 5.

In the context of the original system (5.3) the controls are chosen to be as follows:

$$f^1 = -\dot{w}^2(1) + (w^2)'(1), \; f^2 = \dot{w}^2(0) + (w^2)'(0)$$

which results in the feedback controls

$$(w^1)'(1) = -\dot{w}^1(1), \; (w^3)'(0) = \dot{w}^3(0).$$

It has been shown in [7] that the absorbing boundary feedback, the one with the exact impedance of the half-infinite string, is sensitive to time delays in the velocity feedback. In fact it has subsequently been shown by R. Datko [5], [6] that this phenomenon is not peculiar to strings, and that it is not even circumvented upon the introduction of severe damping. Let us show this instability for our three string model (5.3) in figure 6, and figure 7.

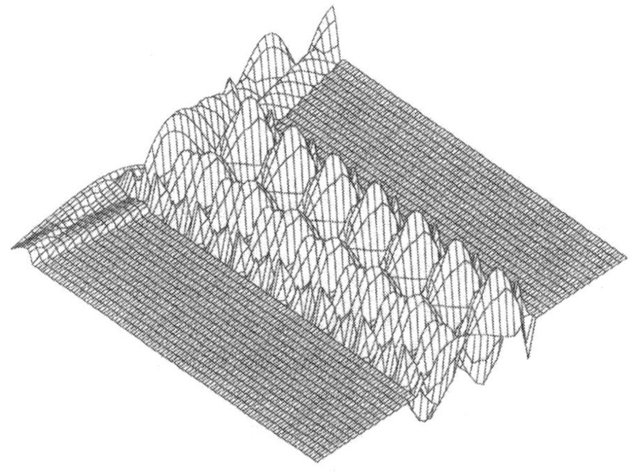

FIGURE 4. Directing controls

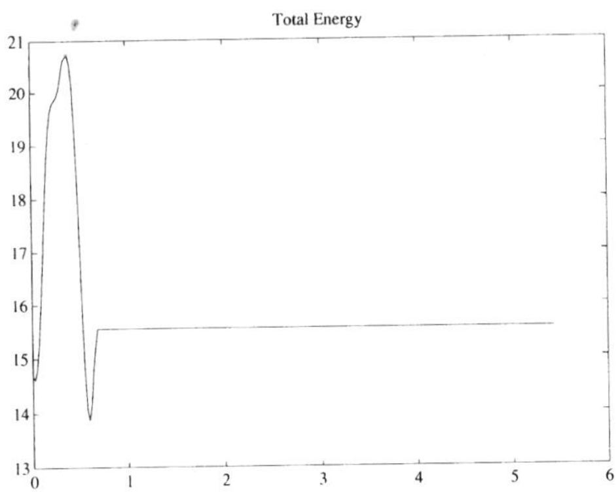

FIGURE 5. Energy of entire system

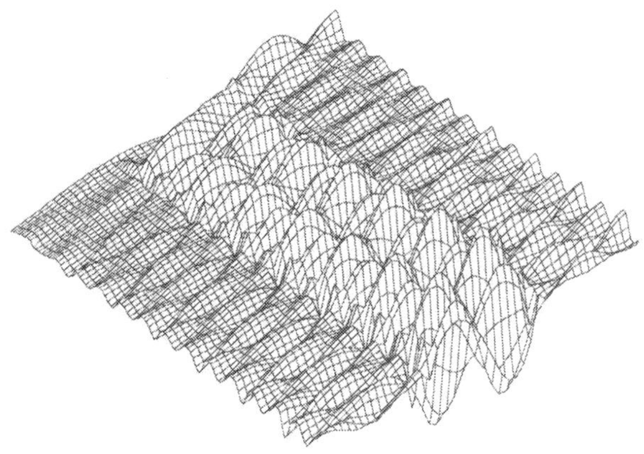

FIGURE 6. Effect of time-delay

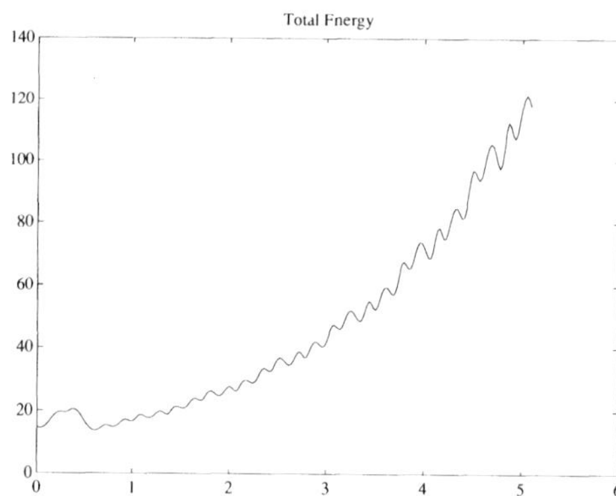

FIGURE 7. Instability due to time delay

Even though this instability is rather discouraging, the situation in a real structure is not as bad as this seems to suggest. First of all, some of the energy absorbing feedback controls can be implemented as passive devices, for which a time lag will not be present. But even if the feedback law is to be realized by numerical computation (in real time), the time-lag due to the action of the microprocessor will not be a constant, but, rather, distributed by a certain stochastic process. There is, however, another remedy to this problem: the growing instability of the solutions under a feedback with time-lag is caused by feeding back an increasing amount of control energy to the system, once it is out-of-phase. In a real structure, however, a controller will have to respect certain a priori given bounds. There is an extensive literature devoted to the problem of controllability/stabilizability by bounded controls. We do not want dwell on this question here. We refer the reader to a recent paper by Rao [16] and to the classical work of Haraux [10], [11] and also to Cabannes [2], [3]. Without going into the theoretical detail, we present the response of the system (5.3) without time delay but with respect to the nonlinear boundary controls:

$$f^1 = -sign(\dot{w}^2(1)) + (w^2)'(1), \; f^2 = sign(\dot{w}^2(0)) + (w^2)'(0)$$

See figure 8, figure8. It is apparent that these controls lead to considerably improved decay to a residual motion in the second string. It is interesting to observe the pattern of the strain in that string. If we now take these nonlinear boundary controls and if we account for a small time delay in the control (say one time step) then we observe no such instability. We dispense with the mesh plot and diplay the energy plot only; see figure 8.

The proof that this observed robusteness is actually true for the infinite dimensional system appears to be an open problem. □

The kind of nonlinearity of boundary controls just considered is in fact derived from a physical principle, namely *dry friction*. We come to this point later, when we are going to give an example of implicit Runge- Kutta methods applied to differential algebraic equations.

In the next example we want to demonstrate how directing controls can be used to improve the system response by 'directing' the energy to a 'safe' part of the structure. This is of great practical importance because a large flexible structure usually contains many uncontrollable circuits. Even if one puts the classical dissipative controls at each node, there will be a residual vibration in the system which will not decay to zero. Usage of 'directing' controls, however, can serve to 'channel' the energy through the network to end up at a set of nodes where either the energy influx is not important or where incoming energy is more easily absorbed.

Example 5.3. *Five strings containing a circuit.*

This model consists of a unit square (labels 1,2,3,4) with another string (#5) attached to the node connecting string #2 and string #3. We have an initial

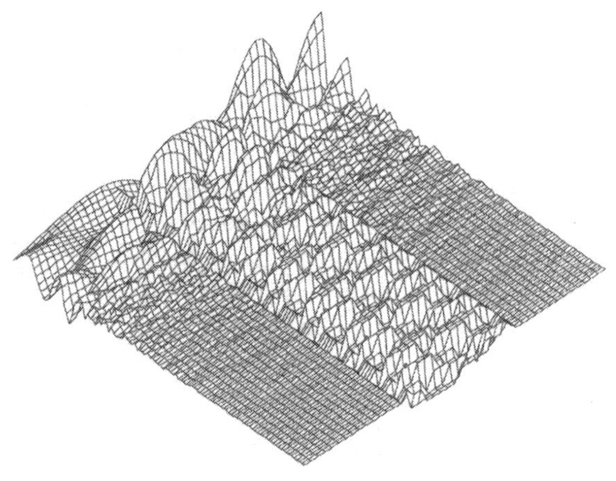

FIGURE 8. bang-bang control

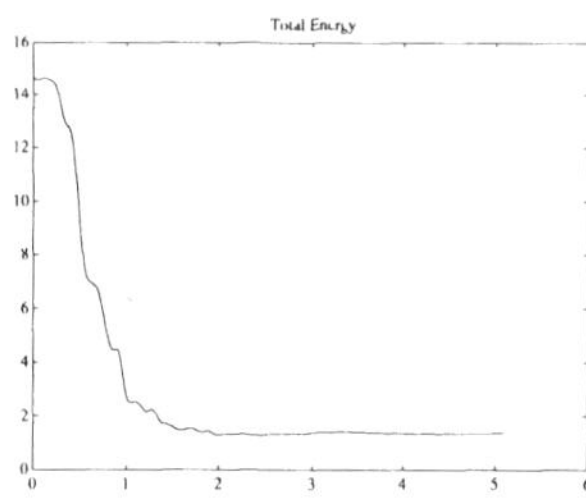

FIGURE 9
Improved decay

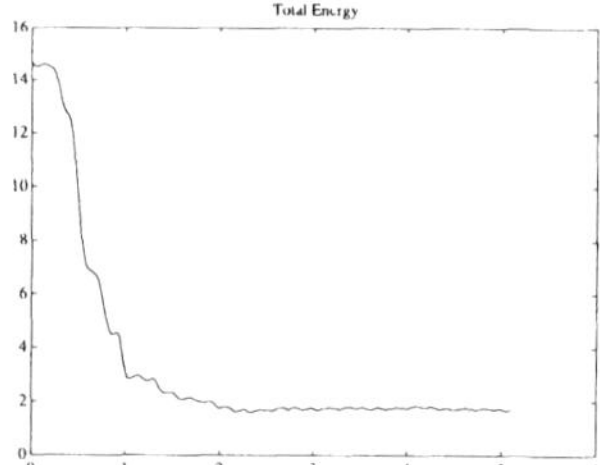

FIGURE 10
Robusteness

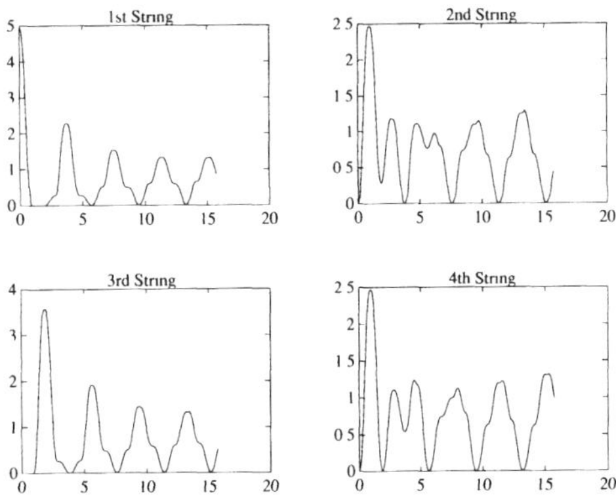

FIGURE 11. Energy in individual strings

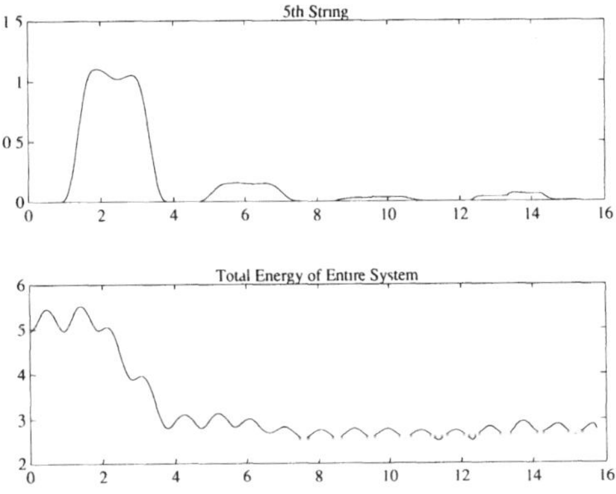

FIGURE 12. Energy in 5th. string and entire system

displacement in the first string only (as usual). In the first experiment we apply an absorbing control only at the free end of the attached string.

Figures 11,12 reveal what one expects.

Energy is taken out of the circuit until a residual motion settles. In particular, the plot of the total energy of the 5th string shows that with each cycle less energy can be absorbed from the circuit because of the gradually vanishing of the nodal excitation. The situation completely changes, once 'directing' controls are introduced around the vertex connecting strings #2, #3 and #5. See figures 13 and 14.

□

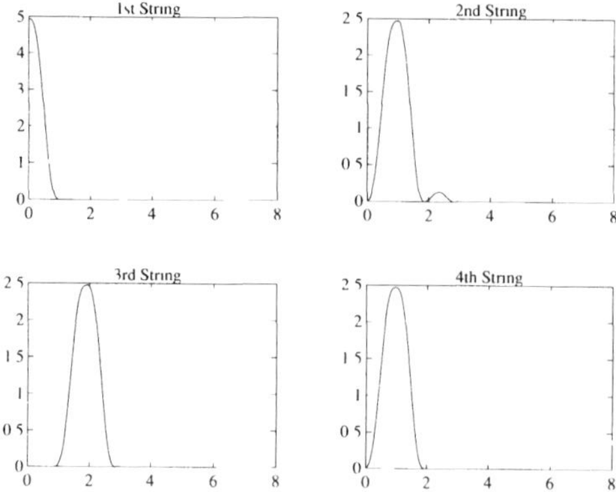

FIGURE 13. Directing controls

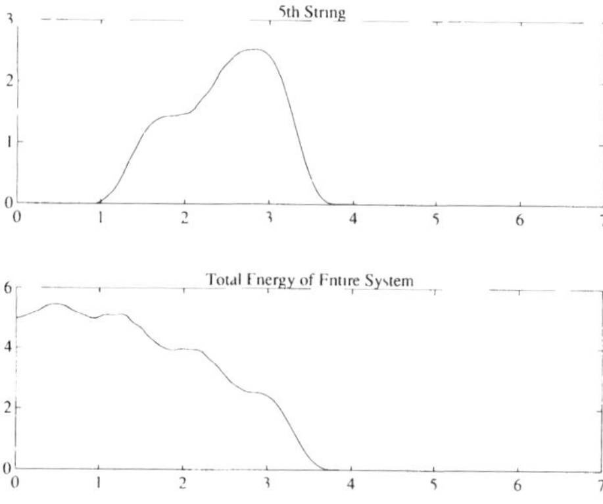

FIGURE 14

5.3. Finite elements. As mentioned above, the previous calculations have been based on string models. In dealing with networks of Timoshenko beams the numerical simulation becomes more involved, and it appears less appealing to use finite difference shemes. In the case of networks of Timoshenko beams and also for more complex string networks, the finite-element method seems best suited. Because of space limitations, we have to be very brief and refer the reader to Hughes [12]. Our concern in this section is to give numerical evidence for the lack of approximate controllability / stabilizability of a network of planar Timoshenko beams containing a circuit.

Let **K**, **C**, **M** denote the global stiffness, global damping and global mass matrix, respectively. Then the system of equations governing the evolution of the time dependent nodal variables can be written down as follows.

$$\mathbf{M}\ddot{\mathbf{U}} + \mathbf{C}\dot{\mathbf{U}} + \mathbf{K}\mathbf{U} = \mathbf{F}, \tag{5.6}$$

where we use $\mathbf{U}$ as the vector containing the nodal variables of all element nodes ($\mathbf{F}$ represents all applied forces). We recall the family of Newmark schemes which are actually of the Predictor- Corrector type. Again, we follow Hughes [12]. To this end we put

$$\mathbf{a}_n = \ddot{\mathbf{U}}(t_n),\ \mathbf{v}_n = \dot{\mathbf{U}}(t_n), \mathbf{d}_n = \mathbf{U}(t_n), \tag{5.7}$$

for the acceleration, velocity and displacement, respectively. We then define the *predictors* by

$$\begin{aligned} \tilde{\mathbf{d}}_{n+1} &= \mathbf{d}_n + \Delta t \mathbf{v}_n + \Delta\frac{t^2}{2}(1-2\beta)\mathbf{a}_n, \\ \tilde{\mathbf{v}}_{n+1} &= \mathbf{v}_n + (1-\gamma)\Delta t \mathbf{a}_n. \end{aligned} \tag{5.8}$$

As for starting values, we have

$$\mathbf{M}\mathbf{a}_0 = \mathbf{F} - \mathbf{C}\mathbf{v}_0 - \mathbf{K}\mathbf{d}_0.$$

This equation can be solved for $\mathbf{a}_0$. In the step $n+1$ the acceleration $\mathbf{a}_{n+1}$ is calculated from previous quantities and a priori known data by the recursion

$$(\mathbf{M} + \gamma\Delta t\mathbf{C} + \beta\Delta t^2\mathbf{K})\mathbf{a}_{n+1} = \mathbf{F}_{n+1} - \mathbf{C}\tilde{\mathbf{v}}_{n+1} - \mathbf{K}\tilde{\mathbf{d}}_{n+1}. \tag{5.9}$$

Once $\mathbf{a}_{n+1}$ is known from (5.9), we update the predictors via socalled *correctors*

$$\begin{aligned} \mathbf{d}_{n+1} &= \tilde{\mathbf{d}}_{n+1} + \beta\Delta t^2\mathbf{a}_{n+1}, \\ \mathbf{v}_{n+1} &= \tilde{\mathbf{v}}_{n+1} + \gamma\Delta t\mathbf{a}_{n+1}. \end{aligned} \tag{5.10}$$

The result of (5.10) is then inserted into (5.8) which, in turn, will then be used in (5.9) and so on. The two Newmark parameters can be chosen in various ways resulting in various different properties of the scheme. We will always make the choice: $\beta = \frac{1}{4}$, $\gamma = \frac{1}{2}$. This choice is known to lead to conservation of the energy above.

In our calculations we confine ourselves to planar frames of Timoshenko beams.

In that case the local stiffness matrix $\mathbf{k}^e$ is a 6×6 matrix as we have vertical and longitudinal displacement and one rotation in the plane for each of the two nodes of one element. For the sake of completeness we write it out explicitly

$$
\mathbf{k}^e = \begin{pmatrix}
\frac{EA}{\ell} & 0 & 0 & -\frac{EA}{\ell} & 0 & 0 \\
0 & \frac{A^s}{\ell} & \frac{A^s}{2} & 0 & -\frac{A^s}{\ell} & \frac{A^s}{2} \\
0 & \frac{A^s}{\ell} & \frac{A^s\ell}{4} + \frac{EI}{2} & 0 & -\frac{A^s}{2} & \frac{A^s\ell}{4} - \frac{EI}{2} \\
-\frac{EA}{2} & 0 & 0 & \frac{EA}{\ell} & 0 & 0 \\
0 & -\frac{A^s}{\ell} & -\frac{A^s}{2} & 0 & \frac{A^s}{\ell} & -\frac{A^s}{2} \\
0 & \frac{A^s}{\ell} & \frac{A^s\ell}{4} - \frac{EI}{2} & 0 & -\frac{A^s}{2} & \frac{A^s\ell}{4} + \frac{EI}{2}
\end{pmatrix} \tag{5.11}
$$

As for the lumped mass matrix we take

$$
\mathbf{m}^e = \frac{m_0 A\ell}{2} \begin{pmatrix}
1 & 0 & 0 & 0 & 0 & 0 \\
0 & 1 & 0 & 0 & 0 & 0 \\
0 & 0 & \frac{I}{A} & 0 & 0 & 0 \\
0 & 0 & 0 & 1 & 0 & 0 \\
0 & 0 & 0 & 0 & 1 & 0 \\
0 & 0 & 0 & 0 & 0 & \frac{I}{A}
\end{pmatrix} \tag{5.12}
$$

See [12] p. 515 for the planar beam without longitudinal displacement. With the bar wave velocity $c = \sqrt{E/m_0}$, and the beam shear wave velocity $c_s = \frac{GA^s}{m_0 A}$, the critical time step Δt can be estimated as

$$
\Delta t \leq \min\{\frac{h}{c}, (\frac{h}{c_s})\big(1 + \frac{A}{I}(\frac{h}{2})^2\big)^{-\frac{1}{2}}\}.
$$

Example 5.4. *The unit square of planar Timoshenko beams*

Here we want to show two things: at first we consider the uncontrolled system. Again, we have an initial displacement only in the bottom beam (#1), which is horizontal. The difficulty in showing all the displacements and rotations is obvious. At the nodes local vertical motion of beam # 1 converts to local longitudinal motion of the upright beams #2 and #4, which, in turn, is converted to local vertical motion in the top (horizontal) beam(#3). Because in the finite-element approach we automatically have the local rotations Γ_i available, we can easily go back to the global reference frame. In this way the locally longitudinal motion of the 2nd and 4th beam is, indeed, globally vertical. See figure 15. This plot clearly shows the collision of waves in the 3rd beam after travelling safely through beams #2 and #4. The initial bump next to the corner at the bottom indicates the location of the first beam. Figure 16 shows the conservation of energy.

In the second experiment we apply dissipative controls at all four nodes. See figure 17 and figure 18. Figure 17 clearly shows how the motion settles at a residual oscillation with nodes at the vertices. Accordingly, the energy settles at a constant value; see figure18.

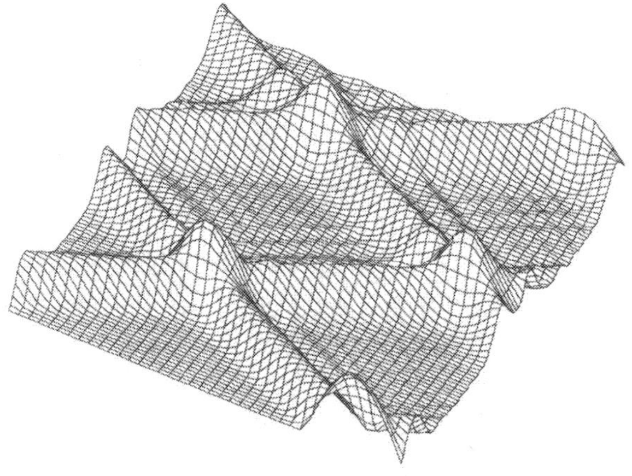

FIGURE 15. Unit square: uncontrolled motion

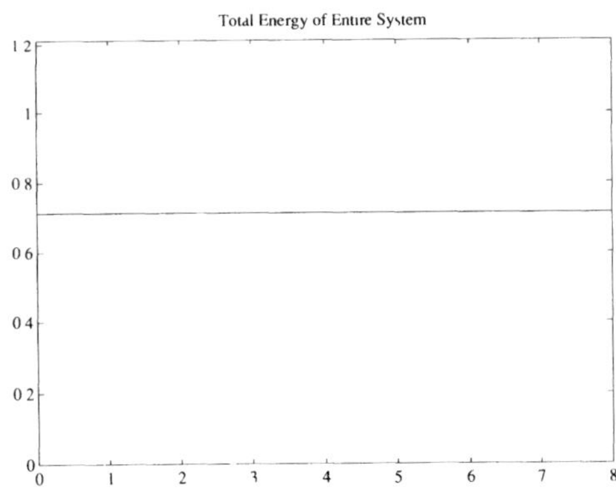

FIGURE 16. Unit square: conservation of energy

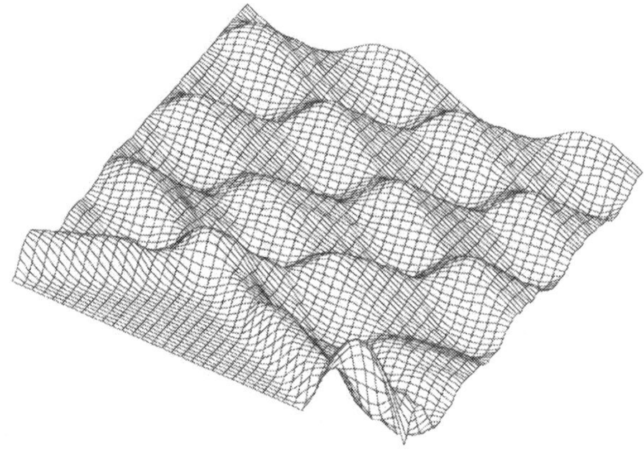

FIGURE 17. Unit square: controlled motion

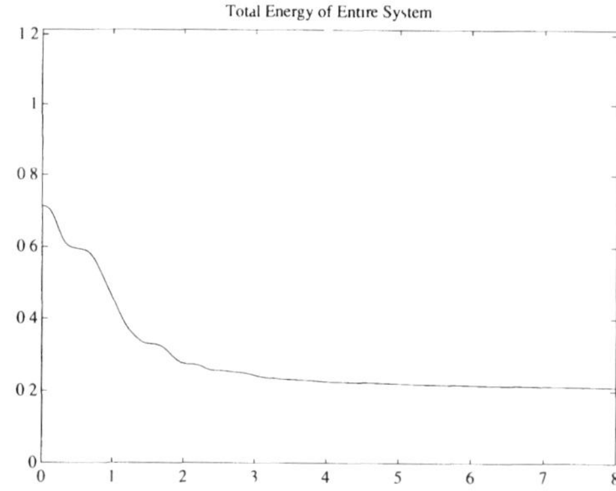

FIGURE 18. Unit square: saturation of energy

6. Implicit Runge-Kutta method applied to differential algebraic systems in the context of dry friction at the joints

We consider two identical strings of length 1 put together at the end point where $x = 1$ for both strings. We assume clamped boundaries at $x = 0$ for both strings, and we assume dry friction at the common point $x = 1$. The resulting model is as follows:

$$
\begin{aligned}
&\ddot{w}^i = (w^i)'', \quad i = 1, 2,\\
&w^1(0) = w^2(0) = 0,\ w^1(1) = w^2(1),\\
&(w^1)'(1) + (w^2)'(1) = -\rho_s z,\ z = \begin{cases} \operatorname{sign}(\dot{w}^1(1)), & \text{if } \dot{w}^1(1) \neq 0 \\ \in [-\frac{\rho_a}{\rho_s}, \frac{\rho_a}{\rho_s}], & \text{else} \end{cases},\\
&w^1(x,0) = w_0^1(x),\ w^2(x,0) = 0,\ \dot{w}^1(x,0) = \dot{w}^2(x,0) = 0.
\end{aligned}
\tag{6.1}
$$

Existence and uniqueness of solutions to this and much more general systems can be treated along the lines of Haraux [10]. In this special case we use the methods of characteristics. We perform the familiar transformations

$$
\begin{aligned}
u_1 &:= \dot{w}^1,\ u_2 := \dot{w}_2,\ u_3 := (w^1)',\ u_4 := (w^2)',\\
v_1 &:= \frac{1}{\sqrt{2}}(-u_1 + u_3),\ v_3 := \frac{1}{\sqrt{2}}(u_1 + u_3),\\
v_2 &:= \frac{1}{\sqrt{2}}(-u_2 + u_4),\ v_4 := \frac{1}{\sqrt{2}}(u_2 + u_4).
\end{aligned}
$$

We then have the system

$$
\dot{\mathbf{v}} = \mathbf{\Gamma}\mathbf{v}',\ \mathbf{\Gamma} = \operatorname{diag}([-1, -1, 1, 1]),
$$

with the boundary and node conditions

$$
\begin{gathered}
(v_3 - v_1)(0) = 0,\ (v_3 - v_1)(1) = (v_4 - v_2)(1),\\
(v_4 - v_2)(0) = 0,\ (v_3 + v_1)(1) + (v_4 + v_2)(1) = -\rho_s\sqrt{2}z,\\
\text{with}\\
z = \begin{cases} \operatorname{sign}(v_3 - v_1)(1) & \text{if } (v_3 - v_1)(1) \neq 0 \\ \in [-\frac{\rho_a}{\rho_s}, \frac{\rho_a}{\rho_s}], & \text{else.} \end{cases}
\end{gathered}
$$

These can be rewritten as

$$
\begin{pmatrix} v_1 \\ v_2 \end{pmatrix}(0) = \begin{pmatrix} v_3 \\ v_4 \end{pmatrix}(0),\ \begin{pmatrix} v_3 \\ v_4 \end{pmatrix}(1) = -\begin{pmatrix} 0 & 1 \\ 1 & 0 \end{pmatrix}\begin{pmatrix} v_1 \\ v_2 \end{pmatrix}(1) - \rho_s\frac{1}{\sqrt{2}}z\begin{pmatrix} 1 \\ 1 \end{pmatrix}.
$$

As a consequence of our special choice of initial conditions (we could do more general but we want to treat this problem only as an exemplaric one), we have $v_2(\cdot,0) = v_4(\cdot,0) = 0$.

We will consider two cases:

i.) $|(w^1)'(\cdot,0)| \leq \rho_a$
ii.) $|(w^1)'(\cdot,0)| > \rho_a$

We claim that in case i.) no motion takes place on the second string for $t \geq 0$.

Proof. We have

$$(v_3 - v_1)(1) = 0 \text{ at } t = 0.$$

Take

$$z = -\frac{\sqrt{2}}{\rho_s} v_1(1,t) \in [-\frac{\rho_a}{\rho_s}, \frac{\rho_a}{\rho_s}], \ (t \in [0,1]).$$

We follow the reflections at $x = 1$ and subsequently at $x = 0$, i.e.

$$v_3(1,t) = -v_{20}(1-t) + v_{10}(1-t) = v_{10}(1-t),$$
$$v_4(1,t) = 0, \qquad t \in [0,1],$$

$$v_1(0,1+t) = v_{10}(1-t),$$
$$v_2(0,1+t) = 0.$$

Hence, upon reflection, v_1 does not change magnitude, and with z as above, v_4 is kept equal to zero, causing also v_2 to be equal to zero.

The analogous case starting with v_{30}, v_{40} and travelling towards the boundary at $x = 0$ first is settled in a similar way. The main point is that the variable z (representing the adhesive force) always compensates the shear force ('kills' the outgoing signal $v_4(1,t)$) at the boundary $x = 1$ which represent the multiple node in our original model. □

We also claim that in case ii.) motion does spill over into the second string

Proof. We have initially $(v_3 - v_1)(1,0) = 0$. Assume now

$$(v_3 - v_1)(1,t) = 0, \ t \in (0,\varepsilon).$$

$$z(t) \in [-\frac{\rho_a}{\rho_s}, \frac{\rho_a}{\rho_s}] \Rightarrow z(t) \neq -\frac{\sqrt{2}}{\rho_s} v_1(1,t), \text{ on } (0,\varepsilon)$$
$$\Rightarrow v_4(1,t) \neq 0, \text{ on } (0,\varepsilon) \Rightarrow v_2(0,1+t) \neq 0.$$

The latter inclusion says that there is non zero motion in the second string. The other cases are handled in a similar way. □

We are going to give numerical evidence to these remarks. We consider the mesh points $x_i = \frac{i}{N}$, $N = 40$ and use the classical finite-difference approximation to the second order operator (which, incidentally, equals the corresponding finite element stiffness matrix). Indeed, we can view this process happening in one string with dry friction at the midpoint. We have the following system:

$$\ddot{w}_i - N^2\{w_{i-1}(t) - 2\,w_i(t) + w_{i+1}(t)\} = \begin{cases} -\rho_s z & \text{if } i = \frac{N}{2} \\ 0 & \text{else} \end{cases} \tag{6.2}$$

where now z is given by

$$\begin{cases} \text{if} \quad \dot{w}_{\frac{N}{2}}(t) \neq 0 & \Rightarrow z = \operatorname{sign}(\dot{w}_{\frac{N}{2}}(t)) \\ \text{if} \quad \dot{w}_{\frac{N}{2}}(t) = 0 & \Rightarrow \rho_s z \in \rho_a[-1,1]. \end{cases} \tag{6.3}$$

We also have boundary and initial conditions:

$$w_0(t) = 0,\; w_N(t) = 0,\; w_i(0) = w_i(0),\; \dot{w}_i = 0. \tag{6.4}$$

In this system, the variable z can be viewed as a control variable which forces the the solution of the system (6.2) to satisfy $\dot{w}_{\frac{N}{2}} = 0$. In order to do that, this variable z, which physically represents the adhesive force, will vary in time. In order to make the rôle of z more transparent, we consider a general differential algebraic equation of the following kind

$$\begin{cases} \dot{\mathbf{y}} &= \mathbf{f}(\mathbf{y}, \mathbf{z}), \\ 0 &= \mathbf{g}(\mathbf{y}), \end{cases}$$

where $\mathbf{f} : \mathbf{R}^n \times \mathbf{R}^m \longrightarrow \mathbf{R}^n$, $\mathbf{g} : \mathbf{R}^n \longrightarrow \mathbf{R}^n$ are C^1, C^2-functions, respectively. Denote the Jacobian of $\mathbf{f}$ with respect to $\mathbf{y}$, $\mathbf{z}$ by $\mathbf{f}_\mathbf{y}$, $\mathbf{f}_\mathbf{z}$. Then we may differentiate $(6.5)_2$ with repsect to the variable x we obtain

$$0 = \mathbf{g}_\mathbf{y}(\mathbf{y})\mathbf{f}(\mathbf{y}, \mathbf{z}).$$

Another differentiation with respect to x yields

$$0 = (\mathbf{g}_{\mathbf{yy}}(\mathbf{y})\mathbf{f}(\mathbf{y}, \mathbf{z}) + \mathbf{g}_\mathbf{y}(\mathbf{y})\mathbf{f}_\mathbf{y}(\mathbf{y}, \mathbf{z}))\mathbf{f}(\mathbf{y}, \mathbf{z}) + \mathbf{g}_\mathbf{y}(\mathbf{y}\mathbf{f}_\mathbf{z}(\mathbf{y}), \mathbf{z})\mathbf{z}'. \tag{6.5}$$

Obviously, if $\mathbf{g}_\mathbf{y}(\mathbf{y})\mathbf{f}_\mathbf{z}(\mathbf{y}, \mathbf{z})$ is invertible in a neighbourhood of a solution, say $(\bar{\mathbf{y}}, \bar{\mathbf{z}})$ then (6.5) is equivalent to an ordinary differential equation. This situation is said to have index 2. Also, in this case for the problem

$$\mathbf{G}(\mathbf{y}, \mathbf{z}) := \mathbf{g}_\mathbf{y}(\mathbf{y})\mathbf{f}(\mathbf{y}, \mathbf{z}) = 0, \quad \mathbf{G}(\bar{\mathbf{y}}, \bar{\mathbf{z}}) = 0,$$

by the implicit function, there exists a neighbourhood $\mathcal{U}(\bar{\mathbf{y}}, \bar{\mathbf{z}})$ and a solution $\mathbf{z}(\mathbf{y}, x)$ of that equation such that (6.5) holds. See Hairer and Wanner [8] for the theory of differential algebraic equations. With this remark in mind we go back to our system, where the case just considered corresponds to $(6.2),(6.3)_2$.

If, now, z is in the mode 2 of (6.3) and reaches the value $\frac{\rho_a}{\rho_s}$, one has to switch to the first case in (6.3), which together with (6.2) then reduces to an ordinary differential equation. As a result, we have to switch between a differential

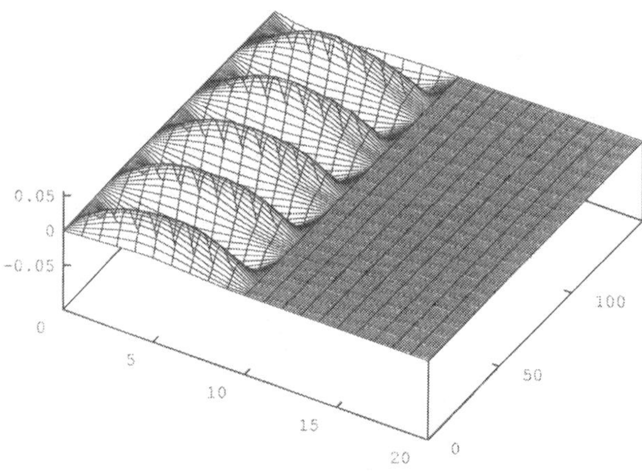

FIGURE 19. Dry friction: locking

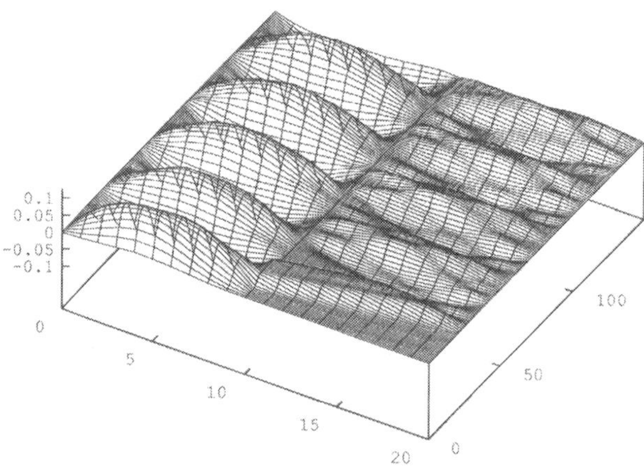

FIGURE 20. Dry friction: spill over

algebraic system (6.2),$(6.3)_2$ and (6.4) and an ordinary differential equation (initial value problem) (6.2), $(6.3)_1$ and (6.4). Switching problems of this kind have been treated by Brasey and Hairer [1]. They used the half- explicit 5 -stage Runge-Kutta method of order 4. We used the Radau5- implementation of the Radau IIa fully implicit Runge-Kutta method of order 5 also given by Hairer and Wanner in [8]. The corresponding driver was originally written by J. Verscht (Bayreuth) for a different problem . We made the appropriate changes to use it also for the problem under consideration and also for beam applications (which we do not reproduce here). The problem of switching between the two models is a major difficulty which is usually solved using 'dense output formulae' and some kind of extrapolation to find the switching manifold $\mathbf{g}(\mathbf{y}) = 0$. See the figures 20,19.

References

1 V Brasey and E Haırer, *Half-explıcıt Runge-Kutta methods for dıfferentıal-algebraıc systems of ındex 2*, SIAM J Numer Anal **30** (1993), no 2, 538–552

2 H Cabannes, *Mouvement d'une corde vıbrante soumıse à un fottement solıde*, C R Acad Sc Parıs **287** (1978), 671–673

3 ______, *Propagatıon des dıscontınuıtés dans les cordes vıbrante soumıses à un frottement solıde*, C R Acad Sc Parıs **289** (1979), 127–130

4 G Chen and Coleman, M and H H West, *Poıntwıse stabılızatıon ın the mıddle of the span for second order systems, nonunıform and unıform exponentıal decay of solutıons*, SIAM J Control Optım **47** (1987), no 4, 751–779

5 R Datko, *Not all feedback stabılızed hyperbolıc systems aer robust wıth respect to small tıme delays ın theır feedbacks*, SIAM J Control Optım **26** (1988) 697–713

6 ______, *A rank-one perturbatıon result on the spectra of certaın operators* Proceedıngs of the Royal Socıety of Edınburgh **111A** (1989), 337–346

7 R Datko and Lagnese, J E and M P Polıs, *An example of the effect of tıme delays ın boundary feedback stabılızatıon of wave equatıons* SIAM J Control Optım **24** (1986), 152–156

8 E Haırer and G Wanner, *Solvıng ordınary dıfferentıal equatıons II Stıff and dıfferentıal algebraıc systems*, Sprınger-Verlag, Berlın, 1987

9 L Halpern, *Absorbıng boundary condıtıons for the dıscretızatıon of the one-dımensıonal wave equatıon*, Mathematıcs of Computatıon **38** (1982), no 158, 415–429

10 A Haraux, *Comportement à l'ınfinı pour une equatıon d'ondes non lınéaıre dıssıpatıve*, C R Acad Sc Parıs **287** (1978), 507–509

11 ______, *Nonlınear evolutıon equatıons - global behavıor of solutıons*, Lect Notes ın Mathematıcs, vol 841, Sprınger-Verlag, Berlın, 1981

12 T J R Hughes, *The finıte element method Lınear statıc and dynamıc finıte element analysıs*, Prentıce-Hall, New Yersey, 1987

13 J E Lagnese and Leugerıng, G and E J P G Schmıdt, *Controllabılıty of planar networks of Tımoshenko beams*, SIAM J Control and Optım **31** (1993), 780–811

14 ______, *Modellıng, analysıs and control of multı-lınk flexıble structures*, Bırkhauser Verlag, 1994

15 G Leugerıng and E J P G Schmıdt, *On the control of networks of vıbratıng strıngs and beams*, Proceedıngs of the 28th IEEE Conference on Decısıon and Control, vol 3, IEEE, 1989, pp 2287–2290

16 Bopeng Rao, *Stabılısatıon du modéle SCOLE par un contrôle a prıorı borné*, C R Acad Scı Parıs, **316** (1993), 1061–1066

17. A. Y. Le Roux, *Stability of numerical schemes solving quasi-linear wave equations*, Mathematics of Computation **36** (1981), no.153, 93–105.
18. D. L. Russell, *Controllability and stabilizability theory for linear partial differential equations: recent progress and open questions*, SIAM Review **20** (1978), 639–739.
19. E. J. P. G. Schmidt, *On the modelling and exact controllability of networks of vibrating strings*, SIAM J. of Control and Optimization **30** (1992), 229–245.
20. J. Schmidt, *Entwurf von Reglern zur aktiven Schwingungsdämpfung an flexiblen mechanischen Strukturen*, Ph.D. thesis, 1987.

Günter Leugering
Mathematisches Institut
Universität Bayreuth
D 95440 Bayreuth, Germany

E-mail: leugering@uni-bayreuth.de

VARIOUS RELAXATIONS IN OPTIMAL CONTROL OF DISTRIBUTED PARAMETER SYSTEMS

TOMÁŠ ROUBÍČEK

Institute of Information Theory and Automation
Czech Academy of Sciences

ABSTRACT. The so-called convex-compactification theory is applied to an extension (=relaxation) of optimal control problems involving evolution distributed parameter systems. An infinite number of relaxed problems and corresponding Pontryagin maximum principles are thus obtained, including those described in the literature. A comparison and an abstract unifying viewpoint is thus made possible.

1991 *Mathematics Subject Classification.* 49J20, 49K20

Key words and phrases. Convex compactifications, optimal control, relaxation, Young measures, Pontryagin maximum principle, parabolic problems..

0. Introduction.

Optimal control problems, game-theoretical problems, or variational problems in their usual formulation typically do not admit any solution because of the absence of any compact topology making the data of the problem continuous. This is related in general with a lack of smoothness, or oscillation or concentration phenomena. Especially the oscillation phenomenon (i.e. the "ε-optimal" controls necessarily oscillate more and more rapidly when $\varepsilon \searrow 0$) appears typically in nonconvex and/or nonlinear optimal control problems. Therefore, an extension (=compactification) of the original space of admissible controls is urgently needed. This procedure, called "relaxation", has many variants especially in the cases where controls are functions of several variables (e.g. space and time), which appears typically in control of evolution distributed parameter systems. The aim of this paper is to present briefly a general theory, which covers by a unified way various relaxations that appeared in the literature, and to compare them together with the resulting Pontryagin maximum principle which is intimately related with the chosen relaxation.

1. Convex compactification theory.

Let us first consider an abstract topological space U of "original controls". The compactification of U (i.e. a continuous dense imbedding of U into a compact space K) is an inappropriately general approach for our purposes, so that we impose a restrictive requirement that K is to be a convex subset of a locally convex space Z. Then we will speak about a convex compactification. More precisely, a triple (K, Z, i) will be called a convex compactification of U if

(1a) K is a convex compact subset of Z,
(1b) Z is a locally convex space,
(1c) $i : U \to K$ is continuous,
(1d) $i(U)$ is dense in K.

Mostly, we will deal with Hausdorff convex compactifications (i.e. i injective) and then we will often identify U with $i(U)$ for simplicity.

To compare various convex compactifications of U, we introduce a natural ordering: for (K_1, Z_1, i_1) and (K_2, Z_2, i_2) two convex compactifications of U, we will say that (K_1, Z_1, i_1) is finer than (K_2, Z_2, i_2) if there is a continuous affine surjection $\psi : K_1 \to K_2$ such that $\psi \circ i_1 = i_2$. Then we will also say that (K_2, Z_2, i_2) is coarser than (K_1, Z_1, i_1), and write $(K_1, Z_1, i_1) \succeq (K_2, Z_2, i_2)$, or briefly $K_1 \succeq K_2$. If (K_1, Z_1, i_1) is simultaneously finer and coarser than (K_2, Z_2, i_2), then we will say that they are equivalent to each other, and write $K_1 \cong K_2$. If $K_1 \succeq K_2$ but $K_1 \not\cong K_2$, then we will say that K_1 is strictly finer than K_2, and write $K_1 \succ K_2$.

Having a convex compactification (K, Z, i), we will sometimes omit Z and i when clear for the context. Examples of convex compactifications will be given later.

One "canonical" construction is always possible. Let us denote by $C(U)$ a Banach space of continuous bounded functions on U, and let us consider a linear subspace $\mathcal{F} \subset C(U)$ containing constants and satisfying: $\forall u_1, u_2 \subset U \ \exists$ a net $\{u_\alpha\}$ $\forall f \in \mathcal{F}$: $\lim f(u_\alpha) = \frac{1}{2} f(u_1) + \frac{1}{2} f(u_2)$; such subspaces will be called "convexifying". Besides, let $e : U \to \mathcal{F}^* : u \mapsto (f \mapsto f(u))$ denote the evaluation mapping, let the dual space $\mathcal{F}^*$ be endowed with the weak* topology, and let

$$M(\mathcal{F}) = \{\mu \in \mathcal{F}^*;\ \|\mu\|_{\mathcal{F}^*} = 1 \ \&\ \langle \mu, 1 \rangle = 1\} .$$

The elements of $M(\mathcal{F})$ are usually called means on $\mathcal{F}$. The following assertion summarizes some results from [20, 24].

Theorem 1. *Let $\mathcal{F}$ be a subspace of $C(U)$ containing constants. Then $(M(\mathcal{F}), \mathcal{F}^*, e)$ is a convex compactification of U if and only if $\mathcal{F}$ is convexifying. Moreover, every convex compactification is equivalent with $M(\mathcal{F})$ for some closed convexifying subspace $\mathcal{F} \subset C(U)$. If $\mathcal{F}_1, \mathcal{F}_2$ are two convexifying subspaces of $C(U)$ containing constants, then $M(\mathcal{F}_1) \succeq M(\mathcal{F}_2)$ provided $\mathcal{F}_1 \supset \mathcal{F}_2$. If the closures of $\mathcal{F}_1$ and $\mathcal{F}_2$ in $C(U)$ coincide with each other, then $M(\mathcal{F}_1) \cong M(\mathcal{F}_2)$. Conversely, if $\mathcal{F}_1 \supset \mathcal{F}_2$ but their closures in $C(U)$ do not coincide, then $M(\mathcal{F}_1) \succ M(\mathcal{F}_2)$.*

Finally, every convexifying subspace $\mathcal{F} \subset C(U)$ is contained in some maximal convexifying subspace.

For a convex compactification K of U, the equivalent convex compactification $M(\mathcal{F})$ will be referred to as a canonical form of K.

2. An abstract relaxation pattern.

If one neglects a concrete structure, every optimal control problem eventually takes the form of an abstract minimization problem:

$$\text{(P)} \qquad \text{Minimize } \Phi(u) \text{ over } u \in U \, ,$$

where $\Phi : U \to \mathbb{R}$ is a cost function. As mentioned in Sect. 0, (P) need not have any solution and the need of its relaxation by a compactification of U and an extension of Φ immediately arises. The reader can certainly anticipate that we will admit only a convex compactification of U, and the canonical form will be used. Therefore, the relaxed problem takes the form

$$\text{(RP)} \qquad \text{Minimize } \bar{\Phi}(\mu) \text{ over } \mu \in M(\mathcal{F}) \, ,$$

where $(M(\mathcal{F}), \mathcal{F}^*, e)$ is some convex compactification of U and $\bar{\Phi} : M(\mathcal{F}) \to \mathbb{R}$ a l.s.c. (=lower semicontinuous) extension of Φ; i.e. $\bar{\Phi} \circ e = \Phi$.

It is obvious that (RP) possesses always a solution, which is to be considered as a generalized solution of (P). Indeed, the relation between (RP) and (P) is very intimate: every cluster point in $\mathcal{F}^*$ of $\{e(u_k)\}_{k \in \mathbb{N}}$, where $\{u_k\}_{k \in \mathbb{N}}$ is a minimizing sequence for (P), solves (RP) and conversely every solution of (RP) can be reached by a net $\{e(u_\alpha)\}$ such that $\lim_\alpha \Phi(u_\alpha) = \inf \Phi(U)$. In particular, every solution of (P) solves (after being imbedded via e) the relaxed problem (RP).

The convex structure enables to set up the first-order necessary optimality condition. For this reason, we will say that $\bar{\Phi}$ has at $\mu_0 \in M(\mathcal{F})$ the Gâteaux differential $\bar{\Phi}'(\mu_0) \in \mathcal{F}$ if

$$(2) \quad \forall \mu \in M(\mathcal{F}) : \qquad \langle \mu - \mu_0, \bar{\Phi}'(\mu_0) \rangle \; = \; \lim_{\varepsilon \searrow 0} \frac{1}{\varepsilon} \left(\bar{\Phi}(\mu_0 + \varepsilon(\mu - \mu_0)) - \bar{\Phi}(\mu_0) \right) \, .$$

Let us only remind that, since $\mathcal{F}^*$ is endowed with the weak* topology, every linear continuous functional on $\mathcal{F}^*$ has the form $\mu \mapsto \langle \mu, f \rangle$ for some $f \in \mathcal{F}$, which explains why $\bar{\Phi}'(\mu_0)$ lives in $\mathcal{F}$ and not in $\mathcal{F}^{**} \setminus \mathcal{F}$. Besides it is known [22] that $\bar{\Phi}'(\mu_0) \in \mathcal{F}$ is determined by (2) uniquely up to constants on U.

If μ_0 solves (RP) and $\bar{\Phi}$ has the Gâteaux differential at μ_0, then it must belong to the negative normal cone to $M(\mathcal{F})$; this means

$$(3) \qquad \bar{\Phi}'(\mu_0) \; \in \; -N_{M(\mathcal{F})}(\mu_0) \, ,$$

where $N_{M(\mathcal{F})}(\mu_0)$ denotes the normal cone $\{f \in \mathcal{F};\ \forall \mu \in M(\mathcal{F}) : \ \langle \mu - \mu_0, f \rangle \le 0\}$. If $N_{M(\mathcal{F})}(\mu_0) = \{\text{constants}\}$, then (3) turns out to the condition $\bar{\Phi}'(\mu_0)$ to be constant, which is essentially the standard Euler-Lagrange equation $\bar{\Phi}'(\mu_0) = 0$ because the constants are always included in $N_{M(\mathcal{F})}(\mu_0)$ so that they are factually irrelevant.

However, in nonlinear optimal control problems where a finer relaxation is generally inevitable the normal cone $N_{M(\mathcal{F})}(\mu_0)$ is typically much larger, cf. also Section 4 below.

In view of the density of $e(U)$ in $M(\mathcal{F})$, we can obviously rewrite (2)–(3) as

$$\langle \mu_0, \bar{\Phi}'(\mu_0) \rangle = \sup_{u \in U} \left[\bar{\Phi}'(\mu_0)\right](u) . \tag{4}$$

We will refer to (4) as an *abstract maximum principle.*

Let us only outline the situation when (P) comes from an optimal control problem having an abstract structure

$$(\mathsf{OCP}) \qquad \begin{cases} \text{Minimize } \varphi(u, y) \\ \text{subject to } u \in U,\ y \in Y,\ A(u, y) = 0, \end{cases}$$

where $\varphi : U \times Y \to \mathbb{R}$ is a cost function, Y a Banach space of states, and $A : U \times Y \to \Lambda$ with Λ another Banach space determines a state operator $\pi : U \to Y$ such that $y = \pi(u)$ if and only if $A(u, y) = 0$. We will suppose that the convex compactification $M(\mathcal{F})$ of U is so fine that both A and φ admit continuous extensions $\bar{\varphi} : M(\mathcal{F}) \times Y \to \mathbb{R}$ and $\bar{A} : M(\mathcal{F}) \times Y \to \Lambda$ and, moreover, for every $\mu \in M(\mathcal{F})$ the equation $\bar{A}(\mu, y) = 0$ has a unique solution $y = \bar{\pi}(\mu)$ and the state operator $\bar{\pi} : M(\mathcal{F}) \to Y$ thus determined is continuous (then $\bar{\pi}$ is the continuous extension of the original operator $\pi : U \to Y$ in the sense that $\bar{\pi} \circ e = \pi$). We can then define the following relaxed optimal control problem:

$$(\mathsf{ROCP}) \qquad \begin{cases} \text{Minimize } \bar{\varphi}(\mu, y) \\ \text{subject to } \mu \in M(\mathcal{F}),\ y \in Y,\ \bar{A}(\mu, y) = 0, \end{cases}$$

The existence of a solution (ROCP) (considered as a generalized solution of (OCP)) and stability of the set of all these solutions is again ensured by the standard compactness and continuity arguments.

The problem (ROCP) can be equivalently written in the form of (RP) by putting $\bar{\Phi}(\mu) = \bar{\varphi}(\mu, \bar{\pi}(\mu))$. Let us suppose that $\bar{\Phi} : M(\mathcal{F}) \to R$ is Gâteaux differentiable. By using the adjoint-equation technique developed essentially in [16] we can evaluate the Gâteaux differential $\bar{\Phi}'(\mu_0)$ as follows: Let us assume that $\bar{\varphi}(\mu_0, .)$ has the continuous Fréchet derivative $\bar{\varphi}'_y : M(\mathcal{F}) \times Y \to Y^*$ and $\bar{\varphi}(., y)$ has a Gâteaux derivative $\bar{\varphi}'_u \in \mathcal{F}$, and the same holds for $\bar{A}$ weakly; this means there is $\bar{A}'_y : M(\mathcal{F}) \times Y \to \mathcal{L}(Y, \Lambda)$ such that $\langle \lambda, \bar{A}'_y(\mu_0, y) \rangle \in Y^*$ is the Fréchet derivative of $\langle \lambda, \bar{A}(\mu_0, .) \rangle : Y \to \mathbb{R}$ at $y \in Y$ and $\langle \lambda, \bar{A}'_y \rangle : M(\mathcal{F}) \times Y \to Y^*$ is continuous for all $\lambda \in \Lambda^*$, and there is $\bar{A}'_u : M(\mathcal{F}) \times Y \to \mathcal{L}(\mathcal{F}^*, \Lambda)$ such that $\langle \lambda, \bar{A}'_u(\mu_0, y) \rangle \in \mathcal{F}$ is the Gâteaux derivative of $\langle \lambda, \bar{A}(., y) \rangle$ at $\mu_0 \in M(\mathcal{F})$ for all λ; $\mathcal{L}(., .)$ denotes the space of all continuous, linear operators. Moreover, let the state operator $\bar{\pi} : M(\mathcal{F}) \to Y$

be directionally Lipschitz continuous, i.e. $\|\bar{\pi}(\mu_0 + h(\mu - \mu_0)) - \bar{\pi}(\mu_0)\| \leq L_{\mu_0,\mu}h$. If the adjoint equation

$$[\bar{A}'_y(\mu_0, \bar{\pi}(\mu_0))]^*\lambda \ + \ \bar{\varphi}'_y(\mu_0, \bar{\pi}(\mu_0)) \ = \ 0 \tag{5}$$

has a solution $\lambda \in \Lambda^*$ (=the so-called adjoint state), then we can evaluate $\bar{\Phi}'(\mu_0) = \bar{\varphi}'_u(\mu_0, \bar{\pi}(\mu_0)) + [\bar{A}'_u(\mu_0, \bar{\pi}(\mu_0))]^*\lambda$+const. Obviously, (4) is then equivalent to (5) completed with the maximum principle

$$\langle \mu_0, f_{\lambda,y,\mu_0} \rangle \ = \ \sup_{u \in U} f_{\lambda,y,\mu_0}(u) \ , \tag{6}$$

where $f_{\lambda,y,\mu_0} \in \mathcal{F}$ is an "abstract Hamiltonian" defined by

$$f_{\lambda,y,\mu_0} \ = \ \bar{\varphi}'_u(\mu_0, y) \ + \ [\bar{A}'_u(\mu_0, y)]^*\lambda \ + \ \text{const.} \qquad \text{with} \quad y = \pi(\mu_0) \ . \tag{7}$$

We will see in Section 4 that in concrete relaxed problems the "Hamiltonian" from (7) actually gives the standard Hamiltonians which typically do not depend explicitly on μ_0 provided $\mathcal{F}$ is large enough so that both $\varphi(\cdot, y) \in \mathcal{F}$ and $\langle \lambda, A(\cdot, y) \rangle \in \mathcal{F}$ for any $\lambda \in \Lambda$.

3. Concrete convex compactifications of U.

In applications, relaxed controls are typically the so-called Young measures (i.e. parameterized probability measures) or some generalizations of them. It is related with a concrete form of the set of controls U. For simplicity, let us consider here only bounded controls in the form

$$U \ = \ \{u \in L^\infty(Q; \mathbb{R}^k); \ \forall \xi \in Q : \ u(\xi) \in S\} \tag{8}$$

with $Q \subset \mathbb{R}^n$ and $S \subset \mathbb{R}^k$ compact. Such U appears quite typically in optimal control of distributed parameter systems. Besides, let us suppose that $n = n_P + n_R$ and Q admits a decomposition $Q = P \times R$, with $P \subset \mathbb{R}^{n_P}$ and $R \subset \mathbb{R}^{n_R}$, $n_P \geq 1$, $n_R \geq 0$. In the case $n_R = 0$, we put simply $Q = P$. Roughly speaking, P denotes the "directions" in Q where the "rapid" oscillations of the controls appear. Let us put

$$B_R \ = \ \{R \to S \text{ measurable}\} \ , \tag{9}$$

which is a bounded subset of $L^\infty(R; \mathbb{R}^k)$; for $n_R = 0$ we put simply $B_R = S \subset \mathbb{R}^k$. Let us choose a linear subspace $H \subset L^1(P; C(B_R))$ and define the mapping $\Psi_P : H \to C(U)$ by $\Psi_P(h) = \big(u \mapsto \int_P h(\zeta, u(\zeta, \cdot))d\zeta\big)$; note that $\Psi_P(h)$ is nothing else than the integral over P of the Nemytskii operator generated by h. Furthermore, let us define the imbedding $i_H : U \to H^* : u \mapsto [\Psi_P(\cdot)](u)$ and denote by $Y_H(P; B_R)$ the weak* closure of $i_H(U)$ in H^*.

Lemma 1. (Cf. [20, 24]) *Let H be a linear subspace of $L^1(P; C(B_R))$ and $\mathcal{F} = \mathcal{F}_H = \Psi_P(H)$+const. Then $\mathcal{F}_H$ is a convexifying subspace of $C(U)$ with U from*

(8), and the convex compactification $(M(\mathcal{F}_H), \mathcal{F}_H^*, e)$ of U is equivalent with $(Y_H(P; B_R), H^*, i_H)$. More precisely, the affine homeomorphism $M(\mathcal{F}_H) \leftrightarrow Y_H(P; B_R)$ fixing U is the adjoint operator $\Psi_P^* : \mathcal{F}^* \to H^*$ restricted to $M(\mathcal{F})$.

Supposing that $C(P) \cdot H = H$ (this means $g \in C(P)$ and $h \in H$ implies $g \cdot h \in H$ for $[g \cdot h](\zeta, \cdot) = g(\zeta) \cdot h(\zeta, \cdot)$), we can define the bilinear mapping $\bullet : H \times Y_H(P; B_R) \to L^1(P)$ by $h \bullet \eta =$w-$\lim_\alpha h \circ u_\alpha$ for $i_H(u_\alpha) \to \eta$, where $[h \circ u](\zeta) = h(\zeta, u(\zeta, \cdot))$. Note that the net $\{|h \circ u_\alpha|\}$ has an integrable majorant so that its weak limit in $L^1(\Omega)$ does exist at least if a subnet is taken, but it exists even for the whole net because this limit is unique and equals $h \bullet \nu \in L^1(\Omega)$ defined alternatively by $\langle h \bullet \nu, g \rangle = \langle \eta, g \cdot h \rangle$ for every $g \in C(P)$.

Let us come back to Sect. 2 and suppose that $\bar{\Phi} : M(\mathcal{F}_H) \to \mathbb{R}$ is Gâteaux differentiable at μ_0. By Lemma 1, we can transfer our considerations to $Y_H(P; B_R)$: Let us put $\tilde{\Phi} = \bar{\Phi} \circ (\Psi_P^*)^{-1} : Y_H(P; B_R) \to \mathbb{R}$. Then also $\tilde{\Phi}$ is Gâteaux differentiable at $\eta_0 = \Psi_P^* \mu_0 \in Y_H(P; B_R)$ and $\tilde{\Phi}'(\eta_0) \in H$ is determined by $\Psi_P^{-1}(\bar{\Phi}'(\mu_0)) + h_0$ with arbitrary $h_0 \in H$ such that $h_0(\zeta, \cdot)$ is constant on B_R for a.a. $\zeta \in P$. We will call $\mathcal{H} = \tilde{\Phi}'(\eta_0) \in H$ the "Hamiltonian"; obviously, $\mathcal{H}$ is determined uniquely up to $H \cap [L^1(P) \otimes 1_R]$ where 1_R denotes the function $R \to \{1\}$.

Theorem 2. *The abstract maximum principle (4) is equivalent*

$$\int_P [\mathcal{H} \bullet \eta_0](\zeta) d\zeta \;=\; \sup_{u \in U} \int_P \mathcal{H}(\zeta, u(\zeta, \cdot)) d\zeta \;. \tag{10}$$

If $H \subset L^1(P; C(B_R^\tau))$ where B_R^τ denotes the set B_R endowed by a topology τ which makes it compact or a complete metrizable separable space, then (10) is also equivalent with

$$[\mathcal{H} \bullet \eta_0](\zeta) \;=\; \sup_{v \in B_R} \mathcal{H}(\zeta, v) \qquad \text{for a.a. } \zeta \in P \;. \tag{11}$$

Sketch of the proof. As for the equivalence of (4) with (10), it suffices to realize the indentity $\int_P [\mathcal{H} \bullet \eta](\zeta) d\zeta = \langle \eta, \mathcal{H} \rangle = \langle \Psi_P^* \mu, \mathcal{H} \rangle = \langle \mu, \Psi_P \mathcal{H} \rangle = \langle \mu, \Phi'(\mu_0) \rangle$ for any $\eta = \Psi_P^* \mu \in Y_H(P; B_R)$.

As for the localization of (10) to (11), we only need to make a measurable selection from a multivalued mapping of the type $\zeta \mapsto \{v \in B_R;\ \mathcal{H}(\zeta, v) \geq c(\zeta)\}$ with a suitable $c \in L^1(P)$, which needs the assumptions on τ as imposed above; for details we refer to [23, Theorem 2]. □

In accord with usual terminology, we will call (10) and (11) the integral and the Pontryagin maximum principles, respectively; cf. [14, 19]. Let us point out that the Hamiltonian $\mathcal{H}$ (determined uniquely only up to integrands $c \in H$ such that $c(\zeta, \cdot)$ is constant) resulting from our theory coincides for the choice $c = 0$ with the usual "guessed" Hamiltonian; cf. [23, Sect. 4] for the case of a finite-dimensional system or (21) below for the parabolic system. Of course, the constancy (on P) of the Hamiltonian along optimal trajectories, which is usually included into the Pontryagin principle, is irrelevant here because our Hamiltonian is unique only up to integrable functions on P. On the other hand, the Hamiltonian has a definite meaning by our derivation, namely it represents the Gâteaux differential with respect to the geometry coming from H^*.

After the transformation of a particular $H \subset L^1(P; C(B_R))$ via Ψ_P, we get the subspace $\mathcal{F}_H$ of $C(U)$ without any explicit reference to the decomposition $Q = P \times R$, which makes possible a comparison between various H even if the decomposition $Q = P \times R$ vary. The following assertion is a consequence of Theorem 1 with Lemma 1. Note that it identifies the topology on $L^1(P; C(B_R))$ induced projectively via Ψ_P from $C(U)$ as the decisive topology in the sense that the closure of H in this topology does not change the resulting convex compactification but any further enlargement does refine it.

Theorem 3. *Let $Q = P_1 \times R_1$ and $Q = P_2 \times R_2$ be two admissible decompositions of Q and let the linear subspaces $H_1 \subset L^1(P_1; C(B_{R_1}))$ and $H_2 \subset L^1(P_2; C(B_{R_2}))$ be given such that $\mathcal{F}_{H_1} \subset \mathcal{F}_{H_2}$. Then $M(\mathcal{F}_{H_1}) \preceq M(\mathcal{F}_{H_2})$, which means just that $Y_{H_1}(P_1, B_{R_1})$ is a coarser convex compactification of U than $Y_{H_2}(P_2, B_{R_2})$. Moreover, $cl_{C(U)}\mathcal{F}_{H_1} = cl_{C(U)}\mathcal{F}_{H_2}$ if and only if $M(\mathcal{F}_{H_1}) \cong M(\mathcal{F}_{H_2})$.*

We can see that there is, in fact, a large freedom in the choice of the decomposition $Q = P \times R$ and then in the choice of the particular subspace $H \subset L^1(P; C(B_R))$. It is obvious that the choice of the decomposition of Q has an immediate impact on the character of the Pontryagin maximum principle (11). Namely, the larger the component P (which contains directions in which "rapid" oscillations in controls are allowed), the more local the Pontryagin maximum principle that can be obtained. From this point of view, the best choice of the decomposition is $Q = P$, $n = n_P$, $n_R = 0$. On the other hand, another decomposition with $n_R > 0$ can yield finer convex compactifications, which might be sometimes inevitable when special problems are to be treated; cf. Sect. 4. As far as the choice of H is concerned, we can say that a larger H makes the resulting convex compactification finer (cf. also Theorem 3) and thus the class of mappings which admit a continuous extension to this compactification is larger. Also we can say that a "limit" information about oscillations of ε-optimal controls contained in the relaxed optimal controls is then greater but, of course, the implementation of such more informative solutions on computers is harder. Taking H smaller has naturally just opposite effects. This is summarized in the following tables.

	H	
	larger	smaller
convex compactification	finer	coarser
information in a solution	greater	lesser
implementation	harder	easier
a class of continuously extendable mappings	more	less

TABLE 1. *The influence of the choice of the linear subspace H.*

	P	
	larger	smaller
optimality conditions	more local	less local
directions for "rapid" oscillations in controls	wider	narrower

TABLE 2. *The influence of the choice of the decomposition $Q = P \times R$.*

Given a class of problems to be relaxed, an attempt for a general recommendation might look like: n_P and P should be chosen as large as possible to localize the maximum principle as much as possible, and simultaneously H should be taken as small as possible (with respect to a given class of problems that are to admit a continuous extension on the resulted convex compactification) to get a convex compactification as coarse as possible (which makes its implementation easier). Also, unnecessarily finer convex compactifications deteriorate a chance that, at least for some special problems from a given class, the solution of the relaxed problem is unique. Therefore, minimal convex compactifications satisfying some prescribed conditions are of a particular interest, cf. also [21, Theorem 3].

4. Examples.

We want to deal with the particular case $Q = (0,T) \times \Omega$ with Ω a domain in $\mathbb{R}^m$. It appears typically in distributed control of evolution partial differential equations, say parabolic or hyperbolic. (Quite equally we could treat the case $Q = (0,T) \times \partial\Omega$ with $\partial\Omega$ an $(m-1)$-dimensional boundary of Ω, which appears in a boundary control of such equations.) Then time t ranges the interval $(0,T)$ while x ranges the spatial domain Ω. In the previous notation, $\xi = (t,x)$, so that the set of controls, cf. (8), is now

$$(12) \quad U = \{u \in L^\infty((0,T) \times \Omega; \mathbb{R}^k);\ u(t,x) \in S \ \text{ for a.a. } t \in (0,T),\ x \in \Omega\}\ .$$

There are now two reasonable decompositions of $Q \equiv (0,T) \times \Omega = P \times R$, namely

1. $P = (0,T) \times \Omega$, $n = n_P = m+1$, $n_R = 0$; then $B_R = S$,

2. $P = (0,T)$, $R = \Omega$, $n_P = 1$, $n_R = m$; then $B_R \equiv B_\Omega = \{u \in L^\infty(\Omega;\mathbb{R}^k);\ u(x) \in S \text{ for a.a. } x \in \Omega\}$.

Of course, we could also imagine other decompositions of $Q = (0,T) \times \Omega$ but the corresponding relaxations do not seem to be reported in the literature.

Let us start with the first possibility. The resulting Pontryagin maximum principle is then formulated for a.a. $t \in (0,T)$ and a.a. $x \in \Omega$, and the supremum in (11) is taken over S. Such kind of maximum principles has been treated by Ahmed, Teo [3], Sadigh-Esfandiari, Sloss, Bruch Jr., Sadek [25, 26], Zolezzi [27], etc.

Though there are uncountably many convex compactifications of the form $M(\mathcal{F}_H)$ with $H \subset L^1((0,T) \times \Omega; C(S))$, we will mention only two of them as examples. The first one is the finest convex compactification in this class, which takes

$$H = L^1((0,T) \times \Omega; C(S)) , \tag{13}$$

while the second one uses

$$H = L^1((0,T) \times \Omega) \otimes \mathcal{L}(S) , \tag{14}$$

where $\mathcal{L}(S)$ denotes the set of all linear functions $\mathbb{R}^k \to \mathbb{R}$ restricted to S; in other words, H from (14) contains just the functions of the form $(t,x,s) \mapsto \sum_{l=1}^k g_l(t,x)s_l$ with $g_l \in L^1((0,T) \times \Omega)$ and $s = (s_1, ..., s_k) \in S \subset \mathbb{R}^k$.

In the case (13), it is known that $H^* = L^1((0,T) \times \Omega; C(S))^*$ is isometrically isomorphic with $L^\infty_w((0,T) \times \Omega; rca(S))$ via the mapping $\psi : L^\infty_w((0,T) \times \Omega; rca(S)) \to L^1((0,T) \times \Omega; C(S))^*$ defined by

$$\nu = \{\nu_{t,x}\}_{t\in(0,T),x\in\Omega} \mapsto \left(h \mapsto \int_0^T \int_\Omega \int_S h(t,x,s)\nu_{t,x}(ds)dxdt \right) \tag{15}$$

where $L^\infty_w((0,T) \times \Omega; rca(S))$ is the space of weakly* measurable essentially bounded mappings from $(0,T) \times \Omega$ to the space $rca(S)$ of regular bounded σ-additive set functions (=Borel measures) on S. According to the usual notation, we wrote $\nu_{t,x}$ instead of $\nu(t,x)$, which is to emphasize that ν is a parameterized measure. Moreover, ψ^{-1} maps $Y_H((0,T) \times \Omega; S)$ onto

$$\begin{aligned} \mathcal{Y}((0,T) \times \Omega; S) \;=\; & \{\nu \in L^\infty_w((0,T) \times \Omega; rca(S));\ \nu_{t,x} \in rca_1^+(S) \\ & \text{for a.a. } t \in (0,T),\ x \in \Omega\} , \end{aligned}$$

where $rca_1^+(S) = \{\nu \in rca(S);\ \nu \geq 0 \ \&\ \nu(S) = 1\}$ denotes the set of all regular probability measures of S. The elements of $\mathcal{Y}((0,T) \times \Omega; S)$ are just the so-called Young measures, parameterized by $(t,x) \in (0,T) \times \Omega$ and supported on S.

In the case (14), it is easy to show that $Y_H((0,T) \times \Omega; S)$ is affinely homeomorphic with

$$\{u \in L^\infty((0,T) \times \Omega; \mathbb{R}^k);\ u(t,x) \in \overline{co}(S) \text{ for a.a. } t \in (0,T),\ x \in \Omega\}$$

endowed with the weak* topology of $L^\infty((0,T)\times\Omega;\mathbb{R}^k)$. The homeomorphism is via the mapping which assigns each $u\in L^\infty((0,T)\times\Omega;\mathbb{R}^k)$ the linear functional $[\Psi_Q(\cdot)](u)\in H^*$. Note that this mapping is linear since all the integrands from H from (14) are linear in terms of $s\in\mathbb{R}^k$. This gives basically the standard relaxation which uses more or less the original controls equipped with the L^∞-weak* topology. However, such coarse relaxation is suitable only for problems having a convex/linear structure, possibly with slightly nonlinear perturbations.

Let us go on to the second mentioned decomposition of $(0,T)\times\Omega$, namely $P=(0,T)$ and $R=\Omega$. The resulting Pontryagin maximum principle is then formulated for a.a. $t\in(0,T)$, and the supremum in (11) is taken over $B_\Omega=\{u\in L^\infty(\Omega;\mathbb{R}^k);\ u(x)\in S$ for a.a. $x\in\Omega\}$. Such kind of relaxations and/or maximum principles has been treated by Ahmed [1, 2], Basile, Mininni [4], Fattorini [10], Frankowska [12, 13], Kampowsky [15], Papageorgiou [17], etc.

Again there are uncountably many convex compactifications of the form $M(\mathcal{F}_H)$ with $H\subset L^1(0,T;C(B_\Omega^\tau))$ with τ some Hausdorff topology on B_Ω, but we will mention only two of them as examples. The first one is

$$(16)\qquad H\ =\ L^1(0,T;C(B_\Omega^\tau))\ ,\quad \tau=\ \text{the norm topology from } L^p(\Omega;\mathbb{R}^k)\ ,$$

while the second one uses

$$(17)\qquad H\ =\ L^1(0,T;C(B_\Omega^\tau))\ ,\quad \tau=\ \text{the weak* topology from } L^\infty(\Omega;\mathbb{R}^k)\ .$$

Note that both (16) for $1\le p<+\infty$ and (17) for $S\subset\mathbb{R}^k$ convex satisfy the assumptions of Theorem 2 needed to establish the Pontryagin maximum principle.

As to the case (16), B_Ω^τ is not compact and $H^*=L^1(0,T;C(B_\Omega^\tau))^*$ is isometrically isomorphic (again via ψ defined like in (15)) with $L_w^\infty(0,T;rba(B_\Omega^\tau))$, where $rba(B)\cong C(B)^*$ denotes the set of all regular bounded additive set functions on a normal topological space B. Then elements of $Y_H(0,T;B_\Omega^\tau)$ can be interpreted as probability regular bounded additive set functions, that means elements of $L_w^\infty(0,T;rba(B_\Omega^\tau))$ valued in $rba_1^+(B_\Omega^\tau)$. In the context of optimal-control theory, elements of $rba(B)$ are also called "finite additive measures" to distinguish it from the usual measures which are σ-additive. Finitely additive measures were used in this context in the works by Fattorini [8, 9, 10, 11], and in general optimization theory also by Chentsov [6, 7], Pashaev [18], etc. Let us only mention that, since $rba(B_\Omega^\tau)\cong rca(\underline{\mathrm{B}}_\Omega^\tau)$ with $\underline{\mathrm{B}}_\Omega^\tau$ denoting the Stone-Čech compatification of B_Ω^τ, the finitely additive measures on a normal space B_Ω^τ can be equally understood as standard σ-additive measures of a (non-metrizable) compact space $\underline{\mathrm{B}}_\Omega^\tau$.

Let us suppose, for simplicity, that $S\subset\mathbb{R}^k$ is convex. Since B_Ω^τ is compact provided τ is the weak* topology of $L^\infty(\Omega;\mathbb{R}^k)$, it is now evident that H^* with H from (17) is isometrically isomorphic with $L_w^\infty(0,T;rca(B_\Omega^\tau))$, and $Y_H(0,T;B_\Omega^\tau)$ is mapped by this isomorphism onto

$$\mathcal{Y}(0,T;B_\Omega^\tau)\ =\ \{\nu\in L_w^\infty(0,T;rca(B_\Omega^\tau));\ \nu_t\in rca_1^+(B_\Omega^\tau)\ \text{for a.a. } t\in(0,T)\}\ .$$

The elements of $\mathcal{Y}((0.T) \times \Omega; S)$ are just the Young measures, parameterized by $t \in (0,T)$ and supported on B_Ω^τ. Such Young measures has been basically used by Papageorgiou [17].

The relations between the mentioned concrete convex compactifications of U are stated in the following assertion. For brevity, we will denote by $H_{(\cdot)}$ and $\mathcal{F}_{(\cdot)}$ respectively the subspace H from $(\cdot)$ and $\mathcal{F}_H \subset C(U)$ created by $H_{(\cdot)}$; for example $\mathcal{F}_{(16)}$ stands for $\mathcal{F}_H$ with H from (16), denoted by $H_{(16)}$.

Theorem 4. *The relations between the convex compactifications $M(\mathcal{F}_{(13)})$, $M(\mathcal{F}_{(14)})$, $M(\mathcal{F}_{(16)})$, and $M(\mathcal{F}_{(17)})$ are given by the following diagram*

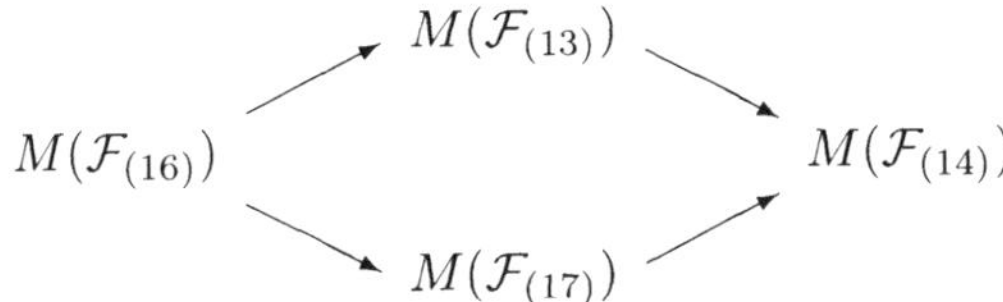

where each arrow indicates the existence of an affine homeomorphism; in other words, each arrow goes from a finer convex compactification to a coarser one. Besides, no arrow can be reversed if S is not a singleton; this means the original convex compactification is strictly finer than the terminal one. Finally, the convex compactifications $M(\mathcal{F}_{(13)})$ and $M(\mathcal{F}_{(17)})$ are actually not comparable.

Sketch of the proof. The fact that $M(\mathcal{F}_{(13)}) \succeq M(\mathcal{F}_{(14)})$ follows immediately from $H_{(13)} \supset H_{(14)}$. Taking two points $s_1, s_2 \in S$ and two sequences $\{u_k^1\}$ and $\{u_k^2\}$, the former one taking the values s_1 and s_2 and converging weakly* to the constant $\frac{1}{2}s_1 + \frac{1}{2}s_2 \in S$ while the latter one being constant $u_k^2 \equiv \frac{1}{2}s_1 + \frac{1}{2}s_2$. Now it is clear that the limit of $\{e(u_k^j)\}_{k\in\mathbb{N}}$ in $M(\mathcal{F}_{(14)})$ exists and is the same for $j = 1$ and 2, while these sequences can be separated on $M(\mathcal{F}_{(13)})$ by, for example, a function $f = \Psi_Q(1_Q \otimes v)$ with $v \in C(S)$ such that $v(s_1) = v(s_2) = 0$ and $v(\frac{1}{2}s_1 + \frac{1}{2}s_2) = 1$ because obviously $\langle e(u_k^1), f\rangle = \int_Q v(u_k^1(t,x))dtdx = 0$ while $\langle e(u_k^2), f\rangle = |Q| > 0$ for any $k \in \mathbb{N}$. This shows that $M(\mathcal{F}_{(13)}) \succ M(\mathcal{F}_{(14)})$.

By analogous arguments we can show also $M(\mathcal{F}_{(16)}) \succ M(\mathcal{F}_{(17)})$. As for the two sequences which have the same limits on $M(\mathcal{F}_{(17)})$ but can be separated on $M(\mathcal{F}_{(16)})$, we can take $u_k^1(t,x) = v_k(x)$ and u_k^2 as previously, where $v_k : \Omega \to S$ converges to $\frac{1}{2}s_1 + \frac{1}{2}s_2$ weakly* in $L^\infty(\Omega; \mathbb{R}^k)$ but not strongly in $L^p(\Omega; \mathbb{R}^k)$.

As for $M(\mathcal{F}_{(17)}) \succeq M(\mathcal{F}_{(14)})$, we cannot show directly $H_{(17)} \supset H_{(14)}$ because the partition $Q = P \times R$ is not the same. Nevertheless, we can show that $\mathcal{F}_{(17)} \supset \mathcal{F}_{(14)}$ because for any $\sum_{l=1}^k g_l \otimes s_l \in H_{(14)}$ there is $h \in H_{(17)}$ such that $\Psi_Q(\sum_{l=1}^k g_l \otimes s_l) = \Psi_{(0,T)}(h)$, namely $h(t,v) = \int_\Omega \sum_{l=1}^k g_l(t,x)v_l(x)dx$ for $t \in (0,T)$ and $v \in B_\Omega$; note that the following estimate holds: $\|h\|_{L^1(0,T;C(B_\Omega^\tau))} \leq \sup_{s\in S} |s| \sum_{l=1}^k \|g_l\|_{L^1((0,T)\times\Omega)}$.

Analogous arguments can be also used for $M(\mathcal{F}_{(16)}) \succ M(\mathcal{F}_{(13)})$. Indeed, having $h \in H_{(13)}$ we can find $h_1 \in H_{(16)}$ such that $\Psi_{(0,T)}(h_1) = \Psi_Q(h)$, namely $h_1(t,v) = \int_\Omega h(t,x,v(x))dx$. Note that the continuity of $h_1(t,\cdot) : B_\Omega^\tau \to \mathbb{R}$ follows by the standard properties of the Nemytskii mappings, while the estimate $\|h_1\|_{L^1(0,T,C(B_\Omega^\tau))} \leq \|h\|_{L^1((0,T)\times\Omega;C(S))}$ is obvious. As for the two sequences, let us take $\{u_k^1\}$ as in the first case but with $u_k^1(t,\cdot)$ constant (and equal to either s_1 or s_2) on Ω, and $\{u_k^2\}$ such that $u_k^2(t,x) = u_k^1(t,x)$ if $x \in \Omega'$ and $u_k^2(t,x) = s_2$ if $u_k^1(t,x) = s_1$ and $u_k^2(t,x) = s_1$ if $u_k^1(t,x) = s_2$ provided $x \in \Omega''$, where Ω' and Ω'' are disjoint parts of Ω of positive measure. Roughly speaking, both sequences converges to the Young measure $\nu \in L_w^\infty(Q; rca(S))$ given by $\nu_{t,x} = \frac{1}{2}\delta_{s_1} + \frac{1}{2}\delta_{s_2}$ in the representation of $M(\mathcal{F}_{(13)})$. On the other hand, on the representation of $M(\mathcal{F}_{(16)})$, the first one converges to the Young measure $\nu^1 \in L_w^\infty(0,T; rba(B_\Omega^\tau))$ given by $\nu_t = \frac{1}{2}\delta_{v_1^1} + \frac{1}{2}\delta_{v_2^1}$ with $v_j^1(x) = s_j$ for $j = 1,2$ while the second one converges to $\nu^2 \in L_w^\infty(0,T; rba(B_\Omega^\tau))$ given by $\nu_t = \frac{1}{2}\delta_{v_1^2} + \frac{1}{2}\delta_{v_2^2}$ with $v_1^2(x) = s_1$ if $x \in \Omega'$ and $v_1^2(x) = s_2$ if $x \in \Omega''$, and $v_2^2(x) = s_2$ if $x \in \Omega'$ and $v_2^2(x) = s_1$ if $x \in \Omega''$.

By this construction we can also show that $M(\mathcal{F}_{(17)}) \not\succeq M(\mathcal{F}_{(13)})$, while the fact that $M(\mathcal{F}_{(17)}) \not\preceq M(\mathcal{F}_{(13)})$ can be proved similarly as the fact that $M(\mathcal{F}_{(16)}) \not\cong M(\mathcal{F}_{(17)})$. $\square$

Finally, let us illustrate the above theory by one model example of an optimal control of a nonlinear parabolic equation. We will take (OCP) from Section 2 with the following data: $k = 1$, U from (12) with $S = [s_1, s_2]$, $Y = H^1(Q) \cap L^2(0,T; H_0^1(\Omega))$, $\Lambda = L^2(0,T; H^{-1}(\Omega)) \times L^2(\Omega)$, the cost function

$$(18) \quad \varphi(u,y) = \int_0^T \int_\Omega \left(u(t,x) - \frac{1}{|\Omega|} \int_\Omega u(t,\zeta)d\zeta \right)^2 dxdt + \int_\Omega (y(T,x) - y_d(x))^2 \, dx \,,$$

with the desired terminal profile $y_d \in L^2(\Omega)$, and the state equation $A(y,u) = 0$ in the form

$$(19) \qquad \frac{\partial y}{\partial t} - \Delta y + \gamma(y) = \gamma(u) \,, \quad y(0,\cdot) = y_0 \,,$$

with $\gamma : \mathbb{R} \to \mathbb{R}$ smooth and nondecreasing, the inital data $y_0 \in H_0^1(\Omega)$ and, for simplicity, the Dirichlet boundary conditions $y(t,\cdot) = 0$ on $\partial\Omega$. More precisely, the operator $A : U \times Y \to \Lambda$ is determined by the standard weak formulation of (19) with the boundary and initial conditions, which means that $A(u,y) = (z_1, z_0) \in L^2(0,T; H^{-1}(\Omega)) \times L^2(\Omega)$ with $z_0 = y(T,\cdot) - y_d$ and $z_1 = \partial y/\partial t - \Delta y + \gamma(y) - \gamma(u)$ understood in the distributional sense. Note that $A(u,y) = 0$ has actually the only weak solution $y = \pi(u) \in H^1(Q) \cap L^2(0,T; H_0^1(\Omega))$. The growth of the nonlinearity γ at infinity is irrelevant because L^∞-estimates are at our disposal thanks to the comparison principle.

The interpretation of the problem (OCP) with (18)–(19) might be the following: the optimal control should drive the parabolic system from a given initial state y_0 as close to the desired terminal state y_d as possible and, simultaneously, the spatial profiles of the control should be "as constant as possible". For example, if (19) has a heat-transfer interpretation (i.e. y is a temperature) and $\gamma(y) = y^4$, then we can say that the control acts in (19) via a Stefan-Boltzmann radiation term (which is of the 4th power).

It is obvious that, supposing γ nonlinear, such (OCP) does not bear a relaxation by means of H from (14) because (19) could not be extended continuously. For the same reason, also (17) is unsatisfactory. As for the choice (13), it would enable a continuous extension of (19) but not of (18). This can be seen easily by taking the two sequences contructed in the proof of Theorem 4 (the part $M(\mathcal{F}_{(16)}) \succ M(\mathcal{F}_{(13)})$). On the other hand, the choice (16) is satisfactory both for (18) and (19) provided $p \geq 2$. Thus we demonstrated a nontrivial situation when we must inevitably choose the finest convex compactification from those investigated by Theorem 4. Let us also remark that the transformation of control variable $\gamma(u) \to \tilde{u}$ would not change the situation because then the failure of the weak continuity, that appears on coarser convex compactifications, would be only transferred from (19) to (18) provided γ is actually nonlinear.

If one evaluates (5) in our case, one gets after integration by parts (cf. [23] for this procedure) the equations for the adjoint state $\lambda = (\lambda_1, \lambda_0) \in \Lambda^* = L^2(0,T; H_0^1(\Omega)) \times L^2(\Omega)$:

$$(20)\quad \frac{\partial \lambda_1}{\partial t} + \Delta\lambda_1 + \gamma'(y)\lambda_1 = 0\,, \quad \lambda_1(T,\cdot) = 2(y(T,\cdot) - y_d)\,, \quad \lambda_0 = \lambda_1(0,\cdot)\,.$$

Of course, (20) is to be understood again in the weak sense.

To evaluate the Hamiltonian, let us observe that both $A(\cdot, y)$ and $\varphi(\cdot, y)$ are affine with respect to the geometry of $\mathcal{F}^*_{(16)}$, so that the Gâteaux differential of the extended mappings does not explicitly depend on particular relaxed controls and equals just $A(\cdot, y)$ and $\varphi(\cdot, y)$, respectively. Then the abstract Hamiltonian $f_{\lambda,y,\mu_0} \in \mathcal{F}_{(16)}$ of (7), after the transformation via $\Psi^{-1}_{(0,T)}$ to $\mathcal{H}_{\lambda,y,\eta_0} \equiv \mathcal{H}_{\lambda_1} \in L^1(0,T; C(B^\tau_\Omega))$, looks like

$$(21)\quad \mathcal{H}_{\lambda_1}(t,v) = \int_\Omega \left(v(x) - \frac{1}{|\Omega|}\int_\Omega v(\zeta)d\zeta\right)^2 dx + \int_\Omega \lambda_1(t,x)\gamma(v(x))dx + c(t)\,,$$

where $t \in (0,T)$ and $v \in B_\Omega$, and $c \in L^1(0,T)$ is arbitrary. The Pontryagin maximum principle for the relaxed problem then looks like

$$\int_{B_\Omega} \mathcal{H}_{\lambda_1}(t,v)\nu_t(dv) = \sup_{v\in B_\Omega} \mathcal{H}_{\lambda_1}(t,v) \qquad \text{for a.a. } t \in (0,T)\,,$$

where $\nu = \{\nu_t\}_{t\in(0,T)} \in L^\infty_w(0,T; rba(B^\tau_\Omega))$ is an optimal relaxed control.

Thus we have shown that our fairly abstract theory coincides with standard approaches in concrete cases.

References

1. N.U. Ahmed, *Optimal control of a class of strongly nonlinear parabolic systems,* J. Math. Anal. Appl. **61** (1977), 188–207.
2. N.U. Ahmed, *Properties of relaxed trajectories for a class of nonlinear evolution equations on a Banach space,* SIAM J. Control Optim. **21** (1983), 953–967.
3. N.U. Ahmed and K.L. Teo, *Necessary conditions for optimality of Cauchy problems for parabolic partial differential systems,* SIAM J. Control **13** (1975), 981–993.
4. N. Basile and M. Mininni, *An extension of the maximum principle for a class of optimal control problems in infinite-dimensional spaces,* SIAM J. Control Optim. **28** (1990), 1113-1135.
5. A.G. Butkovskii, *Distributed Control Systems,* Elsevier, New York, 1969.
6. A.G. Chentsov, *Finitely additive measures and minimum problems,* (In Russian) Kibernetika No.3 (1988), 67–70.
7. A.G. Chentsov, *On the construction of solution to nonregular problems of optimal controls,* Problems Control Inf. Th. **20** (1991), 129–143.
8. H.O. Fattorini: *Relaxed controls in infinite dimensional control systems,* In: Proc. 5th Conf. on Control and Estimation of Distributed Parameter Systems, ISNM **100**, Birkhäuser Verlag, Basel, 1991
9. H.O. Fattorini: *Relaxation theorems, differential inclusions and Filippov's theorem for relaxed controls in semilinear infinite dimensional systems,* J. Diff. Equations (to appear).
10. H.O. Fattorini: *Existence theory and the maximum principle for relaxed inifinite dimensional optimal control problems.* SIAM J. Control Optim. (to appear).
11. H.O. Fattorini: *Relaxation in semilinear infinite dimensional control systems,* Markus Festschrift Volume, Lecture Notes in Pure and Appl. Math. **152**, 1993.
12. H.O. Fattorini and H. Frankowska, *Necessary conditions for infinite dimensional control problems.* (to appear).
13. H. Frankowska, *The maximum principle for differential inclusions with end point constraints.* SIAM J. Control **25** (1987).
14. R.V. Gamkrelidze: *Principles of Optimal Control Theory,* (in Russian), Tbilisi Univ. Press, Tbilisi, 1977. Engl. Transl.: Plenum Press, New York, 1978.
15. W. Kampowsky, *Optimalitätsbedingungen für Prozesse in Evolutionsgleichungen 1. Ordnung,* ZAMM **61** (1981), 501–512.
16. D.G.Luenberger: *Optimization by Vector Space Methods.* J.Wiley, New York, 1969.
17. N.S. Papageorgiou, *Optimal control of nonlinear evolution inclusions,* J. Optim. Th. Appl. **67** (1990), 321–354.
18. A.B. Pashaev and A.G. Chentsov, *Generalized control problem in a class of finitely-additive measures,* (in Russian) Kibernetika No.2 (1986), 110–112.
19. L.S. Pontryagin, V.G. Boltjanski, R.V. Gamkrelidze, E.F. Miscenko: *Mathematical Theory of Optimal Procceses,* (in Russian), Moscow, 1961.
20. T. Roubíček, *Convex compactifications and special extensions of optimization problems,* Nonlinear Analysis, Th., Methods, Appl. **16**, 1117–1126 (1991).
21. T. Roubíček, *A general view to relaxation methods in control theory,* Optimization **23** (1992), 261–268.
22. T. Roubíček, *Minimization on convex compactifications and relaxation of nonconvex variational problems,* Advances in Math. Sciences and Appl. **1**, 1–18 (1992).
23. T. Roubíček, *Convex compactifications in optimal control theory,* In: Proc. 15th IFIP Conf. " Modelling and Optimization of Systems" (Ed. P. Kall), Lecture Notes in Control and Inf. Sci. **180** (1992), Springer, Berlin, pp. 433–439.
24. T. Roubíček, *Relaxation in Optimization Theory and Variational Calculus,* In preparation.
25. R. Sadigh-Esfandiari, J.M. Sloss, and J.C. Bruch Jr., *Maximum principle for optimal control of distributed parameter systems with singular mass matrix,* J. Optim. Th. Appl. **66** (1990),

211-226.
26. J.M. Sloss, J.C. Bruch Jr. and I.S. Sadek, *A maximum principle for nonconservative self-adjoint systems,* IMA J. Math. Control Inf. **6** (1986), 199-216.
27. T. Zolezzi, *Necessary conditions for optimal controls of elliptic or parabolic problems,* SIAM J. Control **10** (1972), 594–602.

Tomáš Roubíček
Institute of Information Theory and Automation
Academy of Sciences of Czech Republic
Pod vodárenskou věží 4
CZ-182 08 Praha 8,
Czech Republic

CONVERGENCE OF AN SQP–METHOD FOR A CLASS OF NONLINEAR PARABOLIC BOUNDARY CONTROL PROBLEMS

FREDI TROLTZSCH

Department of Mathematics
Technical University of Chemnitz-Zwickau

ABSTRACT We investigate local convergence of an SQP method for an optimal control problem governed by a parabolic equation with nonlinear boundary condition Sufficient conditions for local quadratic convergence of the method are discussed

1991 *Mathematics Subject Classification* 49M05, 49M40, 49K24

Key words and phrases Sequential quadratic programming, optimal control, parabolic equation, nonlinear boundary condition, control constraints

1. Introduction

In this paper, we investigate the behaviour of a Sequential Quadratic Programming (SQP)–method applied to the following very simplified model problem (P):

(P) Minimize

$$\frac{1}{2}\int_0^T\int_\Gamma \left\{(w(t,\xi)-q(t,\xi))^2+\lambda u(t,\xi)^2\right\} dS_\xi dt \tag{1.1}$$

subject to

$$\begin{aligned} w_t(t,\xi) &= (\Delta_\xi w - w)(t,\xi) && \text{in } \Omega \\ w(0,\xi) &= 0 && \text{in } \Omega \\ \frac{\partial w}{\partial n}(t,\xi) &= b(w(t,\xi))+u(t,\xi) && \text{on } \Gamma \end{aligned} \tag{1.2}$$

and to the constraints on the control

$$|u(t,\xi)| \le 1, \tag{1.3}$$

$t \in [0,T]$. The control u is looked upon in $L_\infty((0,T)\times\Gamma)$, while the state w is defined as mild solution of (1 2) (cf. section 2).

In this setting, a bounded domain $\Omega \subset R^n (n \geq 2)$ with C^∞–boundary Γ, positive constants λ, T, and functions $b \in C^2(I\!R), q \in L_\infty((0,T) \times \Gamma)$ are given. By n and dS the outward normal vector and the surface measure on Γ, respectively, are denoted.

Pointwise constraints of the type (1.3) are often imposed for a correct modelling of the underlying process. They reflect technical limitations to the possible choice of the control and cannot be realized by a smooth penalization. The term $\lambda\|u\|^2$ in (1.1) is to enhance continuity of the optimal control (it may express the cost for the control, too).

It is known since several years that the SQP algorithm, applied to mathematical programming problems in finite-dimensional spaces, exhibits local quadratic convergence. The method can be easily extended to infinite-dimensional optimization problems such as optimal control problems. We refer, for instance, to the works by Alt [1], [2], Alt and Malanowski [3], Kelley and Wright [7], or Levitin and Polyak [10]. In the context of nonlinear parabolic control problems without control constraint we mention the numerical work by Kupfer and Sachs [8].

Recently, Alt, Sontag and Tröltzsch [4] proved the local quadratic convergence of the SQP method for the optimal control of a weakly singular Hammerstein integral equation with pointwise constraints on the control. In the author's paper [12], the proof of convergence was transferred to the one-dimensional heat equation with nonlinear boundary condition. The aim of this note is to extend the convergence result to a parabolic equation in a domain of dimension n.

We assume that b and its derivatives up to the order 2 are uniformly bounded and Lipschitz: There are constants c_B, c_l:

$$|b^{(i)}(w)| \leq c_B \qquad |b^{(i)}(w_1) - b^{(i)}(w_2)| \leq c_l, \tag{1.4}$$

for all $w, w_1, w_2 \in I\!R, i = 0, 1, 2$. We may weaken the conditions (1.4) to local ones. However, this would lead to difficulties, as the solution w of (1.2) could blow up in finite time. For convenience we consider only the very special type of boundary condition in (1.2). This enables us to work out the principal behaviour of the SQP–method and to avoid tedious technical estimates. The case of more general boundary conditions having the form $\partial w/\partial n = b_1(w) + b_2(w)\, u$ is discussed in [4].

2. Integral equation method

Let us define $A : L_2(\Omega) \supset D(A) \to L_2(\Omega)$ by $D(A) = \{w \in W_2^2(\Omega) : \partial w/\partial n = 0 \text{ on } \Gamma\}$, $Aw = -\Delta w + w$ for $w \in D(A)$. $-A$ is known to generate an analytic semigroup $\{S(t)\}, t \geq 0$, of continuous linear operators in $L_2(\Omega)$. Moreover, we introduce the Neumann operator $N : L_2(\Gamma) \to W_2^s(\Omega)(s < 1 + 1/2)$ by $N : g \mapsto w, \Delta w - w = 0, \partial w/\partial n = g$. Next, we fix $\sigma, p \in I\!R$ by $p > n + 1$ and $n/p < \sigma < 1 + 1/p$. The last two inequalities have a non–void intersection for

$p > n-1$, while $p > n+1$ is needed to work with states being continuous w.r. to t. In the same way we may introduce operators $A_r, S_r(t), N_r$ just by substituting above the order of integrability r for 2, where $1 < r < \infty$. Restricting $AS(t)N$ to $L_r(\Gamma)(r \geq 2)$, we obtain $A_r S_r(t) N_r$. Therefore, we shall use in the paper the same symbol $AS(t)N$ regarded in different spaces L_r. To continue our preparations we define a Nemytskiĭ operator $\mathcal{B} : C(\Gamma) \to C(\Gamma)$ by $(\mathcal{B}x)(\xi) = b(x(\xi))$.

A function $w \in C([0,T], W_p^\sigma(\Omega))$ is said to be a *mild solution* of (1.2), if the Bochner integral equation

$$w(t) = \int_0^t AS(t-s)N(\mathcal{B}(\tau w(s)) + u(s))ds \tag{2.1}$$

holds on $[0,T]$ (τ: trace operator). The expression on the right hand side makes sense, as $u \in L_\infty((0,T)\times\Gamma) \subset L_p((0,T)\times\Gamma) = L_p(0,T; L_p(\Gamma))$ and $W_p^\sigma(\Omega) \subset C(\overline{\Omega})$ by $n/p < \sigma$ (here we regard $AS(t)N$ as operator from $L_p(\Gamma)$ to $W_p^\sigma(\Omega), t > 0$). Owing to the strong assumption (1.4), to each $u \in L_\infty((0,T)\times\Gamma)$ a unique global solution w of (2.1) exists (cf. Tröltzsch [11]). Turning over to the trace $x(t) = \tau w(t)$ in (2.1) we arrive at the integral equation

$$x(t) = \int_0^t \tau AS(t-s)N(\mathcal{B}(x(s)) + u(s))ds \tag{2.2}$$

for $x \in C([0,T], C(\Gamma))$.

The estimate (see Amann [4])

$$\|AS(t)N\|_{L_r(\Gamma)\to W_r^\sigma(\Omega)} \leq c\, t^{-\alpha}, \tag{2.3}$$

where $\alpha = 1 - (\sigma' - \sigma)/2$ and $0 < \sigma < \sigma' < 1 + 1/r$, turns out to be essential for investigating properties of the integral operator K,

$$(Kz)(t) = \int_0^t \tau AS(t-s)Nz(s)ds. \tag{2.4}$$

Let us briefly discuss (2.3) for $r := p$: Taking $\sigma = n/p + \varepsilon, \sigma' = 1 + 1/p - \varepsilon (\varepsilon > 0)$ we find that (2.3) holds for all $\alpha > 0.5 + (n-1)/2p$. K maps continuously $L_p(0,T; L_p(\Gamma))$ into $C([0,T], C(\Gamma))$ provided that $p > 1/(1-\alpha)$. This holds together with the last inequality for α, if $p > n+1$. For $p \downarrow n+1$ we may take $\alpha \downarrow n/(n+1)$.

For convenience we regard K between different spaces: K may be viewed as operator in $L_r(0,T; L_r(\Gamma))$ for all $1 < r < \infty$. Let its adjoint K^* be defined for $r = 2$.

It is known that

$$(K^*z)(t) = \int_t^T \tau A S(s-t) N z(s) ds,$$

hence K^* has the same transformation properties as K. By means of these prerequisites we are able to write (P) as

$$f(x,u) = \frac{1}{2} \int_0^T \{\|x(t) - q(t)\|^2_{L_2(\Gamma)} + \lambda \|u(t)\|^2_{L_2(\Gamma)}\} dt = min! \tag{P}$$

subject to

$$x = K(B(x) + u), \quad u \in U^{ad}. \tag{2.5}$$

Here we have introduced B in $C([0,T], C(\Gamma))$ by $(Bx)(t,\cdot) = \mathcal{B}(x(t,\cdot))$ and $U^{ad} := \{u \in L_\infty((0,T) \times \Gamma) : |u(t,\xi)| \leq 1\}$. The equation in (2.5) is well defined, as K maps $L_p(0,T;L_p(\Gamma))$ into $C([0,T], C(\Gamma))$.
In the paper, the following notation is used: We write $L_r = L_r(0,T;L_r(\Gamma)), 1 \leq r < \infty, L_\infty = L_\infty((0,T) \times \Gamma), C = C([0,T], C(\Gamma)) = C([0,T] \times \Gamma)$ and endow the spaces with their natural norms $\|\cdot\|_r$ and $\|\cdot\|_\infty$, respectively. The natural norm of $L_\alpha(0,T;L_\beta(\Gamma)), 1 < \alpha, \beta < \infty$, is denoted by $\|\cdot\|_{L_{\alpha,\beta}}$. For $\alpha = \beta = \infty$ we set $\|\cdot\|_{L_{\infty,\infty}} := \|\cdot\|_\infty$. In product spaces of this type, the norm is defined as the sum of the corresponding norms. In $C \times L_r$, $\|(x,u)\|_{\infty,r} = \|x\|_\infty + \|u\|_r, 1 \leq r \leq \infty$, and $\|(x,u)\|_r := \|(x,u)\|_{r,r}$, in $C \times L_r \times L_r$: $\|(x,u,y)\|_{\infty,r} = \|x\|_\infty + \|(u,y)\|_r, \|(x,u,y)\|_\infty := \|(x,u,y)\|_{\infty,\infty}$. An "inner product" is defined formally by

$$\langle x, y \rangle = \int_0^T \int_\Gamma x(t,\xi) y(t,\xi) \, dS_\xi dt$$

just denoting integration of xy over $[0,T] \times \Gamma$.

3. Known optimality conditions

The functional $f : C \times L_p \to I\!R$ and the mapping $(x,u) \mapsto B(x) + u$ from $C \times L_p$ to L_p are twice continuously Fréchet differentiable. This enables us to apply lateron second order methods to (P). Owing to the convexity of f and the linear appearance of u in (1.2), standard methods show the existence of at least one optimal control u_0 for (P). Let x_0 be the corresponding state. For $y \in L_\infty$ the *Lagrange function*

$$\mathcal{L}(x,u,y) = f(x,u) - \langle y, x - K(B(x) + u) \rangle$$

is defined. From $\mathcal{L}_x = 0$ we obtain formally the equation $y_0 = f_x + B'(x_0)^* K^* y_0$ for the Lagrange multiplier y_0. A careful discussion (taking derivatives in $C \times L_\infty$

and regarding the derivatives as linear operators in L_p) justifies this (cf. Tröltzsch [11]):

$$y_0(t) = x_0(t) - q(t) + B'(x_0(t)) \int_t^T \tau AS(s-t)N\, y_0(s)ds. \tag{3.1}$$

Here, we took advantage of $(\mathcal{B}'(x_0(t,\cdot))h(t,\cdot))(\xi) = b'(x_0(t,\xi)) \cdot h(t,\xi)$, hence $B'(x_0)$ is formally self–adjoint. The variational inequality $\langle \mathcal{L}_u, u - u_0 \rangle \geq 0\ \forall u \in U^{ad}$ yields $< \lambda u_0 + K^* y_0, u - u_0 > \geq 0$. After a standard pointwise discussion we arrive at

$$u_0(t,\xi) = P_{[-1,1]}\{-\lambda^{-1}(K^* y_0)(t,\xi)\}, \tag{3.2}$$

where $P_{[-1,1]} : I\!R \to [-1,1]$ denotes projection onto $[-1,1]$. We assume that in addition to the first order necessary conditions (3.1), (3.2) the following *second order sufficient optimality condition* is satisfied: There is a $\delta > 0$ such that

$$\mathcal{L}_{vv}(x_0, u_0, y_0)[v - v_0, v - v_0] \geq \delta \|v - v_0\|_2^2 \tag{3.3} \qquad \text{(SSC)}$$

for all $v = (x, u)$ satisfying the linearized equation

$$x - x_0 = K(B'(x_0)(x - x_0) + u - u_0). \tag{3.4}$$

In (3.3), $\mathcal{L}_{vv}$ denotes the second order F-derivative of $\mathcal{L}$ w.r. to $v = (x,u)$ in $C \times L_p$ at $v_0 := (x_0, u_0)$. The sufficiency of (SSC) was discussed in [6]. We finish this section by the following very useful Lemma:

Lemma 3.1. *Let $1 \leq r, \rho \leq \infty$, D be the linear continuous operator in $L_{r,\rho}$ defined by $(Dx)(t,\xi) = \beta(t,\xi)x(t,\xi)$, where $\beta \in L_\infty$. Then $(I - DK)^{-1}$ and $(I - KD)^{-1}$ exist as linear continuous operators in $L_{r,\rho}$ and their norm is bounded by a constant $c_{r\rho}$ which depends only $\|\beta\|_\infty$, r, and ρ.*

Proof. The invertibility follows from a standard application of the Banach fixed point theorem. Let $z \in L_{r,\rho}$ be given. We estimate the solution of the equation $x = BKx + z$. Then

$$\begin{aligned}\|x(t)\|_{L_\rho(\Gamma)} &\leq \int_0^t \|\beta\|_\infty \|\tau AS(t-s)N\|_{L_\rho(\Gamma)\to L_\rho(\Gamma)} \|x(s)\|_{L_\rho(\Gamma)} ds + \|z(t)\|_{L_\rho(\Gamma)} \\ &\leq c \int_0^t (t-s)^{-\alpha} \|x(s)\|_{L_\rho(\Gamma)} ds + \|z(t)\|_{L_\rho(\Gamma)}.\end{aligned}$$

Here, $c > 0$ depends only on ρ, $\|\beta\|_\infty$, and $\alpha \in (0,1)$. $\|x(t)\|$ is majorized by the real function $\varphi(t)$ solving the corresponding weakly singular integral equation. Therefore

$$\|x\|_{L_{r,\rho}} = (\int_0^T \|x(t)\|^r_{L_\rho(\Gamma)} dt)^{1/r} \le \|\varphi\|_{L_r(0,T)} \le c(\int_0^T \|z(t)\|^r_{L_\rho(\Gamma)} dt)^{1/r} = c\|z\|_{L_{r,\rho}}.$$

In this way the Lemma is shown for $I-DK$. The arguing for $I-KD$ is identical. □

4. The SQP method, Hölder estimates

Initiating from a starting point (x_1, u_1, y_1) in $C \times L_\infty \times L_\infty$ the (full) SQP method generates sequences $\{x_n\}, \{u_n\}, \{y_n\}$ by solving certain quadratic programs. Adopting the notation by Alt [1], one step of the method can be described as follows: Let $w = (x_w, u_w, y_w)$ be the result of the last iteration. As before, we write $v_w = (x_w, u_w), v = (x, u)$. The next iterate $\overline{v}_w = (\overline{x}_w, \overline{u}_w)$ is obtained as the solution of

$$(4.1)\quad F(v; w) = f'(v_w)(v - v_w) + \frac{1}{2}\mathcal{L}_{vv}(v_w, y_w)[v - v_w, v - v_w] = min! \qquad (QP)_w$$

$$(4.2)\qquad x = K(B'(x_w)(x - x_w) + B(x_w) + u), \quad u \in U^{ad}.$$

$\overline{y}_w$ is the corresponding Lagrange multiplier.

Remark. Define $g(v) = g(x, u) = x - K(B(x) + u)$. Then the state equation (2.5) reads $g(v) = 0$. (4.2) is its linearization $g(v_w) + g'(v_w)(v - v_w) = 0$, which simplifies by linearity w.r. to u.

The Lagrange function $\tilde{\mathcal{L}}$ to $(QP)_w$ is

$$\tilde{\mathcal{L}}(v, y) = F(v; w) - \langle y, x - K(B'(x_w)(x - x_w) + B(x_w) + u)\rangle.$$

$\tilde{\mathcal{L}}_x = 0$ leads to the (adjoint) equation

$$(4.3)\qquad \begin{aligned} y &= B'(x_w)K^*y + f_x(v_w) + f_{xx}(v_w)[\overline{x}_w - x_w, \cdot] \\ &\quad + B''(x_w)[\overline{x}_w - x_w, \cdot]K^* y_w \end{aligned}$$

for $y = \overline{y}_w$. After a simple calculation we find

$$(4.4)\qquad \begin{aligned} \overline{y}_w(t) &= \mathcal{B}'(x_w(t)) \int_t^T \tau A S(s-t) N \overline{y}_w(s) ds + \overline{x}_w(t) - q(t) \\ &\quad + \mathcal{B}''(x_w(t))[\overline{x}_w(t) - x_w(t), \cdot] \int_t^T \tau A S(s-t) N y_w(s) ds. \end{aligned}$$

The variational inequality determining $\overline{u}_w$, $\langle\tilde{\mathcal{L}}_u, u - \overline{u}_w\rangle \geq 0 \quad \forall u \in U^{ad}$, gives

$$\overline{u}_w(t,\xi) = P_{[-1,1]}\{-\lambda^{-1}(K^*\overline{y}_w)(t,\xi)\}. \tag{4.5}$$

A straightforward calculation by means of (SSC) yields

$$F(v;w_0) \geq \delta\|v - v_0\|_2^2 = F(v_0;w_0) + \delta\|v - v_0\|_2^2 \tag{4.6}$$

for all v being admissible for $(QP)_{w_0}$ (where $w_0 = (v_0, y_0) = (x_0, u_0, y_0)$).

Lemma 4.1. *For all w in a sufficiently small $C \times L_p \times L_p$-neighbourhood $N_1(w_0)$, $(QP)_w$ admits a unique solution $(\overline{x}_w, \overline{u}_w) = \overline{v}_w$ with associated Lagrange multiplier $\overline{y}_w$. There is a constant c_H not depending on w, such that*

$$\|\overline{v}_w - v_0\|_2 \leq c_H\|w - w_0\|_2^{1/2} \tag{4.7}$$

$$\|\overline{y}_w\|_\infty \leq c_H \tag{4.8}$$

for all $w \in N_1(w_0)$.

Proof. We shall only briefly sketch the proof, which is along the lines of Alt [1] or Alt, Sontag and Tröltzsch [3]. A first step initiates from the simple observation $F(\overline{v}_w;w) \leq F(\tilde{v}_w;w)$, where $\tilde{v}_w = (\tilde{x}_w, u_w)$ and $\tilde{x}_w$ is the state obtained from (4.2) for $u = u_w$. By means of Lemma 3.1 the *upper estimate*

$$F(\overline{v}_w;w) \leq c\|w - w_0\|_2 \tag{4.9}$$

can be derived. A *lower estimate*

$$F(\overline{v}_w;w) \geq \frac{\delta}{2}\|\overline{v}_w - v_0\|_2^2 - c\|w - w_0\|_2 \tag{4.10}$$

follows from re-writing terms such as $f'(v_w)$ or $\mathcal{L}_{vv}(v_w, y_w)$ in terms of $f'(v_0)$, $\mathcal{L}_{vv}(v_0, y_0)$ etc., estimating the correction parts and exploiting (SSC) and Lemma 3.1. We omit the details of the lengthy computations. (4.9) and (4.10) yield (4.7). As regards (4.8), we first observe that $\overline{u}_w$ is uniformly bounded, hence $\|\overline{x}\|_\infty$ is bounded, too (this is a consequence of (4.2) for $u = \overline{u}_w$: $\|x_w\|_\infty$ is bounded, as $w \in N_1(w_0)$; apply Lemma 3.1 to $x = \overline{x}_w$ in (4.2)). Now the uniform boundedness of $\overline{y}_w$ is an immediate conclusion of (4.4) and Lemma 3.1. □

Corollary 4.2. *The estimate (4.7) holds true in the form*

$$\|\overline{v}_w - v_0\|_{\infty,p} \leq c'_H\|w - w_0\|_{\infty,p}^{1/p} \tag{4.11}$$

for all $w \in N'_1(w_0) \subset N_1(w_0)$.

Proof. (4.7) means in particular $\|\overline{u}_w - u_0\|_2 \le c_H \|w - w_0\|_2^{1/2}$. Exploiting $|\overline{u}_w - u_0| \le 2$ it is easy to show that

$$\|\overline{u}_w - u_0\|_p \le c\|w - w_0\|_2^{1/p} \le c'\|w - w_0\|_{\infty,p}^{1/p}. \tag{4.12}$$

Subtracting the equations for $\overline{x}_2$ and x_0 we get

$$(\overline{x}_w - x_0) - KB'(x_w)(\overline{x}_w - x_0) = K(B(x_w) - B(x_0) + B'(x_w)(x_0 - x_w) + \overline{u}_w - u_0).$$

Therefore, the L_∞–norm of the left–hand side is less or equal $\|K\|_{L_p \to C}(c_1\|x_w - x_0\|_\infty + c_2\|\overline{u}_w - u_0\|_p) \le c\|w - w_0\|_{\infty,p}^{1/p}$ by (4.12) (provided that $\|w - w_0\|_{\infty,p} \le 1$). Now Lemma 3.1 applies to the left–hand side,

$$\|\overline{x}_w - x_0\|_\infty \le c''\|w - w_0\|_{\infty,p}^{1/p},$$

implying (4.11). □

In the same way, subtraction of the equations for $y_0, \overline{y}_w$ yields

Corollary 4.3. *There is a constant $c_H'' > 0$ such that*

$$\|\overline{y}_w - y_0\|_\infty \le c_H''\|w - w_0\|_{\infty,p}^{1/p} \tag{4.13}$$

for all $w \in N_1''(w_0) \subset N_1(w_0)$.

We omit the proof. In what follows let $N_1(w_0)$ denote the intersection $N_1(w_0) \cap N_1'(w_0) \cap N_1''(w_0)$.

5. Right hand side perturbations, Lipschitz estimate

Following Alt [1], [2], we consider now the close relationship between the stability of $(QP)_w$ and certain perturbations of $(QP)_{w_0}$. We discuss the perturbed problem

$$f'(v_0)(v - v_0) + \tfrac{1}{2}\mathcal{L}_{vv}(v_0, y_0)[v - v_0, v - v_0] - \langle d, v - v_0\rangle = min! \qquad (QS)_\pi$$

$$x = e + K(B'(x_0)(x - x_0) + B(x_0) + u), \; u \in U^{ad} \tag{5.1}$$

belonging to the perturbation $\pi = (d, e) = (d_x, d_u, e) \in L_\infty \times L_\infty \times C$. For $\pi = 0$ this problem has the unique solution $v_0 = (x_0, u_0)$. In $(QS)_\pi$ we regard x, u in L_2, although the constraint $u \in U^{ad}$ automatically generates only L_∞–solutions.

$(QS)_\pi$ is a linear–quadratic parabolic control problem, where the theory is already widely investigated. Owing to (SSC), the following result is therefore standard: There is a neighbourhood $N_2(0)$, and a positive constant $c_h > 0$ such that for all $e \in N_2(0)$ and all $d \in L_\infty \times L_\infty$ problem $(QS)_\pi$ admits a unique solution $v_\pi = (x_\pi, u_\pi)$ and

$$\|v_\pi - v_0\|_2 \leq c_h \|\pi\|_2^{1/2} \tag{5.2}$$

for all $\pi = (d, e)$ such that $e \in N_2(0)$. We are able to improve this estimate in Theorem 5.3.

Lemma 5.1. *Let y_π be the Lagrange multiplier belonging to v_π and $2 \leq \alpha, \beta \leq \infty$. Then there is a constant $c_{\alpha\beta} > 0$ such that*

$$\|y_\pi - y_0\|_{L_{\alpha,\beta}} \leq c_{\alpha\beta} \left(\|x_\pi - x_0\|_{L_{\alpha,\beta}} + \|\pi\|_{L_{\alpha,\beta}}\right) \tag{5.3}$$

for all $\pi \in L_\infty \times L_\infty \times C$.

Proof. The adjoint equations defining y_0, y_π are (3.1) and

$$y_\pi = (x_\pi - q) - d_x + B'(x_0)K^* y_\pi + B''(x_0)[x_\pi - x_0, \cdot]K^* y_0. \tag{5.4}$$

Subtraction of these equations yields after some estimations

$$\|(y_0 - y_\pi) - B'(x_0)K^*(y_0 - y_\pi)\|_{L_{\alpha,\beta}} = \|d_x + x_\pi - x_0 + B''(x_0)[x_\pi - x_0, \cdot]K^* y_0\|_{L_{\alpha,\beta}}.$$

Applying Lemma 3.1

$$\|y_0 - y_\pi\|_{L_{\alpha,\beta}} \leq c_1 \|d_x\|_{L_{\alpha,\beta}} + c_2 \|x_\pi - x_0\|_{L_{\alpha,\beta}}$$

is obtained. This implies (5.3). □

One of the decisive steps for showing quadratic convergence is the following Lipschitz estimate improving (5.2):

Theorem 5.2. *There is a constant $c_L > 0$ such that*

$$\|v_\pi - v_0\|_2 \leq c_L \|\pi\|_2 \tag{5.5}$$

for all $\pi \in L_\infty \times L_\infty \times C$.

Proof. We outline the main steps of the proof. The first order condition for v_π as a solution of $(QS)_\pi$ is

$$0 \leq \langle \tilde{\mathcal{L}}_v(v_\pi, y_\pi),\ v - v_\pi \rangle \ \ \forall v \in L_2 \times U^{ad},$$

i.e.

$$\begin{aligned} 0 \ \leq \ & f'(v_0)(v - v_\pi) + \mathcal{L}_{vv}(v_0, y_0)[v_\pi - v_0, v - v_\pi] \\ & -\langle d, v - v_\pi \rangle - \langle y_\pi, (x - x_\pi) - K(B'(x_0)(x - x_\pi) + u - u_\pi) \rangle \end{aligned}$$

for all $x \in L_2, u \in U^{ad}$. Now we insert $x = x_0, v = (x_0, u_0), u = u_0$ and find after exploiting the first order necessary optimality conditions for v_0 as a solution for (P)

$$(5.6)\qquad \begin{aligned} \mathcal{L}_{vv}(v_0, y_0)[v_\pi - v_0, v_\pi - v_0] &\le -\langle e, y_0 - y_\pi > - < d, v_0 - v_\pi\rangle \\ &\le \|y_0 - y_\pi\|_2 \, \|e\|_2 + \|d\|_2 \, \|v_0 - v_\pi\|_2 \\ &\le c\, \|\pi\|_2^2 + c\, \|\pi\|_2 \, \|v_0 - v_\pi\|_2 \end{aligned}$$

by Lemma 5.1. The difference $\xi = v_\pi - v_0 = (x_\pi - x_0, u_\pi - u_0) = (\xi_x, \xi_u)$ solves $\xi_x - K(B'(x_0)\xi_x + \xi_u) = e$, hence ξ does not satisfy the linearized equation (3.4), where (SSC) applies. Define $\hat{\xi} = (\hat{\xi}_x, \xi_u)$, where $\hat{\xi}_x$ is the solution of $\hat{\xi}_x - K(B'(x_0)\hat{\xi}_x + \xi_u) = 0$. Then $\xi_x = \hat{\xi}_x + \Delta$, and $\|\Delta\|_2 \le c\, \|\pi\|_2$ (apply Lemma 3.1). (SSC) is valid for $\hat{\xi}_x$, hence simple estimations yield

$$(5.7)\qquad \begin{aligned} \mathcal{L}_{vv}(v_0, y_0)[\xi, \xi] &\ge \delta \|\hat{\xi}\|_2^2 - 2c\, \|\hat{\xi}\|_2 \, \|\Delta\|_2 - c\, \|\Delta\|_2^2 \\ &\ge \delta\, \|\xi\|_2^2 - c(\|\xi\|_2 \, \|\Delta\|_2 + \|\Delta\|_2^2). \end{aligned}$$

Taking into account (5.6) and $\|\Delta\|_2 \le c\, \|\pi\|_2$ we easily find

$$\|\xi\|_2^2 \le c(\|\xi\|_2 \, \|\pi\|_2 + \|\pi\|_2^2) \le c\, \|\xi\|_2 \, \|\pi\|_2,$$

if $\|\pi\| \le \|\xi\|$. This implies (5.5), if $\|\xi\| \ge \|\pi\|$. Thus (5.5) holds for $c_L := max(1, c)$. □

The estimate (5.5) in the L_2–norm is not sufficient for our purposes. However, we are able to show

Theorem 5.3. *There is a constant $c_L' > 0$ such that*

$$(5.8)\qquad \|v_\pi - v_0\|_\infty \le c_L' \|\pi\|_{\infty,p}$$

for all $\pi \in L_\infty \times L_\infty \times C$.

Proof. We start with the equation for $x_\pi - x_0$,

$$(5.9)\qquad x_\pi - x_0 - KB'(x_0)(x_\pi - x_0) = e + K(u_\pi - u_0).$$

We have $K(u_\pi - u_0) = \tau w$, where w solves the PDE (1.2) with boundary condition $\partial w/\partial n = u_\pi - u_0$. By L_2–regularity,

$$\|w\|_{L_2(0,T,H^{3/2-\varepsilon}(\Omega))} \le c\, \|u_\pi - u_0\|_2,$$

($\varepsilon > 0$ fixed sufficiently small), hence

$$(5.10)\qquad \|K(u_\pi - u_0)\|_{L_2(0,T,H^{1-\varepsilon}(\Gamma))} \le c\|u_\pi - u_0\|_2.$$

Sobolev embedding theorems yield

$$H^{1-\varepsilon}(\Gamma) \subset L_{\frac{2(n-1)}{n-1-2(1-\varepsilon)}}(\Gamma) = L_{\frac{n-1}{(n-1)/2-(1-\varepsilon)}}(\Gamma) =: L_{p_1}(\Gamma).$$

Denote the left hand side of (5.9) by E. Thus

$$\|E\|_{L_{2\,p_1}} \leq \|e\|_{L_{2\,p_1}} + c\|u_\pi - u_0\|_2 \leq \|e\|_{L_{2\,p_1}} + c\|\pi\|_2$$
$$\leq c\|\pi\|_{L_{2\,p_1}} \text{ (Theorem 5.2).}$$

By Lemma 3.1,

$$\|x_\pi - x_0\|_{L_{2\,p_1}} \leq c\,\|E\|_{L_{2\,p_1}} \leq c\,\|\pi\|_{L_{2\,p_1}}. \tag{5.11}$$

Invoking the first order necessary conditions for u_π, $u_\pi = P_{[-1,1]}\{-\lambda^{-1}(K^* y_\pi - d_u)\}$, a simple estimation yields

$$\begin{aligned} \|u_\pi - u_0\|_{L_{2\,p_1}} &\leq \lambda^{-1}\|K^*\|\,\|y_\pi - y_0\|_{L_{2\,p_1}} + \lambda^{-1}\|d_u\|_{L_{2\,p_1}} \\ &\leq c\,\|x_\pi - x_0\|_{L_{2\,p_1}} + c\,\|\pi\|_{L_{2\,p_1}} \leq c\,\|\pi\|_{L_{2\,p_1}} \end{aligned} \tag{5.12}$$

by Lemma 5.1 and (5.11). In this way, we have already extended (5.5) to the $L_2(0,T;L_{p_1}(\Gamma))$–norm performing one step of a bootstrapping argument. Now we continue estimating (5.9) by means of the L_{p_1}–regularity of parabolic equations. The solution w can be estimated in the $L_2(0,T;W_{p_1}^{1+1/p_1-\varepsilon}(\Omega))$–norm, hence we have for its trace $\tau w = K(u_\pi - u_0)$

$$\|K(u_\pi - u_0)\|_{L_2(0,T,W_{p_1}^{1-\varepsilon}(\Gamma))} \leq c\,\|u_\pi - u_0\,\|_{L_{2\,p_1}}.$$

Embedding $W_{p_1}^{1-\varepsilon}(\Gamma) \subset L_{\frac{n-1}{(n-1)/p_1-(1-\varepsilon)}}(\Gamma) = L_{\frac{n-1}{(n-1)/2-2(1-\varepsilon)}}(\Gamma) = L_{p_2}(\Gamma)$

$$\|x_\pi - x_0\|_{L_{2\,p_2}} \leq c\,\|\pi\|_{L_{2\,p_2}}$$

is obtained as above. Proceeding in the same way we arrive after at most $[(n-1)/2]+1$ steps at the case $(n-1)/2-k(1-\varepsilon) < 0$, while $(n-1)/2-(k-1)(1-\varepsilon) > 0$ (provided $\varepsilon > 0$ is sufficiently small). Here we end up with the possibility of an estimate in the norm of $L_2(0,T;C(\Gamma))$. However, we use only

$$\begin{aligned} \|K(u_\pi - u_0)\|_{L_{2\,p}} &\leq c\|K(u_\pi - u_0)\|_{L_2(0,T,W_{p_{k-1}}^{1-\varepsilon}(\Gamma))} \\ &\leq c\|\pi\|_{L_{2\,p_{k-1}}} \leq c\|\pi\|_{L_{2\,p}}. \end{aligned}$$

provided that $p_{k-1} \leq p$. If $p_{k-1} > p$, then we use the argument

$$\|K(u_\pi - u_0)\|_{L_{2\,p}} \leq c\|K(u_\pi - u_0)\|_{L_{2\,p_{k-1}}} \leq c\|\pi\|_{L_{2\,p_{k-2}}} \leq c\|\pi\|_{L_{2\,p}}$$

(note that $p_{k-2} < p$ must hold, as $L_{2,p_{k-2}}$ is still not transformed into C).

Thus finally (invoking (5.9) and the optimality conditions for u_π, u_0)

$$\|v_\pi - v_0\|_{L_{2\,p}} \leq c\|\pi\|_{L_{2\,p}} \tag{5.13}$$

can be derived.

It remains to lift the regularity with respect to the time t. From Krasnosel'skiĭ a.o. [9] it is known that a weakly singular integral operator with weak singularity $\alpha \in (0,1)$ maps continuously $L_p(0,T)$ into $L_{p'}(0,T)$, if $1/p' > 1/p + \alpha - 1$. Put

$\delta = 1 - \alpha > 0$ and take $\lambda \in (0,1)$. Then K transforms $L_2(0,T;L_p(\Gamma))$ into $L_{\beta_1}(0,T;L_p(\Gamma))$, where $1/\beta_1 = 1/2 - \lambda\delta$. Arguing as in the first part of the proof,

$$\|u_0 - u_\pi\|_{L_{\beta_1,p}} \le c\|\pi\|_{L_{\beta_1,p}} \tag{5.14}$$

is obtained. K transforms $L_{\beta_1}(0,T;L_p(\Gamma))$ into $L_{\beta_2}(0,T;L_p(\Gamma))$, provided that $1/\beta_2 = 1/\beta_1 - \lambda\delta = 1/2 - 2\lambda\delta$. Therefore, the estimate (5.14) can be derived in the norm of $L_{\beta_2}(0,T;L_p(\Gamma))$. Proceeding in this way, (5.14) is seen to hold in the $\|\cdot\|_p$–norm after finitely many steps. (5.8) follows easily, as K transforms L_p into C. □

The next two results are standard. We refer to the proofs given by Alt in [1], which simplify considerably for our model problem (P).

Lemma 5.4. *There is a $C \times L_p \times L_p$–neighbourhood $N_3(w_0)$, such that for all $w \in N_3(w_0)$ the following equivalence holds true: If $w \in N_3(w_0)$, then the solution $\overline{v}_w = (\overline{x}_w, \overline{u}_w)$ is also the unique solution of $(QS)_\pi$ for the following choice of $\pi = (d,e) = (d_x, d_u, e) : d_u = 0$,*

$$\begin{aligned} d_x &= B''(x_0)[\overline{x}_w - x_0, \cdot]K^* y_0 - B''(x_w)[\overline{x}_w - x_w, \cdot]K^* y_w \\ &\quad -(B'(x_w) - B'(x_0))K^*\overline{y}_w \end{aligned} \tag{5.15}$$

$$e = K(B'(x_w)(\overline{x}_w - x_w) - B'(x_0)(\overline{x}_w - x_0) + B(x_w) - B(x_0)). \tag{5.16}$$

Proof. We know that $(QS)_\pi$ is a convex problem with a solution determined uniquely by the conditions (5.1), (5.4), and

$$u_\pi = P_{[-1,1]}\{-\lambda^{-1}(K^* y_\pi - d_u)\}. \tag{5.17}$$

Thus it suffices to show that the triplet $(\overline{x}_w, \overline{u}_w, \overline{y}_w)$ fulfils these relations for an appropriate π and $y_\pi := \overline{y}_w$. As regards $\overline{x}_w$, it is a solution of

$$\overline{x}_w = K(B'(x_w)(\overline{x}_w - x_w) + B(x_w) + \overline{u}_w).$$

In order to comply with (5.1), it must hold

$$\overline{x}_w = K(B'(x_0)(\overline{x}_w - x_0) + B(x_0) + \overline{u}_w) + e.$$

Subtracting the last equations we end up with (5.16). The adjoint state $\overline{y}_w$ is defined by (4.4). Comparing this with (5.4),

$$\overline{y}_w = \overline{x}_w - q - d_x + B'(x_0)K^*\overline{y}_w + B''(x_0)[\overline{x}_w - x_0, \cdot]K^* y_0,$$

we easily arrive at formula (5.15). Obviously, $\overline{u}_w$ satisfies (4.5) together with (5.17) iff $d_u = 0$. □

Lemma 5.5. *Define d and e according to (5.15), (5.16). Then for all $w \in N_4(w_0)$*

$$\|e\|_\infty \le c_T(\|x_0 - x_w\|_\infty^2 + \|x_w - x_0\|_\infty \|\overline{x}_w - x_0\|_\infty) \tag{5.18}$$

$$\begin{aligned} \|d\|_\infty \le c_T(&\|\overline{y}_w\|_p \|x_0 - x_w\|_\infty^2 + \|\overline{x}_w - x_0\|_\infty(\|x_w - x_0\|_\infty \\ &+ \|y_w - y_0\|_p) + \|x_w - x_0\|_\infty(\|y_w - y_0\|_p + \|\overline{y}_w - y_0\|_p)) \end{aligned} \tag{5.19}$$

with a certain constant c_T not depending on w.

The proof follows completely analogous to [4] from re-arranging and estimating (5.15) – (5.16).

6. Quadratic convergence of the SQP–method

Theorem 6.1. *There is a $C \times L_p \times L_p$–neighbourhood $N_5(w_0)$, and a positive constant ν such that for all $w \in N_5(w_0)$ the solution $\overline{v}_w$ of $(QP)_w$ and the corresponding Lagrange multiplier $\overline{y}_w$ satisfy*

$$\|(\overline{v}_w, \overline{y}_w) - (v_0, y_0)\|_{\infty,p} \le \nu \|w - w_0\|_{\infty,p}^2. \tag{6.1}$$

Proof. We take at first $N(w_0) \subset N_1(w_0) \cap N_4(w_0)$ such that the radius of $N(w_0)$ is less than 1. According to Corollary 4.2., $\|\overline{x}_w - x_0\|_\infty$ and $\|\overline{y}_w\|_p$ remain bounded by a constant $c > 0$ for all $w \in N(w_0)$. From (5.15) - (5.16)

$$\begin{aligned} max(\|e\|_\infty, \|d\|_\infty) &\le c(\|v_0 - v_w\|_\infty^2 + \|w_0 - w_w\|_{\infty,p}) \\ &\le c\|w - w_0\|_{\infty,p} \end{aligned} \tag{6.2}$$

as the diameter of $N(w_0)$ is less than 1.

Thus on $N(w_0)$,

$$\|\pi\|_\infty \le c\|w - w_0\|_{\infty,p}. \tag{6.3}$$

On the other hand, Lemma 5.4. and Theorem 5.3. yield now $y_\pi = y_{\overline{w}}$ and

$$\|\overline{v}_w - v_0\|_{\infty,p} \le c\|\pi\|_{\infty,p} \le c\|w - w_0\|_{\infty,p}. \tag{6.4}$$

Analogously we find

$$\|\overline{y}_w - y_0\|_p \le c\|w - w_0\|_{\infty,p} \tag{6.5}$$

by Lemma 5.1. and (6.3), (6.4). Inserting (6.4) - (6.5) in (5.18) - (5.19) we obtain

$$\|\pi\|_{\infty,p} \le c\|\pi\|_\infty = \|e\|_\infty + \|d\|_\infty \le c\|w - w_0\|_{\infty,p}^2, \tag{6.6}$$

implying together with (6.4), (5.3) the relation (6.1). □

Now we reformulate the SQP–method and state the result on its local convergence. The SQP–method runs as follows.

(SQP) Choose a starting point $w_1 = (v_1, y_1)$ Having $w_k = (v_k, y_k)$ compute $w_{k+1} = (v_{k+1}, y_{k+1})$ to be the solution and the associated Lagrange multiplier of the quadratic optimization problem $(QP)_{w_k}$.

Using Theorem 6 1. it follows now by standard proof techniques that this method converges quadratically to $w_0 = (x_0, u_0, y_0)$, if the starting point w_1 is chosen sufficiently close to w_0 (see [2], Theorem 5 1). Let ν be defined by Theorem 6.1. Denote by $B_\delta(w_0)$ the open ball around w_0 with radius r in the sense of $C \times L_p \times L_p$

Theorem 6.2. *Suppose that Assumptions (1 4) and (SSC) are satisfied. Choose $p > n+1$ and let $\gamma > 0$ be such that $\delta = \nu\gamma < 1$, and $B_{\gamma\delta}(w_0) \subset N_5(w_0)$. Then for any starting point $w_1 \in B_{\gamma\delta}(w_0)$ the SQP method computes a unique sequence w_k with*

$$\|w_k - w_0\|_{\infty,p} \leq \gamma\, \delta^{2^k - 1},$$

and $w_k \in B_{\gamma\delta}(w_0)$ for $k \geq 2$.

The proof is identical to that given in [2]

Thus we have local quadratic convergence of the SQP method in (x, u, y) More precisely, Theorem 6 2 expresses r-quadratic convergence, while Theorem 6.1 shows q-quadratic convergence of the method

References

1 Alt, W The Lagrange–Newton method for infinite–dimensional optimization problems Numer Funct Anal and Optimization 11 (1990), 201–224

2 Alt, W Sequential quadratic programming in Banach spaces In W Oettli and D Pallaschke, editors, Advances in Optimization, number 382 in Lecture Notes in Economics and Mathematical Systems, 281–301 Springer Verlag, 1992

3 Alt, W and K Malanowski The Lagrange–Newton method for nonlinear optimal control problems Computational Optimization and Applications, 2 (1993), 77–100

4 Alt,W , Sontag, R , and F Troltzsch An SQP method for optimal control of weakly singular Hammerstein integral equations DFG–Schwerpunktprogramm ,,Anwendungsbezogene Optimierung und Steuerung ", Report No 423 (1993), to appear in Appl Math Optimization

5 Amann, H Parabolic evolution equations with nonlinear boundary conditions In F E Browder, editor, Nonlinear Functional Analysis Proc Sympos Pure Math , Vol 45, Part I, 17–27, 1986

6 Goldberg, H and F Troltzsch Second order optimality conditions for nonlinear parabolic boundary control problems SIAM J Contr Opt 31 (1993), 1007–1025

7 Kelley, C T and S J Wright Sequential quadratic programming for certain parameter identification problems Math Programming 51 (1991) , 281–305

8 Kupfer, S -F and Sachs, E W Numerical solution of a nonlinear parabolic control problem by a reduced sqp method Computational Optimization and Applications 1 (1992), 113–135

9 Krasnoselskiĭ, M A et al Linear operators in spaces of summable functions (in Russian) Nauka, Moscow, 1966

10 Levitin, E S and B T Polyak Constrained minimization methods USSR J Comput Math and Math Phys 6 (1966), 1–50

11 Troltzsch, F On the semigroup approach for the optimal control of semilinear parabolic equations including distributed an boundary control Z fur Analysis und Anwendungen 8 (1989), 431–443
12 Troltzsch, F An SQP Method for Optimal Control of a Nonlinear Heat Equation (1993), to appear in Control Cybernetics

Fredi Troltzsch
Department of Mathematics
Technical University of Chemnitz-Zwickau
PSF 964
09009 Chemnitz, Germany

CONDITIONAL STABILITY IN DETERMINATION OF DENSITIES OF HEAT SOURCES IN A BOUNDED DOMAIN

MASAHIRO YAMAMOTO

Department of Mathematical Sciences
University of Tokyo

ABSTRACT We consider the heat equation in a bounded domain $\Omega \subset \mathbb{R}^r$

$$\frac{\partial u}{\partial t}(x,t) = \Delta u(x,t) + \sigma(t)f(x) \quad (x \in \Omega,\ 0 < t < T),$$

$$u(x,0) = 0 \quad (x \in \Omega), \qquad \frac{\partial u}{\partial n}(x,t) = 0 \quad (x \in \partial\Omega,\ 0 < t < T)$$

Assuming that σ is a known function with $\sigma(0) \neq 0$, we prove (1) $f(x)$ $(x \in \Omega)$ can be uniquely determined from the boundary data $u(x,t)$ $(x \in \partial\Omega,\ 0 < t < T)$ (2) If f is restricted to a compact set in a Sobolev space, then we get an estimate

$$\|f\|_{L^2(\Omega)} = O\left(\left(\log \frac{1}{\eta}\right)^{-\beta}\right) \quad \text{as} \quad \eta \equiv \|u(\ ,\ )\|_{H^1(0\ T\ L^2(\partial\Omega))} \downarrow 0$$

Here the exponent β is given by the order of the Sobolev space which is assumed to contain the set of f's

1991 *Mathematics Subject Classification* 35K05, 35R25, 35R30, 93B30

Key words and phrases Conditional stability, a priori bound, biorthogonal function, boundary observation, density of heat source

1. Introduction

We consider an initial/boundary value problem for the heat flow equation

$$\text{(1.1)} \quad \left\{ \begin{aligned} \frac{\partial u}{\partial t}(x,t) &= \Delta u(x,t) + \sigma(t)f(x) \qquad (x \in \Omega,\ 0 < t < T) \\ u(x,0) &= 0 \qquad (x \in \Omega) \\ \frac{\partial u}{\partial n}(x,t) &= 0 \qquad (x \in \partial\Omega,\ 0 < t < T). \end{aligned} \right\}$$

Here $\Omega \subset \mathbb{R}^r$ is a bounded domain with smooth boundary $\partial\Omega$, Δ is the Laplacian and $\frac{\partial}{\partial n}$ is a trace operator (e.g. Adams [1]). In particular, if u is sufficiently smooth on $\bar{\Omega}$, then $\frac{\partial u}{\partial n}(x) = \sum_{i=1}^r \nu_i(x)\frac{\partial u}{\partial x_i}(x) \quad (x \in \partial\Omega)$, where $\nu(x) = (\nu_1(x), \ \ ,\nu_r(x))$

is the outward unit normal to $\partial\Omega$ at x. Throughout this paper, $T > 0$ is arbitrarily fixed.

The term $\sigma(t)f(x)$ is considered a heat source. We assume that σ is a known non-zero C^1-function and is independent of the space variable x, and $f \in L^2(\Omega)$ is unknown to be determined from boundary data:

$$u(x,t) \equiv h(x,t) \qquad (x \in \partial\Omega,\ 0 < t < T). \tag{1.2}$$

In (1.2), we regard $u(x,t)$ $(x \in \partial\Omega,\ 0 < t < T)$ as a trace of $u(\cdot,t)$ to the boundary. In fact, since $f \in L^2(\Omega)$ and $\sigma \in C^1[0,T]$, there exists a unique *strong solution* $u \in C^0([0,T]; L^2(\Omega)) \cap C^1((0,T]; L^2(\Omega))$ such that $\Delta u \in C^0((0,T]; L^2(\Omega))$ and $u(\cdot,t) \in X_0 \quad (0 < t \le T)$ (e.g. Pazy [13]). Here we set $X_0 = \{u \in H^2(\Omega)\,;\ \frac{\partial u}{\partial n} = 0\}$ where $H^2(\Omega) = W^{2,2}(\Omega)$ is a Sobolev space (e.g. [1]). Therefore the left hand side of (1.2) can be considered as the trace.

In this paper, we consider

Inverse Problem. (I) (Uniqueness) Do the boundary data $h(x,t) = u(x,t)$ $(x \in \partial\Omega,\ 0 < t < T)$ determine f uniquely ? (II) (Stability) Establish an actual stability estimate for the correspondence $h \longrightarrow f$.

This inverse problem is related to determining the density of radioactive heat sources by the thermal radiation on the surface, for example, if $\sigma(t) = \exp(-\lambda_0 t)$ (e.g. Lavrent'ev, Romanov and Vasil'ev [11, pp.49–50], Romanov [14, pp.208–210]).

To the uniqueness, we can get the positive answer (Theorem 1 in §2).

To the stability property, it seems that we cannot generally expect any stability in usual norms such as $\|f\|_{L^2(\Omega)}$ and $\|h(\cdot,\cdot)\|_{H^1(0,T,L^2(\partial\Omega))}$. This can be conjectured by the regularity property of the parabolic equation (e.g. [13]). Here and henceforth we set

$$\|h(\cdot,\cdot)\|_{H^1(0,T,L^2(\partial\Omega))} = \left(\int_0^T \int_{\partial\Omega} \left(|h(x,t)|^2 + \left|\frac{\partial h}{\partial t}(x,t)\right|^2\right) dS_x dt\right)^{\frac{1}{2}}.$$

However, since the uniqueness result can be established and the map $f \longrightarrow h$ is proved to be continuous from $L^2(\Omega)$ to $H^1(0,T; L^2(\partial\Omega))$, by Tikhonov's theorem (e.g. Lavrent'ev, Romanov and Shishat·skiĭ[10, p.28] or [14, §1.2]), we can see the continuity of the map $h \longrightarrow f$ provided that we restrict an admissible set of f to any compact set $\mathcal{U}$ in $L^2(\Omega)$. That is, for any $\epsilon > 0$, there exists some $\delta(\epsilon) > 0$ such that

$$\|u(f) - u(g)\|_{H^1(0,T,L^2(\partial\Omega))} < \delta(\epsilon) \quad \text{and} \quad g,h \in \mathcal{U}$$

imply $\|g - h\|_{L^2(\Omega)} \le \epsilon$. This is what is called a conditional stability. Here $\delta(\epsilon)$ depends also on the choice of a compact set $\mathcal{U}$ and, in general, we cannot specify the order of $\delta(\epsilon)$.

The purpose of this paper is to show the uniqueness in our inverse problem and to give the modulus $\delta(\epsilon)$ of continuity if we take a bounded set in some Sobolev space as an admissible set of unknown densities f's.

For determination of spatially non-homogeneous terms in parabolic equations, we can refer, for example, to Cannon [3], Cannon and Esteva [4], [11], [14] and Yamamoto [18].

This paper is composed of four sections and an appendix. In §2 we will state Theorem 1 (uniqueness) and Theorem 2 (conditional stability) as our main results. In §3 and §4, we will prove them. The appendix is devoted to the proof of another uniqueness result in the case where the observation time length T is infinite.

2. Main Results

Henceforth we assume

$$\sigma(0) \neq 0, \qquad \sigma \in C^1[0,T]. \tag{2.1}$$

Let $L^2(\Omega)$ be the space of real-valued square integrable functions with the norm $\|\cdot\|$ and the inner product $(\cdot,\cdot)$.

For stating our results, we introduce some notations and definitions. Let A be the realization of $-\Delta$ in $L^2(\Omega)$ with Neumann boundary condition:$Au(x) = -\Delta u(x)$ and $\mathcal{D}(A) = \{\, u \in H^2(\Omega);\ \frac{\partial u}{\partial n}_{|\partial\Omega} = 0\,\}$. Let us number the eigenvalues of A repeatedly according to their multiplicities:

$$0 = \lambda_1 < \lambda_2 \le \lambda_3 \le ....$$

That is, if the multiplicity of λ_i is m, then λ_i appears in the above sequence m-times. Let ϕ_k be an eigenfunction for the eigenvalue λ_k of A $(k \ge 1)$. We can choose $\{\phi_k\}_{k\ge 1}$ such that

$$(\phi_k, \phi_l) = \delta_{kl}. \tag{2.2}$$

Here we set $\delta_{kl} = 1\,(k = l), = 0$ (otherwise).

Let us fix $\alpha > 0$ such that

$$\alpha > \frac{r-1}{2}. \tag{2.3}$$

For any $M > 0$, let us define an admissible set of unknown terms f's by

$$\mathcal{U}_{M,\alpha} = \{\, f \in L^2(\Omega);\ \sum_{k=1}^{\infty} |(f,\phi_k)|^2 k^{\frac{2\alpha}{r}} \le M^2\,\}. \tag{2.4}$$

For $\frac{\alpha}{2} \in \mathbb{N}$, by using the asymptotic behaviour of λ_n (see (3.1) below), we can replace the definition of $\mathcal{U}_{M,\alpha}$ by

$$\mathcal{U}_{M,\alpha} = \{f \in H^\alpha(\Omega);\ \frac{\partial}{\partial n}(A^j f) = 0,\ 0 \le j \le \frac{\alpha}{2} - 1,\ \|f\|_{H^\alpha(\Omega)} \le M\}. \tag{2.5}$$

Remark 1. We can define the fractional power $(A+\gamma)^\beta$ for $\gamma > 0$ and $\beta \in \mathbb{R}$ (e.g. [13]). By $(A+\gamma)^{\frac{\alpha}{2}}$, we can rewrite the definition of $\mathcal{U}_{M,\alpha}$:

$$\mathcal{U}_{M,\alpha} = \{\, f \in L^2(\Omega);\ f \in \mathcal{D}\left((A+\gamma)^{\frac{\alpha}{2}}\right),\quad \|\,(A+\gamma)^{\frac{\alpha}{2}} f\| \le M \,\}.$$

Now we are ready to state

Theorem 1 (Uniqueness): *Let f satisfy*

$$\sum_{k=1}^{\infty} |(f,\phi_k)|^2 k^{\frac{2\alpha}{r}} < \infty.$$

Let $u(f)(x,t)$ be the strong solution to

$$\text{(2.6)}\qquad \left\{ \begin{array}{rcl} \dfrac{\partial u}{\partial t}(x,t) & = & \Delta u(x,t) + \sigma(t) f(x) \qquad (x \in \Omega,\ 0 < t < T) \\ u(x,0) & = & 0 \qquad (x \in \Omega) \\ \dfrac{\partial u}{\partial n}(x,t) & = & 0 \qquad (x \in \partial\Omega,\ 0 < t < T). \end{array} \right\}$$

If

$$\text{(2.7)}\qquad u(f)(x,t) = 0 \qquad (x \in \partial\Omega,\ 0 < t < T),$$

then $f(x) = 0$ for almost all $x \in \Omega$.

Theorem 2 (Conditional Stability). *For any fixed $\alpha > \frac{r-1}{2}$ and any $M > 0$, we a priori assume that $f \in \mathcal{U}_{M,\alpha}$. If (2.6) holds, then there exists a constant $C > 0$ such that*

$$\text{(2.8)}\qquad \|f\| \le \frac{M}{N^{\frac{\alpha}{r}}} + C\exp(CN^{\frac{1}{r}})\|u(f)\|_{H^1(0,T;L^2(\partial\Omega))}$$

for any $N \in \mathbb{N}$. Moreover, as $\eta \equiv \|u(f)\|_{H^1(0,T;L^2(\partial\Omega))}$ tends to zero, we have the estimate:

$$\text{(2.9)}\qquad \|f\| = O\left(\left(\frac{1}{\log\frac{1}{\eta}}\right)^{\beta}\right) \qquad \textit{for any}\quad \beta < \alpha.$$

Remark 2. By (4.3) in the proof of Theorem 2 in §4, we can see: As the norm for densities f's, if we adopt $|||f||| = \sup_{k\ge 1} |(f,\phi_k)|\exp(-C_1 k^{\frac{1}{r}})$ for a sufficiently large $C_1 > 0$ and we define an admissible set of f's by $\tilde{X} = \{\, f \in L^2(\Omega);\ |||f||| < \infty \,\}$, then the map $u(x,t)\quad (x \in \partial\Omega,\ 0 < t < T) \longrightarrow f(x)\quad (x \in \Omega)$ is Lipschitz continuous from $H^1(0,T;L^2(\partial\Omega))$ to $\tilde{X}$. However, the topology in $\tilde{X}$ is too weak in the sense that we can not generally get the continuity of the map $f \longrightarrow u(f)$ from $\tilde{X}$ to $H^1(0,T;L^2(\partial\Omega))$.

In [3], a similar problem is considered in the case of $\sigma(t) \equiv -1$ and the maximum principle for the analytic function is a key. On the other hand, in this paper we reduce our inverse problem to some "moment problem" by which we can estimate Fourier coefficients of f. Our technique is control-theoretical, and is derived from Russell [15], where a boundary control problem is considered for heat equation.

If Ω is an r-dimensional parallelepipedon $\{(x_1, ..., x_r);\, 0 < x_i < L_i \quad (1 \le i \le r)\}$, $(L_i > 0 : 1 \le i \le r)$, then we can reach similar results for uniqueness and conditional stability even though we restrict observations on the whole boundary, $u(x,t) \quad (x \in \partial\Omega,\, 0 < t < T)$ to a part of $\partial\Omega$ ([18]). The method in [18] is based on Fattorini [6]. However for a general bounded domain, it is not certain whether or not boundary data restricted to a part of $\partial\Omega$ can guarantee conditional stability such as Theorem 2. As is mentioned in §1, in [11] and [14], a similar problem is considered in the case of $\Omega = \{(x_1, ..., x_{r-1}, x_r);\, -\infty < x_1, ..., x_{r-1} < \infty,\, x_r > 0\}$ and $T = \infty$, where the key technique is the Laplace transform with respect to time t and so it seems essential that the observation is over any $t > 0$, not over a finite time interval $0 < t < T\,(< \infty)$.

Remark 3. For discussing parabolic inverse problems, another method by Reznitskaja is applicable (e.g. Lavrent'ev, Reznitskaja and Yakhno [9, pp.6–8], and Romanov [14, pp.213–218]). That is, if we can observe boundary values over $[0, \infty)$ of the t-axis, then we can replace our inverse problem for a parabolic equation (1.1) by a determination problem of f in a hyperbolic equation $\frac{\partial^2 u}{\partial t^2} = \Delta u + \sigma(t)f(x) \quad (x \in \Omega,\, t > 0)$. Then we can apply results in hyperbolic inverse problems (e.g. [10, Chapter VII] and [14, Chapter 4]). In our paper, however, by constructing biorthogonal functions, we directly discuss a parabolic inverse problem.

If we assume that $T = \infty$, then we can weaken the condition (2.7) for the uniqueness. That is,

Proposition (Uniqueness). *Let us assume that f and σ are continuous and bounded respectively in Ω and $[0, \infty)$ and that σ does not vanish identically. If $u(f)(x,t)$ satisfies*

$$(2.10) \qquad \left\{ \begin{array}{rcl} \frac{\partial u}{\partial t}(x,t) & = & \Delta u(x,t) + \sigma(t)f(x) \qquad (x \in \Omega,\, t > 0) \\ u(x,0) & = & 0 \qquad (x \in \Omega) \\ \frac{\partial u}{\partial n}(x,t) & = & 0 \qquad (x \in \partial\Omega,\, t > 0). \end{array} \right\}$$

and

$$(2.11) \qquad u(f)(x,t) = 0 \quad (x \in \textit{any open set of } \partial\Omega,\, t > 0),$$

then $f(x) = 0 \quad (x \in \Omega)$.

We can prove a uniqueness result of this type in the case where Δ in (2.10) is replaced by any uniformly elliptic differential operator of the second order, but we will omit it. For convenience, the proof of this proposition is given in Appendix.

3. Proof of Theorem 1

We will divide the proof into the following three steps.

1) In this step, we will prove that the trace of the series at the boundary

$$\sum_{k=1}^{\infty}\left(\int_0^t \exp(-\lambda_k(t-s))\sigma(s)ds\right)(f,\phi_k)\phi_k(x)$$

is convergent in $H^1(0,T;L^2(\partial\Omega))$. First, by a theorem of asymptotic distribution of eigenvalues (e.g. Agmon [2], Courant and Hilbert [5]), we have

$$\lambda_k = c_2 k^{\frac{2}{r}} + o(k^{\frac{2}{r}}) \tag{3.1}$$

as $k \longrightarrow \infty$, where c_2 is a positive constant.

For simplicity, we set

$$\psi_k(t) = \int_0^t \exp(-\lambda_k(t-s))\sigma(s)ds \qquad (k \ge 1,\ 0 \le t \le T). \tag{3.2}$$

By $\sigma \in C^1[0,T]$, we have

$$|\psi_k(t)| = \left|\int_0^t \exp(-\lambda_k s)\sigma(t-s)ds\right| \le C_3' \int_0^t \exp(-\lambda_k s)ds$$

and

$$\left|\frac{d\psi_k}{dt}(t)\right| \le C_3'\left(\exp(-\lambda_k t) + \int_0^t \exp(-\lambda_k s)ds\right) \qquad (k \ge 1).$$

Therefore we get

$$\|\psi_k\|_{H^1(0,T)} \le \frac{C_3}{\sqrt{\lambda_k+1}} \qquad (k \ge 1). \tag{3.3}$$

On the other hand, we have $\mathcal{D}((A+1)^{\frac{1}{4}}) = H^{\frac{1}{2}}(\Omega)$ and $\|u\|_{H^{1/2}(\Omega)} \le C_3'\|(A+1)^{\frac{1}{4}}u\|$ (e.g. Fujiwara [7]) and $\|u\|_{L^2(\partial\Omega)} \le C_3'\|u\|_{H^{1/2}(\Omega)}$ (e.g. Adams [1]). Consequently by (2.2) we see

$$\|\phi_k\|_{L^2(\partial\Omega)} \le C_3'^2\|(A+1)^{\frac{1}{4}}\phi_k\| = C_3'^2\|(\lambda_k+1)^{\frac{1}{4}}\phi_k\| = C_3'^2(\lambda_k+1)^{\frac{1}{4}},$$

so that we have

$$\begin{aligned}
&\sum_{k=1}^{\infty}\left\|\left(\int_0^t \exp(-\lambda_k(t-s))\sigma(s)ds\right)(f,\phi_k)\phi_k\right\|_{H^1(0,T;L^2(\partial\Omega))}\\
&=\sum_{k=1}^{\infty}\|(f,\phi_k)\psi_k\phi_k\|_{H^1(0,T;L^2(\partial\Omega))}\\
&\le \sum_{k=1}^{\infty}\frac{C_3}{(\lambda_k+1)^{1/2}}\times|(f,\phi_k)|\times C'^2_3(\lambda_k+1)^{\frac14} \qquad \text{(by (3.3))}\\
&= C_3\sum_{k=1}^{\infty}|(f,\phi_k)|(\lambda_k+1)^{-\frac14}\\
&= C_3\sum_{k=1}^{\infty}|(f,\phi_k)|(\lambda_k+1)^{\frac{\alpha}{2}}(\lambda_k+1)^{-\frac{\alpha}{2}-\frac14}\\
&\le C_3\left(\sum_{k=1}^{\infty}|(f,\phi_k)|^2(\lambda_k+1)^{\alpha}\right)^{\frac12}\left(\sum_{k=1}^{\infty}(\lambda_k+1)^{-\alpha-\frac12}\right)^{\frac12}\\
&\quad \text{(by Schwarz's inequality)}\\
&\le C''_3\left(\sum_{k=1}^{\infty}|(f,\phi_k)|^2k^{\frac{2\alpha}{r}}\right)^{\frac12}\left(\sum_{k=1}^{\infty}k^{-\frac{2\alpha+1}{r}}\right)^{\frac12} \qquad \text{(by (3.1))}\\
&<\infty \qquad \left(\text{by } f\in\mathcal{U}_{M,\alpha} \text{ and } \alpha>\frac{r-1}{2}\right).
\end{aligned}$$

Thus, by Itô [8], for almost all $x\in\partial\Omega$ and almost all $t\in[0,T]$, we can obtain

$$u(f)(x,t)=\sum_{k=1}^{\infty}(f,\phi_k)\psi_k(t)\phi_k(x) \tag{3.4}$$

and we see that the series is convergent in $H^1(0,T;L^2(\partial\Omega))$.

2) In this step, on the basis of Russell [15], in $H^1(0,T;L^2(\partial\Omega))$ we will construct a biorthogonal system $\{\theta_k(x,t)\}_{k\ge1}$ to $\{\psi_k(t)\phi_k(x)\}_{k\ge1}$.

By [15], there exists a system $\{\eta_k(x,t)\}_{k\ge1}$ in $L^2(0,T;L^2(\partial\Omega))$ such that

$$\langle\exp(-\lambda_k t)\phi_k,\eta_l\rangle_{L^2}=\delta_{kl} \qquad (k,l\ge1). \tag{3.5}$$

Here and henceforth we denote the inner products in $L^2(0,T;L^2(\partial\Omega))$ and in $H^1(0,T;L^2(\partial\Omega))$ respectively by $\langle\cdot,\cdot\rangle_{L^2}$ and $((\cdot,\cdot))_{H^1}$:

$$\langle g,h\rangle_{L^2}=\int_0^T\int_{\partial\Omega}g(x,t)h(x,t)dS_x dt$$

and

$$((g,h))_{H^1} = \int_0^T \int_{\partial\Omega} \left(g(x,t)h(x,t) + \frac{\partial g}{\partial t}(x,t)\frac{\partial h}{\partial t}(x,t) \right) dS_x dt.$$

Let us define an operator $K : L^2(0,T;L^2(\partial\Omega)) \longrightarrow H^1(0,T;L^2(\partial\Omega))$ by

$$(Kg)(x,t) = \int_0^t \sigma(t-s)g(x,s)ds \qquad (0 < t < T,\ x \in \partial\Omega).$$

We can easily see that K is bounded from $L^2(0,T;L^2(\partial\Omega))$ to $H^1(0,T;L^2(\partial\Omega))$. By the assumption (2.1) on σ, we have $\sigma(0) \neq 0$, so that we can prove that there exists a constant $C_4 > 0$ such that

$$C_4 \|Kg\|_{H^1(0,T,L^2(\partial\Omega))} \geq \|g\|_{L^2(0,T,L^2(\partial\Omega))} \tag{3.6}$$

for any $g \in L^2(0,T;L^2(\partial\Omega))$. In fact, let

$$h(x,t) = (Kg)(x,t) = \int_0^t \sigma(t-s)g(x,s)ds \qquad (0 < t < T,\ x \in \partial\Omega).$$

Then we have

$$\frac{\partial h}{\partial t}(x,t) = \sigma(0)g(x,t) + \int_0^t \frac{d\sigma}{dt}(t-s)g(x,s)ds \qquad (0 < t < T,\ x \in \partial\Omega),$$

which is a Volterra integral equation in t of the second kind, so that we can construct the real-valued resolvent kernel $\Gamma(t,s)$ for $\frac{1}{\sigma(0)}\frac{d\sigma}{dt}(t-s) \quad (0 \leq s \leq t \leq T)$ and we have

$$g(x,t) = \frac{1}{\sigma(0)}\frac{\partial h}{\partial t}(x,t) - \frac{1}{\sigma(0)}\int_0^t \Gamma(t,s)\frac{\partial h}{\partial t}(x,s)ds$$

(e.g. Tricomi [17]). This means (3.6).

Now (3.6) implies that the range of K is closed in $H^1(0,T;L^2(\partial\Omega))$. Therefore we see that $(K^{-1})^*$ exists and $(K^{-1})^* = (K^*)^{-1}$. Here and henceforth we identify the adjoint space X^* of a Hilbert space with X and $\cdot^*$ denotes the adjoint of an operator under consideration.

We define a system $\{\theta_l\}_{l\geq 1} \subset H^1(0,T,L^2(\partial\Omega))$ by

$$\theta_l(x,t) = ((K^{-1})^*\eta_l)(x,t) \qquad (l \geq 1,\ x \in \partial\Omega,\ 0 < t < T) \tag{3.7}$$

This system $\{\theta_l\}_{l\geq 1}$ is the desired one, that is,

$$((\psi_k\phi_k, \theta_l))_{H^1} = \delta_{kl} \qquad (k \geq 1, l \geq 1). \tag{3.8}$$

Here let us recall that ψ_k is given by (3.2). The relation (3.8) can be proved as follows:

$$\begin{aligned} ((\psi_k\phi_k, \theta_l))_{H^1} &= ((K(\exp(-\lambda_k t)\phi_k),\ (K^{-1})^*\eta_l))_{H^1} \\ &= \langle (K^{-1})^{**}K(\exp(-\lambda_k t)\phi_k),\ \eta_l \rangle_{L^2} \\ &= \delta_{kl}, \end{aligned}$$

by (3.5) and (3.7).

3) We will complete the proof of Theorem 1. By the assumption, we have

$$u(f)(x,t) = 0 \quad (x \in \partial\Omega,\ 0 < t < T).$$

Therefore (3.4) implies

$$\sum_{k=1}^{\infty}(f,\phi_k)\psi_k(t)\phi_k(x) = 0 \qquad \text{in} \quad H^1(0,T;L^2(\partial\Omega)).$$

Since, as is proved in the first step 1), this series is convergent in $H^1(0,T;L^2(\partial\Omega))$, we can take the $H^1(0,T;L^2(\partial\Omega))$-inner product with θ_l in termwise, so that

$$(f,\phi_l) = 0 \quad (l \ge 1)$$

by the biorthogonality (3.8). Thus the proof is complete.

4. Proof of Theorem 2

First we will estimate norms of θ_k constructed by (3.7) in $H^1(0,T;L^2(\partial\Omega))$. By Russell [15], there exists a constant $C_5' > 0$ such that

$$\|\eta_l\|_{L^2(0,T;L^2(\partial\Omega))} \le C_5' \exp(C_5'\sqrt{\lambda_l}) \qquad (l \ge 1).$$

Consequently by (3.1) we can obtain

$$\|\eta_l\|_{L^2(0,T;L^2(\partial\Omega))} \le C_5 \exp(C_5 l^{\frac{1}{r}}) \qquad (l \ge 1). \tag{4.1}$$

By (3.6) and (3.7), we have

$$\begin{aligned}\|\theta_l\|_{H^1(0,T;L^2(\partial\Omega))} &\le \|(K^{-1})^*\| \times \|\eta_l\|_{L^2(0,T;L^2(\partial\Omega))} \\ &\le C_5'C_5 \exp(C_5 l^{\frac{1}{r}}) \equiv C_6 \exp(C_6 l^{\frac{1}{r}}) \ (l \ge 1).\end{aligned} \tag{4.2}$$

Here $\|(K^{-1})^*\|$ is the operator norm of $(K^{-1})^*$ and $C_6 > 0$ is a constant which is independent of $l \ge 1$.

Next we will estimate the Fourier coefficients (f,ϕ_k) by using (4.2) and (3.4), and complete the proof of Theorem 2. Since the series in (3.4) converges in $H^1(0,T;L^2(\partial\Omega))$, we can take the inner product $((\cdot,\cdot))_{H^1}$ with θ_l in termwise. Consequently by (3.8) we can reach

$$((u(f),\theta_l))_{H^1} = (f,\phi_l) \qquad (l \ge 1),$$

and so by (4.2) we obtain

$$|(f,\phi_l)| \le C_6 \|u(f)\|_{H^1(0,T;L^2(\partial\Omega))} \exp(C_6 l^{\frac{1}{r}}) \quad (l \ge 1). \tag{4.3}$$

First let us prove (2.8). We have

$$\begin{aligned}
\|f\|^2 &= \sum_{l=1}^{\infty} |(f,\phi_l)|^2 = \sum_{l=1}^{N-1} |(f,\phi_l)|^2 + \sum_{l=N}^{\infty} \frac{|(f,\phi_l)|^2 l^{\frac{2\alpha}{r}}}{l^{\frac{2\alpha}{r}}}\\
&\le \sum_{l=1}^{N-1} |(f,\phi_l)|^2 + N^{-\frac{2\alpha}{r}} M^2 \qquad (\text{by } f \in \mathcal{U}_{M,\alpha})\\
&\le C_6^2 \|u(f)\|^2_{H^1((0,T),L^2(\partial\Omega))} \sum_{l=1}^{N-1} \exp(2C_6 l^{\frac{1}{r}}) + N^{-\frac{2\alpha}{r}} M^2 \quad (\text{by } (4.3))\\
&\le C_6^2 \|u(f)\|^2_{H^1((0,T),L^2(\partial\Omega))} (N-1) \exp(2C_6 (N-1)^{\frac{1}{r}}) + N^{-\frac{2\alpha}{r}} M^2
\end{aligned}$$

which is (2.8).

Next let us consider (2.9). For $\beta < \alpha$, we fix the natural number N such that

$$M^{\frac{r}{\alpha}} \left(\log \frac{1}{\eta}\right)^{\frac{\beta r}{\alpha}} \le N < 1 + M^{\frac{r}{\alpha}} \left(\log \frac{1}{\eta}\right)^{\frac{\beta r}{\alpha}}.$$

Then we can get

$$\frac{M}{N^{\frac{\alpha}{r}}} \le \left(\frac{1}{\log \frac{1}{\eta}}\right)^{\beta}. \tag{4.4}$$

Moreover we have

$$\begin{aligned}
C\eta \exp(CN^{\frac{1}{r}}) &\le C\eta \exp\left(C\left(1 + M^{\frac{r}{\alpha}} \left(\log \frac{1}{\eta}\right)^{\frac{\beta r}{\alpha}}\right)^{\frac{1}{r}}\right)\\
&\le C'\eta \exp\left(CM^{\frac{1}{\alpha}} \left(\log \frac{1}{\eta}\right)^{\frac{\beta}{\alpha}}\right).
\end{aligned}$$

Therefore, in order to prove (2.9), by (2.8) and (4.4), it is sufficient to verify that

$$\eta \exp\left(CM^{\frac{1}{\alpha}} \left(\log \frac{1}{\eta}\right)^{\frac{\beta}{\alpha}}\right) = O\left(\left(\frac{1}{\log \frac{1}{\eta}}\right)^{\beta}\right)$$

as $\eta \downarrow 0$ for $\beta < \alpha$, which is seen from

$$\lim_{\eta\downarrow 0} \eta \exp\left(CM^{\frac{1}{\alpha}} \left(\log \frac{1}{\eta}\right)^{\frac{\beta}{\alpha}}\right) \left(\log \frac{1}{\eta}\right)^{\beta} = 0 \qquad (\beta < \alpha).$$

Appendix. Proof of Proposition

Let $U(t, x, y)$ be the fundamental solution of $\frac{\partial\cdot}{\partial t} - \Delta\cdot$ with the zero Neumann boundary condition. Then under the assumption on regularity of σ and f, we can express the solution to (2.10) as

$$u(f)(x, t) = \int_0^t \sigma(s) \left(\int_\Omega U(t-s, x, y) f(y) dy \right) ds \qquad (x \in \bar{\Omega},\ t > 0)$$

(Itô [8]). We set

$$u_x(t) = \int_\Omega U(t, x, y) f(y) dy$$

for any fixed $x \in \bar{\Omega}$. Let us assume that $u(f)(x, t) = 0$ for $t > 0$ and any $x \in \Gamma$: an open set of $\partial\Omega$. Therefore we have

$$\int_0^t \sigma(s) u_x(t-s) ds = 0 \qquad (t > 0,\ x \in \Gamma),$$

and since σ does not vanish identically and σ, $u_x(\cdot)$ are continuous, by Titchmarsh's theorem on the convolution (e.g. Titchmarsh [16, p.322]), we can obtain

$$u_x(t) \equiv \int_\Omega U(t, x, y) f(y) dy = 0 \qquad (t > 0,\ x \in \Gamma).$$

This implies that for any $0 < \epsilon < T$, the classical solution u to

$$\left\{ \begin{array}{lll} \frac{\partial u}{\partial t}(x, t) & = & \Delta u(x, t) \qquad (x \in \Omega,\ \epsilon < t < T) \\ u(x, 0) & = & f(x) \qquad (x \in \Omega) \\ \frac{\partial u}{\partial n}(x, t) & = & 0 \qquad (x \in \partial\Omega,\ \epsilon < t < T) \end{array} \right\}$$

vanishes in an open set of $\partial\Omega$. Therefore it follows from the unique continuation theorem (e.g. Mizohata [12]) that $u(x, t) = 0 \quad (x \in \Omega,\ t > \epsilon)$. Consequently by analyticity of u in t (e.g. [13]), we see $u(x, t) = 0\,(x \in \Omega,\ t \geq 0)$. Thus we can see $f(x) = 0 \quad (x \in \Omega)$.

Acknowledgements. The author would like to express his hearty thanks to referees for valuable comments, and sincerely thanks the Alexander von Humboldt Foundation for the financial support. Moreover he expresses his hearty gratitude to Professor Dr. Karl-Heinz Hoffmann of Technische Universität München.

References

1. R. A. Adams, *Sobolev Spaces*, Academic Press, New York, 1975.
2. S. Agmon, *Lectures on Elliptic Boundary Value Problems*, Van Nostrand, Princeton, N.J., 1965.
3. J. R. Cannon, *Determination of an unknown heat source from overspecified boundary data*, SIAM J. Numer. Anal. **5** (1968), 275–286.
4. J. R. Cannon and S. P. Esteva, *An inverse problem for the heat equation*, Inverse Problems **2** (1986) 395–403.
5. R. Courant and D. Hilbert, *Methods of Mathematical Physics*, Volume I, Interscience, New York, 1953.
6. H. O. Fattorini, *Boundary control of temperature distributions in a parallelepipedon*, SIAM J. Control **13** (1975) 1–13.
7. D. Fujiwara, *Concrete characterization of the domains of fractional powers of some elliptic differential operators of the second order*, Proc. Japan Acad. **43** (1967), 82–86.
8. S. Itô, *Diffusion Equations*, (English translation), American Mathematical Society, Providence, Rhode Island, 1992.
9. M. M. Lavrent'ev, K. G. Reznitskaja and V. G. Yakhno, *One-dimensional Inverse Problems of Mathematical Physics*, (English translation), American Mathematical Society, Providence, Rhode Island, 1986.
10. M. M. Lavrent'ev, V. G. Romanov and S. P. Shishat·skiĭ, *Ill-posed Problems of Mathematical Physics and Analysis*, (English translation), American Mathematical Society, Providence, Rhode Island, 1986.
11. M. M. Lavrent'ev, V. G. Romanov and V. G. Vasil'ev, *Multidimensional Inverse Problems for Differential Equations*, (Lecture Notes in Mathematics, No.167), Springer-Verlag, Berlin, 1970.
12. S. Mizohata, *The Theory of Partial Differential Equations*, Cambridge University Press, London, 1973.
13. A. Pazy, *Semigroups of Linear Operators and Applications to Partial Differential Equations* Springer-Verlag, Berlin, 1983.
14. V. G. Romanov, *Inverse Problems of Mathematical Physics*, (English translation), VNU Science Press, Utrecht, 1987.
15. D. L. Russell, *A unified boundary controllability theory for hyperbolic and parabolic partial differential equations*, Studies in Applied Mathematics **52** (1973), 189–211.
16. E. C. Titchmarsh, *Introduction to the Theory of Fourier Integrals*, Oxford University Press, London, 1937.
17. F. G. Tricomi, Integral Equations, Dover, New York, 1985.
18. M. Yamamoto, *Conditional stability in determination of force terms of heat equations in a rectangle*, Mathematical and Computer Modelling **18** (1993), 79–88.

Masahiro Yamamoto
Department of Mathematical Sciences
University of Tokyo
3-8-1 Komaba, Meguro, Tokyo 153
Japan

E-mail: myama@tansei.cc.u-tokyo.ac.jp

BOUNDARY STABILIZATION OF THE KORTEWEG-DE VRIES EQUATION

BING-YU ZHANG

Department of Mathematical Sciences
University of Cincinnati

ABSTRACT Consider herein is the initial-boundary value problem of the KdV equation posed on the bounded interval $(0,1)$

$$(*)\qquad \begin{cases} u_t + uu_x + u_{xxx}, & u(x,0) = \phi(x) \\ u(0,t) = 0, \quad u(1,t) = 0, & u_x(1,t) = \alpha u_x(0,t) \end{cases}$$

where $|\alpha| < 1$

It is shown that (i) the system $(*)$ is globally well-posed in the space

$$H^3_\alpha = \left\{\phi \in H^3(0,1), \quad \phi(0,t) = 0, \quad \phi(1,t) = 0, \quad \phi_x(1,t) = \alpha\phi_x(0,t)\right\}$$

and (ii) if $\alpha \neq 0$, then the system $(*)$ is locally well-posed in the space $H^1_0(0,1)$, but its small amplitude solutions exist globally and decay exponentially to zero as $t \to \infty$

1991 *Mathematics Subject Classification* 35Q20, 93D15, 93C20, 93B05

Key words and phrases Korteweg-de Vries equation, stabilization, boundary control

1. Introduction

In this paper we consider a boundary control system described by the Korteweg-de Vries (KdV) equation posed on a bounded interval $(0,1)$

$$(1.1)\qquad \begin{cases} u_t + uu_x + u_{xxx} = 0, & u(x,0) = \phi(x) \\ u(0,t) = 0, \quad u(1,t) = 0, & u_x(1,t) = h(t) \end{cases}$$

for $x \in (0,1)$, $t \geq 0$, where the subscripts denote partial derivatives and the boundary value function $h(t)$ is considered as a control input The goal is to find an appropriate boundary feedback control law so that the resulting closed loop system is stabilized

The paper was completed while the author held a postdoctoral position in industrial mathematics at the Institute for Mathematics and its Applications at the University of Minnesota

It is readily verified that, for any smooth solution of (1.1), one has

$$\frac{d}{dt}\int_0^1 u^2(x,t) = u_x^2(1,t) - u_x^2(0,t) = h^2(t) - u_x^2(0,t).$$

This suggests to choose a feedback control law

$$h(t) = \alpha u_x(0,t), \qquad \text{with } |\alpha| < 1\ , \tag{1.2}$$

in order to stabilize the system (1.1). Indeed, by choosing h as in (1.2), one would see that the identity

$$\frac{d}{dt}\int_0^1 u^2(x,t) = -(1-\alpha^2)u_x^2(0,t) \tag{1.3}$$

holds for smooth solutions of the system (1.1). Hence the "energy" $\int_0^1 u^2(x,t)dx$ is decreasing as long as $u_x(0,t)$ does not vanish. Plugging (1.2) into (1.1) one obtains a closed-loop system

$$\begin{cases} u_t + uu_x + u_{xxx} = 0, \qquad u(x,0) = \phi(x), \\ u(0,t) = 0, \quad u(1,t) = 0, \quad u_x(1,t) = \alpha u_x(0,t). \end{cases} \tag{1.4}$$

According to (1.3), it possesses a dissipative mechanism introduced through the boundary. Our main concern is whether the introduced dissipative mechanism is strong enough to stabilize the system.

A similar problem has been considered in [16] for the KdV equation posed on the periodic domain:

$$\begin{cases} v_t + vv_x + v_{xxx} = 0, \ v(x,0) = \phi(x), \\ v(1,t) = v(0,t), \ v_{xx}(1,t) = v_{xx}(0,t), \ v_x(1,t) - v_x(0,t) = h(t) \end{cases} \tag{1.5}$$

where a dissipative mechanism is introduced to (1.5) by choosing

$$h(t) = -k(v_x(1,t) + v_x(0,t)), \qquad k > 0.$$

The resulting closed loop system is of the form

$$\begin{cases} v_t + vv_x + v_{xxx} = 0, \qquad v(x,0) = \phi(x), \\ v(0,t) = v(1,t), \ v_{xx}(1,t) = v_{xx}(0,t), \ v_x(1,t) = \alpha v_x(0,t) \end{cases} \tag{1.6}$$

where $\alpha = (1-k)/(1+k)$. It was shown in [16] that there exists a $\delta > 0$ and a $\beta > 0$ such that for any $\phi \in H^1(0,1)$ with $\|\phi\|_{H^1(0,1)} \le \delta$, the system (1.6) has a unique solution $v \in L^\infty(0,\infty; H^1(0,1))$ and

$$\|v(\cdot,t) - [\phi]\|_{L^2(0,1)} \le ce^{-\beta t}\|\phi - [\phi]\|_{L^2(0,1)}$$

for any $t \geq 0$ where $[\phi]$ is the mean value of $\phi(x)$ over the interval $(0, 1)$, i.e.,

$$[\phi] = \int_0^1 \phi(x)dx.$$

In this paper we are going to show that a similar result is also true for the system (1.4).

We will first consider the well-posedness problem of the system (1.4), which is an interesting initial-boundary value problem for the KdV equation in its own right. In the past several decades there have been many literatures discussing the initial-boundary value problem for the KdV equation. For a beginning of collection of references, we refer to [2], [8] – [10] and [18] for the equation posed on the whole real line R, or on a periodic domain and refer to [3] for the equation posed on on a half real line R^+. As for the equation posed on a finite interval we refer to [4]. In particular, the following initial boundary value problem has been considered in [4]

$$(1.7) \qquad \begin{cases} u_t + uu_x + u_{xxx} = f, \qquad u(x, 0) = 0 \\ \alpha_1 u_{xx}(0, t) + \alpha_2 u_x(0, t) + \alpha_3 u(0, t) = 0 \\ \beta_1 u_{xx}(1, t) + \beta_2 u_x(1, t) + \beta_3 u(1, t) = 0 \\ \xi_1 u_x(1, t) + \xi_2 u(1, t) = 0 \end{cases}$$

where α_i, β_i, $\xi_j \in R$, $i = 1, 2, 3$; $j = 1, 2$ and some assumptions are imposed so that the 'energy' of the solutions of (1.7) is decreasing. It was shown in [4] that (1.7) has a unique solution

$$u \in L^2(0, T; H^3(0, 1)), \quad u_t \in L^\infty(0, T; L^2(0, 1)) \cap L^2(0, T; H^1(0, 1))$$

for f, $f_t \in L^2(0, T; L^2(0, 1))$. The result is local in the sense that T depends on f. A counter example is given in [4] to show that one cannot expect to have global result for (1.7) in general. We point out here that the boundary conditions of the system (1.4) is not covered by (1.7) except the case that $\alpha = 0$. In this paper we will show that the system (1.4) including the case $\alpha = 0$ is globally well-posed in the space $H^3_\alpha(0, 1)$ where

$$H^3_\alpha(0, 1)) := \{g \in H^3(0, 1) : \ g(0, t) = 0, \quad g(1, t) = 0, \quad g_x(1, t) = \alpha g_x(0, t)\}.$$

The notion of well-posedness includes existence, uniqueness, persistence property (i.e., the solution describes a continuous curve in the space H^3_α) and continuous dependence of the solution upon the initial data. We first establish the local well-posedness result using contraction principle and the smoothing properties of the associated linear system. The global result is then obtained by finding global *a priori* estimates for smooth solutions of (1.4). It is interesting to note that while the L^2 estimate of solutions is easy to establish, the global H^1 and H^2 estimates of solutions are very difficult to obtain. What we do is to find an L^2 estimate of the time derivative of the solution, u_t which, in turn, would provide us the needed global H^3 estimate. Our argument relies heavily on the boundary conditions $u(0,t) = u(1,t) = 0$. The approach used here does not apply to the system (1.6) considered in [16]. If fact, only local well-posedness result was obtained in [16] for (1.6). It remains open whether the solutions of the system (1.6) exist globally in time or blows up in finite time (see [17]).

After showing the closed loop system (1.4) is well-posed, we show that the disspative mechanism introduced through the boundary is strong enough to enable the small amplitude solutions of the system (1.4) ($\alpha \neq 0$) decay exponentially. The approach we will use is similar to that used in [16]. Following the lines outlined there, we will begin with studying spectral properties of the operator A_α defined by

$$(A_\alpha g)(x) = -g'''(x) \tag{1.8}$$

with domain $\mathcal{D}(A_\alpha) = H^3_\alpha$. In particular, we show that the operator A_α is a discrete spectral operator whose eigenfunctions form a Riesz basis for the space $L^2(0,1)$ which would provide us a working ground to apply nonharmonic analysis techniques (cf. [14]). Then some smoothing properties of the associated linear system will be presented and will be used to show that the system (1.4) is locally well-posed in the space $H^1_0(0,1)$, but small amplitude solutions exist globally. As in [16] the decay property of small amplitude solutions is obtained by using an infinite dimension version of the second method of Lyapounov. It is not clear how large amplitude solutions of the system (1.4) behave as $t \to \infty$.

The paper is organized as follows. In section 2, we present the global well-posedness results for the system (1.4). In section 3, we study spectral properties of the operator A_α. The decay property of small amplitude solutions of (1.4) is established in section 4.

2. Global Well-posedness

In this section we show that the closed loop system (1.4) is globally well-posed in the space H^3_α. To begin, we consider the associated inhomogeneous linear system

$$\begin{cases} u_t + u_{xxx} = f, \qquad u(x,0) = \phi(x), \\ u(0,t) = 0, \quad u(1,t) = 0, \quad u_x(1,t) = \alpha u_x(0,t), \end{cases} \tag{2.1}$$

for $x \in (0,1)$ and $t \geq 0$. Using the notation of the operator A_α (cf. (1.8)) we rewrite the system (2.1) as an abstract evolution equation

$$\frac{du}{dt} = A_\alpha u + f, \qquad u(0) = \phi.$$

Since the operator A_α generates a continuous semigroup $S_\alpha(t)$ in the space $L^2(0,1)$, it follows from standard semigroup theory (cf. [16]) that the system (2.1) has a unique solution $u \in C([0,T]; H^3_\alpha)$ for any $\phi \in \mathcal{D}(A_\alpha) = H^3_\alpha$ and $f \in L^1(0,T; H^3_\alpha)$. The following two propositions present some estimates of these solutions that are needed for establishing the well-posedness of the nonlinear system (1.4).

Proposition 2.1. *Let $T > 0$ be given and $u(x,t)$ is a solution of the system* (2.1). *Then,*

$$\sup_{[0,T]} \|u(\cdot,t)\|_{L^2(0,1)} \le \|\phi\|_{L^2(0,1)} + \int_0^T \|f(\cdot,\tau)\|_{L^2(0,1)} d\tau, \tag{2.2}$$

$$\int_0^T u_x^2(0,\tau)d\tau \le \frac{1}{1-\alpha^2}\left(\|\phi\|_{L^2(0,1)} + \int_0^T \|f(\cdot,\tau)\|_{L^2(0,1)} d\tau\right)^2 \tag{2.3}$$

and

$$\|u\|_{L^2(-T,T;H^1(0,1))} \le c\left(\|\phi\|_{L^2(0,1)} + \int_0^T \|f(\cdot,\tau)\|_{L^2(0,1)} d\tau\right) \tag{2.4}$$

Proof: The estimates (2.2)-(2.3) and (2.4) are established using the usual energy estimate method with the multiplier $2u$ and $e^{\lambda x}u$, respectively.

Proposition 2.2. *Let $T > 0$ be given. If $f \in W^{1,1}(0,T; L^2(0,1))$ and $\phi \in H^3_\alpha$, then the system* (2.1) *has a unique solution $u \in C([0,T]; H^3_\alpha)$ for which,*

$$u_t \in L^\infty(0,T; L^2(0,1)) \cap L^2(0,T; H^1(0,1)), \qquad u_{xt}(0,t) \in L^2(0,T)$$

and

$$\begin{aligned} &\|u_t\|_{L^\infty(0,T;L^2(0,1))} + \|u_{xt}\|_{L^2(0,T;L^2(0,1))} + \|u_{xt}(0,\cdot)\|_{L^2(0,T)} \\ &\qquad \le c\left(\|\phi\|_{H^3_\alpha} + \|f\|_{W^{1,1}(0,T;L^2(0,1))}\right). \end{aligned} \tag{2.5}$$

In addition, if $f \in L^2(0,T; H^1(0,1))$, then $u \in L^2(0,T; H^4(0,1))$ and

$$\begin{aligned} &\|u\|_{L^2(0,T;H^4(0,1))} \\ &\qquad \le c\left(\|\phi\|_{H^3_\alpha} + \|f\|_{W^{1,1}(0,T;L^2(0,1))} + \|f\|_{L^2(0,T;H^1(0,1))}\right). \end{aligned} \tag{2.6}$$

Proof: It is easy to see that $v = u_t$ solves

$$\begin{cases} v_t + v_{xxx} = f_t, & v(x,0) = \psi(x), \\ v(0,t) = 0, \quad v(1,t) = 0, & v_x(1,t) = \alpha v_x(0,t) \end{cases}$$

where

$$\psi(x) = f(x,0) - \phi'''(x).$$

Then estimate (2.5) is established in the same way as that to obtain (2.2)-(2.4). If $f \in L^2(0,T; H^1(0,1))$, from the equation $u_{xxx}(x,t) = f(x,t) - u_t(x,t)$ and the fact that $u_t \in L^2(0,T;H^1(0,1))$ it follows that $u \in L^2(0,T;H^4(0,1))$. Then estimate (2.6) follows from (2.5).

Now we turn to consider that the nonlinear system (1.4) and show that it is locally well-posed using a contraction principle based on the estimates provided above.

Proposition 2.3. *For any $\phi \in H^3_\alpha$, there exists a $T = T(\|\phi\|_{H^3_\alpha}) > 0$ such that (1.4) has a unique solution*

$$u \in C([0,T);H^3_\alpha) \cap L^2(0,T;H^4(0,1)). \tag{2.7}$$

Moreover, for any $T' < T$ there exists a neighborhood U of ϕ in H^3_α such that the map from the initial data ϕ to the solution u is Lipschitz continuous from U to the space class defined by (2.7).

Proof: Let

$$\begin{aligned} Y_T = \{v \in & C([0,T];H^3_\alpha) \cap L^2(0,T;H^4(0,1)), \\ & v_t \in L^\infty(0,T;L^2(0,1) \cap L^2(0,T;H^1(0,1))\} \end{aligned}$$

which is a Banach space equipped with the norm

$$\|v\|_{Y_T} := \|v\|_{C([0,T]\ H^3_\alpha)} + \|v\|_{L^2(0,T\ H^4(0,1))} + \|v_t\|_{L^2(0,T\ H^1(0,1))} + \|v_t\|_{L^\infty(0,T,L^2(0,1))}.$$

For given $\phi\ H^3_\alpha$,

$$\begin{cases} u_t + u_{xxx} = -vv_x, & v(x,0) = \phi(x) \\ u(0,t) = 0, \quad u(1,t) = 0, & u_x(1,t) = \alpha u_x(0,t). \end{cases} \tag{2.8}$$

defines a map Γ from the space Y_T to Y_T: $u = \Gamma(v)$ for any $v \in Y_T$ where u is the corresponding solution of (2.8). Applying Proposition 2.1-2.2 to (2.8) yields that

$$\begin{aligned} \|\Gamma(v)\|_{Y_T} &\leq c\|\phi\|_{H^3_\alpha} + c\|vv_x\|_{W^{1\,1}(0,T,L^2(0,1))} + c\|vv_x\|_{L^2(0,T,H^1(0,1))} \\ &\leq c\|\phi\|_{H^3_\alpha} + cT^{1/2}\|v\|^2_{Y_T}. \end{aligned}$$

Choose

$$M = 2c\|\phi\|_{H^3_\alpha} \quad \text{and} \quad T = (2cM)^{-2}. \tag{2.9}$$

Then Γ maps the ball $S_M = \{v \in Y_T;\ \ \|v\|_{Y_T} \leq M\}$ into itself. In addition, a similar argument shows that

$$\|\Gamma(v_1) - \Gamma(v_2)\|_{Y_T} \leq \frac{1}{2}\|v_1 - v_2\|_{Y_T}$$

for any $v_1,\ v_2 \in S_M$. Hence the map Γ is a contraction in the ball S_M if T and M are as chosen in is the needed solution of (1.4). The proof is complete. □

Next we consider regularity problem of (1.4). Roughly speaking, we show that if ϕ is a smooth function, so is the solution. This would be a consequence of the following proposition to the linear problem

$$\text{(2.10)} \qquad \begin{cases} u_t + (wu)_x + u_{xxx} = f, \quad u(x,0) = \phi(x), \\ u(0,t) = 0, \quad u(1,t) = 0, \quad u_x(1,t) = \alpha u_x(0,t). \end{cases}$$

Proposition 2.4. *Let $T > 0$ be given and $w \equiv w(x,t) \in Y_T$. Then (2.10) has a unique solution $u \in Y_T$ for any $\phi \in H^3_\alpha$ and $f \in W^{1,1}(0,T;L^2) \cap L^2(0,T;H^1)$. Moreover*

$$\text{(2.11)} \qquad \|u\|_{Y_T} \le c\left(\|\phi\|_{H^3_\alpha} + \|f\|_{W^{1,1}(0,T;L^2(0,1))} + \|f\|_{L^2(0,T;H^1(0,1))}\right)$$

where $c > 0$ depends continuously on $\|w\|_{Y_T}$.

Proof: For given ϕ and f,

$$\text{(2.12)} \qquad \begin{cases} u_t + u_{xxx} = -(vw)_x + f, \quad u(x,0) = \phi \\ u(0,t) = 0, \quad u(1,t) = 0, \quad u_x(1,t) = \alpha u_x(0,t) \end{cases}$$

defines a map Γ from Y_{T^*} to Y_{T^*}: $u = \Gamma(v)$ for any $v \in Y_{T^*}$ where u is the corresponding solution of (2.12). Using Proposition 2.1 and Proposition 2.2,

$$\|\Gamma(v)\|_{Y_{T^*}} \le c\left(\|\phi\|_{H^3_\alpha} + \|f\|_{W^{1,1}(0,T^*;L^2(0,1))}\right) + c(T^*)^{1/2}\|w\|_{Y_{T^*}}\|v\|_{Y_{T^*}}$$

Thus if one chooses $T^* > 0$ such that

$$\text{(2.13)} \qquad \begin{array}{l} c(T^*)^{1/2}\|w\|_{Y_T^*} = 1/2 \quad \text{and} \\ b = 2c(\|\phi\|_{H^3_\alpha} + \|f\|_{W^{1,1}(0,T^*;L^2(0,1))} + \|f\|_{L^2(0,T^*;H^1(0,1))}), \end{array}$$

then for any $v \in Y_T^*$ with $\|v\|_{Y_T^*} \le b$,

$$\|\Gamma(v)\|_{Y_{T^*}} \le b.$$

Thus Γ maps the ball $S_b := \{v \in Y_{T^*} : \quad \|v\|_{Y_{T^*}} < b\}$ to itself. Similarly, one can also show that Γ is a contraction in the ball S_b. Its fixed point is the solution of (2.10). Note that T^* determined by (2.13) only depends on w and, in particular, not on ϕ and f. Thus a standard argument can be used to show that T^* may be extended to any large previously given number and (2.11) holds. The proof is complete. □

Define a series of differential operators Q_k for $k = 0, 1, 2, \ldots$,

$$\text{(2.14)} \qquad \begin{cases} Q_0\phi = \phi \\ Q_{k+1}\phi = -\partial_x^3(Q_k\phi) - \frac{1}{2}\partial_x\left(\sum_{j=0}^k \begin{pmatrix} j \\ k \end{pmatrix} Q_j\phi Q_{k-j}\phi\right) \end{cases}$$

Proposition 2.5. *Let $k \geq 0$ be a given integer. Assume that*

$$Q_j \phi \in H^3_\alpha, \qquad j = 0, 1, ..., k$$

Then there exists a $T = T(\|\phi\|_{H^3_\alpha}) > 0$ such that the system (1.4) has a unique solution

$$\partial_t^j u \in Y_T, \qquad j = 0, 1, 2, ..., k.$$

As a consequence,

$$u \in C([0, T]; H^{3(k+1)}(0, 1)) \cap L^2(0, T; H^{3(k+1)+1}(0, 1)).$$

Proof: It is Proposition 2.3 if $k = 0$. We only provide a proof here for the case $k = 1$. The general case follows easily then from induction.

Let $v = u_t$. It solves

$$(2.15) \qquad \begin{cases} v_t + (uv)_x + v_{xxx} = 0, \quad v(x, 0) = Q_1 \phi, \\ v(0, t) = 0, \quad v(1, t) = 0, \quad v_x(1, t) = \alpha v_x(0, t). \end{cases}$$

It is from Proposition 2.4 that $u_t = v \in Y_T$. Then from $u_{xxx} = -u_t - uu_x$ it follows

$$u \in C([0, T]; H^6(0, 1)) \cap L^2(0, T; H^7(0, 1)).$$

The proof is complete. □

Now we turn to show globally well-posedness of (1.4) by establishing the following global *a priori* estimates for solutions of (1.4).

Proposition 2.6. *Let $T > 0$ be given and u be a smooth solution of (1.4). Then*

$$(2.16) \qquad \int_0^1 u^2(x, t) dx + (1 - \alpha^2) \int_0^T u_x^2(0, \tau) d\tau \leq \int_0^1 \phi^2(x) dx$$

and

$$(2.17) \qquad \int_0^T \|u(., t)\|_{H^1(0,1)} dt \leq g_1(\|\phi\|_{L^2(0,1)}, T)$$

for any $0 \leq t \leq T$, where $g_1(\cdot, \cdot)$ is a nonnegative continuous function.

Proof: This is a nonlinear version of Proposition 2.1 but with a similar proof.

It seems natural that the next step should be finding a global H^1 estimate. However, it turns out being very difficult, if it is not possible, to obtain a global H^1 or H^2 estimate of the solutions of the system (1.4). Instead, we directly obtain the H^3 estimate by estimating the L^2 norm of u_t. The following two technique lemmas are standard (cf. [1] and [12]).

Lemma 2.1. *There exists a constant $c > 0$ such that for any $g \in H^1(0,1)$ satisfying either $\int_0^1 g(x)dx = 0$ or $g(0) = g(1) = 0$, one has*

$$\sup_{x\in(0,1)} |g(x)| \leq c\|g\|^{1/2}_{L^2(0,1)}\|g\|^{1/2}_{H^1(0,1)}.$$

Lemma 2.2. *For given positive integer m, there is a constant $c > 0$ such that*

$$\|g\|_{H^j(0,1)} \leq c\|g\|^{j/m}_{H^m(0,1)}\|g\|^{(m-j)/m}_{L^2(0,1)}$$

for any $g \in H^m(0,1)$ and $0 \leq j \leq m$.

Proposition 2.7. *Let $T > 0$ be given and assume that u is a smooth solution of (1.4). Then*

$$\sup_{[0,T]} \|u(\cdot,t)\|_{H^3_\alpha} + \|u\|_{L^2(0,T;H^4(0,1))} \leq g_2(\|\phi\|_{H^3_\alpha}, T) \tag{2.18}$$

where $g_2(\cdot,\cdot)$ is a nonnegative continuous function.

Proof: Let $v = u_t$. It solves (2.15). Multiply both sides of the equation (2.15) by $2e^{\lambda x}v$ and integrate over the domain $(0,1)\times(0,t)$. Integration by parts leads to

$$\begin{aligned}
&\int_0^1 e^{\lambda x}v^2dx + (1 - e^\lambda\alpha^2)\int_0^t v_x^2(0,\tau)d\tau + 3\lambda\int_0^t\int_0^1 e^{\lambda x}v_x^2(x,\tau)dxd\tau \\
&+\int_0^t\int_0^1 e^{\lambda x}v^2(x,\tau)(u_x(x,\tau) - \lambda u(x,\tau) - \lambda^3)dxd\tau \\
&= \int_0^1 e^{\lambda x}\psi^2(x)dx
\end{aligned} \tag{2.19}$$

where $\psi = Q_1\phi$. Taking $\lambda = 0$ in (2.19) yields

$$\int_0^1 v^2(x,t)dx+(1-\alpha^2)\int_0^t v_x^2(0,\tau)d\tau\leq\int_0^t\int_0^1 v^2|u_x|dxd\tau+\int_0^1\psi^2dx. \tag{2.20}$$

Using Lemma 2.1 and Lemma 2.2,

$$\begin{aligned}
\int_0^1 v^2|u_x|dx &\leq \|u_x\|_{L^\infty(0,1)}\|v\|^2_{L^2(0,1)} \\
&\leq c\|u\|^{1/2}_{H^2(0,1)}\|u\|^{1/2}_{H^1(0,1)}\|v\|^2_{L^2(0,1)} \\
&\leq c\|u\|^{3/8}_{H^4(0,1)}\|v\|^2_{L^2(0,1)},
\end{aligned}$$

and similarly,

$$\int_0^1 v^2|u|dx \leq c\|u\|^{1/8}_{H^4(0,1)}\|v\|^2_{L^2(0,1)}.$$

Noticing that $v = u_t = -uu_x - u_{xxx}$, one obtains

$$\begin{aligned}\int_0^1 v^2 dx &= \int_0^1 (uu_x + u_{xxx})^2 dx \\ &\leq \int_0^1 u^2 u_x^2 dx + \int_0^1 u_{xxx}^2 dx + 2\int_0^1 |u||u_x||u_{xxx}|dx \\ &\leq c\|u\|_{H^4(0,1)}^{12/8}.\end{aligned}$$

Consequently,

$$\int_0^1 v^2(|u_x| + |u|)dx \leq c\|u\|_{H^4(0,1)}^{15/8}. \tag{2.21}$$

In addition, since $v_x = u_{xt} = -u_x^2 - uu_{xx} - u_{xxxx}$,

$$\begin{aligned}\int_0^1 (v_x^2)dx &= \int_0^1 (u_x^2 + uu_{xx} + u_{xxxx})^2 dx \\ &= \int_0^1 (u_{xxxx})^2 dx + \int_0^1 u^2(u_{xx})^2 dx + \int_0^1 u_x^4 dx + 2\int_0^1 u_{xxxx}u_{xx}u dx \\ &\quad +2\int_0^1 uu_x^2 u_{xx} dx + 2\int_0^1 u_x^2 u_{xxxx} dx.\end{aligned}$$

Using Lemma 2.1 and Lemma 2.2,

$$\begin{aligned}\left|\int_0^1 u_{xxxx}uu_{xx}dx\right| &\leq \|u_{xx}\|_{L^\infty(0,1)}\|u\|_{L^2(0,1)}\|u\|_{H^4(0,1)} \\ &\leq c\|u\|_{H^3(0,1)}^{1/2}\|u\|_{H^2(0,1)}^{1/2}\|u\|_{H^4(0,1)} \\ &\leq c\|u\|_{H^4(0,1)}^{13/8}.\end{aligned}$$

Similarly,

$$\left|\int_0^1 uu_x^2 u_{xx}dx\right| \leq c\|u\|_{H^4(0,1)}^{7/4}, \qquad \left|\int_0^1 u_x^2 u_{xxxx}dx\right| \leq c\|u\|_{H^4(0,1)}^{15/8},$$

and

$$\int_0^1 u^2 u_{xx}^2 dx \leq c\|u\|_{H^4(0,1)}^{5/4}, \qquad \int_0^1 u_x^4 dx \leq c\|u\|_{H^4(0,1)}^{5/4}.$$

Thus

$$\int_0^1 v_x^2 dx \geq \int_0^1 u_{xxxx}^2 dx - c\|u\|_{H^4(0,1)}^{15/8}. \tag{2.22}$$

We obtain

$$\int_0^1 v^2 dx + 3\lambda \int_0^t \int_0^1 u_{xxxx}^2 dx d\tau$$

$$\leq \lambda^3 e^\lambda \int_0^t \int_0^1 v^2 dx d\tau + c(\lambda) \int_0^t \left(\int_0^1 u_{xxxx}^2 dx \right)^{15/8} d\tau + \|\phi\|_{H_\alpha^3}^2 \tag{2.23}$$

for any $0 \leq t \leq T$, which implies (2.18). The proof is complete. □

Remark 2.1. *As a consequence of continuous dependence of solution on its initial data, (2.21) is not only true for smooth solutions but also true for any solution in the class of $C([0,T]; H^s(S))$ for $s \geq 3$.*

The following theorem now follows from a standard argument.

Theorem 2.1. *Let $T > 0$ be given. For any $\phi \in H_\alpha^3$, there exists a unique solution $u \in C([0,T]; H_\alpha^3)$ to (1.4). Moreover, the map from the initial data ϕ to the solution u is Lipschiz continuous from H_α^3 to the space $C([0,T]; H_\alpha^3) \cap L^2(0,T; H^4(0,1))$.*

Remark 2.2. *In fact, as in [19], we can show the map K is analytic from the space H_α^3 to the space $C([0,T]; H_\alpha^3) \cap L^2(0,T; H^4(0,1))$.*

As a corollary of the above theorem and the Proposition 2.4, we have the following regularity result.

Corollary 2.1. *Let $T > 0$ be given and k be a positive integer. If $Q_j\phi \subset H_\alpha^3$, $j = 1, 2, \cdots, k$, then (1.4) has a unique solution $\partial_t^j u \in Y_T$ for $j = 0, 1, ..., k$. and consequently,*

$$u \in C([0,T]; H^{3(k+1)}(0,1)) \cap L^2(0,T; H^{3(k+1)+1}(0,1)).$$

3. Spectral Properties of the Operator A_α

The operator A_α defined in section 1 generates a strongly continuous semigroup $S_\alpha(t)$ on $L^2 \equiv L^2[0,1]$ for $t \geq 0$. For the work to be done in this paper we require very detailed information about the eigenvalues of A_α and its adjoint A_α^*, which agrees with (1.8) except for removal of the "-" sign there and reversal of the roles played by 0 and 1 in the boundary conditions. Since A_α for $|\alpha| < 1$ is a dissipative differential operator, it has compact resolvent and therefore its spectra consists of only countable many eigenvalues λ_k with $Re\lambda_k \leq 0$ for $k = 0, \pm 1, \pm 2, ...$

Proposition 3.1. *For any $|\alpha| < 1$, the corresponding eigenvalues of A_α have the asymptotic form*

$$\lambda_k = \begin{cases} \frac{-8\pi^3|k|^3}{3\sqrt{3}} + O(k^2) & \text{if } \alpha = 0 \\ (12\pi^2k^2 \ln|\alpha| + O(k)) + (8\pi^3k^3 + O(k^2))i & \text{if } \alpha \neq 0. \end{cases} \tag{3.1}$$

as $k \to \pm\infty$. Moreover, there exists an $\eta < 0$ such that

$$Re\lambda_k < \eta, \qquad \forall k. \tag{3.2}$$

Proof: Assume that $Re\lambda \leq 0$. By symmetry we only need to consider the case that $Im\lambda \leq 0$. We denote the three cube roots of $-\lambda$ by μ_0, μ_1, μ_2. These must have distinct real parts; we let μ_0 be the unique root such that $0 \leq arg(\mu_0) \leq \pi/6$ and

$$\mu_1 = e^{\frac{2\pi i}{3}}\mu_0 \equiv \rho\mu_0, \qquad \mu_2 = \rho^2\mu_0. \tag{3.3}$$

The general solution of the characteristic equation

$$\phi'''(x) + \lambda\phi(x) = 0 \tag{3.4}$$

is then

$$\phi(x) = c_0e^{\mu_0 x} + c_1e^{\mu_1 x} + c_2e^{\mu_2(x)} \tag{3.5}$$

where c_0, c_1 and c_2 are arbitrary coefficients. Substituting (3.5) into the boundary conditions

$$\phi(1) = \phi(0) = 0, \quad \phi'(1) = \alpha\phi'(0) \tag{3.6}$$

we arrive at the system of equations

$$\begin{bmatrix} 1 & 1 & 1 \\ e^{\mu_0} & e^{\mu_1} & e^{\mu_2} \\ (e^{\mu_0} - \alpha)\mu_0 & (e^{\mu_1} - \alpha) - \mu_1 & (e^{\mu_2} - \alpha)\mu_2 \end{bmatrix} \begin{bmatrix} c_0 \\ c_1 \\ c_2 \end{bmatrix} = 0. \tag{3.7}$$

Setting the determinant of the matrix equal to zero we have

$$\begin{aligned} &(\mu_2 - \mu_1)e^{\mu_2+\mu_1} + (\mu_0 - \mu_2)e^{\mu_0+\mu_2} + (\mu_1 - \mu_0)e^{\mu_0+\mu_1} + \alpha\mu_0(e^{\mu_1} - e^{\mu_2}) + \\ &\qquad \alpha\mu_1(e^{\mu_2} - e^{\mu_0}) + \alpha\mu_2(e^{\mu_0} - e^{\mu_1}) = 0. \end{aligned} \tag{3.8}$$

By our assumption, $Re\mu_0 > 0$, $Re\mu_1 < 0$, $Re\mu_2 \leq 0$, and $Re\,\mu_0 \to +\infty$, $Re\,\mu_1 \to -\infty$ as $\lambda \to \infty$.

Multiplying both sides of (3.8) by $e^{-\mu_0}$ and neglecting the terms $(\mu_2 - \mu_1)e^{\mu_2-\mu_1-\mu_0}$, $\alpha\mu_0(e^{\mu_1}-e^{\mu_2})e^{-\mu_0}$, $\alpha\mu_1 e^{\mu_2-\mu_0}$ and $\alpha\mu_2 e^{\mu_1-\mu_0}$, which are very small for large $|\lambda|$, we arrive at the equation

$$(\mu_0 - \mu_2)e^{\mu_2} + (\mu_1 - \mu_0)e^{\mu_1} - \alpha\mu_1 + \alpha\mu_2 = 0$$

which is, after simplification,

$$(1+\rho)e^{\mu_2} - e^{\mu_1} = -\alpha\rho \tag{3.9}$$

If $\alpha = 0$, (3.9) leads to $e^{\sqrt{3}i\mu_0} = e^{i\pi/3}$. Thus $\mu_{0,k} = \frac{1}{\sqrt{3}}(\pi/3 + 2k\pi)$, and

$$\lambda_k = -\mu_{0,k}^3 = -(\frac{8\pi^3}{3\sqrt{3}}k^3 + O(k^2)) \tag{3.10}$$

as $k \to +\infty$. If $\alpha \neq 0$, write μ_2 as $\mu_2 = y + iz$. Separating the real and imaginary parts of the equation (3.9), we obtain

$$\begin{cases} e^{-\frac{1}{2}y+\frac{\sqrt{3}}{2}z}\cos(-\frac{\sqrt{3}}{2}y - \frac{1}{2}z) = \frac{\alpha}{2} + \frac{1}{2}e^{y}(\cos(z) - \sqrt{3}\sin(z)), \\ e^{-\frac{1}{2}y+\frac{\sqrt{3}}{2}z}\sin(-\frac{\sqrt{3}}{2}y - \frac{1}{2}z) = -\frac{\sqrt{3}}{2}\alpha + \frac{1}{2}e^{y}(\sqrt{3}\cos(z) + \sin(z)), \end{cases} \tag{3.11}$$

from which, by squaring both sides of the equations and adding together,

$$e^{-y+\sqrt{3}z} = \left(\frac{\alpha}{2} + \frac{1}{2}e^{y}(\cos(z) - \sqrt{3}\sin(z))\right)^2 + \left(-\frac{\sqrt{3}}{2}\alpha + \frac{1}{2}e^{y}(\sqrt{3}\cos(z) + \sin(z))\right)^2. \tag{3.12}$$

Note that $\mu_2 \to \infty$ in the third quarter of the plane as $\lambda \to \infty$ by our assumption. There are only three possible cases.

(i). y is bounded and $y \to 0$, $z \to -\infty$;
(ii). $y \to -\infty$, $z \to -\infty$;
(iii). $z \to -\infty$ and $-N < y < -\beta$ for some $N > 0$ and $\beta > 0$.

Case (i): letting $y \to 0$ and $z \to -\infty$ in (3.11),

$$\lim_{z\to-\infty}(\cos(z) - \sqrt{3}\sin z) = -\alpha, \qquad \lim_{z\to-\infty}(\sqrt{3}\cos(z) + \sin(z)) = \sqrt{3}\alpha.$$

Thus $\sin(z) = \frac{\sqrt{3}}{2}\alpha$ and $\cos(z) = \frac{\alpha}{2}$ that lead a contradiction

$$1 = \sin^2(z) + \cos^2(z) = \alpha^2 < 1.$$

Case (ii): if $y \to -\infty$, then the term e^{μ_2} in (3.9) is smaller than other terms and we obtain approximately

$$e^{\mu_1} = -\alpha\rho.$$

It yields that

$$\mu_{1,k} = \ln|\alpha| + \frac{5\pi}{3}i + 2k\pi i + o(k).$$

Case (iii): letting $z \to -\infty$ in (3.12), we have that

$$\lim_{z\to-\infty}\left(\cos(z) - \sqrt{3}\sin(z) + e^{-y}\alpha\right) = 0,$$
$$\lim_{z\to-\infty}\left(\sqrt{3}\cos(z) + \sin(z) - e^{-y}\sqrt{3}\alpha\right) = 0,$$

or equivalently,

$$\lim_{z\to-\infty} e^{y}\sin(z) = \frac{\sqrt{3}}{2}\alpha, \qquad \lim_{z\to-\infty} e^{y}\cos(z) = \frac{1}{2}\alpha.$$

It leads to

$$\lim_{z\to-\infty}\tan(z) = \sqrt{3}, \qquad \lim_{z\to-\infty} y = \ln|\alpha|.$$

Thus

$$\mu_{2,k} = \ln|\alpha| + i(-\frac{5\pi}{3} + 2k\pi) + o(k)$$

as $k \to -\infty$ and A_α has eigenvalues

$$\lambda_k = -\mu_2^3 = -\mu_1^3 = [8\pi^3k^3 + O(k^2)]i + 12\pi^2k^2\ln|\alpha| \tag{3.13}$$

as $k \to -\infty$. (3.10) and (3.13) give the formula (3.1) where the neglected terms referred to following (3.9) are corrected using the implicit function theorem; the one to one relationship between the eigenvalues and the indices k can be established using Rouché's theorem. An entirely similar argument gives the eigenvalues λ_{-k} as the conjugates of the λ_k shown.

To show (3.2), it suffices to prove that A_α has no eigenvalue on the imaginary axis. Suppose that A_α has an eigenvalue $i\xi$ with ξ a real number and a corresponding eigenfunction $g \in H^3_\alpha$, i.e.

$$A_\alpha g = i\xi g. \tag{3.14}$$

From the identity

$$(A_\alpha g, g)_{L^2(0,1)} = -\frac{1}{2}(1-\alpha^2)|g'(0)|^2,$$

one has $g'(0) = 0$ and therefore $g'(1) = 0$ since $(A_\alpha g, g)_{L^2(0,1)} = i\xi\|g\|_{L^2(0,1)}$ and $g'(1) = \alpha g'(0)$. Write g as

$$g = g_1 + ig_2$$

with both g_1 and g_2 being real functions. Then it is from (3.14) that

$$g_1''' = \xi g_2, \qquad g_2''' = -\xi g_1 \tag{3.15}$$

with boundary conditions

$$g_k(1) = g_k(0) = g_k'(0) = g_k'(1) = 0, \quad k = 1, 2, \tag{3.16}$$

which force $g \equiv 0$. Hence $i\xi$ is not an eigenvalue of A_α. The proof is complete. □

Proposition 3.2. *Assume that $\alpha \neq 0$. Then the operator A_α is a discrete spectral operator, all but a finite number of whose eigenvalues λ corresponds to one-dimensional projections $E(\lambda; T)$. The operators A_α, A_α^* with the indicated domains, have complete sets of eigenvectors, respectively,*

$$\{\phi_k \mid -\infty < k < +\infty\}, \qquad \{\psi_k \mid -\infty < k < +\infty\}$$

which, normalized so that (with $\delta_{k,j}$, the Kronecker delta)

$$(\phi_k, \psi_j)_{L^2(0,1)} = \delta_{k,j} \tag{3.17}$$

form dual Riesz basis for $L^2[0,1]$.

Proof: The proof is almost same as those of Proposition 2.1 and 2.2 in [16] after the needed asymptotic form of the eigenvalues of A_α has been established in Proposition 3.1.

Remark 3.1. *The $\alpha = 0$ is an exceptional case as that of $\alpha = -1/2$ in [16] It is not clear whether the corresponding eigenvectors form a Riesz basis in the space $L^2(0,1)$ in this exceptional case although its eigenvalues has faster grow rate.*

Definition 3.1. *We denote by H_α^n the Hilbert space consisting of functions w in the Sobolev space $H^n[0,1]$ which obey boundary conditions of the form*

$$w^{(3j)}(1) = w^{(3j)}(0) = 0, \quad w^{(3j+1)}(1) = \alpha w^{(3j+1)}(0) \tag{3.18}$$

as long as the indicated derivatives are of order $\leq n-1$. The norm and inner product in H_α^n are $\| \; \|_n$, $(\,\cdot\,)_n$, inherited from $H^n[0,1]$.

The space H_α^n can be characterized as follows using eigenvectors of A_α.

Proposition 3.3. *For $\alpha \neq 0$, a function $w \in L^2$ also lies in H_α^n if and only if, represented in the form*

$$w = \sum_{k=-\infty}^{\infty} c_k \phi_k$$

we have

$$\sum_{k=-\infty}^{\infty} |k^n c_k|^2 < \infty. \tag{3.19}$$

Moreover, $\|w\|^2_{H_\alpha^n}$ is a equivalent to the sum

$$\sum_{k=-\infty}^{\infty} \left(|c_k|^2 + |k^n c_k|^2\right),$$

i.e., there are positive constants e_n, E_n (these will, in general, depend on α) such that, uniformly for $w \in H^n_\alpha$,

$$e_n^2 \sum_{k=-\infty}^{\infty} \left[|c_k|^2 + |k^n c_k|^2\right] \leq \|w\|^2_{H^n_\alpha} \leq E_n^2 \sum_{k=-\infty}^{\infty} \left(|c_k|^2 + |k^n c_k|^2\right). \tag{3.20}$$

Proof: The proof is similar to that of Proposition 2.4 in [16] and therefore is omitted here.

4. Asymptotic Stability

In this section we consider asymptotic stability of solutions of the system (1.4) and show that small amplitude solutions decay exponentially in the space $L^2(0,1)$. This is of course due to the dissipative mechanism introduced through the boundaries (cf. 1.3. Our approach is entirely same as the one used in [16] within the framework already established in the previous section.

We assume $|\alpha| < 1$ and $\alpha \neq 0$. According the previous section, the resolution of the identity associated with the operator A_α is the sum, strongly convergent in $\mathcal{L}(L^2, L^2)$,

$$I = \sum_{k=-\infty}^{\infty} P_k$$

where P_k, $-\infty < k < \infty$, is the (generally non-orthogonal) projection defined by

$$P_k f = (f, \psi_k)_{L^2(0,1)} \phi_k \qquad \text{for any } f \in L^2(0,1).$$

The corresponding strongly convergent representation of the semigroup generated by A_α is

$$S_\alpha(t) = \sum_{k=-\infty}^{\infty} e^{\lambda_k t} P_k. \tag{4.1}$$

where $f_k = P_k f$. From the representation (4.1) and the spectral properties of the operator A_α it is easy to get the following proposition.

Proposition 4.1. *For any $n \geq 0$, there exists a $B_n > 0$ and a $\beta > 0$ such that*

$$\|S_\alpha(t) w_0\|_n \leq B_n e^{-\beta t} \|w_0\|_n \tag{4.2}$$

and

$$\int_0^\infty \|S_\alpha(t) w_0\|^2_{n+1} dt \leq B_n^2 \|w_0\|^2_n \tag{4.3}$$

for any $w_0 \in H^n_\alpha$ *and* $t > 0$. *Besides,*

$$(4.4) \qquad \sup_{0\le t\le\infty} \| \int_0^t S_\alpha(t-\tau)f(\cdot,\tau)d\tau\|_{n+1} \le B_n \left(\int_0^\infty \|f(\cdot,\tau)\|_n^2 d\tau\right)^{\frac{1}{2}}$$

for any $f \in L^2(0,\infty; H^n_\alpha)$, *and*

$$(4.5) \qquad \sup_{0\le t<\infty} \| \int_0^t S_\alpha(t-\tau)f(\cdot,\tau)d\tau\|_{n+1} \le B_n \sup_{0\le t<+\infty} \|f(\cdot,t)\|_n$$

for any $f \in L^\infty(0,\infty; H^n_\alpha)$.

Now we turn to consider the well-posedness of the nonlinear system (1.4). The global well-posedness of (1.4) in the space H^n_α for $n \ge 3$ has been established in section 2. However the approach there cannot be used to obtain the well-posedness of (1.4) in the space H^n_α for $n < 3$. Here, based on the estimates established in Proposition 4.1, we are able to show that that (1.4) is locally well-posed in the space H^n_α for $n = 1$ or 2.

Theorem 4.1. *Let* $n = 1$ *or* 2. *Then for any* $\phi \in H^n_\alpha$ *there exists a* $T = T(\|\phi\|_n) > 0$ *such that the system* (1.4) *has a unique solution*

$$u \in X_T := C(0,T; H^1_\alpha) \cap L^\infty(0,T; H^1_\alpha)$$

where $T \to \infty$ *as* $\|\phi\|_1 \to 0$. *For any* $T' < T$, *there exists a neighborhood* U *of* ϕ *in* H^n_α *such that the map* $K : \phi \to u(\cdot,t)$ *from* U *to* $X_{T'}$ *is Lipschitz continuous.*

Proof: It uses the same argument as that in [16].

Since the a priori global H^1 or H^2 estimates of solutions are not available, it is not clear whether (1.4) is global well-posed in the space H^1_α or H^2_α. However the small amplitude solutions of (1.4) do exists globally as shown by the following theorem.

Theorem 4.2. *Let* $n = 1$ *or* 2. *There exists a* $\beta > 0$ *such that for any* $\phi \in H^n_\alpha$ *with* $\|\phi\|_n \le \beta$, (1.4)) *has a unique solution*

$$u \in C(0,\infty; H^n_\alpha) \cap L^\infty(0,\infty; H^n_\alpha).$$

Proof: see [16].

Then, using an infinite dimensional version of the second methods of Lyapounouv as in [16] we may show that a small amplitude solution decays exponentially to 0.

Theorem 4.3. *There exists a* $\delta > 0$ *such that for any* $\phi \in H^1_\alpha$ *with* $\|\phi\|_1 < \delta$, *the corresponding unique solution of* (1.4) *satisfies*

$$(4.6) \qquad \|u(\cdot,t)\|_{L^2(0,1)} \le ce^{-\rho t}\|\phi\|_{L^2(0,1)}, \quad t \ge 0,$$

where $c > 0$ *and* $\rho > 0$ *are independent of* ϕ.

Proof: see [16]

Finally we end this paper by presenting the following problem

Problem: *As we have known from Theorem 2 1 that the solution of* (1 4) *exists globally if its initial data is in the space* H^3_α *What is long time behavior of those solutions with large size of initial states?*

The long time behavior of the solutions would reflect competition among the nonlinearity due to the term uu_x, dispersive effect due to the term u_{xxx} and the dissipative mechanism introduced through the boundary conditions For the associated linear system, all solutions decay exponentially (see Proposition 4 1) which shows that the system possesses a rather strong dissipative mechanism For the nonlinear system (1 4), Theorem 4 3 shows that a weak nonlinearity is not enough to overcome the introduced dissipative mechanism for the nonlinear term uu_x is even smaller comparing to the solution u if its initial data is small However, for large amplitude solutions, the nonlinear term uu_x is expected to play a much more important role It would be very interesting to see how a stronger nonlinearity to influence the long time behavior of the solutions of system (1 4)

References

1 R A Adams, *Sobolev Spaces*, Academic Press, New York, 1975

2 J L Bona and R Smith, The initial value problem for the Korteweg-de Vries equation, *Philos Trans Roy Soc London A* **278** (1978), 555 601

3 J L Bona and R Winther, The Korteweg-de Vries equation in a quarter-plane, continuous dependence results, *Differential and Integral Equations* **2** (1989), 228 250

4 B A Bubnov, Generalized boundary value problems for the Korteweg de Vries equation in bounded domain, *Differential Equations* **15** (1979) 17 - 21

5 B A Bubnov, Solvability in the large of nonlinear boundary-value problem for the Korteweg de Vries equations, *Differential Equations* **16** (1980), 24 30

6 N Dunford and J T Schwartz *Linear Operators,* part III, Wiley-interscience, 1971

7 J P LaSalle and S Lefschetz, *Stability by Lyapounov's direct method with applications,* Academic Press, New York, 1961

8 T Kato, On the Cauchy problem for the (generalized) Korteweg-de Vries equations, *Advances in Mathematics Supplementary Studies, Studies in Applied Math* , **8** (1983), 93 128

9 C E Kenig, G Ponce and L Vega, Well-posedness of the initial value problem for the Korteweg-de Vries, *J Amer Math Soc* **4** (1991), 323 - 347

10 ————-, Well posedness and scattering results for the generalized Korteweg-de Vries equation via the contraction principle, *Comm Pure Appl Math* **46** (1993), 527 - 620

11 V Komornik, D L Russell and B -Y Zhang, Stabilisation de l'equation de Korteweg-de Vries, *C R Acad Sci Paris* **312** (1991), 841 843

12 O A Ladyzenskaja, *Linear and Quasilinear Equations of Parabolic Type*, Translations of Mathematics Monograph, Vol 23, AMS

13 A Pazy, *Semigroup of Linear Operators and Applications to Partial Differential Equations*, Applied Mathematical Sciences, Vol 44, Springer-Verlag, 1983

14 D L Russell, Nonharmonic Fourier series in the control theory of distributed parameter systems, *J Math Anal Appl* **18** (1967), 542 - 560

15 D L Russell and B -Y Zhang, Controllability and stabilizability of the third order linear dispersion equation on a periodic domain *SIMA Contrl & Opt* **3** (1993), 659 - 676

16 D L Russell and B -Y Zhang, Smoothing properties of solutions of the Korteweg-de Vries equation on a periodic domain with point dissipation, *IMA preprint*, series # 1085, December, 1992 (to appear in *J Math Anal Appl*)

17 D L Russell and B -Y Zhang, Stabilization of the Korteweg-de Vries equation on a periodic domain, *Control and Optimal Design of Distributed Parameter Systems*, edited by J Lagnese, D L Russell and L White, the IMA Volumes in Mathematics and Its Applications, Springer-Verlag, to appear

18 J C Saut and R Temam, Remarks on the Korteweg-de Vries equation, *Israel J Math* **24** (1976), 78 – 87

19 B -Y Zhang, Taylor series expansion for solutions of the Koteweg-de Vries equation with respect to their initial values, *IMA preprint*, series # 1015, August 1992 (to appear in *J Func Anal*)

Bing-Yu Zhang
Department of Mathematical Sciences
University of Cincinnati
Cincinnati, Ohio 45221-0025, USA

CONTROLLABILITY OF THE LINEAR SYSTEM OF THERMOELASTICITY: DIRICHLET-NEUMANN BOUNDARY CONDITIONS

ENRIQUE ZUAZUA

Departamento de Matemática Aplicada
Universidad Complutense

ABSTRACT We prove that the linear system of thermoelasticity with various boundary conditions is controllable in the following sense If the control time is large enough and we act in the equations of displacement by means of a control supported in a neighborhood of the boundary of the thermoelastic body, then we may control exactly the displacement and simultaneously the temperature in an approximate way We consider the following two cases a) The displacement satisfies Dirichlet boundary conditions and the temperature takes Neumann zero boundary value, b) The displacement satisfies Neumann boundary conditions and the temperature vanishes at the boundary The method of proof is inspired in our earlier work where the same result was proved for the case where both displacement and temperature satisfy Dirichlet boundary conditions

1991 *Mathematics Subject Classification* 93B05, 73C05, 35B37

Key words and phrases Linear system of thermoelasticity, exact controllability, approximate controllability, decoupling, observability inequalities

1. Introduction: First type of boundary conditions

Let us consider an isotropic and homogeneous thermoelastic body occupying an open and bounded set Ω of $\mathbb{R}^n (n \geq 1)$ with boundary $\Gamma = \partial\Omega$ of class C^2

We denote by $x = (x_1, \ldots, x_n)$ a point of Ω while t stands for the time variable. The displacement-vector is denoted by $u = (u_1, \ldots, u_n)(u_i = u_i(x,t), i = 1,..,n)$ and the temperature by $\theta = \theta(x,t)$.

We fix a control time $T > 0$ and a control region ω; an open and non-empty subset of Ω. We are allowed to act on the system through the equations of displacement by means of a control function

$$f = f(x,t) \in \left(L^2(\Omega \times (0,T))\right)^n$$

that represents an exterior force. The support of the control is restricted to the control region ω In the sequel, by χ_ω we denote the characteristic function of the set ω.

The linear system of thermoelasticity with Dirichlet boundary conditions for the displacement and Neumann boundary conditions for the temperature and in the presence of this type of control f is as follows:

$$(1.1) \quad \begin{cases} u_{tt} - \mu\Delta u - (\lambda + \mu)\nabla \text{ div } u + \alpha\nabla\theta = f\chi_\omega & \text{in } \Omega \times (0, \infty) \\ \theta_t - \Delta\theta + \beta \text{ div } u_t = 0 & \text{in } \Omega \times (0, \infty) \\ u = 0, \quad \partial\theta/\partial\nu = 0 & \text{on } \Gamma \times (0, \infty) \\ u(x, 0) = u^0(x),\ u_t(x, 0) = u^1(x),\ \theta(x, 0) = \theta^0(x) & \text{in } \Omega \end{cases}$$

where $\mu, \lambda > 0$ are Lamé's constants, $\alpha, \beta \in \mathbb{R}$ $(\alpha/\beta > 0)$ the coupling parameters and ν the outward unit normal to Ω in Γ.

When $f = 0$ system (1.1) is well-posed in

$$H = (H_0^1(\Omega))^n \times (L^2(\Omega))^n \times L^2(\Omega).$$

More precisely, for every initial data $(u^0, u^1, \theta^0) \in H$ there exists a unique solution

$$(u, u_t, \theta) \in C([0, \infty); H).$$

This solution is given by

$$(u(t), u_t(t), \theta(t)) = S(t)(u^0, u^1, \theta^0), t > 0$$

where $S(t) : H \to H, t > 0$, is the strongly continuous semigroup generated by system (1.1). We will denote by $S_i(t), i = 1, 2, 3$ the three components of $S(t)$.

On the other hand, the energy

$$(1.2) \qquad E(t) = \frac{1}{2}\int_\Omega \left[|u_t|^2 + \mu|\nabla u|^2 + (\lambda + \mu)|\text{ div } u|^2 + \frac{\alpha}{\beta}\theta^2\right] dx$$

satisfies

$$\frac{dE(t)}{dt} = -\frac{\alpha}{\beta}\int_\Omega |\nabla\theta|^2 dx.$$

If $f \in L^1(0, T; (L^2(\Omega))^n)$ system (1.1) possesses an unique solution (u, u_t, θ) in $C([0, T]; H)$ given in terms of the semigroup S by the variation of constants formula.

We consider the following controllability problem: Given (u^0, u^1, θ^0) and (v^0, v^1, η^0) in H and $\varepsilon > 0$, to find a control f such that the solution of (1.1) satisfies

$$(1.3) \qquad \begin{cases} u(T) = v^0, u_t(T) = v^1 \\ \|\theta(T) - \eta^0\|_{L^2(\Omega)} \le \varepsilon. \end{cases}$$

In other words, we request the exact controllability of the displacement and the approximate controllability of the temperature.

In [Z3] we gave a positive answer to this question in the case where both displacement and temperature satisfy Dirichlet boundary conditions. However, in the present case, the approximate controllability does not hold in any time T and whatever the support of the controls ω is. This is due to the fact that the quantity

$$F(t) = \int_\Omega \theta(x,t)dx \tag{1.4}$$

remains constant in time along solution trajectories of (1.1). This can be checked easily integrating the second equation of (1.1) in Ω.

The fact that $F(\cdot)$ is preserved along trajectories suggests the following controllability problem: Given $(u^0, u^1, \theta^0), (v^0, v^1, \eta^0) \in H$ such that

$$\int_\Omega \theta^0(x)dx = \int_\Omega \eta^0(x)dx \tag{1.5}$$

and $\varepsilon > 0$, to find a control f such that the solution of (1.1) satisfies (1.3).

In other words, we request the exact controllability of the displacement and the approximate controllability of the temperature at a prescribed level of total amount of heat.

In the sequel, if this property holds we will say that system (1.1) is exact-approximately controllable.

We have the following result.

Theorem 1.1. *Let ω be a neighborhood of the boundary Γ in Ω, i. e. $\omega = \Omega \cap \Theta$ where Θ is a neighborhood of Γ in $\mathbb{R}^n$. Suppose that $T > diam(\Omega \setminus \omega)/\sqrt{\mu}$. Then, system (1.1) is exact-approximately controllable in time T.*

Remark 1.1. Theorem 1.1 is a variant of the result proved in [Z3] in the case where both displacement and temperature satisfy Dirichlet boundary conditions. For the proof of Theorem 1.1 we will use basically the method developed in [Z3] (see also [Z1] and [Z2]). However, considering that we have different boundary conditions and that the constraint (1.5) is now present, some additional developments will be necessary. ∎

Remark 1.2. In one space dimension the same result can be proved for any open subset ω of Ω. We refer to [Z3] for a detailed proof in the case where both displacement and temperature satisfy Dirichlet boundary conditions. ∎

A natural way of addessing the presence of the constraint (1.5) is to decompose the temperature θ so that

$$\theta(x,t) = \frac{1}{|\Omega|}\int_\Omega \theta^0(x)dx + \tilde{\theta}(x,t)dx \tag{1.6}$$

where $\tilde{\theta}$ is such that $(u, \tilde{\theta})$ satisfy (1.1) with the following modified initial data for $\tilde{\theta}$:

$$\tilde{\theta}(x,0) = \tilde{\theta}^0(x) = \theta^0(x) - \frac{1}{|\Omega|}\int_\Omega \theta^0(x)dx. \tag{1.7}$$

We then introduce the Hilbert space $\mathcal{H} = \{f \in L^2(\Omega) : \int_\Omega f(x)dx = 0\}$ and the new phase space for the system of thermoelasticity: $\mathcal{V} = (H_0^1(\Omega))^n \times (L^2(\Omega))^n \times \mathcal{H}$.

Since the space $\mathcal{V}$ is invariant under the action of the semigroup $S(t)$ even if an external force is added in the first equation of (1.1) (i. e. when $f \neq 0$), then our controllability problem can be formulated in the following way: Given $(u^0, u^1, \theta^0), (v^0, v^1, \eta^0) \in \mathcal{V}$ and $\varepsilon > 0$, to find a control f such that the solution of (1.1) satisfies (1.3).

One of the main ingredients of the proof of Theorem 1 is the following observability inequality for the adjoint system of thermoelasticity:

$$(1.8) \quad \begin{cases} \varphi_{tt} - \mu\Delta\varphi - (\lambda+\mu)\nabla \text{ div } \varphi + \beta\nabla\psi_t = 0 & \text{in } \Omega \times (0,T) \\ -\psi_t - \Delta\psi - \alpha \text{ div } \varphi = 0 & \text{in } \Omega \times (0,T) \\ \varphi = 0, \ \partial\psi/\partial\nu = 0 & \text{on } \partial\Omega \times (0,T) \\ \varphi(x,T) = \varphi^0(x), \varphi_t(x,T) = \varphi^1(x) & \text{in } \Omega \\ \psi(x,T) = \psi^0(x) & \text{in } \Omega \end{cases}$$

where $\varphi = (\varphi_1, \dots, \varphi_n)$ is the adjoint displacement variable and ψ the temperature.

Proposition 1.1. *Under the assumptions of Theorem 1.1, for every bounded set B of $\mathcal{H}$ there exists $\delta = \delta(B) > 0$ such that*

$$(1.9) \qquad \delta \leq \int_0^T \int_\omega |\varphi|^2 dxdt$$

holds for every solution of (1.8) *with initial data such that*

$$\|(\varphi^0, \varphi^1 + \beta\nabla\psi^0)\|_{(L^2(\Omega))^n \times (H^{-1}(\Omega))^n} \geq 1, \psi^0 \in B.$$

By $P : (L^2(\Omega))^n \to (L^2(\Omega))^n$ we denote the continuous and linear operator defined as $Pf = \nabla g$ where $g \in H^1(\Omega) \cap \mathcal{H}$ is the unique solution of the following variational problem:

$$(1.10) \qquad \int_\Omega \nabla g \cdot \nabla \rho dx = \int_\Omega f \cdot \nabla \rho dx, \qquad \text{for all } \rho \in H^1(\Omega) \cap \mathcal{H}.$$

Obviously, the solution g of (1.10) in $H^1(\Omega) \cap \mathcal{H}$ exits and it is unique. Note that when $f \in (H_0^1(\Omega))^n$, (1.10) is the weak formulation of the equation

$$(1.11) \qquad \Delta g = \text{div } f \ \text{ in } \Omega; \ \int_\Omega g dx = 0; \ \partial g/\partial\nu = 0 \ \text{ on } \partial\Omega.$$

To prove Proposition 1.1 we combine multiplier techniques, compactness arguments, Hölmgren's Uniqueness Theorem and the following deep result that can be easily derived from the decoupling technique developed by D. Henry, O. Lopes and A. Perissinitto [HeLP].

Theorem 1.2. [HeLP]. *Let P be as above and let us denote by $\{S^0(t)\}_{t\geq 0}$ the strongly continuous semigroup in $\mathcal{V}$ associated to the following decoupled system:*

$$\text{(1.12)} \quad \begin{cases} u_{tt} - \mu\Delta u - (\lambda + \mu)\nabla \text{ div } u + \alpha\beta P u_t = 0 & \text{in } \Omega \times (0, \infty) \\ \theta_t - \Delta\theta + \beta \text{ div } u_t = 0 & \text{in } \Omega \times (0, \infty) \\ u = 0, \ \partial\theta/\partial\nu = 0 & \text{on } \Gamma \times (0, \infty) \\ u(0) = u^0, u_t(0) = u^1, \theta(0) = \theta^0 & \text{in } \Omega. \end{cases}$$

Then, $S(t) - S^0(t) : \mathcal{V} \to C([0,T]; \mathcal{V})$ is continuous and compact (recall that $S(t)$ is the semigroup associated to the system of thermoelasticity (1.1)).

We will denote by $S_i^0(t), i = 1, 2, 3$ the three components of $S^0(t)$.

To our knowledge, the first results on controllability of thermoelastic systems are due to K. Narukawa [N]. But in [N] and in the more recent works by J. Lagnese [La], J. Lagnese and J. L. Lions [LaLi] and J. L. Lions [Li2], only "partial controllability" results are proved. Indeed, in these papers various models of thermoelasticity are considered but the results that are obtained are, roughly, that one may control exactly the displacement by means of one control acting in the equations of displacement when the coupling parameters are small enough. Note that nothing is said about the controllability of the temperature. More recently, S. Hansen [H] has proved the controllability of both the displacement and the temperature in one space dimension by means of one sole boundary-control, for various boundary conditions and without restrictions in the size of the coupling parameters. The methods of [H] are based on moment problems and nonharmonic Fourier series and they do not seem to extend to several space dimensions.

The results of [Z3] and of the present paper prove that in several space dimensions, without any restriction on the size of the coupling parameters, one may achieve simultaneously the exact controllability of the displacement and the approximate controllability of the temperature by means of one sole control.

The extension of these results to the case where both displacement and temperature satisfy Neumann boundary conditions is to be done.

The rest of the paper is organized as follows. In Section 2 we prove Theorem 1.1 following the method of [Z3]. In Section 4, we consider the case where the displacement satisfies Neumann boundary conditions and the temperature takes zero Dirichlet value.

2. Proof of Theorem 1.1

To prove Theorem 1.1 we need to use Proposition 1.1 , Theorem 1.2 and a uniqueness or unique-continuation result for the adjoint system (1.8) as well. In the first three subsections we will state those results and give an outline of its prove. The proof of the Theorem 1.1 is concluded in the fourth subsection where the tools developed in [Li] and [FPZ1, 2, 3, 4] are applied.

2.1. A uniqueness result

Proposition 2.1. *Suppose that $T > \mathrm{diam}(\Omega \setminus \omega)/\sqrt{\mu}$. Let (φ, ψ) be a solution of system (1.8) such that $\varphi = 0$ in $\omega \times (0,T)$ and $\int_\Omega \psi(x,t)dx = 0$ for all $t \in [0,T]$. Then $\varphi \equiv \sigma \equiv 0$ in $\Omega \times (0,T)$.*

This result is basically a consequence of Hölmgren's Uniqueness Theorem (see for instance [J]). Note however that we do not do any assumption about the set of zeros of the temperature ψ. But this is not necessary because of the very strong coupling between displacement and temperature. Indeed, since $\varphi = 0$ in $\omega \times (0,T)$, then from the first equation in (1.8) we deduce that $\nabla \psi_t = 0$ in $\omega \times (0,T)$. The arguments of [Z3] allow to conclude that $\varphi \equiv 0$ and $\psi \equiv c$ for some real constant $c \in \mathbb{R}$. The fact that $c = 0$ is deduced from the hypothesis $\int_\Omega \psi(x,t)dx = 0$. For the technical details we refer to [Z3] where the case where both φ and ψ satisfy Dirichlet boundary conditions is treated in detail.

2.2. The observability inequality

This section is devoted to the proof of Proposition 1.1. We will use the notation $\tilde{\mathcal{V}} = (L^2(\Omega))^n \times (H^{-1}(\Omega))^n \times \mathcal{H}$.

First we observe that if (φ, ψ) solves (1.8), then (ϕ, ψ), where

$$\phi(x,t) = -\int_t^T \varphi(x,s)\, ds + \chi(x)$$

with

$$\begin{cases} -\mu\Delta\chi - (\lambda+\mu)\nabla \operatorname{div} \chi = -\varphi^1 - \beta\nabla\psi^0 & \text{in } \Omega \\ \chi = 0 & \text{on } \partial\Omega, \end{cases}$$

solves

$$(2.1) \begin{cases} \phi_{tt} - \mu\Delta\phi - (\lambda+\mu)\nabla \operatorname{div} \phi + \beta\nabla\psi = 0 & \text{in } \Omega \times (0,T) \\ -\psi_t - \Delta\psi - \alpha \operatorname{div} \phi_t = 0 & \text{in } \Omega \times (0,T) \\ \phi = 0, \partial\psi/\partial\nu = 0 & \text{on } \partial\Omega \times (0,T) \\ \phi(x,T) = \chi(x), \phi_t(x,T) = \phi^0(x), \psi(x,T) = \psi^0(x) & \text{in } \Omega. \end{cases}$$

Taking into account that $\|\chi\|_{(H_0^1(\Omega))^n}$ and $\|\varphi^1 + \beta\nabla\psi^0\|_{(H^{-1}(\Omega))^n}$ are equivalent norms, we see that Proposition 1 is equivalent to the following one:

Proposition 2.2. *Under the assumptions of Theorem 1.1, for every bounded set B of $\mathcal{H}$ there exists $\delta = \delta(B) > 0$ such that*

$$(2.2) \qquad \delta \le \int_0^T \int_\omega |\phi_t|^2 dxdt$$

holds for every solution of (2.1) with initial data such that

$$(2.3) \qquad \|(\chi, \varphi^0)\|_{(H_0^1(\Omega))^n \times (L^2(\Omega))^n} \ge 1, \psi^0 \in B.$$

Thus, it is sufficient to prove Proposition 2.2.

To prove Proposition 2.2, first we introduce the decoupled system associated to (2.1):

$$(2.4)\begin{cases} \tilde\phi_{tt} - \mu\Delta\tilde\phi - (\lambda+\mu)\nabla \text{ div } \tilde\phi - \alpha\beta P\tilde\phi_t = 0 & \text{in } \Omega\times(0,T) \\ -\tilde\psi_t - \Delta\tilde\psi - \alpha \text{ div } \tilde\phi_t = 0 & \text{in } \Omega\times(0,T) \\ \tilde\phi = 0, \partial\tilde\psi/\partial\nu = 0 & \text{on } \partial\Omega\times(0,T) \\ \tilde\phi(x,T) = \chi(x), \tilde\phi_t(x,T) = \varphi^0(x), \tilde\psi(x,T) = \psi^0(x) & \text{in } \Omega \end{cases}$$

and consider the subsystem that $\tilde\phi$ satisfies:

$$(2.5)\quad\begin{cases} \tilde\phi_{tt} - \mu\Delta\tilde\phi - (\lambda+\mu)\nabla \text{ div } \tilde\phi - \alpha\beta P\tilde\phi_t = 0 & \text{in } \Omega\times(0,T) \\ \tilde\phi = 0 & \text{on } \partial\Omega\times(0,T) \\ \tilde\phi(x,T) = \chi(x), \tilde\phi_t(x,T) = \varphi^0(x) & \text{in } \Omega. \end{cases}$$

We have the following observability inequality for system (2.5).

Proposition 2.3. *Suppose that $T > \text{diam}(\Omega\setminus\omega)/\sqrt{\mu}$. Then, there exists a constant $C > 0$ and a semi-norm $X : (H_0^1(\Omega))^n \times (L^2(\Omega))^n \to I\!R^+$ such that*

$$(2.6)\qquad \|\chi\|^2_{(H_0^1(\Omega))^n} + \|\varphi^0\|^2_{(L^2(\Omega))^n} \le C\Big[\int_0^T\int_\omega |\tilde\phi_t|^2 dxdt + X^2(\chi,\varphi^0)\Big]$$

holds for every solution of (3.5), *$X : (H_0^1(\Omega))^n \times (L^2(\Omega))^n \to I\!R^+$ being continuous and compact.*

Proof. The proof is the same as in Proposition 2.3 of [Z3]. Indeed, the only properties of the operator P that are used in [Z3] are the following ones:

$$\text{curl}\, Pf = 0;\ \text{div}\, Pf = \text{div}\, f \quad \text{for all } f \in (L^2(\Omega))^n$$

and these conditons are fulfilled for P as defined in (1.10). ∎

Let us now conclude the proof of Proposition 2.2.

We decompose the solution of (2.1) as $(\phi,\psi) = (\tilde\phi,\tilde\psi) + (\xi,\eta)$ where $(\tilde\phi,\tilde\psi)$ solves (2.4) and (ξ,η) satisfies

$$(2.7)\quad\begin{cases} \xi_{tt} - \mu\Delta\xi - (\lambda+\mu)\nabla \text{ div } \xi = -\alpha\beta P\tilde\phi_t - \beta\nabla\psi & \text{in } \Omega\times(0,T) \\ -\eta_t - \Delta\eta - \alpha \text{ div } \xi_t = 0 & \text{in } \Omega\times(0,T) \\ \xi = 0, \partial\eta/\partial\nu = 0 & \text{on } \partial\Omega\times(0,T) \\ \xi(x,T) = \xi_t(x,T) = 0, \eta(x,T) = 0 & \text{in } \Omega. \end{cases}$$

As a consequence of Proposition 2.3 we have

$$(2.8)\ \|\chi\|^2_{(H_0^1(\Omega))^n} + \|\varphi^0\|^2_{(L^2(\Omega))^n} \le C\Big[\int_0^T\int_\omega (|\phi|_t^2 + |\xi_t|^2)dxdt + X^2(\chi,\varphi^0)\Big].$$

We argue by contradiction. Suppose that Proposition 2.2 does not hold. Then, there exists a bounded set B of $\mathcal{H}$ and a sequence of initial data $(\chi_j, \varphi_j^0, \psi_j^0)$ satisfying (2.3) such that

$$\int_0^T \int_\omega |\phi_{j,t}|^2 dxdt \to 0 \text{ as } j \to \infty. \tag{2.9}$$

In view of (2.8) and taking into account (2.9) and that

$$\|(\chi_j, \varphi_j^0)\|_{(H_0^1(\Omega))^n \times (L^2(\Omega))^n} \geq 1$$

holds, we deduce that

$$\liminf_{j\to\infty}[\int_0^T \int_\omega |\xi_{j,t}|^2 dxdt + X^2(\chi_j, \varphi_j^0)] > 0. \tag{2.10}$$

We introduce the normalized data

$$(\hat{\chi}_j^0, \hat{\varphi}_j^0, \hat{\psi}_j^0) = (\chi_j^0, \varphi_j^0, \psi_j^0)/[\|\xi_{j,t}\|_{(L^2(\omega\times(0,T)))^n} + X^2(\chi_j, \varphi_j^0)]^{1/2}$$

and the corresponding solutions $(\hat{\phi}_j, \hat{\psi}_j)$ and $(\hat{\xi}_j, \hat{\eta}_j)$ of (2.1) and (2.7). We have then

$$\int_0^T \int_\omega |\hat{\xi}_{j,t}|^2 dxdt + X^2(\hat{\chi}_j, \hat{\varphi}_j^0) = 1; \quad \int_0^T \int_\omega |\hat{\phi}_{j,t}|^2 dxdt \to 0 \text{ as } j \to \infty. \tag{2.11}$$

In view of (2.8) we deduce that

$$\|(\hat{\chi}_j^0, \hat{\varphi}_j^0)\|_{(H_0^1(\Omega))^n \times (L^2(\Omega))^n} \leq C.$$

On the other hand, by (2.10) we know that $(\hat{\psi}_j^0)$ remains in a bounded set $\hat{B}$ of $\mathcal{H}$. By extracting subsequences we deduce that

$$\begin{cases} (\hat{\chi}_j, \hat{\varphi}_j^0) \longrightarrow (\hat{\chi}, \hat{\varphi}^0) & \text{weakly in } (H_0^1(\Omega))^n \times (L^2(\Omega))^n \\ \hat{\psi}_j^0 \longrightarrow \hat{\psi}^0 & \text{weakly in } L^2(\Omega) \end{cases}$$

and

$$\begin{cases} \hat{\varphi}_{j,t} \longrightarrow \hat{\varphi}_t & \text{weakly in } (L^2(\Omega\times(0,T)))^n \\ \hat{\xi}_{j,t} \longrightarrow \hat{\xi}_t & \text{weakly in } (L^2(\Omega\times(0,T)))^n \end{cases} \tag{2.12}$$

as $j \to \infty$, where $(\hat{\varphi}, \hat{\psi})$ and $(\hat{\xi}, \hat{\eta})$ are the solutions of (1.8) and (2.7) corresponding to the limit initial data.

On the other hand, in virtue of Theorem 1.2 we know that $(\hat{\xi}_{j,t})$ is relatively compact in $C([0,T]; (L^2(\Omega))^n)$ and therefore

$$\hat{\xi}_{j,t} \longrightarrow \hat{\xi}_t \quad \text{strongly in } (L^2(\Omega\times(0,T)))^n. \tag{2.13}$$

As a consequence of (2.11) and (2.12) we deduce that

$$\hat{\varphi} = \hat{\phi}_t = 0 \text{ in } \omega \times (0,T). \tag{2.14}$$

In view of (2.14) and applying Proposition 2.1 we obtain that $(\hat{\varphi}, \hat{\xi}) \equiv 0$ in $\Omega \times (0,T)$ and therefore

$$(\hat{\varphi}^0, \hat{\varphi}^1, \hat{\psi}^0) \equiv 0. \tag{2.15}$$

This implies that

$$\hat{\xi} \equiv 0. \tag{2.16}$$

However, combining (2.11), (2.13) and the fact that $X : (H_0^1(\Omega))^n \times (L^2(\Omega))^n \to \mathbb{R}^+$ is compact we deduce that

$$\|\hat{\xi}\|^2_{(L^2(\omega\times(0,T)))^n} + X^2(\hat{\chi}, \hat{\varphi}^0) = 1 \tag{2.17}$$

and this contradicts (2.15)-(2.16).

2.3. Proof of the decoupling result

In this subsection we give a sketch of the proof of Theorem 1.2 which follows the arguments of D. Henry, O. Lopes and A. Perissinitto [HeLP].

Let B be a bounded set of $\mathcal{V}$. We set

$$(u(t), u_t(t), \theta(t)) = [S(t)](u^0, u^1, \theta^0),$$

$$(\tilde{u}(t), \tilde{u}_t(t), \tilde{\theta}(t)) = [S^0(t)](u^0, u^1, \theta^0)$$

and

$$(v(t), v_t(t), \eta(t)) = [S(t) - S^0(t)](u^0, u^1, \theta^0)$$

for any $(u^0, u^1, \theta^0) \in B$.

We have:

$$\begin{cases} v_{tt} - \mu\Delta v - (\lambda+\mu)\nabla \text{ div } v + \alpha\nabla\eta = \alpha[\beta P\tilde{u}_t - \nabla\tilde{\theta}] & \text{in } \Omega\times(0,T) \\ \eta_t - \Delta\eta + \beta \text{ div } v_t = 0 & \text{in } \Omega\times(0,T) \\ v = 0, \ \partial\eta/\partial\nu = 0 & \text{on } \Gamma\times(0,T) \\ v(0) = v_t(0) = 0, \eta(0) = 0 & \text{in } \Omega. \end{cases} \tag{2.18}$$

It is sufficient to check that $\beta P\tilde{u}_t - \nabla\tilde{\theta}$ is bounded in $L^1(0,T;(H^s(\Omega))^n)$ for some $s > 0$ when (u^0, u^1, θ^0) varies in B.

Let us decompose $\beta P\tilde{u}_t - \nabla\tilde{\theta}$ as follows:

$$\beta P\tilde{u}_t - \nabla\tilde{\theta} = \nabla w_1 + \nabla w_2$$

where w_1 satisfies

$$\begin{cases} w_{1,t} - \Delta w_1 = 0 & \text{in } \Omega\times(0,T) \\ \partial w_1/\partial\nu = 0 & \text{on } \Gamma\times(0,T) \\ w_1(0) = \beta g - \theta^0 & \text{in } \Omega \end{cases} \tag{2.19}$$

with $g \in V$ the solution of (1.10) for $f = u^1$ and w_2 verifies

$$
(2.20) \qquad \begin{cases} w_{2,t} - \Delta w_2 = \beta h_{tt} & \text{in } \Omega \times (0,T) \\ \partial w_2 / \partial \nu = 0 & \text{on } \Gamma \times (0,T) \\ w_2(0) = 0 & \text{in } \Omega \end{cases}
$$

where, for each $t \in [0,T]$, $h(x,t)$ solves (1.10) with $f(x) = \tilde{u}(x,t)$.

Since $\beta g - \theta^0$ is bounded in $L^2(\Omega)$, because of the regularizing effect of the heat equation (2.19), we deduce that w_1 is bounded in $L^1(0,T;H^{1+s}(\Omega))$ for any $0 < s < 1$.

On the other hand, by elliptic regularity it is easy to check that h_{tt} is bounded in $L^\infty(0,T;L^2(\Omega))$. Therefore, w_2 is bounded in $L^1(0,T;H^2(\Omega))$. This concludes the proof of Theorem 1.2.

2.4. Conclusion

In this subsection we conclude the proof of Theorem 1.1.

First we observe that it is sufficient to consider the case where $u^0 \equiv u^1 \equiv 0$ and $\theta^0 \equiv 0$. Therefore, in the sequel we will assume that $u^0 \equiv u^1 \equiv 0$ and $\theta^0 \equiv 0$.

Given any $(v^0, v^1, \eta^0) \in \mathcal{V}$ and $\varepsilon > 0$ we introduce the functional $J : \tilde{\mathcal{V}} \to \mathbb{R}$ defined as follows

$$
(2.21) \quad J(\varphi^0, \varphi^1, \psi^0) = \frac{1}{2} \int_0^T \int_\omega |\varphi|^2 dx dt - \int_\Omega v^1 . \varphi^0 \mathrm{dx} + \langle v^0, \varphi^1 \rangle + \varepsilon \|\psi^0\|_{L^2(\Omega)} - \int_\Omega (\eta^0 + \beta \text{ div } v^0) \psi^0 dx
$$

where $\langle \cdot, \cdot \rangle$ denotes the duality product between $(H_0^1(\Omega))^n$ and $(H^{-1}(\Omega))^n$ and (φ, ψ) the solution of (1.8).

The functional J is coercive in $\tilde{\mathcal{V}}$. More precisely, we have the following result:

Lemma 2.1. *Under the assumptions of Theorem 1.1,*

$$
(2.22) \qquad \liminf_{\|(\varphi^0, \varphi^1 + \beta \nabla \psi^0)\|_{\tilde{\mathcal{V}}} \to \infty} \frac{J(\varphi^0, \varphi^1, \psi^0)}{\|(\varphi^0, \varphi^1 + \beta \nabla \psi^0)\|_{\tilde{\mathcal{V}}}} \geq \varepsilon.
$$

In view of Proposition 1.1, the proof of this lemma can be done following the arguments of Lemma 4.1 in [Z4] in a straightforward way.

By using the coercivity property (2.22) it is easy to check that the infimum of J over $\tilde{\mathcal{V}}$ is achieved at some $(\hat{\varphi}^0, \hat{\varphi}^1, \hat{\psi}^0) \in \tilde{\mathcal{V}}$ (see [FPZ1, 2, 3, 4] and [Li] for the details and some other related questions). At this minimizer $(\hat{\varphi}^0, \hat{\varphi}^1, \hat{\psi}^0)$ we have the following optimality conditions:

$$
(2.23) \quad \left| \int_0^T \int_\omega \hat{\varphi} \cdot \rho dx dt - \int_\Omega v^1 \cdot \rho^0 dx + \langle v^0, \rho^1 \rangle - \int_\Omega (\eta^0 + \beta \text{ div } v^0) \xi^0 dx \right| \leq \varepsilon \|\xi^0\|_{L^2(\Omega)}
$$

for all $(\rho^0, \rho^1, \xi^0) \in \tilde{\mathcal{V}}$ where $(\hat{\varphi}, \hat{\psi})$ denotes the solution of (1.8) corresponding to the minimizer $(\hat{\varphi}^0, \hat{\varphi}^1, \hat{\psi}^0)$ and (ρ, ξ) the solution of (1.8) with data (ρ^0, ρ^1, ξ^0).

Observe that if (u, θ) solves (1.1) with $u^0 \equiv u^1 \equiv 0$ and $\theta^0 = 0$ and $f = \hat{\varphi}$ then

$$\int_0^T \int_\omega \hat{\varphi} \cdot \rho dx dt = \int_\Omega u_t(T) \cdot \rho^0 dx - \langle u(T), \rho^1 \rangle + \int_\Omega (\theta(T) + \beta \text{ div } u(T)) \xi^0 dx. \tag{2.24}$$

Combining (2.23) and (2.24) we easily deduce that (u, θ) satisfy (1.3).

3. Boundary conditions of the second type

When the displacement satisfies Neumann boundary conditions and the temperature takes Dirichlet boundary conditions the controlled system of thermoelasticity reads as follows:

$$\begin{cases} u_{tt} - \mu \Delta u - (\lambda + \mu) \nabla \text{ div } u + \alpha \nabla \theta = f \chi_\omega & \text{in } \Omega \times (0, \infty) \\ \theta_t - \Delta \theta + \beta \text{ div } u_t = 0 & \text{in } \Omega \times (0, \infty) \\ \mu \partial u / \partial \nu + (\lambda + \mu) \text{ div } u \; \nu = 0, \quad \theta = 0 & \text{on } \Gamma \times (0, \infty) \\ u(x, 0) = u^0(x), \; u_t(x, 0) = u^1(x), \; \theta(x, 0) = \theta^0(x) & \text{in } \Omega. \end{cases} \tag{3.1}$$

In this situation we have the following controllability result:

Theorem 3.1. *Let ω be a neighborhood of the boundary Γ in Ω. Suppose that $T > diam(\Omega \setminus \omega)/\sqrt{\mu}$. Then, system (3.1) is exact-approximately controllable in time T in the following sense: for any $(u^0, u^1, \theta^0), (v^0, v^1, \eta^0) \in \mathcal{V}_1 = (H^1(\Omega))^n \times (L^2(\Omega))^n \times L^2(\Omega)$ and $\varepsilon > 0$ there exists a control function $f \in (L^2(\Omega \times (0, T)))^n$ such that the solution of (3.1) satisfies (1.3).*

Note that, in this case, there is no restriction like (1.5) on the final data we may reach starting from a given initial data. This is due to the fact that there is no quantity that remains constant in time, this constant being independent of f.

The proof of this theorem is similar to that of Theorem 1.1. However, in this case, the decoupling operator P has to be defined in a different way. In this case, $P : (L^2(\Omega))^n \to (L^2(\Omega))^n$ is given by $Pf = \nabla g$ where $g \in H_0^1(\Omega)$ is the solution of

$$\int_\Omega \nabla g \cdot \nabla \rho dx = \int_\Omega f \cdot \nabla \rho dx, \quad \text{for all } \rho \in H_0^1(\Omega). \tag{3.2}$$

Obviously, the solution g of (3.2) in $H^1(\Omega) \cap \mathcal{H}$ exits and it is unique. Note that (3.2) is the weak formulation of the equation

$$\Delta g = \text{ div } f \text{ in } \Omega; \; g = 0 \text{ on } \partial\Omega. \tag{3.3}$$

Acknowledgements

This work was done while the author was visiting the Department of Computational and Applied Mathematics of Rice University in Houston with the support of the "Ministerio de Educación y Ciencia (Spain)". This research was partially supported by DGICYT Project PB90-0245 and by the EEC Grant SC1*-CT91-0732

References

[FPZ1] C Fabre, J P Puel and E Zuazua, Contrôlabilité approchée de l'équation de la chaleur semilinéaire, C R Acad Sci Paris, 315, série I, 1992, 807-812

[FPZ2] C Fabre, J P Puel and E Zuazua, Approximate controllability of the semilinear heat equation, Proc Royal Soc Edinburgh, to appear

[FPZ3] C Fabre, J P Puel and E Zuazua, Controlabilité approché de l'équation de la chaleur linéaire avec des contrôles de norme L^∞ minimale, C R Acad Sci Paris, 316, série I, 1993, 679-684

[FPZ4] C Fabre, J P Puel and E Zuazua, On the density of the range of the semigroup for semilinear heat equations, IMA Preprint Series, # 1093, 1992

[H] S W Hansen, Boundary control of a one-dimensional, linear, thermoelastic rod, SIAM J Cont Optim , to appear

[HeLP] D Henry, O Lopes and A Perissinitto, On the essential spectrum of a semigroup of thermoelasticity, preprint

[J] F John, *Partial Differential Equations*, Applied Mathematical Sciences 1, 4th edition, Springer Verlag, New York, 1986

[La] J E Lagnese, The reachability problem for thermoelastic plates, Arch Rat Mech Anal , 112 (1990), 223-267

[LaLi] J E Lagnese and J L Lions, *Modelling, Analysis and Control of Thin Plates*, Masson, RMA 6, Paris, 1988

[Li] J L Lions, Remarks on approximate controllability, J Analyse Math , 59, 1992, 103-116

[N] K Narukawa, Boundary value control of thermoelastic systems, Hiroshima Math J , 13 (1983), 227-272

[Z1] E Zuazua, Contrôlabilité du système de la thermoelasticité, C R Acad Sci Paris, 317 (1993), 371-376

[Z2] E Zuazua, Contrôlabilité du systeme de la thermoélasticité sans restrictions sur les paramètres de couplage, C R Acad Sci Paris, 318 (1994), 643-648

[Z3] E Zuazua, Controllability of the linear system of thermoelasticity, preprint

Enrique Zuazua
Departamento de Matemática Aplicada
Universidad Complutense
28040 Madrid, Spain

Zeitfracht Medien GmbH
Ferdinand-Jühlke-Straße 7
99095 Erfurt, Deutschland
produktsicherheit@kolibri360.de